KARL SCHWARZ.

Contaminant Hydrogeology

Contaminant Hydrogeology

C. W. Fetter

Department of Geology
University of Wisconsin—Oshkosh

PRENTICE HALL
Upper Saddle River, NJ 07458

Library of Congress Cataloging-in-Publication Data

Fetter, C. W. (Charles Willard)
   Contaminant hydrogeology/C.W. Fetter.
   p. cm.
   Includes bibliographical references and index.
   ISBN 0-02-337135-8
   1. Water, Underground--Pollution. 2. Water, Underground--Pollution--
United States. 3. Transport theory. 4. Hydrogeology. I. Title.
TD426.F48   1992
628.1'68--dc20   92-17787
                                                                    CIP

Editor: Robert A. McConnin
Production Editor: Sharon Rudd
Art Coordinator: Peter A. Robison
Text Designer: Debra A. Fargo
Cover Designer: Robert Vega
Productiøn Buyer: Pamela D. Bennett
Illustrations: Maryland CartoGraphics Inc.

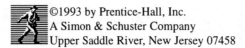
©1993 by Prentice-Hall, Inc.
A Simon & Schuster Company
Upper Saddle River, New Jersey 07458

All rights reserved. No part of this book may be reproduced,
in any form or by any means, without permission in writing
from the publisher.

Printed in the United States of America
10 9 8 7 6 5 4

ISBN 0-02-337135-8

Prentice-Hall International (UK) Limited, *London*
Prentice-Hall of Australia Pty. Limited, *Sydney*
Prentice-Hall Canada Inc., *Toronto*
Prentice-Hall Hispanoamericana, S.A., *Mexico*
Prentice-Hall of India Private Limited, *New Delhi*
Prentice-Hall of Japan, Inc., *Tokyo*
Simon & Schuster Asia Pte. Ltd., *Singapore*
Editora Prentice-Hall do Brasil, Ltda., *Rio de Janeiro*

*This book is dedicated to my parents,
C. Willard Fetter and Grace Fetter.*

Preface

When I completed the second edition of *Applied Hydrogeology*, I realized that it provided only the barest of introductions to what is one of the most fascinating aspects of hydrogeology, the occurrence and movement of dissolved and nonaqueous phase contaminants. Consulting work that I was doing also demonstrated that to understand fully the distribution of contaminants below the water table one must consider the movement of soil moisture and contaminants in the vadose zone. As none of the standard textbooks present advanced topics of solute movement and retardation in both the saturated and vadose zone as well as the occurrence and movement of nonaqueous phase liquids, I think that there is a place for an advanced textbook on contaminant hydrogeology.

In a very real sense this new book is a sequel to *Applied Hydrogeology*. There is almost no overlap between the two books; although some material needed to be repeated to lay the logical foundation for the advanced concepts presented in this book. *Contaminant Hydrogeology* is intended to be a textbook for a graduate-level course in mass transport and ground-water contamination. Such a course might be taught in departments such as geology, civil engineering, geological engineering, or agricultural engineering. In order to obtain the fullest benefit from such a course, the students should have completed a course in geohydrology or hydrogeology. Basic knowledge of physics and chemistry is needed to understand the concepts presented herein.

In addition to its utility as a textbook, *Contaminant Hydrogeology* will be a valuable reference book for the working professional. Both solved example problems and case histories are presented. There is a mixture of the theoretical and the practical. Chapter 1 presents an overview of ground-water contamination and a review of basic mathematics. The theory of mass transport in the saturated zone is presented in Chapter 2. Topics include advective-dispersive theory, stochastic transport theory, and description of solute flow using fractals. Retardation and attenuation of dissolved solutes is covered in Chapter 3, whereas Chapter 4 introduces flow and mass transport in the vadose zone. The distribution and movement of nonaqueous phase liquids both above and below the water table is discussed in Chapter 5. The reactions of inorganic compounds dissolved in ground water is the topic of Chapter 6. Chapter 7 contains an overview of organic chemistry and an exhaustive look at biodegradation of organic compounds in the ground. Chapter 8 contains "how-to" information on conducting field investigations to install borings and monitoring wells as well as collecting soil, soil-water, and ground-water samples. The latest information on site remediation is found in Chapter 9.

In a book of this nature there are a very large number of variables—far more than can be accommodated by the 26 letters of the English and the 24 letters of the Greek alphabets. Many variables are indicated by symbols that are a combination of English and/or Greek letters. A variable is defined where first used in a chapter and then is listed in a table of notation at the end of the chapter. In order to accommodate the large number of variables in the book, the meaning of some symbols changes from chapter to chapter. Although this is not a desirable circumstance, it seemed preferable to such tactics as also utilizing the Hebrew and Russian alphabets. In many cases, if the reader goes to the original literature cited in the text, the notation of the original article will not be the same as that used in this text. This was necessary to have consistency within the text.

Units of measurement have been abbreviated in the text. Appendix E contains a key to these abbreviations.

I am grateful to all who helped with this project. The following individuals provided helpful reviews of chapter drafts: Jean M. Bahr, University of Wisconsin–Madison; Robert A. Griffin, University of Alabama; James I. Hoffman, Eastern Washington University; Martinus Th. van Genuchten, U.S. Department of Agriculture Salinity Laboratory; Stephen Kornder, James River Paper Company; Garrison Sposito, University of California, Berkeley; and Nicholas Valkenburg, Geraghty and Miller, Inc. Peter Wierenga, University of Arizona, provided information on measuring soil-moisture tensions and Shlomo Neuman, University of Arizona, furnished me with a copy of Mualem's Soil Property Catalogue. Mary Dommer prepared the manuscript, and Sue Birch provided some of the figures.

C. W. Fetter

Contents

Chapter One

Introduction

1.1 Ground Water as a Resource   1
1.2 Types of Ground-Water Contaminants   2
1.3 Drinking-Water Standards   11
1.4 Risk and Drinking Water   14
1.5 Sources of Ground-Water Contamination   15
    1.5.1 Category I: Sources Designed to Discharge Substances   16
    1.5.2 Category II: Sources Designed to Store, Treat and/or Dispose of Substances   19
    1.5.3 Category III: Sources Designed to Retain Substances During Transport   25
    1.5.4 Category IV: Sources Discharging Substances as a Consequence of Other Planned Activities   25
    1.5.5 Category V: Sources Providing a Conduit for Contaminated Water to Enter Aquifers   27
    1.5.6 Category VI: Naturally Occurring Sources Whose Discharge is Created and/or Exacerbated by Human Activity   28
1.6 Relative Ranking of Ground-Water–Contamination Sources   29
1.7 Ground-Water Contamination as a Long-Term Problem   31
1.8 Review of Mathematics and the Flow Equation   32
    1.8.1 Derivatives   32
    1.8.2 Darcy's Law   35
    1.8.3 Scaler, Vector, and Tensor Properties of Hydraulic Head and Hydraulic Conductivity   35
    1.8.4 Derivation of the Flow Equation in a Deforming Medium   37
    1.8.5 Mathematical Notation   40
References   41

Chapter Two

Mass Transport in Saturated Media   43

- 2.1   Introduction   43
- 2.2   Transport by Concentration Gradients   43
- 2.3   Transport by Advection   47
- 2.4   Mechanical Dispersion   49
- 2.5   Hydrodynamic Dispersion   51
- 2.6   Derivation of the Advection-Dispersion Equation for Solute Transport   52
- 2.7   Diffusion versus Dispersion   54
- 2.8   Analytical Solutions of the Advection-Dispersion Equation   56
  - 2.8.1   Methods of Solution   56
  - 2.8.2   Boundary and Initial Conditions   56
  - 2.8.3   One-Dimensional Step Change in Concentration (First-Type Boundary)   57
  - 2.8.4   One-Dimensional Continuous Injection into a Flow Field (Second-Type Boundary)   58
  - 2.8.5   Third-Type Boundary Condition   60
  - 2.8.6   One-Dimensional Slug Injection into a Flow Field   61
  - 2.8.7   Continuous Injection into a Uniform Two-Dimensional Flow Field   61
  - 2.8.8   Slug Injection into a Uniform Two-Dimensional Flow Field   63
- 2.9   Effects of Transverse Dispersion   65
- 2.10   Tests to Determine Dispersivity   66
  - 2.10.1   Laboratory Tests   66
  - 2.10.2   Field Tests for Dispersivity   68
  - 2.10.3   Single-Well Tracer Test   69
- 2.11   Scale Effect of Dispersion   71
- 2.12   Stochastic Models of Solute Transport   77
  - 2.12.1   Introduction   77
  - 2.12.2   Stochastic Descriptions of Heterogeneity   78
  - 2.12.3   Stochastic Approach to Solute Transport   81
- 2.13   Fractal Geometry Approach to Field-scale Dispersion   85
  - 2.13.1   Introduction   85
  - 2.13.2   Fractal Mathematics   85
  - 2.13.3   Fractal Geometry and Dispersion   88
  - 2.13.4   Fractal Scaling of Hydraulic Conductivity   90
- 2.14   Deterministic Models of Solute Transport   93

Case Study: Borden Landfill Plume   96

Contents   xi

2.15   Transport in Fractured Media   103
2.16   Summary   107
       Chapter Notation   109
       References   111

Chapter Three

Transformation, Retardation, and Attenuation of Solutes   115

3.1   Introduction   115
3.2   Classification of Chemical Reactions   116
3.3   Sorption Processes   117
3.4   Equilibrium Surface Reactions   117
       3.4.1   Linear Sorption Isotherm   117
       3.4.2   Freundlich Sorption Isotherm   119
       3.4.3   Langmuir Sorption Isotherm   122
       3.4.4   Effect of Equilibrium Retardation on Solute Transport   123
3.5   Nonequilibrium (Kinetic) Sorption Models   129
3.6   Sorption of Hydrophobic (Organic) Compounds   132
       3.6.1   Introduction   132
       3.6.2   Partitioning onto Soil or Aquifer Organic Carbon   132
       3.6.3   Estimating K_{oc} from K_{ow} Data   133
       3.6.4   Estimating K_{oc} from Solubility Data   134
       3.6.5   Estimating K_{oc} from Molecular Structure   138
       3.6.6   Multiple Solute Effects   140
3.7   Homogeneous Reactions   140
       3.7.1   Introduction   140
       3.7.2   Chemical Equilibrium   141
       3.7.3   Chemical Kinetics   141
       3.7.4   Tenads in Chemical Reactions   142
3.8   Radioactive Decay   144
3.9   Biodegradation   144
3.10   Colloidal Transport   149
       Case Study: Large-scale Field Experiment on the Transport of Reactive and Nonreactive Solutes in a Scale Aquifer under Natural Ground-Water Gradients—Borden, Ontario   150
3.11   Summary   157
       Chapter Notation   158
       References   160

Chapter Four

Flow and Mass Transport in the Vadose Zone — 163

- 4.1 Introduction 163
- 4.2 Soil as a Porous Medium 163
- 4.3 Soil Colloids 164
- 4.4 The Electrostatic Double Layer 165
- 4.5 Salinity Effects on Hydraulic Conductivity of Soils 167
- 4.6 Flow of Water in the Unsaturated Zone 168
  - 4.6.1 Soil-Water Potential 168
  - 4.6.2 Soil-Water Characteristic Curves 169
  - 4.6.3 Hysteresis 175
  - 4.6.4 Construction of a Soil-Water-Retention Curve 176
  - 4.6.5 Measurement of Soil-Water Potential 177
  - 4.6.6 Unsaturated Hydraulic Conductivity 180
  - 4.6.7 Buckingham Flux Law 182
  - 4.6.8 Richard Equation 183
  - 4.6.9 Vapor Phase Transport 184
- 4.7 Mass Transport in the Unsaturated Zone 185
- 4.8 Equilibrium Models of Mass Transport 186
- 4.9 Nonequilibrium Models of Mass Transport 188
- 4.10 Anion Exclusion 190
  - Case Study: Relative Movement of Solute and Wetting Fronts 193
- 4.11 Preferential Flowpaths in the Vadose Zone 196
- 4.12 Summary 198
  - Chapter Notation 198
  - References 200

Chapter Five

Multiphase Flow — 202

- 5.1 Introduction 202
- 5.2 Basic Concepts 203
  - 5.2.1 Saturation Ratio 203
  - 5.2.2 Interfacial Tension and Wettability 203
  - 5.2.3 Capillary Pressure 204
  - 5.2.4 Relative Permeability 206
  - 5.2.5 Darcy's Law for Two-Phase Flow 211
  - 5.2.6 Fluid Potential and Head 212

Contents   xiii

5.3   Migration of Light Nonaqueous Phase Liquids (LNAPLs)   217
5.4   Measurement of the Thickness of a Floating Product   225
5.5   Effect of the Rise and Fall of the Water Table on the Distribution of LNAPLs   231
5.6   Migration of Dense Nonaqueous Phase Liquids   231
      5.6.1   Vadose Zone Migration   231
      5.6.2   Vertical Movement in the Saturated Zone   233
      5.6.3   Horizontal Movement in the Saturated Zone   235
5.7   Monitoring for LNAPLs and DNAPLs   238
5.8   Summary   239
      Chapter Notation   240
      References   242

Chapter Six

Inorganic Chemicals in Ground Water   244

6.1   Introduction   244
6.2   Units of Measurement and Concentration   244
6.3   Chemical Equilibrium and the Law of Mass Action   245
6.4   Oxidation-Reduction Reactions   249
6.5   Relationship between pH and Eh   253
      6.5.1   pH   253
      6.5.2   Relationship of Eh and pH   253
      6.5.3   Eh-pH Diagrams   254
      6.5.4   Calculating Eh-pH Stability Fields   257
6.6   Metal Complexes   267
      6.6.1   Hydration of Cations   267
      6.6.2   Complexation   267
      6.6.3   Organic Complexing Agents   269
6.7   Chemistry of Nonmetallic Inorganic Contaminants   270
      6.7.1   Fluoride   270
      6.7.2   Chlorine and Bromine   271
      6.7.3   Sulfur   272
      6.7.4   Nitrogen   272
      6.7.5   Arsenic   274
      6.7.6   Selenium   276
      6.7.7   Phosphorus   276
6.8   Chemistry of Metals   276
      6.8.1   Beryllium   277
      6.8.2   Strontium   277
      6.8.3   Barium   277

xiv Contents

        6.8.4 Vanadium 277
        6.8.5 Chromium 277
        6.8.6 Cobalt 278
        6.8.7 Nickel 279
        6.8.8 Molybdenum 279
        6.8.9 Copper 279
        6.8.10 Silver 279
        6.8.11 Zinc 280
        6.8.12 Cadmium 280
        6.8.13 Mercury 280
        6.8.14 Lead 280

6.9 Radioactive Isotopes 281
        6.9.1 Introduction 281
        6.9.2 Adsorption of Cationic Radionuclides 282
        6.9.3 Uranium 282
        6.9.4 Thorium 285
        6.9.5 Radium 286
        6.9.6 Radon 287
        6.9.7 Tritium 288

6.10 Geochemical Zonation 288
6.11 Summary 292
     Chapter Notation 292
     References 293

Chapter Seven

Organic Compounds in Ground Water   295

7.1 Introduction 295
7.2 Physical Properties of Organic Compounds 295
7.3 Organic Structure and Nomenclature 297
        7.3.1 Hydrocarbon Classes 297
        7.3.2 Aromatic Hydrocarbons 300
7.4 Petroleum Distillates 301
7.5 Functional Groups 305
        7.5.1 Organic Halides 305
        7.5.2 Alcohols 308
        7.5.3 Ethers 308
        7.5.4 Aldehydes and Ketones 311
        7.5.5 Carboxylic Acids 311
        7.5.6 Esters 312
        7.5.7 Phenols 312

Contents

| | 7.5.8 | Organic Compounds Containing Nitrogen | 314 |
| | 7.5.9 | Organic Compounds Containing Sulfur and Phosphorus | 315 |

7.6 Degradation of Organic Compounds 316
    7.6.1 Introduction 316
    7.6.2 Degradation of Hydrocarbons 318
    7.6.3 Degradation of Chlorinated Hydrocarbons 319
    7.6.4 Degradation of Organic Pesticides 323

7.7 Field Examples of Biological Degradation of Organic Molecules 326
    7.7.1 Introduction 326
    7.7.2 Chlorinated Ethanes and Ethenes 327
    7.7.3 Aromatic Compounds 328

7.8 Analysis of Organic Compounds in Ground Water 329

7.9 Summary 334
    References 335

Chapter Eight

Ground Water and Soil Monitoring — 338

8.1 Introduction 338

8.2 Monitoring Well Design 338
    8.2.1 General Information 338
    8.2.2 Monitoring Well Casing 339
    8.2.3 Monitoring Well Screens 345
    8.2.4 Naturally Developed and Filter-Packed Wells 346
    8.2.5 Annular Seal 347
    8.2.6 Protective Casing 348
    8.2.7 Screen Length and Setting 349
    8.2.8 Summary of Monitoring Well Design 351

8.3 Installation of Monitoring Wells 353
    8.3.1 Decontamination Procedures 353
    8.3.2 Methods of Drilling 354
    8.3.3 Drilling in Contaminated Soil 359

8.4 Sample Collection 360

8.5 Installation of Monitoring Wells 364

8.6 Monitoring Well Development 370

8.7 Record Keeping During Monitoring Well Construction 375

8.8 Monitoring Well and Borehole Abandonment 375

8.9 Multiple-level Devices for Ground-Water Monitoring 376

8.10 Well Sampling 378
    8.10.1 Introduction 378

	8.10.2	Well Purging 379
	8.10.3	Well-Sampling Devices 380
8.11	Soil-Gas Monitoring 383
	8.11.1	Introduction 383
	8.11.2	Methods of Soil-Gas Monitoring 384
8.12	Soil-Water Sampling 385
	8.12.1	Introduction 385
	8.12.2	Suction Lysimeters 385
	8.12.3	Installation of Suction Lysimeters 389
8.13	Summary 389
	References 390

Chapter Nine

Site Remediation   392

9.1	Introduction 392
9.2	Source-Control Measures 392
	9.2.1 Solid Waste 392
	9.2.2 Removal and Disposal 393
	9.2.3 Containment 393
	9.2.4 Hydrodynamic Isolation 399
9.3	Pump-and-Treat Systems 401
	9.3.1 Overview 401
	9.3.2 Capture Zones 403
	9.3.3 Computation of Capture Zones 405
	9.3.4 Optimizing Withdrawal-Injection Systems 414
	9.3.5 Permanent Plume Stabilization 416
9.4	Treatment of Extracted Ground Water 416
	9.4.1 Overview 416
	9.4.2 Treatment of Inorganic Contaminants 417
	9.4.3 Treatment of Dissolved Organic Contaminants 417
9.5	Recovery of Nonaqueous Phase Liquids 418
9.6	Removal of Leaking Underground Storage Tanks 424
9.7	Soil-Vapor Extraction 427
9.8	*In Situ* Bioremediation 429
	Case Study: Enhanced Biodegradation of Chlorinated Ethenes 433
9.9	Combination Methods 434
	Case Study: Remediation of a Drinking Water Aquifer Contaminated with Volatile Organic Compounds 438

	Case Study: Ground-Water Remediation Using a Pump-and-Treat Technique Combined with Soil Washing 439
9.10	Summary 442
	Chapter Notation 443
	References 443

Appendix A
Error Function Values   **445**

Appendix B
Bessel Functions   **446**

Appendix C
$W(t, B)$ Values   **448**

Appendix D
Exponential Integral   **450**

Appendix E
Unit Abbreviations   **451**

Index   **452**

Chapter One

Introduction

1.1   Ground Water as a Resource

Ground water is the source for drinking water for many people around the world, especially in rural areas. In the United States ground water supplies 42.4% of the population served by public water utilities. Virtually all the homes that supply their own water have wells and use ground water. In all, more than half of the population (52.5%) of the United States relies upon a ground-water source for drinking water (Solley, Merk, and Pierce 1988).

Table 1.1 shows the ground-water withdrawals by category of use in the United States in 1985 as well as the percentage of total use for that category supplied by ground water. In Table 1.1 public supply refers to water provided by either a public water utility or a private water company and used for residential, commercial, and industrial uses, power-plant cooling, and municipal uses such as fire fighting. All other categories are self-supplied, with the user owning the water system. Many of the self-supplied systems rely upon water wells. From 1980 to 1984 an average of 370,000 water wells were drilled in the United States each year (Hindall and Eberle 1989).

Inasmuch as ground water provides drinking water to so many people, the quality of ground water is of paramount importance. Public water suppliers in the United States are obligated by the Safe Drinking Water Act of 1986 to furnish water to their consumers that meets specific drinking-water standards. If the water does not meet the standards when it is withdrawn from its source, it must be treated. Ground water may not meet the standards because it contains dissolved constituents coming from natural sources. Common examples of constituents coming from natural sources are total dissolved solids, sulfate, and chloride. Ground water also may not meet the standards because it contains organic liquids, dissolved organic and inorganic constituents, or pathogens that came from an anthropogenic source. In such cases the ground water has been contaminated by the acts of humans.

In the case of self-supplied systems, a source of uncontaminated water is of even greater importance. Such systems are typically tested initially for only a very limited range of constituents, such as coliform bacteria, nitrate, chloride, and iron. Most times ground-water contamination cannot be tasted, so that with such limited testing it is possible for

TABLE 1.1 Ground-water usage in the United States, 1985.

Category	Ground-water Use (million gallons/day)	Percent of Total Use Supplied by Ground Water
Public water supply	14,600	40.0
Domestic, self-supplied	3,250	97.9
Commercial, self-supplied	746	60.7
Irrigation	45,700	33.4
Livestock	3,020	67.6
Industrial (fresh)	3,930	17.6
Industrial (saline)	26	0.7
Mining (fresh)	1,410	52.8
Mining (saline)	626	81.9
Power plant cooling	608	0.5

Source: Solley, Merk, and Pierce, 1988.

a user to have a contaminated source and not be aware of it. Additionally, self-supplied systems rarely undergo treatment other than softening and perhaps iron removal. There are limited options available for the homeowner who wishes to treat contaminated ground water so that it can be consumed.

In addition to providing for the sustenance of human life, ground water has important ecological functions. Many freshwater habitats are supplied by the discharge of springs. If the ground water supplying these springs is contaminated, the ecological function of the freshwater habitat can be impaired.

1.2 Types of Ground-Water Contaminants

A wide variety of materials have been identified as contaminants found in ground water. These include synthetic organic chemicals, hydrocarbons, inorganic cations, inorganic anions, pathogens, and radionuclides. Table 1.2 contains an extensive listing of these compounds. Most of these materials will dissolve in water to varying degrees. Some of the organic compounds are only slightly soluble and will exist in both a dissolved form and as an insoluble phase, which can also migrate through the ground. Examples of the uses of these materials are also given on Table 1.2. These uses may provide help in locating the source of a compound if it is found in ground water. The inorganic cations and anions occur in nature and may come from natural as well as anthropogenic sources. Some of the radionuclides are naturally occurring and can come from natural sources as well as mining, milling, and processing ore, industrial uses, and disposal of radioactive waste. Other radionuclides are man-made and come from nuclear weapons production and testing.

Table 1.3 lists the organic contaminants found in ground water at a single hazardous waste site. Almost 80 compounds were detected at this former organic solvent–recycling facility.

TABLE 1.2 Substances known to occur in ground water.

Contaminant	Examples of uses
Aromatic hydrocarbons	
Acetanilide	Intermediate manufacturing, pharmaceuticals, dyestuffs
Alkyl benzene sulfonates	Detergents
Aniline	Dyestuffs, intermediate, photographic chemicals, pharmaceuticals, herbicides, fungicides, petroleum refining, explosives
Anthracene	Dyestuffs, intermediate, semiconductor research
Benzene	Detergents, intermediate, solvents, antiknock gasoline
Benzidine	Dyestuffs, reagent, stiffening agent in rubber compounding
Benzyl alcohol	Solvent, perfumes and flavors, photographic developer inks, dyestuffs, intermediate
Butoxymethylbenzene	NA[a]
Chrysene	Organic synthesis, coal tar by-product
Creosote mixture	Wood preservatives, disinfectants
Dibenz[a.h.]anthracene	NA
Di-butyl-p-benzoquinone	NA
Dihydrotrimethylquinoline	Rubber antioxidant
4,4-Dinitrosodiphenylamine	NA
Ethylbenzene	Intermediate, solvent, gasoline
Fluoranthene	Coal tar by-product
Fluorene	Resinous products, dyestuffs, insecticides, coal tar by-product
Fluorescein	Dyestuffs
Isopropyl benzene	Solvent, chemical manufacturing
4,4'-methylene-bis-2-chloroaniline (MOCA)	Curing agent for polyurethanes and epoxy resins
Methylthiobenzothiazole	NA
Napthalene	Solvent, lubricant, explosives, preservatives, intermediate, fungicide, moth repellant
o-Nitroaniline	Dyestuffs, intermediate, interior paint pigments, chemical manufacturing
Nitrobenzene	Solvent, polishes, chemical manufacturing
4-Nitrophenol	Chemical manufacturing
n-Nitrosodiphenylamine	Pesticides, retarder of vulcanization of rubber
Phenanthrene	Dyestuffs, explosives, synthesis of drugs, biochemical research
n-Propylbenzene	Dyestuffs, solvent
Pyrene	Biochemical research, coal tar by-product
Styrene (vinyl benzene)	Plastics, resins, protective coatings, intermediate
Toluene	Adhesive solvent in plastics, solvent, aviation and high-octane blending stock, dilutent and thinner, chemicals, explosives, detergents
1,2,4-Trimethylbenzene	Manufacture of dyestuffs, pharmaceuticals, chemical manufacturing
Xylenes (m, o, p)	Aviation gasoline, protective coatings, solvent, synthesis of organic chemicals, gasoline
Oxygenated hydrocarbons	
Acetic acid	Food additives, plastics, dyestuffs, pharmaceuticals, photographic chemicals, insecticides
Acetone	Dyestuffs, solvent, chemical manufacturing, cleaning and drying of precision equipment
Benzophenone	Organic synthesis, odor fixative, flavoring, pharmaceuticals
Butyl acetate	Solvent
n-Butyl-benzylphthalate	Plastics, intermediate

Source: Office of Technology Assessment, *Protecting The Nation's Groundwater from Contamination*, 1984, pp. 23–31.

[a] NA: No information in standard sources.

TABLE 1.2 Cont'd

Contaminant	Examples of uses
Oxygenated hydrocarbons (cont'd)	
Di-n-butyl phthalate	Plasticizer, solvent, adhesives, insecticides, safety glass, inks, paper coatings
Diethyl ether	Chemical manufacturing, solvent, analytical chemistry, anesthetic, perfumes
Diethyl phthalate	Plastics, explosives, solvent, insecticides, perfumes
Diisopropyl ether	Solvent, rubber cements, paint and varnish removers
2,4-Dimethyl-3-hexanol	Intermediate, solvent, lubricant
2,4-Dimethyl phenol	Pharmaceuticals, plastics, disinfectants, solvent, dyestuffs, insecticides, fungicides, additives to lubricants and gasolines
Di-n-octyl phthalate	Plasticizer for polyvinyl chloride and other vinyls
1,4-Dioxane	Solvent, lacquers, paints, varnishes, cleaning and detergent preparations, fumigants, paint and varnish removers, wetting agent, cosmetics
Ethyl acrylate	Polymers, acrylic paints, intermediate
Formic acid	Dyeing and finishing, chemicals, manufacture of fumigants, insecticides, solvents, plastics, refrigerants
Methanol (methyl alcohol)	Chemical manufacturing, solvents, automotive antifreeze, fuels
Methylcyclohexanone	Solvent, lacquers
Methyl ethyl ketone	Solvent, paint removers, cements and adhesives, cleaning fluids, printing, acrylic coatings
Methylphenyl acetamide	NA
Phenols (e.g., p-tert-butylphenol)	Resins, solvent, pharmaceuticals, reagent, dyestuffs and indicators, germicidal paints
Phthalic acid	Dyestuffs, medicine, perfumes, reagent
2-Propanol	Chemical manufacturing, solvent, deicing agent, pharmaceuticals, perfumes, lacquers, dehydrating agent, preservatives
2-Propyl-1-heptanol	Solvent
Tetrahydrofuran	Solvent
Varsol	Paint and varnish thinner
Hydrocarbons with specific elements (e.g., with N, P, S, Cl, Br, I, F)	
Acetyl chloride	Dyestuffs, pharmaceuticals, organic preparations
Alachlor (Lasso)	Herbicides
Aldicarb (sulfoxide and sulfone; Temik)	Insecticide, nematocide
Aldrin	Insecticides
Atrazine	Herbicides, plant growth regulator, weed-control agent
Benzoyl chloride	Medicine, intermediate
Bromacil	Herbicides
Bromobenzene	Solvent, motor oils, organic synthesis
Bromochloromethane	Fire extinguishers, organic synthesis
Bromodichloromethane	Solvent, fire extinguisher fluid, mineral and salt separations
Bromoform	Solvent, intermediate
Carbofuran	Insecticide, nematocide
Carbon tetrachloride	Degreasers, refrigerants and propellants, fumigants, chemical manufacturing
Chlordane	Insecticides, oil emulsions
Chlorobenzene	Solvent, pesticides, chemical manufacturing
Chloroform	Plastics, fumigants, insecticides, refrigerants and propellants

TABLE 1.2 Cont'd

Contaminant	Examples of uses
Hydrocarbons with specific elements (cont'd)	
Chlorohexane	NA
Chloromethane (methyl chloride)	Refrigerants, medicine, propellants, herbicide, organic synthesis
Chloromethyl sulfide	NA
2-Chloronaphthalene	Oil: plasticizer, solvent for dyestuffs, varnish gums and resins, waxes wax: moisture-, flame-, acid-, and insect-proofing of fibrous materials; moisture- and flame-proofing of electrical cable; solvent (see oil)
Chlorpyrifos	NA
Chlorthal-methyl (DCPA, or Dacthal)	Herbicide
p-Chlorophenyl methylsulfone	Herbicide manufacture
Chlorophenylmethyl sulfide	Herbicide manufacture
Chlorophenylmethyl sulfoxide	Herbicide manufacture
o-Chlorotoluene	Solvent, intermediate
p-Chlorotoluene	Solvent, intermediate
Cyclopentadine	Insecticide manufacture
Dibromochloromethane	Organic synthesis
Dibromochloropropane (DBCP)	Fumigant, nematocide
Dibromodichloroethylene	NA
Dibromoethane (ethylene dibromide, EDB)	Fumigant, nematocide, solvent, waterproofing preparations, organic synthesis
Dibromomethane	Organic synthesis, solvent
Dichlofenthion (DCFT)	Pesticides
o-Dichlorobenzene	Solvent, fumigants, dyestuffs, insecticides, degreasers, polishes, industrial odor control
p-Dichlorobenzene	Insecticides, moth repellant, germicide, space odorant, intermediate, fumigants
Dichlorobenzidine	Intermediate, curing agent for resins
Dichlorocyclooctadiene	Pesticides
Dichlorodiphenyldichloroethane (DDD, TDE)	Insecticides
Dichlorodiphenyldichloroethylene (DDE)	Degradation product of DDT, found as an impurity in DDT residues
Dichlorodiphenyltrichloroethane (DDT)	Pesticides
1,1-Dichloroethane	Solvent, fumigants, medicine
1,2-Dichloroethane	Solvent, degreasers, soaps and scouring compounds, organic synthesis, additive in antiknock gasoline, paint and finish removers
1,1-Dichloroethylene (vinylidiene chloride)	Saran (used in screens, upholstery, fabrics, carpets, etc.), adhesives, synthetic fibers
1,2-Dichloroethylene (cis and trans)	Solvent, perfumes, lacquers, thermoplastics, dye extraction, organic synthesis, medicine
Dichloroethyl ether	Solvent, organic synthesis, paints, varnishes, lacquers, finish removers, drycleaning, fumigants
Dichloroiodomethane	NA
Dichloroisopropylether (= bis-2-chloroisopropylether)	Solvent, paint and varnish removers, cleaning solutions
Dichloromethane (methylene chloride)	Solvent, plastics, paint removers, propellants, blowing agent in foams
Dichloropentadiene	NA
2,4-Dichlorophenol	Organic synthesis
2,4-Dichlorophenoxyacetic acid (2,4-D)	Herbicides

TABLE 1.2 Cont'd

Contaminant	Examples of uses
Hydrocarbons with specific elements (cont'd)	
1,2-Dichloropropane	Solvent, intermediate, scouring compounds, fumigant, nematocide, additive for antiknock fluids
Dicyclopentadiene (DCPD)	Insecticide manufacture
Dieldrin	Insecticides
Diiodomethane	Organic synthesis
Diisopropylmethyl phosphonate (DIMP)	Nerve gas manufacture
Dimethyl disulfide	NA
Dimethylformamide	Solvent, organic synthesis
2,4-Dinotrophenol (Dinoseb, DNBP)	Herbicides
Dithiane	Mustard gas manufacture
Dioxins (e.g., TCDD)	Impurity in the herbicide 2,4,5-T
Dodecyl mercaptan (lauryl mercaptan)	Manufacture of synthetic rubber and plastics, pharmaceuticals, insecticides, fungicides
Endosulfan	Insecticides
Endrin	Insecticides
Ethyl chloride	Chemical manufacturing, anesthetic, solvent, refrigerants, insecticides
Bis-2-ethylhexylphthalate	Plastics
Di-2-ethylexylphthalate	Plasticizers
Fluorobenzene	Insecticide and larvicide intermediate
Fluoroform	Refrigerants, intermediate, blowing agent for foams
Heptachlor	Insecticides
Heptachlorepoxide	Degradation product of heptachlor, also acts as an insecticide
Hexachlorobicycloheptadiene	NA
Hexachlorobutadiene	Solvent, transformer and hydraulic fluid, heat-transfer liquid
α-Hexachlorocyclohexane (= Benzenehexachloride, or α-BHC)	Insecticides
β-Hexachlorocyclohexane (β-BHC)	Insecticides
γ-Hexachlorocyclohexane (γ-BHC, or Lindane)	Insecticides
Hexachlorocyclopentadiene	Intermediate for resins, dyestuffs, pesticides, fungicides, pharmaceuticals
Hexachloroethane	Solvent, pyrotechnics and smoke devices, explosives, organic synthesis
Hexachloronorbornadiene	NA
Isodrin	Intermediate compound in manufacture of Endrin
Kepone	Pesticides
Malathion	Insecticides
Methoxychlor	Insecticides
Methyl bromide	Fumigants, pesticides, organic synthesis
Methyl parathion	Insecticides
Oxathine	Mustard gas manufacture
Parathion	Insecticides
Pentachlorophenol (PCP)	Insecticides, fungicides, bactericides, algicides, herbicides, wood preservative
Phorate (Disulfoton)	Insecticides
Polybrominated biphenyls (PBBs)	Flame retardant for plastics, paper, and textiles
Polychlorinated biphenyls (PCBs)	Heat-exchange and insulating fluids in closed systems
Prometon	Herbicides

TABLE 1.2 *Cont'd*

Contaminant	Examples of uses
Hydrocarbons with specific elements (cont'd)	
RDX (Cyclonite)	Explosives
Simazine	Herbicides
Tetrachlorobenzene	NA[a]
Tetrachloroethanes (1,1,1,2 and 1,1,2,2)	Degreasers, paint removers, varnishes, lacquers, photographic film, organic synthesis, solvent, insecticides, fumigants, weed killer
Tetrachloroethylene (or perchloroethylene, PCE)	Degreasers, drycleaning, solvent, drying agent, chemical manufacturing, heat-transfer medium, vermifuge
Toxaphene	Insecticides
Triazine	Herbicides
1,2,4-Trichlorobenzene	Solvent, dyestuffs, insecticides, lubricants, heat-transfer medium (e.g., coolant)
Trichloroethanes (1,1,1 and 1,1,2)	Pesticides, degreasers, solvent
1,1,2-Trichloroethylene (TCE)	Degreasers, paints, drycleaning, dyestuffs, textiles, solvent, refrigerant and heat exchange liquid, fumigant, intermediate, aerospace operations
Tricholorfluoromethane (Freon 11)	Solvent, refrigerants, fire extinguishers, intermediate
2,4,6-Trichlorophenol	Fungicides, herbicides, defoliant
2,4,5-Tricholorophenoxyacetic acid (2,4,5-T)	Herbicides, defoliant
2,4,5-Trichlorophenoxypropionic acid (2,4,5-TP or Silvex)	Herbicides and plant growth regulator
Trichlorotrifluoroethane	Dry-cleaning, fire extinguishers, refrigerants, intermediate, drying agent
Trinitrotoluene (TNT)	Explosives, intermediate in dyestuffs and photographic chemicals
Tris-(2,3-dibromopropyl) phosphate	Flame retardant
Vinyl chloride	Organic synthesis, polyvinyl chloride and copolymers, adhesives
Other hydrocarbons	
Alkyl sulfonates	Detergents
Cyclohexane	Organic synthesis, solvent, oil extraction
1,3,5,7-Cyclooctatetraene	Organic research
Dicyclopentadiene (DCPD)	Intermediate for insecticides, paints and varnishes, flame retardants
2,3-Dimethylhexane	NA
Fuel oil	Fuel, heating
Gasoline	Fuel
Jet fuels	Fuel
Kerosene	Fuel, heating solvent, insecticides
Lignin	Newsprint, ceramic binder, dyestuffs, drilling fuel additive, plastics
Methylene blue activated substances (MBAS)	Dyestuffs, analytical chemistry
Propane	Fuel, solvent, refrigerants, propellants, organic synthesis
Tannin	Chemical manufacturing, tanning, textiles, electroplating, inks, pharmaceuticals, photography, paper
4,6,8-Trimethyl-1-nonene	NA
Undecane	Petroleum research, organic synthesis
Metals and cations	
Aluminum	Alloys, foundry, paints, protective coatings, electrical industry, packaging, building and construction, machinery and equipment
Antimony	Hardening alloys, solders, sheet and pipe, pyrotechnics

TABLE 1.2 *Cont'd*

Contaminant	Examples of uses
Metals and cations (cont'd)	
Arsenic	Alloys, dyestuffs, medicine, solders, electronic devices, insecticides, rodenticides, herbicide, preservative
Barium	Alloys, lubricant
Beryllium	Structural material in space technology, inertial guidance systems, additive to rocket fuels, moderator and reflector of neutrons in nuclear reactors
Cadmium	Alloys, coatings, batteries, electrical equipment, fire-protection systems, paints, fungicides, photography
Calcium	Alloys, fertilizers, reducing agent
Chromium	Alloys, protective coatings, paints, nuclear and high-temperature research
Cobalt	Alloys, ceramics, drugs, paints, glass, printing, catalyst, electroplating, lamp filaments
Copper	Alloys, paints, electrical wiring, machinery, construction materials, electroplating, piping, insecticides
Iron	Alloys, machinery, magnets
Lead	Alloys, batteries, gasoline additive, sheet and pipe, paints, radiation shielding
Lithium	Alloys, pharmaceuticals, coolant, batteries, solders, propellants
Magnesium	Alloys, batteries, pyrotechnics, precision instruments, optical mirrors
Manganese	Alloys, purifying agent
Mercury	Alloys, electrical apparatus, instruments, fungicides, bactericides, mildew proofing, paper, pharmaceuticals
Molybdenum	Alloys, pigments, lubricant
Nickel	Alloys, ceramics, batteries, electroplating, catalyst
Palladium	Alloys, catalyst, jewelry, protective coatings, electrical equipment
Potassium	Alloys, catalyst
Selenium	Alloys, electronics, ceramics, catalyst
Silver	Alloys, photography, chemical manufacturing, mirrors, electronic equipment, jewelry, equipment, catalyst, pharmaceuticals
Sodium	Chemical manufacturing, catalyst, coolant, nonglare lighting for highways, laboratory reagent
Thallium	Alloys, glass, pesticides, photoelectric applications
Titanium	Alloys, structural materials, abrasives, coatings
Vanadium	Alloys, catalysts, target material for x-rays
Zinc	Alloys, electroplating, electronics, automotive parts, fungicides, roofing, cable wrappings, nutrition
Nonmetals and anions	
Ammonia	Fertilizers, chemical manufacturing, refrigerants, synthetic fibers, fuels, dyestuffs
Boron	Alloys, fibers and filaments, semiconductors, propellants
Chlorides	Chemical manufacturing, water purification, shrink-proofing, flame-retardants, food processing
Cyanides	Polymer production (heavy duty tires), coatings, metallurgy, pesticides
Fluorides	Toothpastes and other dentrifices, additive to drinking water
Nitrates	Fertilizers, food preservatives
Nitrites	Fertilizers, food preservatives

TABLE 1.2 Cont'd

Contaminant	Examples of uses
Nonmetals and anions (cont'd)	
Phosphates	Detergents, fertilizers, food additives
Sulfates	Fertilizers, pesticides
Sulfites	Pulp production and processing, food preservatives
Microorganisms	
Bacteria (coliform)	
Giardia	
Viruses	
Radionuclides	
Cesium 137	Gamma radiation source for certain foods
Chromium 51	Diagnosis of blood volume, blood cell life, cardiac output, etc.
Cobalt 60	Radiation therapy, irradiation, radiographic testing, research
Iodine 131	Medical diagnosis, therapy, leak detection, tracers (e.g., to study efficiency of mixing pulp fibers, chemical reactions, and thermal stability of additives to food products), measuring film thicknesses
Iron 59	Medicine, tracer
Lead 210	NA
Phosphorus 32	Tracer, medical treatment, industrial measurements (e.g., tire-tread wear and thickness of films and ink)
Plutonium 238, 243	Energy source, weaponry
Radium 226	Medical treatment, radiography
Radium 228	Naturally occurring
Radon 222	Medicine, leak detection, radiography, flow rate measurement
Ruthenium 106	Catalyst
Scandium 46	Tracer studies, leak detection, semiconductors
Strontium 90	Medicine, industrial applications (e.g., measuring thicknesses, density control)
Thorium 232	Naturally occurring
Tritium	Tracer, luminous instrument dials
Uranium 238	Nuclear reactors
Zinc 65	Industrial tracers (e.g., to study wear in alloys, galvanizing, body metabolism, function of oil additives in lubricating oils)
Zirconium 95	NA

The occurrence of the substances found on Tables 1.2 and 1.3 can be detected only if a ground-water sample has been collected and analyzed. In low concentrations most of these substances are colorless, tasteless, and odorless. A specific analytical technique must be employed to determine the presence and concentration of each substance. Unless a sample is collected and a specific test is performed, the presence of a contaminant may not be detected. With so many potential contaminants, it is possible that a sample could be collected and tested and a specific contaminant still not be found because no analysis was done for that compound or element.

TABLE 1.3 Organic compounds detected in ground water at Seymour Recycling Corporation hazardous waste site, Seymour, Indiana.

Extractable Organics	
Phenol	2-Chlorophenol
2,3,6-Trimethylphenol	2,4-Dimethylphenol
2,3-Dimethylphenol	2,6-Dimethylphenol
3,4-Dimethylphenol	3,5-Dimethyl phenol
2-Ethylphenol	2-Methyl phenol
3- and/or 4-Methylphenol	Bis(2-ethylhexyl)phthalate
Di-n-butyl phthalate	Isophorone
Benzo(a)anthracene	Chrysene
2-Butanone	2-Hexanone
4-Methyl-2-pentanone	3,3,5-Trimethylhexanol
2-Hexanol	2-Heptanone
Cyclohexanol	Cyclohexanone
4-Methyl-2-pentanol	4-Hydroxy-4-methyl-2-pentanone
2-Hydroxy-triethylamine	Tri-n-propyl-amine
Alkyl amine	1,4-Dioxane
n-n'-Dimethylformamide	n-n-Dimethylacetamide
Benzoic acid	4-Methylbenzoic acid
3-Methylbenzoic acid	3-Methyl-butanoic acid
Benzenepropionic acid	Benzeneacetic acid
2-Ethyl-hexanoic acid	2-Ethyl butanoic acid
Octanoic acid	Heptanoic acid
Hexanoic acid	Decanoic acid
Nonanoic acid	Pentanoic acid
Cyclohexanecarboxylic acid	1-Methyl-2-pyrrolidinone
1-1'-Oxy bis (2-methoxy ethane)	1,2-Dichlorobenzene
1,1,2-Trichloroethane	Tetrachloroethene
Volatile Organics	
Benzene	Ethyl benzene
Chloroform	Chloromethane
Chloroethane	1,2-Dichloroethane
1,1-Dichloroethane	1,1,1-Trichloroethane
1,1,2-Trichloroethane	1,1-Dichloroethene
Trans-1,2-Dichloroethene	Trichloroethene
Tetrachloroethene	Methylene chloride
Vinyl chloride	Dichlorofluoromethane
Tetrahydrofuran	Acetone
2-Butanone	2-Methyl-2-propanol
2-Methyl-2-butanol	2-Propanol
2-Butanol	2-Hexone
4-methyl-2-pentanol	Ethyl ether
m-Xylene	o- and/or p-Xylene
Toluene	

Note: Some compounds are detected in both the extractable and the volatile fractions and thus appear twice in the list.

Source: C. W. Fetter, Final Hydrogeologic Report, Seymour Recycling Corporation Hazardous Waste Site, Report to U. S. Environmental Protection Agency, Region V, September, 1985.

Introduction

TABLE 1.4 Cost of analysis of a single ground-water sample.

Superfund list of 137 synthetic organic compounds	$965
Twenty-three metals	270
Cyanide	40
Radiological compounds	275
Bacterial analysis (fecal coliform and streptococcus)	36
Chloride	10
Fluoride	18
Nitrate	15
Nitrite	15
Ammonia	15
Phosphorous, total	19
Sulfate	16
pH	6
Total	$1700

A great deal of expense is involved with a water-quality analysis. Table 1.4 lists the cost of an extensive laboratory analysis (at the 1992 list price from one independent, Wisconsin-certified lab). This table does not include the cost of collection of the sample to be analyzed.

The cost of analysis increases as the **detection limit,** the lowest concentration that can be reliably detected, decreases. Ground-water contaminants can be routinely detected at the parts-per-billion level, and with care some compounds can be quantified at the parts-per-trillion level. To put that concentration in perspective, 0.4 mm is one-trillionth of the distance to the moon.

1.3 Drinking-Water Standards

When measured at the parts-per-trillion level, even carefully prepared, triple-distilled, deionized water will be seen to contain some dissolved constituents. What does this mean? We must consider the quality of water with respect to the use to which it will be placed. Water for many industrial purposes need not be as pure as water used for drinking. In the United States the Safe Drinking Water Act and its amendments direct the Environmental Protection Agency to establish maximum contaminant-level goals (MCLGs) and maximum contaminant levels (MCLs) for drinking water supplied by public water agencies.

A maximum contaminant-level goal is a nonenforceable goal set at a level to prevent known or anticipated adverse health effects with a wide margin of safety. The MCLG for a carcinogen is zero, whereas for chronically toxic compounds it is based on an acceptable daily intake that takes into account exposure from air, food, and drinking water. Maximum contaminant levels are enforceable standards that are set as close as feasible to the MCLGs, taking into account water-treatment technologies and cost. Primary MCLs are based on health risk, and secondary MCLs are based on aesthetics. Table 1.5 contains the drinking-water standards promulgated by the U.S. Environmental Protection Agency.

TABLE 1.5 USEPA drinking-water standards and health goals.

Chemical	MCLG (μg/L)	MCL (μg/L)	SMCL (μg/L)
Synthetic organic chemicals			
Acrylamide (1)	0[d]	Treatment technique[d]	
Adipates (di(ethylhexyl)adipate)	500[f]	500[f]	
Alachlor	0[d]	2[d]	
Aldicarb	1[e]	3[e]	
Aldicarb sulfoxide	1[e]	4[e]	
Aldicarb sulfone	1[e]	2[e]	
Atrazine	3[d]	3[d]	
Benzene	0[a]	5[b]	
Benzo[a]anthracene (5)	0[f]	0.1[f]	
Benzo[a]pyrene	0[f]	0.2[f]	
Benzo[b]fluoranthene (5)	0[f]	0.2[f]	
Benzo[k]fluoranthene (5)	0[f]	0.2[f]	
Butylbenzyl phthalate (5)	100[f]	100[f]	
Carbofuran	40[d]	40[d]	
Carbontetrachloride	0[a]	5[b]	
Chlorodane	0[d]	2[d]	
Chrysene (5)	0[f]	0.2[f]	
Dalapon	200[f]	200[f]	
Dibenz[a,h]anthracene (5)	0[f]	0.3[f]	
Dibromochloropropane (DBCP)	0[d]	0.2[d]	
o-Dichlorobenzene (9)	600[d]	600[d]	10
p-Dichlorobenzene (9)	75[b]	75[b]	5
1,2-Dichloroethane	0[a]	5[b]	
1,1-Dichloroethylene	7[a]	7[b]	
cis-1,2-Dichloroethylene	70[a]	70[b]	
trans-1,2-Dichloroethylene	100[d]	100[d]	
1,2-Dichloropropane	0[d]	5[d]	
2,4-Dichlorophenoxyacetic acid (2,4-D)	70[d]	70[d]	
Di(ethylhexyl)phthalate	0[f]	4[f]	

Note: A pCi (picocurie) is a measure of the rate of radioactive disintegrations. Mrem ede/yr is a measure of the dose of radiation received by either the whole body or a single organ.

1. This is a chemical used in treatment of drinking water supply. The USEPA specifies how much may be used in the treatment process.
2. Dual numbers were proposed for aluminum because it is a constituent of a chemical used in the treatment of drinking water and it might not be possible for all treatment systems to meet the lower limit.
3. The total of nitrate plus nitrite cannot exceed 10 mg/L.
4. The proposed rule has two levels being considered.
5. The establishment of MCLGs and MCLs is not required by the Safe Drinking Water Act for these compounds; however, MCLGs and MCLs for them are being considered at the indicated levels.
6. This MCL would replace the current MCL of 5 pCi/L for combined 226 Ra and 228 Ra.
7. There is no MCL for copper and lead. The indicated values are proposed action levels that, under a complicated set of rules, would require treatment of a water supply to reduce potential corrosion of the water mains and pipes. The usual source of these compounds in public water supplies is primarily from the corrosion of copper and lead pipe and solder containing lead.
8. Standard under review as of January 1992.
9. SMCL is a suggested value only. Concentrations above this level may cause adverse taste. See *Federal Register*, January 30, 1991.

[a] Final value. Published in *Federal Register*, April 2, 1986.
[b] Final value. Published in *Federal Register*, July 8, 1987.
[c] Final value. Published in *Federal Register*, June 28, 1989.
[d] Final value. Published in *Federal Register*, January 30, 1991.
[e] Final value. Published in *Federal Register*, July 1, 1991.
[f] Proposed value. Published in *Federal Register*, July 25, 1990.
[g] Proposed value. Published in *Federal Register*, July 18, 1991.
[h] Final value. Published in *Federal Register*, July 7, 1991.
[i] Proposed value. Published in *Federal Register*, Nov. 13, 1985.
[j] Proposed value. Published in *Federal Register*, February, 1978.

TABLE 1.5 Cont'd

Chemical	MCLG (µg/L)	MCL (µg/L)	SMCL (µg/L)
Synthetic organic chemicals (cont'd)			
Diquat	20[f]	20[f]	
Dinoseb	7[f]	7[f]	
Endothall	100[f]	100[f]	
Endrin	2[f]	2[f]	
Epichlorohydrin (1)	0[d]	Treatment technique[d]	
Ethylbenzene (9)	700[d]	700[d]	30
Ethylene dibromide (EDB)	0[d]	0.05[d]	
Glyphosate	700[f]	700[f]	
Heptachlor	0[d]	0.4[d]	
Heptachlor epoxide	0[d]	0.2[d]	
Hexachlorobenzene	0[f]	1[f]	
Hexachlorocyclopentadiene [HEX]	50[f]	50[f]	8[f]
Indenopyrene (5)	0[f]	0.4[f]	
Lindane	0.2[d]	0.2[d]	
Methoxychlor	40[d]	40[d]	
Methylene chloride	0[f]	5[f]	
Monochlorobenzene	100[d]	100[d]	
Oxamyl (vydate)	200[f]	200[f]	
PCBs as decachlorobiphenol	0[d]	0.5[d]	
Pentachlorophenol	0[d]	1[d]	
Picloram	500[f]	500[f]	
Simaze	1[f]	1[f]	
Styrene (9)	100[d]	100[d]	10
2,3,7,8-TCDD (dioxin)	0[f]	5×10^{-5}[f]	
Tetrachloroethylene	0[d]	5[d]	
1,2,4-Trichlorobenzene	9[f]	9[f]	
1,1,2-Trichloroethane	3[f]	5[f]	
Trichloroethylene (TCE)	0[a]	5[b]	
1,1,1-Trichloroethane	200[a]	200[b]	
Toluene (9)	1000[d]	1000[d]	40
Toxaphene	0[d]	3[d]	
2-(2,4,5-Trichlorophenoxy)-propionic acid (2,4,5-TP, or Silvex)	50[d]	50[d]	
Vinyl chloride	0[a]	2[b]	
Xylenes (total) (9)	10,000[d]	10,000[d]	20
Inorganic chemicals			
Aluminum (2)			50–200[d]
Antimony (4)	3[f]	10/5[f]	
Arsenic (8)	50[i]	50[i]	
Asbestos (fibers per liter)	7×10^{6}[d]	7×10^{6}[d]	
Barium	2000[e]	2000[e]	
Beryllium	0[f]	1[f]	
Cadmium	5[d]	5[d]	
Chromium	100[d]	100[d]	
Copper (7)	1,300[h]	1,300[h]	
Cyanide	200[f]	200[f]	
Fluoride (8)	4,000[a]	4,000[a]	2,000[a]
Lead (7)	0[h]	15[h]	

TABLE 1.5 Cont'd

Chemical	MCLG (µg/L)	MCL (µg/L)	SMCL (µg/L)
Inorganic chemicals (cont'd)			
Mercury	2^d	2^d	
Nickel	100^f	100^f	
Nitrate (as N) (3)	$10,000^d$	$10,000^d$	
Nitrite (as N) (3)	$1,000^d$	$1,000^d$	
Selenium	50^d	50^d	
Silver			100^d
Sulfate (4)	$4 \times 10^5 - 5 \times 10^{5f}$	$4 \times 10^5 - 5 \times 10^{5f}$	
Thallium (4)	0.5^f	$2/1^f$	
Microbiological parameters			
Giardia lamblia	0 organisms^c		
Legionella	0 organisms^c		
Heterotrophic bacteria	0 organisms^c		
Viruses	0 organisms^c		
Radionuclides			
Radium 226 (6)	0^g	20 pCi/L^g	
Radium 228 (6)	0^g	20 pCi/L^g	
Radon 222	0^g	300 pCi/L^g	
Uranium	0^g	20 µg/L (30 pCi/L)^g	
Beta and Photon emitters (excluding radium 228)	0^g	4 mrem ede/yr^g	
Adjusted gross alpha emitters (excluding radium 226, uranium, and radon 222)	0^g	15 pCi/L^g	

1.4 Risk and Drinking Water

Cancer-risk levels for varying concentrations of contaminants have been established by toxicologists using extremely conservative methods. These methods are so conservative that some have questioned their validity (Ames, Magaw, and Gold 1987; Lehr 1990b). Such tests are performed by feeding chemicals in large doses to rodents and then extrapolating the effects to humans exposed to low doses by using linear extrapolation rates. However controversial the methods of establishing cancer risks in drinking water are, the MCLs obtained from them have the force of law. The basic cancer-risk level that the Environmental Protection Agency (EPA) uses is the 10^{-6} level—that is, one additional cancer death per million people. The EPA assumes that the person will consume 2 L of drinking water from the same source every day of their lives for 70 yr in arriving at the concentration that has a 10^{-6} cancer risk. The population at large appears to support such conservatism, even though about 25% of the population will eventually contract cancer (Wilson and Crouch 1987). If you are exposed to a carcinogen with a 10^{-6} risk level, your personal chances of contracting cancer are increased from 25% to 25.001% (Lehr 1990a). The cost to society to support this level of conservatism in purifying

Introduction

drinking water is significant. The cost is even greater when one considers the restoration of a large number of sites where the ground water has become contaminated with chemicals believed to be carcinogens. In 1988 the EPA examined the studies that had been performed at 153 Superfund sites. At close to one quarter of the sites, the cleanup costs were more than $10 million dollars, and at one site they were $120 million (Hanmer 1989).

There is an irreducible risk associated with drinking water. In order to protect against pathogenic disease, drinking water is usually chlorinated, especially if the water comes from a surface source. Prior to chlorination of drinking-water supplies, waterborne disease such as typhoid and cholera took many lives. Between 1920 and 1950, a period when the percentage of the population served by safe drinking-water supplies was increasing, there were 1050 deaths in the United States due to waterborne disease, including typhoid fever, gastroenteritis, shigellosis, and amebiasis. Since 1950 there have only been 20 deaths from similar causes (van der Leeden, Troise, and Todd 1990, Table 7-148).

The chlorine reacts with naturally occurring organics in the water to produce trihalomethanes. The average chlorinated tap water in the United States is reported to contain 83 μg/L of chloroform (Ames, Magaw, and Gold 1987). Ames, Magaw, and Gold used this as a base with which to compare other potential cancer risks. Table 1.6 contains cancer risks relative to drinking a liter of chlorinated tap water a day, with tap water having a risk of 1.0. The relative risks were determined as an index obtained by dividing the daily lifetime human exposure in milligrams per kilogram of body weight by the daily dose rate for rodents in milligrams per kilogram of body weight. The dose rate of rodents is the daily dose necessary to give cancer to half the rodents at the end of a standard lifetime. Examination of the table shows that there are numerous cancer risks associated with living and eating. Water from a contaminated well that was closed in Santa Clara County, California (Silicon Valley), had 2800 μg/L of trichloroethylene. Drinking 1 L of this water per day has about half the relative cancer risk (4) as the risk from nitrosamines ingested when one has bacon for breakfast (9). The bacon carries additional risk because high dietary fat is thought to be a possible contributor to colon cancer (Ames, Magaw, and Gold 1987). Water with 2800 μg/L of trichloroethylene has a 10^{-3} cancer risk based on the Environmental Protection Agency's Section 304(1)(1) criteria (Federal Register, November 28, 1983).

However, the consumer of bacon has made a conscious decision to eat it and accept the health risks. Consumers of tap water have the expectation that it is "safe" to drink and are probably not willing to accept even very low cancer risks. Society as a whole places a high value on pure water and is willing to pay to protect it.

1.5 Sources of Ground-Water Contamination

In a 1984 report, *Protecting the Nation's Groundwater from Contamination,* the Office of Technology Assessment (OTA) of the U.S. Congress listed more than 30 different potential sources of ground-water contamination. Although most attention has focused on waste materials as a source of ground-water contamination, there are numerous sources that are not associated with solid or liquid wastes. The OTA report divides the

TABLE 1.6 Risk of getting cancer relative to drinking chlorinated tap water.

Relative Risk	Source/Daily Human Exposure	Carcinogen
	Water	
1.0	Chlorinated tap water, 1 L	Chloroform, 82 µg
4.0	Well water, 1 L (worst well in Silicon Valley)	Trichloroethylene, 2800 µg
	Risks in Food	
30.0	Peanut butter, 1 sandwich	Aflatoxin
100.0	Mushroom, 1, raw	Hydrazines, etc.
2,800.0	Beer, 12 oz	Ethyl alcohol
4,700.0	Wine, 1 glass	Ethyl alcohol
0.3	Coffee, 1 cup	Hydrogen peroxide
30.0	Comfrey herbal tea, 1 cup	Symphytine
400.0	Bread, 2 slices	Formaldehyde
2,700.0	Cola, 1	Formaldehyde
90.0	Shrimp, 100 g	Formaldehyde
9.0	Cooked bacon, 100 g	Dimethylnitrosamine, diethylnitrosamine
60.0	Cooked fish or squid, broiled in a gas oven, 54 g	Dimethylnitrosamine
70.0	Brown mustard, 5 g	Allyl isothiocyanate
100.0	Basil, 1 g of dried leaf	Estragole
20.0	All cooked food, average U.S. diet	Heterocyclic amines
200.0	Natural root beer, 12 oz. (now banned)	Safrole
	Food Additives and Pesticides	
60.0	Diet soft drink, 12 oz.	Saccharin
0.4	Bread and grain products, average U.S. diet	Ethylene dibromide
0.5	Other food with pesticides, average U.S. diet	PCBs, DDE/DDT
	Risks Around the Home	
604.0	Breathing air in a conventional home, 14 hr	Formaldehyde, benzene
2,100.0	Breathing air in a mobile home, 14 hr	Formaldehyde
8.0	Swimming pool, 1 hr (for a child)	Chloroform
	Risks at Work	
5,800.0	Breathing air at work, U.S. average	Formaldehyde
	Commonly Used Drugs	
16,000.0	Sleeping pill (Phenobarbital), 60 mg	Phenobarbital
300.0	Pain-relief pill (Phenacetin), 300 mg	Phenacetin

Source: Jay Lehr, "Toxicological risk assessment distortions: Part III—A different look at environmentalism," Ground Water 28, no. 3 (1990): 330–40. Based on a table and data in Bruce Ames, Renae Magaw, and Lois Gold, "Ranking possible carcinogenic hazards," Science 236 (April 17, 1987): 271–79.

contamination sources into six categories. The following discussion has added some sources not contained in the OTA report. Figure 1.1 illustrates some of these contamination sources.

1.5.1 Category I: Sources Designed to Discharge Substances

Septic tanks and cesspools

Septic tanks and cesspools are designed to discharge domestic wastewater into the subsurface above the water table. Water from toilets, sinks and showers, dishwashers,

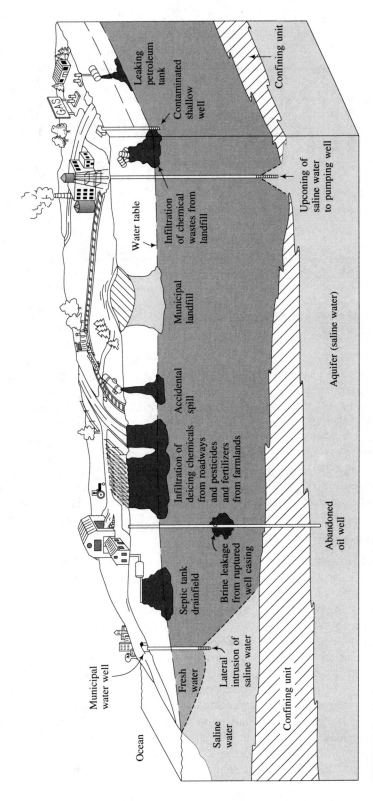

FIGURE 1.1 Mechanisms of ground-water contamination.

TABLE 1.7 Effluent quality from six septic tanks.[a]

Site	Average Flow (g/da)	BOD (mg/L)	COD (mg/L) (unfiltered)	COD (mg/L) (filtered)	TSS (mg/L)	Fecal Coliforms (no./mL)	Fecal Strep (no./mL)	Total N (mg/L)	Ammonia N (mg/L)	Nitrate-Nitrogen (mg/L)	Total P (mg/L)	Ortho P (mg/L)
A	75	131	325	249	69	2907	2.7	50.5	34.1	0.68	12.3	10.8
B	125	176	361	323	44	4127	39.7	57.8	42.5	0.46	14.1	13.6
C	245	272	542	386	68	27,931	1387	76.3	45.6	0.60	31.4	14.0
D	315	127	291	217	52	11,113	184	40.2	33.2	0.35	11.0	10.1
E	860[b]	120	294	245	51	2310	20.7	31.6	20.1	0.16	11.1	10.5
F	150	122	337	281	48	3246	25.3	56.7	38.3	0.83	11.6	10.5

Source: R. J. Otis, W. C. Boyle, and D. K. Sauer, Small-Scale Waste Management Program, University of Wisconsin—Madison, 1973.
[a] All values are means.
[b] Includes 340-g/da sewer flow and 520-g/da from foundation drain.

and washing machines passes from the home into a septic tank, where it undergoes settling and some anaerobic decomposition. It is then discharged to the soil via a drainage system. In 1977 there were an estimated 16.8 million septic systems in use in the United States (Miller 1980). Septic systems discharge a variety of inorganic and organic compounds. Table 1.7 contains an analysis of septic-tank effluent. In addition to the domestic wastewater, septic-tank cleaners containing synthetic organic chemicals such as trichloroethylene, benzene, and methylene chloride are discharged to the subsurface. An estimated 400,000 gal of septic-tank cleaning fluids were used on Long Island, New York, in 1979 (Burmaster and Harris 1982). Shallow ground water on Long Island is known to be contaminated by these same chemicals (Eckhardt and Oaksford 1988).

Injection wells

Injection wells are used to discharge liquid wastes and other liquids into subsurface zones below the water table. Liquids that are injected include (1) hazardous wastes, (2) brine from oil wells, (3) agricultural and urban runoff, (4) municipal sewage, (5) air-conditioning return water, (6) heat-pump return water, (7) liquids used for enhanced oil recovery from oil fields, (8) treated water intended for artificial aquifer recharge, and (9) fluids used in solution mining.

Injection wells can cause ground-water contamination if the fluid being injected accidentally or deliberately enters a drinking-water aquifer. This could happen because of poor well design, poor understanding of the geology, faulty well construction, or deteriorated well casing. Wastewater correctly injected into subsurface zones containing unusable water could still migrate to a usable aquifer by being forced through cracks in a confining layer under unnatural pressures or by flowing through the aquifer to a nearby well that was improperly constructed or abandoned. Injection wells are now regulated under the Underground Injection Control Program of the Safe Drinking Water

Act. The 1984 amendments to the Resource Conservation and Recovery Act prohibit the underground injection of certain hazardous wastes.

Land application

Treated or untreated municipal and industrial wastewater is applied to the land primarily via spray irrigation systems. Exposure to the elements, plants, and microorganisms in the soil can break down the natural organic matter in the wastewater.

Sludge from wastewater-treatment plants is often applied to the soil as a fertilizer, as is manure from farm animals and whey from cheese manufacturing. Oily wastes from refining operations have been applied to the soil so that they could be broken down by soil microbes. Nitrogen, phosphorous, heavy metals, and refractory organic compounds are potential ground-water contaminants that can leach from soil used for land applications of wastes and wastewater.

1.5.2 Category II: Sources Designed to Store, Treat and/or Dispose of Substances

Landfills

Landfills are, by definition, designed to minimize adverse effects of waste disposal (Miller 1980). However, many were poorly designed and are leaking liquids, generically termed **leachate,** which are contaminating ground water. Landfills can contain nonhazardous municipal waste, nonhazardous industrial waste, or hazardous waste as defined by the Resource Conservation and Recovery Act. Peterson (1983) reported that there were 12,991 landfills in the United States, including 2395 open dumps. There are an unknown number of abandoned landfills.

Materials placed in landfills include such things as municipal garbage and trash, demolition debris, sludge from wastewater-treatment plants, incinerator ash, foundry sand and other foundry wastes, and toxic and hazardous materials. Although no longer permitted, liquid hazardous waste was disposed in landfills in the past.

Leachate is formed from the liquids found in the waste as well as by leaching of the solid waste by rainwater. Table 1.8 contains information on the chemical composition of leachate from municipal landfills. To minimize the amount of leachate generated, modern landfills are built in sections, with a low-permeability cover placed over the waste as soon as possible to limit the infiltration of rainwater. Modern landfills also have low-permeability liner systems and collection pipes to remove the leachate that forms so that it can be taken to a wastewater-treatment plant. A modern landfill that is properly sited with respect to the local geology and that has a properly designed and constructed liner, leachate collection system, and low-permeability cover has limited potential to contaminate ground water. However, many landfills do not have liners and leachate collection systems. In the past, landfills tended to be placed in any convenient hole or low spot, such as a sand pit, quarry, or marsh. Ground-water contamination from such landfills is highly probable.

Municipal landfills are usually located near urban areas. The trend is toward large landfills that can handle many thousands of tons of waste per year. Hazardous-waste landfills are now regulated under the Resource Conservation and Recovery Act. There is frequently strong local opposition to the siting of either a municipal or a hazardous-waste landfill. This is referred to as the NIMBY syndrome: **N**ot **I**n **M**y **B**ack **Y**ard!

TABLE 1.8 Overall summary from the analysis of municipal solid-waste leachates in Wisconsin.

Parameter	Overall Range[a]	Typical Range (range of site medians)[a]	Number of Analyses
TDS	584–50,430	2180–25,873	172
Specific conductance	480–72,500	2840–15,485	1167
Total suspended solids	2–140,900	28–2835	2700
BOD	ND–195,000	101–29,200	2905
COD	6.6–97,900	1120–50,450	467
TOC	ND–30,500	427–5890	52
pH	5–8.9	5.4–7.2	1900
Total alkalinity ($CaCO_3$)	ND–15,050	960–6845	328
Hardness ($CaCO_3$)	52–225,000	1050–9380	404
Chloride	2–11,375	180–2651	303
Calcium	200–2500	200–2100	9
Sodium	12–6010	12–1630	192
Total Kjeldahl nitrogen	2–3320	47–1470	156
Iron	ND–1500	2.1–1400	416
Potassium	ND–2800	ND–1375	19
Magnesium	120–780	120–780	9
Ammonia-nitrogen	ND–1200	26–557	263
Sulfate	ND–1850	8.4–500	154
Aluminum	ND–85	ND–85	9
Zinc	ND–731	ND–54	158
Manganese	ND–31.1	0.03–25.9	67
Total phosphorus	ND–234	0.3–117	454
Boron	0.87–13	1.19–12.3	15
Barium	ND–12.5	ND–5	73
Nickel	ND–7.5	ND–1.65	133
Nitrate-nitrogen	ND–250	ND–1.4	88
Lead	ND–14.2	ND–1.11	142
Chromium	ND–5.6	ND–1.0	138
Antimony	ND–3.19	ND–0.56	76
Copper	ND–4.06	ND–0.32	138
Thallium	ND–0.78	ND–0.31	70
Cyanide	ND–6	ND–0.25	86
Arsenic	ND–70.2	ND–0.225	112
Molybdenum	0.01–1.43	0.034–0.193	7
Tin	ND–0.16	0.16	3
Nitrite-nitrogen	ND–1.46	ND–0.11	20
Selenium	ND–1.85	ND–0.09	121
Cadmium	ND–0.4	ND–0.07	158
Silver	ND–1.96	ND–0.024	106
Beryllium	ND–0.36	ND–0.008	76
Mercury	ND–0.01	ND–0.001	111

[a] All concentrations in milligrams per liter except pH (standard units) and specific conductance (μmhos/cm). ND indicates not detected.
Source: Wisconsin Department of Natural Resources.

Introduction

Open dumps

Open dumps are typically unregulated. They receive waste mainly from households but are used for almost any type of waste. Waste is frequently burned, and the residue is only occasionally covered with fill. Such dumps do not have liners and leachate-collection systems and by their nature are highly likely to cause ground-water contamination. The use of open dumps in the United States is no longer possible due to 1991 EPA regulations issued under Subtitle D of the Resource Conservation and Recovery Act, which requires extensive ground-water monitoring at such facilities, requires the placement of daily cover, prohibits burning, and will require engineered liners for future expansions. Most operators of open dumps did not want the expense of such regulations and so closed the dumps.

Residential disposal

Homeowners who are not served by a trash collection service must find alternative ways of disposing of their household waste. Included in the household waste are hazardous substances such as used engine oil and antifreeze and leftover yard and garden chemicals such as pesticides, unused paint, and used paint thinner. In the past these were often taken to the town dump. However, with the closing of most town dumps, the homeowner must find alternative means of disposal.

In Wisconsin virtually all town dumps were closed in 1989 and 1990. Most, but not all, counties offer waste disposal in a secure, engineered landfill. However, in large counties the county landfill may be 10 to 20 mi from some parts of the county and a fee is charged, as opposed to the old town dump, which was close by and free. In some situations the residents must drive to a different county to find an open landfill. Unfortunately, this closing of town dumps has resulted in an increase in illegal dumping in state and national forests and a great increase in trash left at roadside rest areas and parks.

Homeowners may pour waste liquids into ditches or the sanitary sewer; combustibles may be burned in the backyard. These are undesirable practices that can easily result in environmental pollution, including ground-water contamination.

Surface impoundments

Pits, ponds, and lagoons are used by industries, farmers, and municipalities for the storage and/or treatment of both liquid nonhazardous and hazardous waste and the discharge of nonhazardous waste. Prior to the passage of the Resource Conservation and Recovery Act, liquid hazardous wastes were also discharged into pits. These pits may be unlined or lined with natural material, such as clay, or artificial materials, such as plastic sheets, rubber membranes, or asphalt.

Impoundments are used to treat wastewater by such processes as settling of solids, biological oxidation, chemical coagulation and precipitation, and pH adjustment. They may also be used to store wastewater prior to treatment. Water from surface impoundments may be discharged to a receiving water course such as a stream or a lake. Unless a discharging impoundment is lined, it will also lose water by seepage into the subsurface. Nondischarging impoundments release water either by evaporation or seepage into the ground or a combination of both. Evaporation ponds are effective only in arid regions, where potential evapotranspiration far exceeds precipitation. Even evaporation ponds that were originally lined may leak and result in ground-water contamination if the liner deteriorates from contact with the pond's contents.

Impoundments are used for wastewater treatment by municipalities and industries such as paper manufacturing, petroleum refining, metals industry, mining, and chemical manufacturing. They are also used for treatment of agricultural waste, such as farm animal waste from feedlots. Power plants use surface impoundments as cooling ponds. Mining operations use surface ponds for the separation of tailings, which is waste rock from the processing of ore that occurs in a slurry.

Although it is now prohibited, until the 1970s lagoons were used for the disposal of untreated wastewater from manufacturing, ore processing, and other industrial uses into the ground water. Brine pits were used for many years in the oil patch for the disposal of brines pumped up with the oil. Miller (1980) lists 57 cases of ground-water contamination caused by the leakage of wastewater from surface impoundments. In most of the reported cases water-supply wells had been affected; at the time when use of such impoundments was allowed, ground-water monitoring was not required; usually the only way that leakage was detected was by contamination of a supply well.

In one case in Illinois, up to 500,000 gals per day of mineralized wastewater, containing high total dissolved solids (TDS), which included chloride, sulfate, and calcium, from an ore-processing plant were discharged into waste-disposal ponds excavated in a glacial drift aquifer for a period of about 40 yr. Concentrations of chloride, sulfate, TDS, and hardness were elevated in an underlying bedrock aquifer as much as a mile away from the site (U.S. Nuclear Regulatory Commission 1983).

Wastewater from the manufacturing of nerve gas and pesticides at the Rocky Mountain Arsenal at Denver was discharged into unlined evaporation ponds from 1942 until 1956. In 1956 a new pond lined with asphalt was constructed; ultimately that liner failed and the lined pond also leaked. Contamination of nearby farm wells was first detected in 1951 and was especially severe in the drought year of 1954, when irrigated crops died. Ground-water contamination extended at least 8 mi from the ponds and was indicated by high chloride content. Ultimately the ground water under and near the Rocky Mountain Arsenal was found to contain dozens of synthetic organic chemicals, including two that are especially mobile in the subsurface: diisopropylmethylphosphonate (DIMP), a by-product of the manufacture of nerve gas, and dicyclopentadiene (DCPD) a chemical used in the manufacture of pesticides (Konikow and Thompson 1984; Spanggord, Chou, and Mabey 1979). It is estimated that the cleanup of contaminated soil and ground water at the Rocky Mountain Arsenal will ultimately cost more than $1 billion (U.S. Water News, March, 1988).

The Environmental Protection Agency performed a survey of the surface impoundments located in the United States (U.S. EPA 1982). They reported a total of 180,973 impoundments, including 37,185 municipal, 19,437 agricultural, 27,912 industrial, 25,038 mining, 65,688 brine pits for oil and gas, and 5913 miscellaneous. The large number of impoundments provides a significant threat to ground-water resources (OTA 1984).

Mine wastes

Mining can produce spoils, or unneeded soil, sediment, and rock moved during the mining process, and tailings, or solid waste left over after the processing of ore. These wastes may be piled on the land surface, used to fill low areas, used to restore the land to premining contours, or placed in engineered landfills with leachate-collection systems. Mine wastes can generate leachate as rainwater passes through them. If sulfate or sulfide minerals are present, sulfuric acid can be generated, and the resulting drainage water can be acidic. This is likely to occur with coal-mining wastes, copper and gold ores, and ores from massive sulfide mineralization. Mine-waste leachate may also contain heavy metals and, in the case of uranium and thorium mines, radionuclides. Neutralization of the mine wastes can prevent the formation of acidic leachate and prevent the mobilization of many, but not all, metallic ions and radionuclides. The mine-waste disposal issue is a large one, because an estimated 2.3 billion tons of mine wastes are generated annually in the United States. Leachate produced by unneutralized or uncontained mine wastes is a threat to surface and ground water.

Material stockpiles

Many bulk commodities, such as coal, road salt, ores, phosphate rock, and building stone, are stored in outdoor stockpiles. Rainwater percolating through the stockpile can produce leachate similar to that produced by the waste material that resulted from mining the commodities. For example, rainwater draining through a coal pile can become acidic from sulfide minerals contained in the coal. In the northern states road salt is usually stored indoors, although in the past outdoor storage piles were common. Leachate from the road-salt piles was a common source of ground-water contamination that has now been mostly eliminated.

Graveyards

If bodies are buried without a casket or in a nonsealed casket, decomposition will release organic material. Areas of high rainfall with a shallow water table are most susceptible to ground-water contamination from graves. According to Bouwer (1978) contaminants can include high bacterial counts, ammonia, nitrate, and elevated chemical oxygen demand. Nash (1962) reported that hydrogen sulfide gas in a well was the result of a seventeenth-century graveyard for black plague victims. The well had apparently been unwittingly bored through the graveyard.

Animal burials

Unless an animal is a famous Kentucky thoroughbred or a beloved family pet, it is likely to simply be buried in an open excavation. If large numbers of animals are buried in close proximity, ground-water contamination might occur from the decomposing carcasses. If the animals had died due to some type of toxic poisoning, then additional opportunities for ground-water contamination would exist if the toxic chemical were released as the animals decomposed.

Above-ground storage tanks

Petroleum products, agricultural chemicals, and other chemicals are stored in above-ground tanks. Ruptures or leaks in the tanks can release chemicals, which then have the opportunity to seep into the ground. A serious case of ground-water contamination occurred in Shelbyville, Indiana, when one 55-gal tank of perchloroethylene was damaged by vandals and the contents leaked into the ground.

Underground storage tanks

The Office of Technology Assessment estimates that in the United States there are some 2.5 million underground storage tanks used to store fuel and other products (OTA 1984). There are at least two tanks, and frequently more, at every gas station. Many homeowners and farmers have private underground tanks to store heating oil and fuel. Chemicals are also routinely stored in underground tanks at industrial facilities. Liquid hazardous wastes can also be stored in underground tanks. Leachate from landfills with leachate-collection systems may be stored in a tank while it awaits trucking to a treatment facility.

Underground tanks can leak through holes either in the tank itself or in any associated piping. The piping appears to be more vulnerable. Steel tanks are susceptible to corrosion and are being replaced by fiberglass tanks. However, even with fiberglass tanks, the associated pipes can still leak. Fiberglass tanks do not have the strength of steel and may crack. A gas-station owner with a leaking tank can encounter tens of thousands of dollars in costs to remove a leaking tank and associated contaminated soil. Costs can be even higher if extensive ground-water contamination has occurred. In a 1-yr period a small consulting firm made 28 assessments of sites that contained underground fuel storage tanks. Even though none of the sites was known to have contamination prior to the assessments, 22 of the 28 sites (78%) were found to have leaking tanks (Gordon 1990). If one considered the sites being investigated because tanks were known to be leaking, the percentage of leaking tanks would be even higher.

Even the homeowner is at risk. One purchaser of an older home in the town of Black Wolf, Wisconsin, had the misfortune to discover an abandoned fuel-oil tank buried on his property. A total of forty-two 55-gal drums of a mixture of fuel oil and water were removed from the tank and had to be disposed of at considerable expense. Fortunately, as the tank was mostly below the water table, the water had leaked into the tank, rather than the fuel oil leaking out. Had the latter occurred, the costs to remove and dispose of contaminated soil would have been much higher.

Containers

Many chemical and waste products are stored in drums and other containers. Should these leak, there is a potential for ground-water contamination.

Open incineration and detonation sites

Sites for the open incineration of wastes are licensed under RCRA. In 1981 there were 240 such facilities in the United States (OTA 1984). The Department of Defense operates burning grounds and detonation sites for old ammunition. Chemicals released from such sites can leach into the ground with rainwater.

Radioactive-waste–disposal sites

The disposal of civilian radioactive wastes and uranium mill tailings is licensed under the Nuclear Regulatory Commission. High-level radioactive wastes from nuclear power plants are currently in temporary storage but will eventually go into an underground repository excavated into rock. The first repository is planned for Yucca Mountain, Nevada (U.S. Department of Energy 1988). Low-level wastes are buried in shallow landfills. Unless radioactive wastes are properly buried in engineered sites, there is a potential for radionuclides to migrate from the waste into ground water, as happened at Oak Ridge,

Tennessee; Hanford, Washington; Savannah River Facility, Georgia; and the Idaho National Engineering Lab.

1.5.3 Category III: Sources Designed to Retain Substances During Transport

Pipelines

Included in Category III are sewers to transmit wastewater as well as pipelines for the transmission of natural gas, petroleum products, and other liquids such as anhydrous ammonia. Although the pipelines are designed to retain their contents, many leak to a greater or lesser extent. This is particularly true of sewers, especially older sections. Sewers usually have a friction joint that can leak if the pipe shifts position. If the sewer is above the water table, leaking sewage can contaminate the ground water with bacteria, nitrogen, and chloride. Steel pipelines are subjected to corrosion and can also develop leaks. Such pipelines have been known to leak crude oil, gasoline, fuel oil, liquified petroleum gas, natural gas liquids, jet fuel, diesel fuel, kerosene, and anhydrous ammonia (OTA 1984).

Material transport and transfer

Material transport and transfer occurs by the movement of products and wastes via truck and train along transportation corridors and the associated use of loading facilities. Spills may result from accidents, and leaks can occur because of faulty equipment. A wide variety of materials can be released to the environment in this manner. Experienced and well-trained crews with the proper equipment are needed to clean up such spills. Improper actions can result in a spill becoming more severe as a result of a misguided cleanup effort.

1.5.4 Category IV: Sources Discharging Substances as a Consequence of Other Planned Activities

Irrigation

When crops are irrigated, more water is applied to the field than is needed for evapotranspiration. The excess water, called **return flow,** percolates through the soil zone to the water table. In doing so it can mobilize chemicals applied to the fields as fertilizers and pesticides. Soil salinity and salinity of the shallow ground water can also increase, because the evaporation of water concentrates the natural salts carried in the irrigation water. Selenium has been concentrated in irrigation return water that has been discharged to the Kesterson Wildlife Refuge in California's Central Valley.

Pesticide applications

Chemicals are applied to crops to control weeds, insects, fungi, mites, nematodes, and other pests. In addition they are used for defoliation, desiccation, and growth regulation (OTA 1984). Approximately 552 million pounds of active ingredients were applied to crops in the United States in 1982, and there were 280 million acre-treatments with pesticides; some land was treated more than once, so the number of acres treated is actually less than 280 million acres (OTA 1984).

The use of pesticides has extensive potential for contaminating ground water. Pesticides applied to the soil may migrate through the soil to the water table. Pesticides in use today are usually biodegradable to some extent. However, their breakdown products (metabolites) can also be found in ground water. The potential for contamination is higher at sites where pesticides are mixed and application equipment is loaded and then rinsed when its use is finished. Soils under such areas may receive a much greater loading of pesticides than the cropland to which the pesticides are applied. Application of pesticides by aerial spraying may result in uneven distribution. More than 65% of pesticides are applied by aerial spraying, and the cleanup of the planes and disposal of associated wastewater poses a special problem (OTA 1984).

Atrazine has been used extensively for weed control in corn cultivation. In 1985, 3.3 million acres of Wisconsin farmland planted with corn was treated with it. A survey of atrazine in Wisconsin ground waters showed it occurred unevenly in areas where it was used on fields. Highest concentrations, up to 3.5 parts per billion, were associated with mixing sites and sandy river-bottom land (Wollenhaupt and Springman 1990).

Fertilizer application

Farmers and homeowners alike apply fertilizers containing nitrogen, phosphorous, and potassium (potash). Phosphorous is not very mobile in soil and thus does not pose a significant threat to ground water. The rate of potassium application is generally low and, although it is mobile, the literature does not indicate that potassium from fertilizers is a major factor in causing ground-water problems. However, nitrogen from fertilizers can be a major cause of ground-water contamination.

Farm animal wastes

Farm animal wastes have the potential to contaminate ground water with bacteria, viruses, nitrogen, and chloride. Animals that are kept on an open range disperse their wastes over a large area, and the potential for environmental contamination is low. Animals confined to a small area will concentrate their wastes in the barn, barnyard, or feedlot. Rainwater infiltrating these wastes can mobilize contaminants, which can be leached into the soil and eventually into ground water. Manure from farms may be spread onto fields as a fertilizer, whereas large feedlot operations often have wastewater treatment plants. In northern climates manure spread on frozen fields can have a deleterious effect on both surface and ground water during the spring melt. Many farms in northern areas now have concrete storage tanks for holding manure during the winter months.

Salt application for highway deicing

Many states in the snowbelt have a dry-pavement policy that requires the use of highway deicing salts on city streets, rural highways, and interstate highways. The primary deicing salt is rock salt, consisting mainly of sodium chloride. Additives to improve the handling of the salt include ferric ferrocyanide and sodium ferrocyanide. Chromate and phosphate may be added to reduce the corrosiveness of the salt (OTA 1984). The salt and additives eventually are carried from the roadway in runoff and may either wash into surface streams or seep into ground water.

Home water softeners

In areas where the water supply has high calcium and magnesium content, home water softeners are used to reduce the hardness. Home water softeners are recharged with sodium chloride salt. Chlorides from the salt are contained in the backwash water. If the area is not served by sewers, the backwash water is disposed by subsurface drainage via septic tanks or separate drain fields. Chlorides from this source can enter the ground-water reservoir (Hoffman and Fetter 1978).

Urban runoff

Precipitation over urban areas typically results in a greater proportion of runoff and less infiltration than that falling on nearby rural areas because of the greater amount of impervious land surface in the urban area. In addition, the urban runoff contains high amounts of dissolved and suspended solids from auto emissions, fluid leaks from vehicles, home use of fertilizers and pesticides, refuse, and pet feces. For the most part, the urban runoff is carried into surface receiving waters, but it may recharge the water table from leaking storm sewers. This can contribute to degradation of ground-water quality in urban areas.

Percolation of atmospheric pollutants

Atmospheric pollutants reach the land either as dry deposition or as dissolved or particulate matter contained in precipitation. Sources include automobile emissions, power-plant smokestacks, incinerators, foundries, and other industrial processes. Pollutants include hydrocarbons, synthetic organic chemicals, natural organic chemicals, heavy metals, sulfur, and nitrogen compounds. Infiltrating precipitation may carry these compounds into the soil and ground water.

Mine drainage

Surface and underground mining may disrupt natural ground-water flow patterns and expose rocks containing pyrite to oxygenated water. This can result in the production of acid water, which then drains from the mine. The acid mine drainage can result in surface- and ground-water contamination. In one very interesting case in Shullsburg, Wisconsin, a lead and zinc mine was active for 25 yr. In order to work the mine, the ground water table was lowered below the mine levels by pumping. Sulfide minerals in the rock were subjected to biologically mediated oxidation along fractures in the rock and mine workings. Contact of the resulting sulfuric acid with the dolomite host rock neutralized the sulfuric acid and produced highly soluble sulfate minerals. When the mining ceased due to economic factors, the dewatering pumps were shut down and the mine workings were flooded. Ground water in the mine workings dissolved the sulfate minerals and resulted in high sulfate (up to 3500 mg/L), iron (up to 20 mg/L), and zinc (up to 18 mg/L) concentrations. As a result ground-water quality of a number of nearby water supply wells was adversely impacted (Hoffman 1984).

1.5.5 Category V: Sources Providing a Conduit for Contaminated Water to Enter Aquifers

Production wells

Wells are drilled for the production of oil, gas, geothermal energy, and water. Contaminants can be introduced into the ground during the drilling of production wells. Improperly constructed wells, corroded well casings, and improperly abandoned wells can provide a conduit for the flow of contaminated surface water into the ground or the movement of contaminated ground water from one aquifer into another. Homeowners may route drainage water from their roof and basement drains into abandoned water-supply wells. Old dug wells may become receptacles for trash.

Monitoring wells and exploration borings

Many thousands of monitoring wells are being installed in the United States each year. Exploration borings are installed for the purposes of mineral exploration or construction design. These wells and borings have the same potential for cross contamination of aquifers and introduction of contaminated surface water as production wells.

Construction excavation

Construction activities can strip the soil from bedrock, thus removing much of the natural protection of bedrock aquifers from ground-water contamination. Urban runoff water can collect in open foundation excavations, which then provide a conduit to aquifers.

1.5.6 Category VI: Naturally Occurring Sources Whose Discharge is Created and/or Exacerbated by Human Activity

Ground-water–surface-water interactions

Some aquifers are recharged naturally from surface water if the stream stage is higher than the water table (Fetter 1988). If the surface-water body becomes contaminated, then the aquifer being recharged by that water could also become contaminated. An exception to this might occur if the surface-water contamination is by a material that could be adsorbed or removed by filtration when it passes through the alluvium under the stream. Wells located near a stream can induce infiltration from the stream into the ground-water reservoir by development of a cone of depression. Contaminated surface water can thereby be drawn into an aquifer.

Natural leaching

Dissolved minerals occur in ground water due to natural leaching from rocks and soil. Naturally occurring ground water may have total dissolved solids in excess of 10,000 to 100,000 mg/L and may contain undesirable concentrations of various anions and cations. Human activity that results in acid rain may enhance the ability of infiltrating rainwater to leach naturally occurring substances from rock and soil.

Saltwater intrusion

Development of freshwater supplies from coastal aquifers may lower the water table and induce saline ground water that occurs naturally beneath the oceans to move landward into formerly freshwater aquifers. Upconing of the saltwater-freshwater interface may also occur if the well field overlies an aquifer containing saline water (Fetter 1988). Ground-water development in areas susceptible to saltwater intrusion should be undertaken with a clear plan that is designed to maximize the amount of fresh water that can be developed while minimizing the amount of saltwater intrusion and other undesirable effects that can occur (Fetter 1972).

1.6 Relative Ranking of Ground-Water–Contamination Sources

Although there are many potential sources of ground-water contamination, some pose much more of a threat to ground water than others. Section 305(b) of the Federal Clean Water Act requires individual States to submit reports to the Environmental Protection Agency on the sources of ground water contamination in the state and the type of contaminants observed. The data submitted were used to compile *National Water Quality Inventory—1988 Report to Congress* (U.S. EPA 1990).

The states indicated all the ground-water–contamination sources that they considered to be major threats to ground water in their state. Figure 1.2 shows that more than half the states and territories listed underground storage tanks, septic tanks, agricultural activities, municipal landfills, and abandoned hazardous-waste sites as major threats to ground water. Other frequently listed sources include industrial landfills, other landfills, injection wells, regulated hazardous-waste sites, land application, road salt, saltwater intrusion, and brine pits from oil and gas wells.

States and territories could also give a priority ranking from 1 to 5, 1 being the highest priority, for the various sources of ground-water contamination. Figure 1.3 shows these priority rankings for the sources listed on Figure 1.2, whereas Figure 1.4 shows these rankings for the individual sources grouped as "other" on Figure 1.2. The highest-priority ranking was given to underground storage tanks with 15 states listing this as the

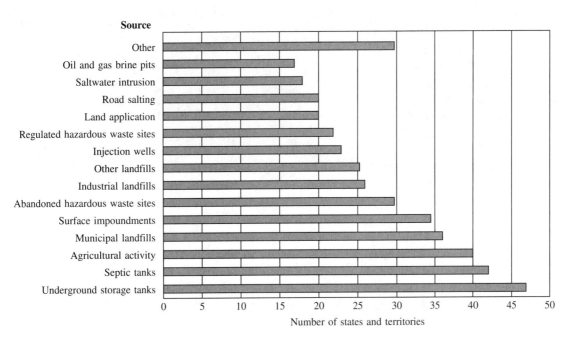

FIGURE 1.2 Frequency of various contamination sources considered by states and territories of the United States to be major threats to ground-water quality. *Source:* National Water Quality Inventory, 1988 Report to Congress, Environmental Protection Agency, 1990.

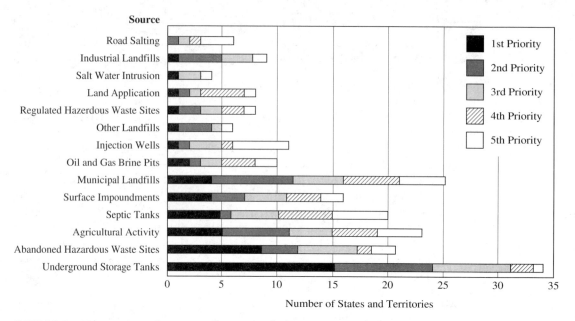

FIGURE 1.3 Priority ranking of contamination sources considered by more than 10 states and territories of the United States to be a major threat to ground-water quality. *Source:* National Water Quality Inventory, 1988 Report to Congress, Environmental Protection Agency, 1990.

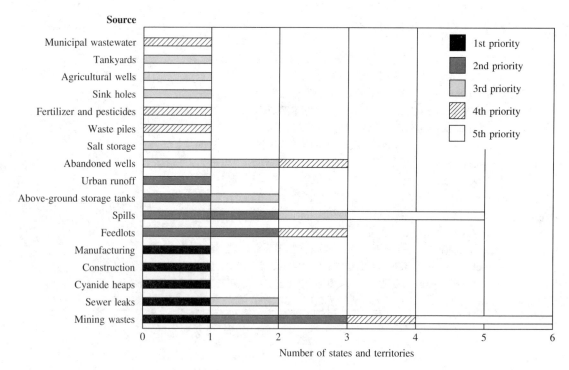

FIGURE 1.4 Priority ranking of contamination sources considered by fewer than 10 states and territories of the United States to be a major threat to ground-water quality. *Source:* National Water Quality Inventory, 1988 Report to Congress, Environmental Protection Agency, 1990.

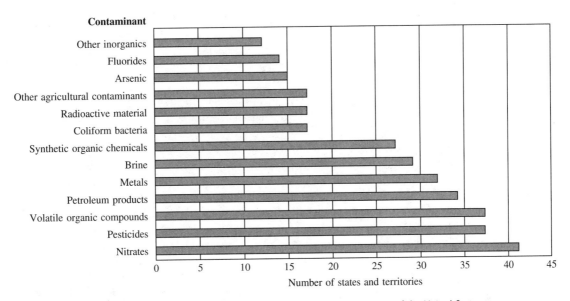

FIGURE 1.5 Frequency of various contaminants considered by states and territories of the United States to be a major threat to ground-water quality. *Source:* National Water Quality Inventory, 1988 Report to Congress, Environmental Protection Agency, 1990.

top problem and nine listing it as the second most severe problem. Rounding out a "dirty half-dozen" are abandoned hazardous-waste sites, agricultural activity, septic tanks, surface impoundments, and municipal landfills. Figure 1.4 shows that some states have unique problems that are severe at the state level but are not particularly problems across the country. Manufacturing, construction, cyanide heaps, and sewer leaks were listed as the number-one priority by one state but, with the exception of sewer leaks, were not even ranked by any other state.

The states also reported the ground-water contaminants of concern. These data are reported in Figure 1.5. The most frequently reported contaminants were nitrates, pesticides, volatile organic compounds, petroleum products, metals, brine, and synthetic organic chemicals. Other parameters of importance were bacteria, radioactive materials, other agricultural contaminants, arsenic, fluoride, and other inorganics.

1.7   Ground-Water Contamination as a Long-Term Problem

One of the factors of ground-water contamination that makes it so serious is its long-term nature. Wastes buried long ago may cause ground-water contamination that takes decades to be discovered. Although many ground-water–contamination sites are small, some of the long-term sites are fairly extensive due to the long time period over which contamination has been migrating away from the source.

In the 1930s poison baits utilizing arsenic were used in the Midwest to counter a grasshopper infestation. Apparently, leftover poison bait was buried when the infestations ended. In 1972 a water-supply well was drilled for a small business. In short order, 11 of 13 employees became ill with arsenic poisoning. Tests of the well showed it contained 21 mg/L of arsenic and soil at the site had 3000 to 12,000 mg/L of arsenic. This was apparently a mixing or burial site for arsenic-laden grasshopper bait (American Water Resources Association 1975).

Beginning in 1910 waste fuel oil and solvents from a railroad yard were discharged into the dry bed of the Mojave River near Barstow, California. A study in 1972 showed that a zone of contaminated ground water extended nearly 4.25 mi from the site and was 1800 ft wide (Hughes 1975).

Starting in 1936 a seepage lagoon was used for the disposal of treated domestic sewage at the Otis Air Force base, Cape Cod, Massachusetts. Over a 50-yr period about 2.5 billion gals of treated sewage was discharged into the rapid-infiltration ponds. The sewage percolated through the unsaturated zone and recharged a shallow sand and gravel aquifer. Because of the high rate of ground-water flow, about 1.0 to 1.5 ft per day, the plume has migrated more than 2 mi downgradient. The plume can be traced by elevated concentrations of chloride, boron, nitrate, detergents, and volatile organic compounds. The plume is narrow and thin due to limited transverse dispersion (Hess 1988).

A coal-tar distillation and wood-preservative plant was operated from 1918 to 1972 at St. Louis Park, Minnesota. Coal tar, which is obtained by heating coal in the absence of air, is a complex mixture of hundreds of organic compounds, including polynuclear aromatic hydrocarbons (PAH). The coal tar was distilled to form creosote, which was then used as a wood preservative. Coal-tar chemicals and creosote entered the environment by spills and drippings at the wood-preservative facility as well as via plant-process discharge water, which went into ponds. Coal tar is denser and more viscous than water and is only slightly soluble. The coal-tar compounds migrated downward into the underlying glacial drift aquifer. Several old, deep wells on the site had defective casings, which allowed coal tar to migrate downward into deep, bedrock aquifers. One 595-ft-deep well on the site was found to contain a column of coal tar 100 ft long. About 150 gal/min of contaminated water was entering this well from the glacial drift aquifer through a leak in the casing. This water then drained downward into the deep bedrock aquifers, carrying contamination with it. After 60 yr of leakage the contamination had spread more than 2 mi from the plant site in several directions. Water supply wells located outside of the area of contamination have drawn contaminated water into the bedrock aquifers up-gradient of the site in terms of the regional ground-water–flow direction (Hult and Stark 1988).

1.8 Review of Mathematics and the Flow Equation

1.8.1 Derivatives

Soil-moisture movement, ground-water flow, and solute transport may be described by means of partial differential equations. Thus, a brief review is in order.

If a bicyclist is traveling down a highway, we can measure the time that it takes the rider, who has a flying start, to go from a starting point ($S(t_1)$, or the location at the starting time, t_1) to a point somewhere down the highway ($S(t_2)$, or the location at elapsed time t_2). If we wish to know the average speed of the rider over this distance, we divide the distance from point $S(t_1)$ to point $S(t_2)$ by the elapsed time, $t_2 - t_1$.

$$\frac{\Delta S}{\Delta t} = \frac{S(t_2) - S(t_1)}{t_2 - t_1} \tag{1.1}$$

The rider will be going more slowly uphill and faster downhill. The average speed will thus include a lot of variation. If we were to measure the rider's speed over a shorter part of the course, there would be less variation in speed. As the length of time over which the distance traveled is measured becomes shorter and shorter, the variation in speed decreases. If the time becomes infinitesimally small—for example, the time that it takes the rider to travel a few microns—we obtain an instantaneous speed. This is known as the **first derivative of distance with respect to time** and is defined by

$$\frac{dS(t_1)}{dt} = \lim_{t \to t_1} \frac{S(t) - S(t_1)}{t - t_1} \tag{1.2}$$

where t is any arbitrary time. Figure 1.6 shows a graph of distance traveled by our bicyclist as a function of time. The slope of the line from time t_1 to time t_2 is the average speed over that part of the highway and is expressed as $\Delta S/\Delta t$. The instantaneous speed

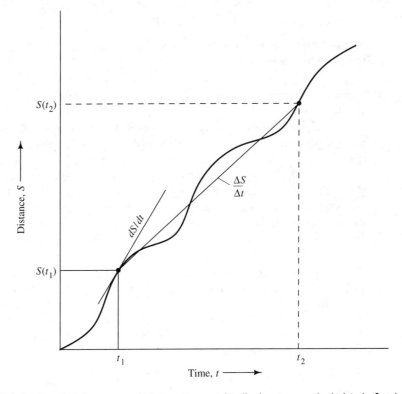

FIGURE 1.6 Graph of distance traveled versus time graphically showing speed, which is the first derivative of distance with respect to time.

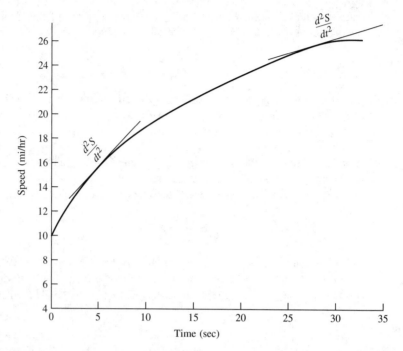

FIGURE 1.7 Graph of speed versus time graphically showing acceleration, which is the second derivative of distance with respect to time.

at time t_1 is the slope of the tangent to the curve at that point, which is expressed as dS/dt.

Note that the slope of distance versus time on Figure 1.6 keeps changing. This reflects the changes in speed that occur as the rider goes up and down hills. As the rider goes over the crest of a hill, he or she will perhaps be going rather slowly. As the rider goes downhill, the velocity will increase. We can compare the crest-of-the-hill velocity with the bottom-of-the-hill velocity and see that it has increased. This is a measure of the acceleration that occurs as gravity and the leg muscles of the bicyclist combine to increase speed. Figure 1.7 shows the speed of the rider as he or she goes over a hill. At $t = 0$ the rider is coming over the crest of the hill and the speed is 10 mi/hr. At $t = 30$ sec, when the rider is near the bottom of the hill, the speed is 26 mi/hr. The average rate of change in speed is (26 mi/hr − 10 mi/hr)/30 sec, or 0.53 mi/hr/sec. The rate of change is faster near the top of the hill where the slope is steeper and there is less wind resistance, since the rider is moving more slowly. From 0 to 5 sec the speed changes from 10 to 15 mi/hr, or 1.0 mi/hr/sec. Acceleration is the rate of change of speed with time, which is a second derivative. It is the slope of a tangent to the curve at a given time. It can be expressed as

$$\frac{d\left(\frac{dS}{dt}\right)}{dt} \quad \text{or} \quad \frac{d^2S}{dt^2}$$

Introduction

The tangent at 5 sec can be seen to be steeper than the tangent at 30 sec, where the rate of change is less.

In hydrogeology we have many parameters that are a function of more than one independent variable. For example, hydraulic head is a function of the three space variables: $h = h(x, y, z)$. We frequently differentiate head with respect to one of the space variables while holding the other two variables constant. Such derivatives of a parameter with respect to a single variable are called **partial derivatives.** The second derivative of hydraulic head with respect to the space variables is

$$\frac{\partial^2 h}{\partial x^2} + \frac{\partial^2 h}{\partial y^2} + \frac{\partial^2 h}{\partial z^2}$$

1.8.2 Darcy's Law

The first experimental study of ground-water flow was performed by Henry Darcy (Darcy 1856). He found that the one-dimensional flow of water through a pipe filled with sand was proportional to the cross-sectional area and the head loss along the pipe and inversely proportional to the flow length. Darcy's law can be expressed as

$$Q = -KA\frac{dh}{dl} \tag{1.3}$$

where

Q = volumetric discharge

K = proportionality constant known as hydraulic conductivity

A = cross-sectional area

dh/dl = gradient of hydraulic head

This equation can also be expressed in terms of **specific discharge,** or **Darcy flux,** q, which is the volumetric flow rate, Q, divided by the cross-sectional area, A.

$$q = -K\frac{dh}{dl} \tag{1.4}$$

Darcy's law was obtained for one-dimensional flow. However, as was previously stated, head is a function of all three dimensions: $h = h(x, y, z)$.

The hydraulic conductivity is the measure of the ability of the fractured or porous media to transmit water. It can have different values, depending upon the actual direction that the water is flowing through the porous media. In such a case the medium is said to be **anisotropic.** The value of the hydraulic conductivity can be measured in three principle directions, K_x, K_y, and K_z. If the hydraulic conductivity is the same in all directions, then $K_x = K_y = K_z = K$ and the medium is said to be **isotropic.**

1.8.3 Scalar, Vector, and Tensor Properties of Hydraulic Head and Hydraulic Conductivity

We first need to define some terms relating to **tensors.** A zero-order tensor, also called **a scalar,** is a quantity characterized only by its size or magnitude. Examples in hydrogeology include hydraulic head, chemical concentration, and temperature. A first-order tensor, or **vector,** is a quantity that has both a magnitude and a direction. Vectors require three components, each having a magnitude and direction. Velocity, specific discharge, mass flux, and heat flux are examples. A second-order tensor—or, simply, **tensor**—acts like the product of two vectors, requiring nine components to account for all possible products of the three components of each vector. Examples in hydrogeology are intrinsic permeability, hydraulic conductivity, thermal conductivity, and hydrodynamic dispersion.

The hydraulic head is a scalar. However, the gradient of the head is a vector as it has both a magnitude and a direction. The gradient of h is designated as grad h:

$$\text{grad } h = \mathbf{i}\frac{\partial h}{\partial x} + \mathbf{j}\frac{\partial h}{\partial y} + \mathbf{k}\frac{\partial h}{\partial z} \tag{1.5}$$

where $\mathbf{i}$, $\mathbf{j}$, and $\mathbf{k}$ are unit vectors in the x, y, and z directions. An equivalent notation is the use of the vector differential operator, del, which has the symbol ∇. This operator is equivalent to

$$\mathbf{i}\frac{\partial}{\partial x} + \mathbf{j}\frac{\partial}{\partial y} + \mathbf{k}\frac{\partial}{\partial z} \tag{1.6}$$

Another vector is the specific discharge, $\mathbf{q}$. It has three components, q_x, q_y, and q_z, when measured along the Cartesian coordinate axes. Associated with any vector is a positive scalar with a value equal to the magnitude of the vector. If q is the magnitude of the vector $\mathbf{q}$, this can be expressed as

$$q = |\mathbf{q}| \tag{1.7}$$

A second-order tensor, such as $\mathbf{K}$, hydraulic conductivity, can be described by nine components. In matrix form they are expressed as:

$$\mathbf{K} = \begin{bmatrix} K_{xx} & K_{xy} & K_{xz} \\ K_{yx} & K_{yy} & K_{yz} \\ K_{zx} & K_{zy} & K_{zz} \end{bmatrix} \tag{1.8}$$

If the tensor is symmetric, $K_{ij} = K_{ji}$; then inspection of (1.8) shows that there are only six independent components of $\mathbf{K}$.

If the coordinate system is oriented along the principal axes, the tensor becomes

$$\mathbf{K} = \begin{bmatrix} K_{xx} & 0 & 0 \\ 0 & K_{yy} & 0 \\ 0 & 0 & K_{zz} \end{bmatrix} \tag{1.9}$$

For the special case of an isotropic media—that is, the value of $\mathbf{K}$ does not depend upon the direction in which it is measured—the tensor becomes

$$\mathbf{K} = \begin{bmatrix} K & 0 & 0 \\ 0 & K & 0 \\ 0 & 0 & K \end{bmatrix} \tag{1.10}$$

The three components of the specific discharge vector, **q**, are

$$q_x = -K_{xx}\frac{\partial h}{\partial x} - K_{xy}\frac{\partial h}{\partial y} - K_{xz}\frac{\partial h}{\partial z}$$

$$q_y = -K_{yx}\frac{\partial h}{\partial x} - K_{yy}\frac{\partial h}{\partial y} - K_{yz}\frac{\partial h}{\partial z} \quad (1.11)$$

$$q_z = -K_{zx}\frac{\partial h}{\partial x} - K_{zy}\frac{\partial h}{\partial y} - K_{zz}\frac{\partial h}{\partial z}$$

For the special case where we orient the axes of the *x*, *y*, and *z* coordinate system with the three principal directions of anisotropy, **K** is the matrix shown in (1.9) and the three components of the specific discharge vector are

$$q_x = -K_{xx}\frac{\partial h}{\partial x}$$

$$q_y = -K_{yy}\frac{\partial h}{\partial y} \quad (1.12)$$

$$q_z = -K_{zz}\frac{\partial h}{\partial z}$$

For an isotropic material, **K** is represented by the matrix in (1.10) and

$$q = -K\frac{\partial h}{\partial x} - K\frac{\partial h}{\partial y} - K\frac{\partial h}{\partial z} \quad (1.13)$$

or

$$q = -K\,\text{grad}\,h \quad (1.14)$$

If we multiply two vectors together and the result is a scalar, then the product is called a **dot product,** or **inner product.** For example, the del operator dotted into a vector yields a scalar, called the **divergence.** Based on grad *h*, we can find a velocity vector **v** such that the magnitude and direction vary throughout the porous media. If we apply the del operator to **v**, we obtain the following:

$$\boldsymbol{\nabla}\cdot\mathbf{v} = \text{div}\,\mathbf{v} = \frac{\partial v_x}{\partial x} + \frac{\partial v_y}{\partial y} + \frac{\partial v_z}{\partial z} \quad (1.15)$$

If we apply the del operator to grad *h*, the result is the second derivative of head:

$$\boldsymbol{\nabla}\cdot\text{grad}\,h = \frac{\partial^2 h}{\partial x^2} + \frac{\partial^2 h}{\partial y^2} + \frac{\partial^2 h}{\partial z^2} \quad (1.16)$$

1.8.4 Derivation of the Flow Equation in a Deforming Medium

The **law of mass conservation** states that there can be no net change in the mass of fluid in a small representative elementary volume (REV) of a porous medium. In other words, the mass entering the REV less the mass leaving the REV is equal to the change in mass storage with time.

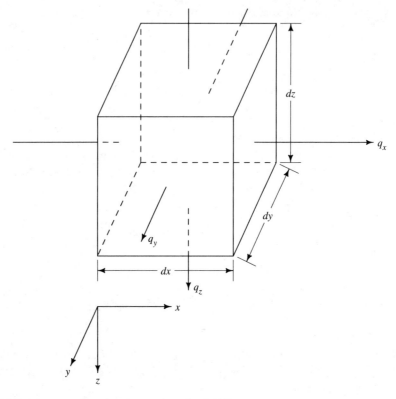

FIGURE 1.8 Representative elementary volume for fluid flow.

The representative elementary volume is shown on Figure 1.8. The three sides have length dx, dy, and dz, respectively. The area of the two faces normal to the x axis is $dy\,dz$, the area of the faces normal to the y axis is $dx\,dz$, and the area of the faces normal to the z axis is $dx\,dy$.

The component of mass flux into the REV parallel to the x axis is the fluid density times the flux rate:

$$\text{Mass influx along } x \text{ axis} = \rho_w q_x \, dy \, dz \tag{1.17}$$

where

ρ_w = fluid density (M/L^3)

q_x = specific discharge or volume of flow per cross-sectional area (L/T)

$dy\,dz$ = cross-sectional area (L^2)

The units of mass inflow are mass per unit time (M/T).*

---

*The units of a variable can be expressed in terms of their fundamental dimensions. These are length, L, mass, M, and time, T. The fundamental dimensions for density are mass per unit volume. Volume is length cubed, so the shorthand expression for the fundamental dimensions of density is M/L^3. Specific discharge has the dimensions of velocity, so the fundamental dimensions are L/T, and area has fundamental dimensions of L^2.

Introduction

The mass outflow rate will be different than the inflow rate and can be given as:

$$\text{Mass outflow rate parallel to } x \text{ axis} = \left[\rho_w q_x + \frac{\partial(\rho_w q_x)\, dx}{\partial x}\right] dy\, dz \quad (1.18)$$

The net mass accumulation within the control volume due to the flow component parallel to the x axis is the mass inflow minus the mass outflow, or

$$\frac{-\partial(\rho_w q_x)\, dx\, dy\, dz}{\partial x}$$

Similar terms exist for the net mass accumulation due to flow components parallel to the y and z axes:

$$\frac{-\partial(\rho_w q_y)\, dy\, dx\, dz}{\partial y}$$

$$\frac{-\partial(\rho_w q_z)\, dz\, dx\, dy}{\partial z}$$

These three terms can be summed to find the total net mass accumulation within the control volume.

$$-\left[\left(\frac{\partial}{\partial x}(\rho_w q_x) + \frac{\partial}{\partial y}(\rho_w q_y) + \frac{\partial}{\partial z}(\rho_w q_z)\right)\right] dx\, dy\, dz \quad (1.19)$$

The mass of water in the REV, M, is the density of water, ρ_w, times the porosity, n, times the volume, $dx\, dy\, dz$. The change in mass with respect to time is

$$\frac{\partial M}{\partial t} = \frac{\partial}{\partial t}(\rho_w n\, dx\, dy\, dz) \quad (1.20)$$

From the law of conservation of mass, Equation 1.19 must equal Equation 1.20.

$$-\left[\frac{\partial}{\partial x}(\rho_w q_x) + \frac{\partial}{\partial y}(\rho_w q_y) + \frac{\partial}{\partial z}(\rho_w q_z)\right] dx\, dy\, dz = \frac{\partial}{\partial t}(\rho_w n)\, dx\, dy\, dz \quad (1.21)$$

We can assume that although density of the fluid may change with time, at any given time it will be the same everywhere in the REV. Under this assumption Equation 1.21 can be simplified to

$$-\left[\frac{\partial q_x}{\partial x} + \frac{\partial q_y}{\partial y} + \frac{\partial q_z}{\partial z}\right] = \frac{1}{\rho_w}\frac{\partial}{\partial t}(\rho_w n) \quad (1.22)$$

We may substitute Darcy's law for the specific discharge components given on the left side. If the xyz coordinate system is aligned with the principal axes of anisotropy, then Equation 1.12 may be used, and the left side of Equation 1.22 becomes

$$\frac{\partial}{\partial x}\left(K_{xx}\frac{\partial h}{\partial x}\right) + \frac{\partial}{\partial y}\left(K_{yy}\frac{\partial h}{\partial y}\right) + \frac{\partial}{\partial z}\left(K_{zz}\frac{\partial h}{\partial z}\right) \quad (1.23)$$

The change in mass within the REV is due to changes in the porosity and the density of water as the head changes with time. Thus the change in the volume of water in storage is proportional to the change in head with time. The right side of Equation

1.22 can be expressed as a proportionality constant, S_s, the specific storage, times the change in head with time.

$$\frac{1}{\rho_w}\frac{\partial}{\partial t}(\rho_w n) = S_s \frac{\partial h}{\partial t} \qquad (1.24)$$

Combining Equations 1.22, 1.23, and 1.24 we obtain the main equation for transient flow in an anisotropic medium when the coordinate system is oriented along the principal axes of anisotropy:

$$\frac{\partial}{\partial x}\left(K_{xx}\frac{\partial h}{\partial x}\right) + \frac{\partial}{\partial y}\left(K_{yy}\frac{\partial h}{\partial y}\right) + \frac{\partial}{\partial z}\left(K_{zz}\frac{\partial h}{\partial z}\right) = S_s \frac{\partial h}{\partial t} \qquad (1.25)$$

1.8.5 Mathematical Notation

In del and tensor notation Equation 1.25 becomes

$$\nabla \cdot \mathbf{K} \cdot \nabla \mathbf{h} = S_s \frac{\partial h}{\partial t} \qquad (1.26)$$

Another form of expression is called Einstein's summation notation. For example, Darcy's law in the familiar, one-dimensional form is

$$q = -\mathbf{K}\frac{dh}{dl} \qquad (1.27)$$

It is implied in the preceding equation that the specific discharge is parallel to the direction of dh/dl and that the medium is isotropic. In a more general form, specific discharge, $\mathbf{q}$, is a vector with components q_1, q_2, and q_3. Grad h is a vector that we will call $\mathbf{h}$. This vector also has components h_1, h_2, and h_3. Hydraulic conductivity, $\mathbf{K}$, is a tensor with nine components. To describe Darcy's law in the most general form, we need three equations.

$$q_1 = K_{11}h_1 + K_{12}h_2 + K_{13}h_3 \qquad (1.28a)$$
$$q_2 = K_{21}h_1 + K_{22}h_2 + K_{23}h_3 \qquad (1.28b)$$
$$q_3 = K_{31}h_1 + K_{32}h_2 + K_{33}h_3 \qquad (1.28c)$$

The inner product can be expressed in index notation as

$$q_i = \sum_j K_{ij}h_j \qquad (i, j = 1, 2, 3) \qquad (1.29)$$

In Einstein's summation notation, the $\sum$ is dropped with the understanding that the summation is over the repeated indices:

$$q_i = K_{ij}h_j \qquad (i, j = 1, 2, 3) \qquad (1.30)$$

In vector notation this can be expressed as either

$$\mathbf{q} = \mathbf{K} \cdot \text{grad } h \qquad (1.31)$$

or

$$\mathbf{q} = \mathbf{K} \cdot \mathbf{h} \qquad (1.32)$$

In del notation this is

$$q = K \cdot \nabla h \qquad (1.33)$$

In general, we will use the standard form of differential equations rather than any of the shorthand notation. However, the literature cited in this text often uses the compact forms and the reader should be aware of them.

References

American Water Resources Association. 1975. Status of waterborne diseases in the U.S. and Canada. *Journal of American Water Works Association* 67, no. 2:95–98.

Ames, B. N., Renae Magaw, and L. S. Gold. 1987. Ranking possible carcinogenic hazards. *Science* 236 (April 17, 1987):271–77.

Bouwer, Herman. 1978. *Groundwater hydrology.* New York: McGraw-Hill Book Company, pp. 423–24.

Burmaster, D. R. and R. H. Harris. 1982. Groundwater contamination, an emerging threat. *Technology Review* 85, no. 5:50–62.

Darcy, Henry. 1856. *Les fontaines publiques de la ville de Dijon.* Paris: Victor Dalmont, 647 pp.

Eckhardt, D. A., and E. T. Oaksford. 1988. "Relation of land use to ground-water quality in the upper glacial aquifer, Long Island, New York." In *National Water Summary 1986—Hydrologic Events and Ground Water Quality.* U.S. Geological Survey Water Supply Paper 2325, pp. 115–121.

Fetter, C. W. 1972. The concept of safe groundwater yield in coastal aquifers. *Water Resources Bulletin* 8, no. 5:1173–76.

———. 1988. *Applied hydrogeology.* 2d ed. Columbus, Ohio: Merrill Publishing Company, 588 pp.

Gordon, James. 1990. OMNI Engineers, personal communication.

Hanmer, Rebecca. 1989. Environmental Protection Agency, Testimony in a hearing on the seriousness and extent of ground-water contamination before the Senate Subcommittee on Superfund, Ocean and Water Protection of the Committee of Environment and Public Works, August 1, 1989.

Hess, K. M. 1988. "Sewage Plume in a sand and gravel aquifer, Cape Cod, Massachusetts." In *National Water Summary, 1986,* 87–92. U.S. Geological Survey Water Supply Paper 2325.

Hindall, S. M., and Michale Eberle. 1989. National and regional trends in water well drilling in the United States, 1964–84. U.S. Geological Survey Circular 1029, 15 pp.

Hoffman, J. I. 1984. "Geochemistry of acid mine drainage on the aquifers of southeastern Wisconsin and regulatory implications." In *Proceedings of the National Water Well Association Conference on the Impact of Mining on Ground Water,* 146–161. August 24–27, 1984, Denver, Colorado, National Water Well Association, Dublin, Ohio.

Hoffman, J. I., and C. W. Fetter. 1978. Water softener salt: a major source of groundwater contamination. Geological Society of America, *Abstracts with Programs* 10, no. 7:423.

Hughes, J. L. 1975. Evaluation of ground-water degradation resulting from waste disposal to alluvium near Barstow, California. U.S. Geological Survey Professional Paper 878.

Hult, M. F., and J. R. Stark. 1988. Coal-tar derivatives in the Prairie du Chien–Jordan Aquifer, St. Louis Park, Minnesota. In *National Water Summary, 1986.* U.S. Geological Survey Water Supply Paper 2325, pp. 87–92.

Konikow, L. F., and D. W. Thompson. 1984. Groundwater contamination and aquifer reclamation at the Rocky Mountain Arsenal, Colorado. In *Groundwater Contamination,* Washington, D.C.: National Academy Press. 93–103.

Lehr, J. H. 1990a. Toxicological risk assessment distortions: Part I. *Ground Water* 28, no. 1:2–8.

———. 1990b. Toxicological risk assessment distortions: Part II—The dose makes the poison. *Ground Water* 28, no. 2:170–5.

Miller, David. 1980. *Waste disposal effects on ground water.* Berkley, Calif.: Premier Press, 512 pp.

Nash, G. J. C. 1962. Discussion of a paper by E. C. Wood. *Proceedings of the Society of Water Treatment and Examination* 11:33.

Office of Technology Assessment. 1984. *Protecting the nation's groundwater from contamination.* Washington, D.C.: U.S. Congress, OTA O-276, two volumes.

Peterson, N. M. 1983. 1983 survey of landfills. *Waste Age* (March 1983):37–40.

Solley, W. B., C. F. Merk, and R. R. Pierce. 1988. *Estimated use of water in the United States in 1985.* U.S. Geological Survey Circular 1004, 82 pp.

Spanggord, R. J., Tsong-Wen Chou, and W. R. Mabey. 1979. *Studies of environmental fates of DIMP and DCPD.* Contract report by SRI International, Menlo Park, Calif., for U.S. Army Medical Research and Development Command, Fort Detrick, Fredrick, Md, 65 pp.

U.S. Department of Energy. 1988. *Site characterization plan overview, Yucca Mountain Site, Nevada Research and Development Area.* Washington, D. C.: Office of Civilian Radioactive Waste Management, 164 pp.

U.S. Environmental Protection Agency. 1982. *Surface impoundment assessment national report.* Office of Drinking Water.

———. 1990. *National water quality inventory.* 1988 Report to Congress, EPA-440-4-90-003, 187 pp.

U.S. Nuclear Regulatory Commission. 1983. *Final environmental statement related to the decommissioning of the Rare Earths Facility, West Chicago, Illinois.* NUREG-0904.

U.S. Water News. 1988. At $1 billion, Colorado arsenal is costliest cleanup yet. 4, no. 9 (March 1988):1.

van der Leeden, Fritz, F. L. Troise, and D. K. Todd. 1990. *The water encyclopedia,* Chelsea, Mich: Lewis Publishers.

Wilson, Richard, and E. A. C. Crouch. 1987. Risk assessment and comparisons: An introduction. *Science,* 236 (April 17, 1987):267–70.

Wollenhaupt, N. C., and R. E. Springman. 1990. Atrazine in groundwater: A current perspective. University of Wisconsin—Extension, Agricultural Bulletin G3525, 17 pp.

Chapter Two
Mass Transport in Saturated Media

2.1 Introduction

In this chapter we will consider the transport of solutes dissolved in ground water. This is known as **mass** or **solute transport.** The methods presented in this chapter are based on partial differential equations for dispersion that have been developed for homogeneous media (Ogata and Banks 1961; Ogata 1970; Bear 1972; Bear and Verruijt 1987). These equations are similar in form to the familiar partial differential equations for fluid flow. In recent years much work has been done on the theories of mass transport in response to the great interest in problems of ground-water contamination. One of the outcomes of this has been the development of what is essentially a new branch of subsurface hydrology, where the flow of fluid and solutes is treated by statistical models; these models can account for the role of varying hydraulic conductivity that accompanies aquifer heterogeneity. Very recently fractal geometry has been used to describe the solute transport based on the concept that aquifer heterogeneities have repeating patterns.

2.2 Transport by Concentration Gradients

A solute in water will move from an area of greater concentration toward an area where it is less concentrated. This process is known as **molecular diffusion,** or **diffusion.** Diffusion will occur as long as a concentration gradient exists, even if the fluid is not moving. The mass of fluid diffusing is proportional to the concentration gradient, which can be expressed as **Fick's first law;** in one dimension, Fick's first law is

$$F = -D_d (dC/dx) \tag{2.1}$$

where

F = mass flux of solute per unit area per unit time

D_d = diffusion coefficient (L^2/T)

C = solute concentration (M/L^3)

dC/dx = concentration gradient ($M/L^3/L$)

The negative sign indicates that the movement is from areas of greater concentration to those of lesser concentration. Values for D_d are well known and range from 1×10^{-9} to 2×10^{-9} m^2/sec at 25°C. They do not vary much with concentration, but they are somewhat temperature-dependent, being about 50% less at 5°C (Robinson and Stokes 1965).

For systems where the concentrations are changing with time, **Fick's second law** applies. In one dimension this is

$$\partial C/\partial t = D_d \, \partial^2 C/\partial x^2 \qquad (2.2)$$

where $\partial C/\partial t$ = change in concentration with time (M/L^3/T).

In porous media, diffusion cannot proceed as fast as it can in water because the ions must follow longer pathways as they travel around mineral grains. To account for this, an effective diffusion coefficient, D^*, must be used.

$$D^* = \omega D_d \qquad (2.3)$$

where ω is a coefficient that is related to the tortuosity (Bear 1972). **Tortuosity** is a measure of the effect of the shape of the flowpath followed by water molecules in a porous media. If L is the straight-line distance between the ends of a tortuous flowpath of length L_e, the tortuosity, T, can be defined as $T = L_e/L$. Tortuosity in a porous media is always greater than 1, because the flowpaths that water molecules take must diverge around solid particles. Flowpaths across a representative sample of a well-sorted sediment would tend to be shorter than those across a poorly sorted sediment in which the smaller grains were filling the voids between the larger grains. Thus the well-sorted sediment would tend to have a lower value for tortuosity than the poorly sorted sediment. (Tortuosity has also been defined as $(L/L_e)^2$ (Carman 1937; Bear 1972). With this definition, tortuosity always has a value less than 1. This definition will not be used in this text.)

The value of ω, which is always less than 1, can be found from diffusion experiments in which a solute is allowed to diffuse across a volume of a porous medium. Perkins and Johnson (1963) found that ω was equal to 0.7 for sand column studies using a uniform sand. According to Freeze and Cherry (1979), ω ranges from 0.5 to 0.01 for laboratory studies using geologic materials.

Diffusion will cause a solute to spread away from the place where it is introduced into a porous medium, even in the absence of ground-water flow. Figure 2.1 shows the distribution of a solute introduced at concentration C_0, at time t_0, over an interval $(x - a)$ to $(x + a)$. At succeeding times t_1 and t_2, the solute has spread out, resulting in a lower concentration over the interval $(x - a)$ to $(x + a)$ but increasing concentrations outside of this interval.

The solute concentration follows a normal, or Gaussian, distribution and can be described by two statistical properties, the mean, $\overline{C}$, and variance, σ_c^2, which are defined in Section 2.12.2.

The effective diffusion coefficient, D^*, can be defined (De Josselin De Jong 1958) as

$$D^* = \frac{\sigma_c^2}{2t} \qquad (2.4)$$

Mass Transport in Saturated Media

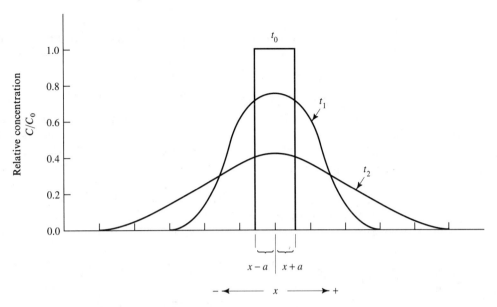

FIGURE 2.1 Spreading of a solute slug with time due to diffusion. A slug of solute was injected into the aquifer at time t_0 with a resulting initial concentration of C_0.

This is an alternative definition of effective diffusion coefficient to the one given in Equation 2.3.

The process of diffusion is complicated by the fact that the ions must maintain electrical neutrality as they diffuse. If we have a solution of NaCl, the Na^+ cannot diffuse faster than the Cl^- unless there is some other negative ion in the region into which the Na^+ is diffusing. If the solute is adsorbed onto the mineral surfaces of the porous medium, the net rate of diffusion will be obviously less that for a nonadsorbed species.

Diffusion can occur when the concentration of a chemical species is greater in one stratum than in an adjacent stratum. For example, solid waste containing a high concentration of chloride ion may be placed directly on the clay liner of a landfill. The concentration of chloride in the leachate contained in the solid waste is so much greater than the concentration of chloride in the pore water of the clay liner that the latter may be considered to be zero as a simplifying assumption in determining a conservative estimate of the maximum diffusion rate. If the solid waste and the clay are both saturated, the chloride ion will diffuse from the solid waste, where its concentration is greater, into the clay liner, even if there is no fluid flow. The concentration of chloride in the solid waste, C_0, will be assumed to be a constant with time, as it can be replaced by dissolution of additional chloride. The concentration of chloride in the clay liner, $C_i(x, t)$, at some distance x from the solid waste interface and some time t after the waste was placed, can be determined from Equation 2.5 (Crank, 1956). This is a solution to Equation 2.2 for the appropriate boundary and initial conditions.

$$C_i(x, t) = C_0 \operatorname{erfc} \frac{x}{2(D^* t)^{0.5}} \tag{2.5}$$

where

C_i = the concentration at distance x from the source at time t since diffusion began

C_0 = the original concentration, which remains a constant

erfc = the complementary error function (Appendix A)

The complementary error function, erfc, is a mathematical function that is related to the normal, or Gaussian, distribution. This means that the solution described by Equation 2.5 is normally distributed, as is expected for a diffusional process. Figure 2.2 shows the profile of relative concentration for a solute diffusing from a region where the concentration is C_0 to a region where it was initially zero. Because the profile is normally distributed, 84% of the values will be less than the value that is one standard deviation more than the mean and 16% of the values will be less than the value that is one standard deviation less than the mean. The standard deviation is the square root of the variance.

The complementary error function is tabulated in Appendix A. It is related to the error function, erf, by

$$\text{erfc}(B) = 1 - \text{erf}(B)$$

The value of erfc(B) is 0 for all positive values of B greater than 3.0 and 1.0 for a B of 0. For some applications it may be necessary to find erfc of a negative number. Appendix A does not give values for erfc(B) for negative values of B. These must be computed from the relationship

$$\text{erfc}(-B) = 1 + \text{erf } B$$

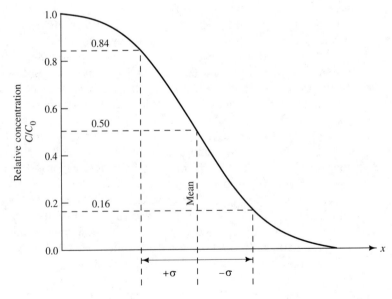

FIGURE 2.2 The profile of a diffusing front as predicted by the complementary error function.

Mass Transport in Saturated Media

Thus values of erfc(B) range from 0 to +2, since the maximum value of erf(B) is 1.0 for 3.0 and all greater numbers.

EXAMPLE PROBLEM

Assume a D of 1×10^{-9} m²/sec and an ω of 0.5, to give a D^* of 5×10^{-10} m²/sec. Find the value of the concentration ratio, C_i/C_0, at a distance of 5 m after 100 yr of diffusion.

1. Convert 100 yr to seconds:

   100 yr × 365 days/yr × 1440 min/day × 60 sec/min = 3.15×10^9 sec

2. Insert values into Equation 2.5:

$$\frac{C_i}{C_0} = \text{erfc} \frac{5}{2(5 \times 10^{-10} \text{ m}^2/\text{sec} \times 3.15 \times 10^9 \text{ sec})^{0.5}}$$

3. Solve:

$$\frac{C_i}{C_0} = \text{erfc}\left(\frac{5}{2.51}\right) = \text{erfc } 1.99 = 0.005$$

In 100 yr, diffusion over a 5-m distance would yield a concentration that is 0.5% of the original.

From the preceding example problem it is obvious that diffusion is not a particularly rapid means of transporting dissolved solutes. Diffusion is the predominant mechanism of transport only in low-permeability hydrogeologic regimes. However, it is possible for solutes to move through a porous or a fractured medium by diffusion even if the ground water is not flowing.

2.3 Transport by Advection

Dissolved solids are carried along with the flowing ground water. This process is called **advective transport,** or **convection.** The amount of solute that is being transported is a function of its concentration in the ground water and the quantity of the ground water flowing. For one-dimensional flow normal to a unit cross-sectional area of the porous media, the quantity of water flowing is equal to the *average linear velocity* times the *effective porosity*. **Average linear velocity**, v_x, is the rate at which the flux of water across the unit cross-sectional area of pore space occurs. It is not the average rate at which the water molecules are moving along individual flowpaths, which is greater than the average linear velocity due to tortuosity. The **effective porosity**, n_e, is the porosity through which flow can occur. Noninterconnected and dead-end pores are not included in the effective porosity.

$$v_x = \frac{K}{n_e} \frac{dh}{dl} \quad (2.6)$$

where

$$v_x = \text{average linear velocity (L/T)}$$
$$K = \text{hydraulic conductivity (L/T)}$$
$$n_e = \text{effective porosity}$$
$$dh/dl = \text{hydraulic gradient (L/L)}$$

The one-dimensional mass flux, F_x, due to advection is equal to the quantity of water flowing times the concentration of dissolved solids and is given by Equation 2.7:

$$F_x = v_x n_e C \tag{2.7}$$

The one-dimensional advective transport equation is

$$\frac{\partial C}{\partial t} = -v_x \frac{\partial C}{\partial x} \tag{2.8}$$

(The derivation of this equation is given in Section 2.6.)

Solution of the advective transport equation yields a sharp concentration front. On the advancing side of the front, the concentration is equal to that of the invading ground water, whereas on the other side of the front it is unchanged from the background value. This is known as **plug flow,** with all the pore fluid being replaced by the invading solute front. The sharp interface that results from plug flow is shown in Figure 2.3. The vertical dashed line at V represents an advancing solute front due to advection alone.

Due to the heterogeneity of geologic materials, advective transport in different strata can result in solute fronts spreading at different rates in each strata. If one obtains a sample of water for purposes of monitoring the spread of a dissolved contaminant from a borehole that penetrates several strata, the water sample will be a composite of the water from each strata. Due to the fact that advection will transport solutes at different rates in each stratum, the composite sample may be a mixture of water containing the transported solute coming from one stratum and uncontaminated ground water coming from a different stratum where the average linear velocity is lower. The concentration of the contaminant in the composite sample would thus be less than in the source.

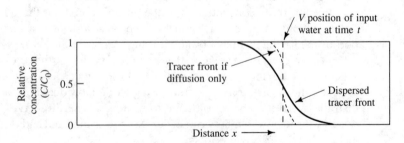

FIGURE 2.3 Advective transport and the influence of longitudinal dispersion and diffusion on the transport of a solute in one-dimensional flow. *Source:* C. W. Fetter, *Applied Hydrogeology,* 2d ed. (New York: Macmillan Publishing Company, 1988).

Mass Transport in Saturated Media

2.4 Mechanical Dispersion

Ground water is moving at rates that are both greater and less than the average linear velocity. At the macroscopic scale—that is, over a domain including a sufficient volume that the effects of individual pores are averaged (Bear 1972)—there are three basic causes of this phenomenon: (1) As fluid moves through the pores, it will move faster in the center of the pores than along the edges. (2) Some of the fluid particles will travel along longer flow paths in the porous media than other particles to go the same linear distance. (3) Some pores are larger than others, which allows the fluid flowing through these pores to move faster. These factors are illustrated in Figure 2.4.

If all ground water containing a solute were to travel at exactly the same rate, it would displace water that does not contain the solute and create an abrupt interface between the two waters. However, because the invading solute-containing water is not all traveling at the same velocity, mixing occurs along the flowpath. This mixing is called **mechanical dispersion,** and it results in a dilution of the solute at the advancing

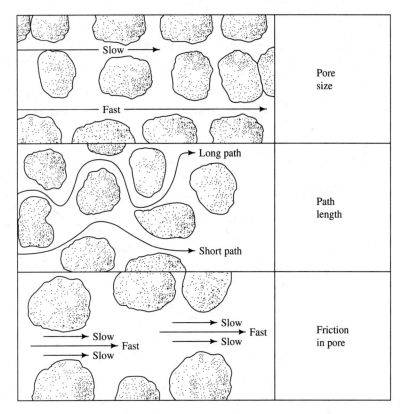

FIGURE 2.4 Factors causing longitudinal dispersion at the scale of individual pores. *Source:* C. W. Fetter, *Applied Hydrogeology*, 2d ed. (New York: Macmillan Publishing Company, 1988).

edge of flow. The mixing that occurs along the direction of the flowpath is called **longitudinal dispersion.**

An advancing solute front will also tend to spread in directions normal to the direction of flow because at the pore scale the flowpaths can diverge, as shown in Figure 2.5. The result of this is mixing in directions normal to the flow path called **transverse dispersion.**

If we assume that mechanical dispersion can be described by Fick's law for diffusion (Equations 2.1 and 2.2) and that the amount of mechanical dispersion is a function of the average linear velocity, then we can introduce a coefficient of mechanical dispersion. This is equal to a property of the medium called *dynamic dispersivity*, or simply *dispersivity*, α, times the average linear velocity. If i is the principle direction of flow, the following definitions apply:

$$\text{Coefficient of longitudinal mechanical dispersion} = \alpha_i \, v_i \qquad (2.9)$$

where

v_i = the average linear velocity in the i direction (L/T)

α_i = the dynamic dispersivity in the i direction (L)

and

$$\text{Coefficient of transverse mechanical dispersion} = \alpha_j \, v_i \qquad (2.10)$$

where

v_i = the average linear velocity in the i direction (L/T)

α_j = the dynamic dispersivity in the j direction (L)

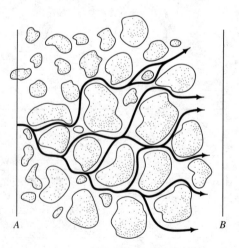

FIGURE 2.5 Flowpaths in a porous medium that cause lateral hydrodynamic dispersion. *Source:* C. W. Fetter, *Applied Hydrogeology*, 2d ed. (New York: Macmillan Publishing Company, 1988).

2.5 Hydrodynamic Dispersion

The process of molecular diffusion cannot be separated from mechanical dispersion in flowing ground water. The two are combined to define a parameter called the **hydrodynamic dispersion coefficient,** D. It is represented by the following formulas:

$$D_L = \alpha_L v_i + D^* \qquad (2.11a)$$

$$D_T = \alpha_T v_i + D^* \qquad (2.11b)$$

where

D_L = hydrodynamic dispersion coefficient parallel to the principal direction of flow (longitudinal)

D_T = hydrodynamic dispersion coefficient perpendicular to the principal direction of flow (transverse)

α_L = longitudinal dynamic dispersivity

α_T = transverse dynamic dispersivity

Figure 2.3 shows the effect of diffusion and mechanical dispersion on the relative concentration (C/C_0) of a solute acting as a tracer that has been injected into a porous medium under one-dimensional flow conditions. The vertical line at V represents the advective transport without dispersion. Effects of diffusion and mechanical dispersion are shown.

The process of hydrodynamic dispersion can be illustrated by Figure 2.6. A mass of solute is instantaneously introduced into the aquifer at time t_0 over the interval $x = 0 + a$. The resulting initial concentration is C_0. The advecting ground water carries the mass of solute with it. In the process the solute slug spreads out, so that the maximum concentration decreases with time, as shown for times t_1 and t_2. The diffusional model of hydrodynamic dispersion predicts that the concentration curves will have a Gaussian distribution that is described by the mean and the variance. With this distribution the

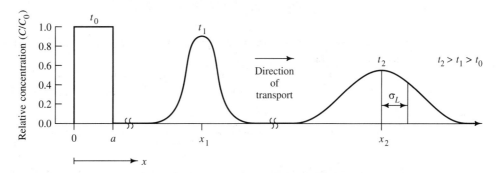

FIGURE 2.6 Transport and spreading of a solute slug with time due to advection and dispersion. A slug of solute was injected at $x = 0 + a$ at time t_0 with a resulting concentration of C_0. The ground-water flow is to the right.

coefficients of longitudinal and transverse hydrodynamic dispersion can be defined as

$$D_L = \frac{\sigma_L^2}{2t} \quad (2.12a)$$

$$D_T = \frac{\sigma_T^2}{2t} \quad (2.12b)$$

where

t = time
σ_T^2 = variance of the transverse spreading of the plume
σ_L^2 = variance of the longitudinal spreading of the plume

2.6 Derivation of the Advection-Dispersion Equation for Solute Transport

This derivation of the advection-dispersion equation is based on work by Freeze and Cherry (1979), Bear (1972), and Ogata (1970). Working assumptions are that the porous medium is homogeneous, isotropic, and saturated with fluid and that flow conditions are such that Darcy's law is valid.

The derivation is based on the conservation of mass of solute flux into and out of a small representative elementary volume (REV) of the porous media. The flow is at a macroscopic scale, which means that it accounts for the differences in flow from pore to pore. A representative elementary volume is illustrated in Figure 1.8.

The average linear velocity, v, has components v_x, v_y, and v_z. The concentration of solute, C, is mass per unit volume of solution. Mass of solute per unit volume of aquifer is the product of the porosity, n_e, and C. Porosity is considered to be a constant because the aquifer is homogeneous.

The solute will be transported by advection and hydrodynamic dispersion. In the i direction the solute transport is given by

$$\text{Advective transport} = v_i n_e C \, dA \quad (2.13)$$

$$\text{Dispersive transport} = n_e D_i \frac{\partial C}{\partial i} \, dA \quad (2.14)$$

where dA is the cross-sectional area of the element and the i direction is normal to that cross-sectional face.

The total mass of solute per unit cross-sectional area transported in the i direction per unit time, F_i, is the sum of the advective and the dispersive transport and is given by

$$F_i = v_i n_e C - n_e D_i \frac{\partial C}{\partial i} \quad (2.15)$$

The negative sign indicates that the dispersive flux is from areas of greater to areas of lesser concentration.

The total amount of solute entering the representative elementary volume is

$$F_x \, dz \, dy + F_y \, dz \, dx + F_z \, dx \, dy$$

The total amount of solute leaving the representative elementary volume is

$$\left(F_x + \frac{\partial F_x}{\partial x}dx\right)dz\,dy + \left(F_y + \frac{\partial F_y}{\partial y}dy\right)dz\,dx + \left(F_z + \frac{\partial F_z}{\partial z}dz\right)dx\,dy$$

The difference between the mass of the solute entering the representative elementary volume and the amount leaving it is

$$\left(\frac{\partial F_x}{\partial x} + \frac{\partial F_y}{\partial y} + \frac{\partial F_z}{\partial z}\right)dx\,dy\,dz$$

The rate of mass change in the representative elementary volume is

$$-n_e\frac{\partial C}{\partial t}dx\,dy\,dz$$

By the law of mass conservation, the rate of mass change in the representative elementary volume must be equal to the difference in the mass of the solute entering and the mass leaving.

$$\frac{\partial F_x}{\partial x} + \frac{\partial F_y}{\partial y} + \frac{\partial F_z}{\partial z} = -n_e\frac{\partial C}{\partial t} \quad (2.16)$$

Equation 2.15 can be used to find the values of F_x, F_y, and F_z. These are substituted in Equation 2.16, which becomes, after cancellation of n_e from both sides,

$$\left[\frac{\partial}{\partial x}\left(D_x\frac{\partial C}{\partial x}\right) + \frac{\partial}{\partial y}\left(D_y\frac{\partial C}{\partial y}\right) + \frac{\partial}{\partial z}\left(D_z\frac{\partial C}{\partial z}\right)\right] \\ - \left[\frac{\partial}{\partial x}(v_xC) + \frac{\partial}{\partial y}(v_yC)\frac{\partial}{\partial z}(v_zC)\right] = \frac{\partial C}{\partial t} \quad (2.17)$$

Equation 2.17 is the three-dimensional equation of mass transport for a *conservative* solute—that is, one that does not interact with the porous media or undergo biological or radioactive decay.

In a homogeneous medium, D_x, D_y, and D_z do not vary in space. However, because the coefficient of hydrodynamic dispersion is a function of the flow direction, even in an isotropic, homogeneous medium, $D_x \neq D_y \neq D_z$. For those domains where the average linear velocity, v_x, is uniform in space, Equation 2.17 for one-dimensional flow in a homogeneous, isotropic porous media is

$$D_L\frac{\partial^2 C}{\partial x^2} - v_x\frac{\partial C}{\partial x} = \frac{\partial C}{\partial t} \quad (2.18)$$

In a homogeneous medium with a uniform velocity field, Equation 2.17 for two-dimensional flow with the direction of flow parallel to the x axis is

$$D_L\frac{\partial^2 C}{\partial x^2} + D_T\frac{\partial^2 C}{\partial y^2} - v_x\frac{\partial C}{\partial x} = \frac{\partial C}{\partial t} \quad (2.19)$$

where

D_L = the longitudinal hydrodynamic dispersion (L²/T)

D_T = the transverse hydrodynamic dispersion (L²/T)

Equation 2.17 for radial flow from a well can be written in polar coordinates (Ogata 1970) as

$$\frac{\partial}{\partial r}\left(D\frac{\partial C}{\partial r}\right) + \frac{D}{r}\frac{\partial C}{\partial r} - u\frac{\partial C}{\partial r} = \frac{\partial C}{\partial t} \quad (2.20)$$

where

r = radial distance to the well

u = average pore velocity of injection, which is found from

$$u = \frac{Q}{2\pi n_e R r^2}$$

where

Q = the rate of injection into the well

n_e = effective porosity

R = length of well screen or open bore hole

2.7 Diffusion versus Dispersion

In the previous section the mass transport equation was derived on the basis of hydrodynamic dispersion, which is the sum of mechanical dispersion and diffusion. It would have been possible to separate the hydrodynamic dispersion term into the two components and have separate terms in the equation for them. However, as a practical matter, under most conditions of ground-water flow, diffusion is insignificant and is neglected.

It is possible to evaluate the relative contribution of mechanical dispersion and diffusion to solute transport. A **Peclet number** is a dimensionless number that can relate the effectiveness of mass transport by advection to the effectiveness of mass transport by either dispersion or diffusion. Peclet numbers have the general form of $v_x d/D_d$ or $v_x L/D_L$, where v_x is the advective velocity, d and L are characteristic flow lengths, D_d is the coefficient of molecular diffusion, and D_L is the longitudinal hydrodynamic dispersion coefficient. The Peclet number, P, which defines the ratio of transport by advection to the rate of transport by molecular diffusion in column studies, is a dimensionless parameter defined as $v_x d/D_d$, where d is the average grain diameter and D_d is the coefficient of molecular diffusion. A plot of the ratio of D_L/D_d versus the Peclet number is given in Figure 2.7a. Shown on this figure are the results of a number of experimental measurements using sand columns and tracers as well as some experimental curves from several investigators (Perkins and Johnson 1963).

At zero flow velocity D_L is equal to D^*, since $D_L = \alpha_L v_x + D^*$. In this manner the value of ω, the tortuosity factor, can be experimentally determined as $D^* = \omega D_d$. At very low velocities, the ratio of D_L/D_d is a constant with a value of about 0.7, which is the experimentally determined value of ω for uniform sand. This shows up on the left side of Figure 2.7(a) as a horizontal line. In this zone diffusion is the predominant force, and dispersion can be neglected. Between a Peclet number of about 0.4 to 6 there is a transition zone, where the effects of diffusion and longitudinal mechanical dispersion are more or less equal.

Mass Transport in Saturated Media

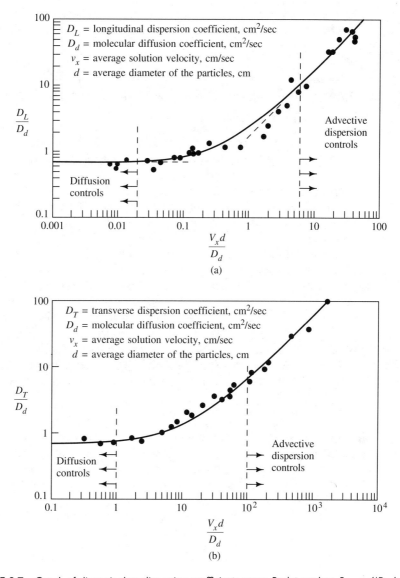

FIGURE 2.7 Graph of dimensionless dispersion coefficients versus Peclet number, $P = v_x d/D_d$. (a) D_L/D_e versus P and (b) D_T/D_e versus P. *Source:* T. K. Perkins and O. C. Johnson, *Society of Petroleum Engineers Journal*, 3 (1963):70–84. Copyright 1963, Society of Petroleum Engineers.

Figure 2.7(b) shows the plot of D_T/D_d as a function of Peclet number. Although the curve has the same shape as in (a), it occurs at Peclet numbers roughly 100 times greater. This means that diffusion has more control over transverse dispersion at higher Peclet numbers than it does for longitudinal dispersion. Higher Peclet numbers occur with higher velocities and/or longer flow paths. At higher Peclet numbers mechanical dispersion is the predominant cause of mixing of the contaminant plume (Perkins and

Johnson 1963; Bear 1972; Bear and Verruijt 1987) and the effects of diffusion can be ignored. Under these conditions D_i can be replaced with $\alpha_i v_i$ in the advection-dispersion equations.

2.8 Analytical Solutions of the Advection-Dispersion Equation

2.8.1 Methods of Solution

The advection-dispersion equations can be solved by either numerical or analytical methods. Analytical methods involve the solution of the partial differential equations using calculus based on the initial and boundary value conditions. They are limited to simple geometry and in general require that the aquifer be homogeneous. A number of analytical solutions are presented in this chapter. They are useful in that they can be solved with a calculator and a table of error functions or even a pencil and paper, if one is so inclined.

Numerical methods involve the solution of the partial differential equation by numerical methods of analysis. They are more powerful than analytical solutions in the sense that aquifers of any geometry can be analyzed and aquifer heterogeneities can be accommodated. However, there can be other problems with numerical models, such as numerical errors, which can cause solutions to show excess spreading of solute fronts or plumes that are not related to the dispersion of the tracer that is the subject of the modeling. Bear and Verruijt (1987) present a good introduction to the use of numerical models to solve mass transport equations. These solutions are normally found by methods of computer modeling, a topic beyond the scope of this text.

2.8.2 Boundary and Initial Conditions

In order to obtain a unique solution to a differential equation it is necessary to specify the initial and the boundary conditions that apply. The **initial conditions** describe the values of the variable under consideration, in this case concentration, at some initial time equal to 0. The **boundary conditions** specify the interaction between the area under investigation and its external environment.

There are three types of boundary conditions for mass transport. The boundary condition of the first type is a **fixed concentration.** The boundary condition of the second type is a **fixed gradient.** A **variable flux** boundary constitutes the boundary condition of the third type.

Boundary and initial conditions are shown in a shorthand form. For one-dimensional flow we need to specify the conditions relative to the location, x, and the time, t. By convention this is shown in the form

$$C(x, t) = C(t)$$

where $C(t)$ is some known function.

For example, we can write

$$C(0, t) = C_0, \quad t \geq 0$$
$$C(x, 0) = 0, \quad x \geq 0$$
$$C(\infty, t) = 0, \quad t \geq 0$$

The first statement says that for all time t equal to or greater than zero, at $x = 0$ the concentration is maintained at C_0. This is a fixed-concentration boundary condition located at $x = 0$ (first-type boundary). The second statement is an initial condition that says at time $t = 0$, the concentration is zero everywhere within the flow domain, that is, where x is greater than or equal zero. As soon as flow starts, solute at a concentration of C_0 will cross the $x = 0$ boundary.

The third condition shows that the flow system is infinitely long and that no matter how large time gets, the concentration will still be zero at the end of the system (first-type boundary condition at $x = \infty$).

We could also have specified an initial condition that within the domain the initial solute concentration was C_i. This would be written as

$$C(x, 0) = C_i, \quad x \geq 0$$

Other examples of concentration (first-type) boundary conditions are exponential decay of the source term and pulse loading at a constant concentration for a period of time followed by another period of time with a different constant concentration.

Exponential decay for the source term can be expressed as

$$C(0, t) = C_0 t^{-it}$$

where i = a decay constant.

Pulse loading where the concentration is C_0 for times from 0 to t_0 and then is 0 for all time more than t_0 is expressed as

$$C(0, t) = C_0 \quad 0 < t \leq t_0$$
$$C(0, t) = 0 \quad t > t_0$$

Fixed-gradient boundaries are expressed as

$$\left.\frac{dC}{dx}\right|_{x=0} = f(t) \quad \text{or} \quad \left.\frac{dC}{dx}\right|_{x=\infty} = f(t)$$

where $f(t)$ is some known function. A common fixed-gradient condition is $dC/dx = 0$, or a no-gradient boundary.

The variable-flux boundary, a third type, is given as

$$-D\frac{\partial C}{\partial x} + v_x C = v_x C(t)$$

where $C(t)$ is a known concentration function. A common variable-flux boundary is a constant flux with a constant input concentration, expressed as

$$\left.\left(-D\frac{dC}{dx} + vC\right)\right|_{x=0} = vC_0$$

2.8.3 One-Dimensional Step Change in Concentration (First-Type Boundary)

Sand column experiments have been used to evaluate both the coefficients of diffusion and dispersion at the laboratory scale. A tube is filled with sand and then saturated with water. Water is made to flow through the tube at a steady rate, creating, in effect, a permeameter. A solution containing a tracer is then introduced into the sand column in place of the water. The initial concentration of the solute in the column is zero, and the concentration of the tracer solution is C_0. The tracer in the water exiting the tube is analyzed, and the ratio of C, the tracer concentration at time t, over C_0, the injected tracer concentration, is plotted as a function of time. This is called a **fixed-step function.**

The boundary and initial conditions are given by

$$C(x, 0) = 0 \qquad x \geq 0 \qquad \text{Initial condition}$$
$$\left. \begin{array}{l} C(0, t) = C_0 \quad t \geq 0 \\ C(\infty, t) = 0 \quad t \geq 0 \end{array} \right\} \quad \text{Boundary conditions}$$

The solution to Equation 2.18 for these conditions is (Ogata and Banks 1961)

$$C = \frac{C_0}{2} \left[\text{erfc}\left(\frac{L - v_x t}{2\sqrt{D_L t}}\right) + \exp\left(\frac{v_x L}{D_L}\right) \text{erfc}\left(\frac{L + v_x t}{2\sqrt{D_L t}}\right) \right] \qquad (2.21)$$

This equation may be expressed in dimensionless form as

$$C_R(t_R, P_e) = 0.5 \left\{ \text{erfc}\left[\left(\frac{P_e}{4 t_R}\right)^{1/2} \cdot (1 - t_R)\right] + \exp(P_e) \text{erfc}\left[\left(\frac{P_e}{4 t_R}\right)^{1/2} (1 + t_R)\right] \right\}$$
$$(2.22)$$

where

$t_R = v_x t / L$
$C_R = C / C_0$
P_e = Peclet number when flow distance, L, is chosen as the reference length ($P_e = v_x L / D_L$)

erfc = complementary error function

2.8.4 One-Dimensional Continuous Injection into a Flow Field (Second-Type Boundary)

In nature there are not many situations where there would be a sudden change in the quality of the water entering an aquifer. A much more likely condition is that there would be leakage of contaminated water into the ground water flowing in an aquifer. For the one-dimensional case, this might be a canal that is discharging contaminated water into an aquifer as a line source (Figure 2.8).

The rate of injection is considered to be constant, with the injected mass of the solute proportional to the duration of the injection. The initial concentration of the solute in the aquifer is zero, and the concentration of the solute being injected is C_0. The solute is free to disperse both up-gradient and down-gradient.

The boundary and initial conditions are

$$C(x, 0) = 0, \qquad -\infty < x < +\infty \qquad \text{Initial condition}$$
$$\left. \begin{array}{l} \int_{-\infty}^{+\infty} n_e C(x, t) \, dx = C_0 n_e v_x t, \quad t > 0 \\ C(\infty, t) = 0 \quad t \geq 0 \end{array} \right\} \quad \text{Boundary conditions}$$

Mass Transport in Saturated Media

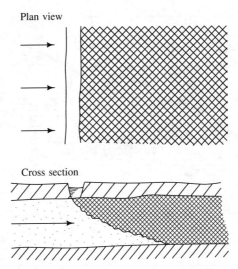

FIGURE 2.8 Leakage from a canal as a line source for injection of a contaminant into an aquifer. *Source:* J. P. Sauty, *Water Resources Research* 16, no. 1 (1980):145–58. Copyright by the American Geophysical Union.

The second boundary condition states that the injected mass of contaminant over the domain from $-\infty$ to $+\infty$ is proportional to the length of time of the injection.

The solution to this flow problem (Sauty 1980) is

$$C = \frac{C_0}{2}\left[\text{erfc}\left(\frac{L - v_x t}{2\sqrt{D_L t}}\right) - \exp\left(\frac{v_x L}{D_L}\right)\text{erfc}\left(\frac{L + v_x t}{2\sqrt{D_L t}}\right)\right] \quad (2.23)$$

In dimensionless form this is

$$C_R(t_R, P_e) = 0.5\left\{\text{erfc}\left[\left(\frac{P_e}{4t_R}\right)^{1/2}(1 - t_R)\right] - \exp(P_e)\,\text{erfc}\left[\left(\frac{P_e}{4t_R}\right)^{1/2}(1 + t_R)\right]\right\} \quad (2.24)$$

It can be seen that Equations 2.21 and 2.23 are very similar, the only difference being that the second term is subtracted rather than added in 2.23.

Sauty (1980) gives an approximation for the one-dimensional dispersion equation as

$$C = \frac{C_0}{2}\left[\text{erfc}\left(\frac{L - v_x t}{2\sqrt{D_L t}}\right)\right] \quad (2.25)$$

In dimensionless form this is

$$C_R(t_R, P_e) = 0.5\,\text{erfc}\left[\left(\frac{P_e}{4t_R}\right)^{1/2}(1 - t_R)\right] \quad (2.26)$$

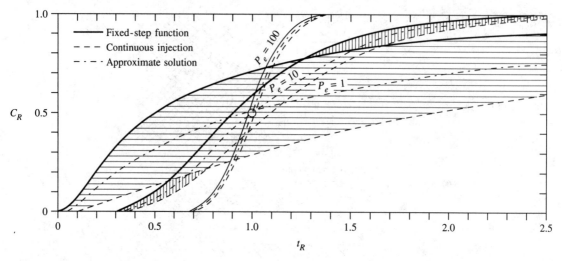

FIGURE 2.9 Dimensionless-type curves for the continuous injection of a tracer into a one-dimensional flow field. *Source:* J. P. Sauty, *Water Resources Research* 16, no. 1 (1980):145–58. Copyright by the American Geophysical Union.

This approximation comes about because for large Peclet numbers, the second term of Equations 2.21 and 2.23 is much smaller than first term and can be neglected. Figure 2.9 demonstrates under what conditions this approximation is valid. In Figure 2.9 the dimensionless concentration, C_R, is plotted as a function of dimensionless time, t_R, for continuous tracer injection using the fixed-step function, Equation 2.22, the continuous-injection function, Equation 2.24, and the approximate solution, Equation 2.26. Curves are plotted for three Peclet numbers, 1, 10, and 100. Sauty (1980) defined a Peclet number as $P_e = v_x L / D_L$, where L is the distance from the point of injection of the solute to the point of measurement and D_L is the coefficient of hydrodynamic dispersion. This Peclet number defines the rate of transport by advection to the rate of transport by hydrodynamic dispersion. For Peclet number 1, the fixed-step function and the continuous-injection function give quite different results, whereas for Peclet number 100 they are almost identical. The approximate solution lies midway between the other two. This figure suggests that for Peclet numbers less than about 10, the exact solutions need to be considered, whereas for Peclet numbers greater than 10, the approximate solution is probably acceptable, especially as the Peclet number approaches 100. This Peclet number increases with flow-path length as advective transport becomes more dominant over dispersive transport. Thus for mass transport near the inlet boundary, it is important to use the correct equation, but as one goes away from the inlet boundary, it is less important that the correct form of the equation is employed.

2.8.5 Third-Type Boundary Condition

A solution for Equation 2.18 for the following boundary condition was given by van Genuchten (1981).

Mass Transport in Saturated Media

$$C(x, 0) = 0 \quad \text{Initial condition}$$

$$\left.\left(-D\frac{\partial C}{\partial x} + v_x C\right)\right|_{x=0} = v_x C_0$$

$$\left.\frac{\partial C}{\partial x}\right|_{x \to \infty} = \text{(finite)} \quad \text{Boundary conditions}$$

The third condition specifies that as x approaches infinity, the concentration gradient will still be finite. Under these conditions the solution to Equation 2.18 is:

$$C = \frac{C_0}{2}\left[\operatorname{erfc}\left[\frac{L - v_x t}{2\sqrt{D_L t}}\right] + \left(\frac{v_x^2 t}{\pi D_L}\right)^{1/2} \exp\left[-\frac{(L - v_x t)^2}{4 D_L t}\right]\right.$$

$$\left. - \frac{1}{2}\left(1 + \frac{v_x L}{D_L} + \frac{v_x^2 t}{D_L}\right)\exp\left(\frac{v_x L}{D_L}\right)\operatorname{erfc}\left[\frac{L - v_x t}{2\sqrt{D_L t}}\right]\right] \quad (2.27)$$

This equation also reduces to the approximate solution, Equation 2.25, as the flow length increases.

2.8.6 One-Dimensional Slug Injection into a Flow Field

If a slug of contamination is instantaneously injected into a uniform, one-dimensional flow field, it will pass through the aquifer as a pulse with a peak concentration, C_{max}, at some time after injection, t_{max}. The solution to Equation 2.18 under these conditions (Sauty 1980) is in dimensionless form:

$$C_R(t_R, P_e) = \frac{E}{(t_R)^{1/2}} \exp\left(-\frac{P_e}{4 t_R}(1 - t_R)^2\right) \quad (2.28)$$

with

$$E = (t_{Rmax})^{1/2} \cdot \exp\left(\frac{P_e}{4 t_{Rmax}}(1 - t_{Rmax})^2\right) \quad (2.29)$$

where

$t_{Rmax} = (1 + P_e^{-2})^{1/2} - P_e^{-1}$ (dimensionless time at which peak concentration occurs)

$C_R = C/C_{max}$

In Figure 2.10, C_R (C/C_{max}) for a slug injected into a uniform one-dimensional flow field is plotted against dimensionless time, t_R, for several Peclet numbers. It can be seen that the time for the peak concentration (C_{max}) to occur increases with the Peclet number, up to a limit of $t_R = 1$. Breakthrough becomes more symmetric with increasing P_e.

2.8.7 Continuous Injection into a Uniform Two-Dimensional Flow Field

If a tracer is continuously injected into a uniform flow field from a single point that fully penetrates the aquifer, a two-dimensional plume will form that looks similar to Figure

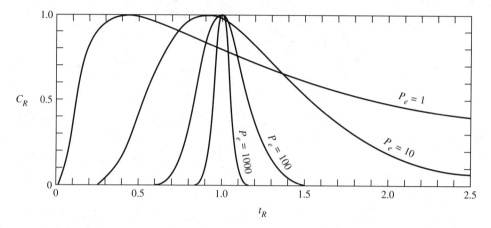

FIGURE 2.10 Dimensionless-type curve for the injection of a slug of a tracer into a one-dimensional flow field. *Source:* J. P. Sauty, *Water Resources Research* 16, no. 1 (1980):145–58. Copyright by the American Geophysical Union.

2.11. It will spread along the axis of flow due to longitudinal dispersion and normal to the axis of flow due to transverse dispersion. This is the type of contamination that would occur due to leakage of liquids from a landfill or lagoon.

The mass transport equation for two-dimensional flow, Equation 2.19, has been solved for several boundary conditions. The well is located at the origin ($x = 0$, $y = 0$) and there is a uniform velocity at a rate v_x parallel to the x axis. There is a continuous injection of a solute of concentration, C_0, at a rate Q at the origin.

Bear (1972) gives the solution to Equation 2.19 for the condition where the growth of the plume has stabilized—that is, as time approaches infinity—as

$$C(x, y) = \left(\frac{C_0 Q}{2\pi (D_L D_T)^{1/2}}\right) \exp\left(\frac{v_x x}{2 D_L D_T}\right) K_0 \left[\left(\frac{v_x^2}{4 D_L}\left(\frac{x^2}{D_L} + \frac{y^2}{D_T}\right)\right)^{1/2}\right] \quad (2.30)$$

where

K_0 = the modified Bessel function of the second kind and zero order (values are tabulated in Appendix B)

Q = rate at which a tracer of concentration C_0 is being injected

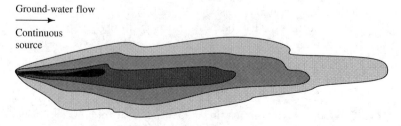

FIGURE 2.11 Plume resulting from the continuous injection of a tracer into a two-dimensional flow field. *Source:* C. W. Fetter, *Applied Hydrogeology*, 2nd ed. (New York: Macmillan Publishing Company, 1988).

The two-dimensional growth of a plume from a continuous source can be tracked through time using a solution to Equation 2.19 developed by Y. Emsellem (see Fried 1975). The solution has the form

$$C(x, y, t) = \frac{C_0 Q}{4\pi (D_L D_T)^{1/2}} \exp\left(\frac{v_x x}{2D_L}\right) [W(0, B) - W(t, B)] \quad (2.31)$$

where

$$B = \left[\frac{(v_x x)^2}{4D_L^2} + \frac{(v_x y)^2}{4D_L D_T}\right]^{1/2}$$

t = time

$W[t, B]$ = a function derived by Hantush and tabulated in Appendix C (In well hydraulics this function is known as the leaky well function $W[u, r/b]$.)

2.8.8 Slug Injection into a Uniform Two-Dimensional Flow Field

If a slug of contamination is injected over the full thickness of a two-dimensional uniform flow field in a short period of time, it will move in the direction of flow and spread with time. This result is illustrated by Figure 2.12 and represents the pattern of contamination at three increments that result from a one-time spill. Figure 2.12 is based on the results of a laboratory experiment conducted by Bear (1961). Figure 2.13 shows the spread of a plume of chloride that was injected into an aquifer as a part of a large-scale field test (Mackay et al. 1986). The plume that resulted from the field test is more complex than the laboratory plume due to the heterogeneities encountered in the real world and the fact the plume may not be following the diffusional model of dispersion.

De Josselin De Jong (1958) derived a solution to this problem on the basis of a statistical treatment of lateral and transverse dispersivities. Bear (1961) later verified it experimentally. If a tracer with concentration C_0 is injected into a two-dimensional flow field over an area A at a point (x_0, y_0), the concentration at a point (x, y), at time t after the injection is

$$C(x, y, t) = \frac{C_0 A}{4\pi t (D_L D_T)^{1/2}} \exp\left[-\frac{(x - (x_0 - v_x t))^2}{4 D_L t} - \frac{(y - y_0)^2}{4 D_T t}\right] \quad (2.32)$$

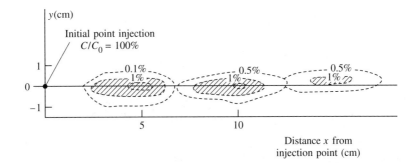

FIGURE 2.12 Injection of a slug of a tracer into a two-dimensional flow field shown at three time increments. Experimental results from J. Bear, *Journal of Geophysical Research* 66, no. 8 (1961):2455–67. Copyright by the American Geophysical Union.

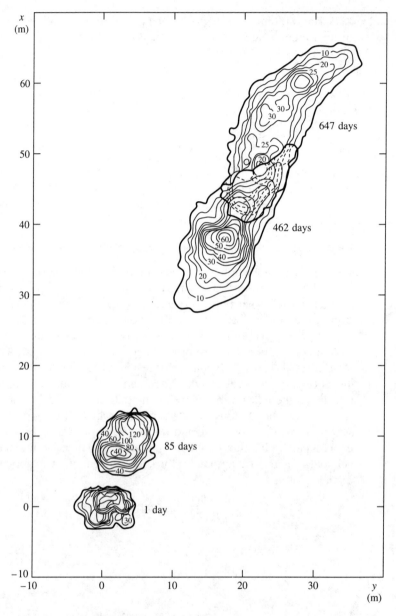

FIGURE 2.13 Vertically averaged chloride concentration at 1 day, 85 days, 462 days, and 647 days after the injection of a slug into a shallow aquifer. *Source:* D. M. Mackay et al. *Water Resources Research* 22, no. 13 (1986): 2017–29. Copyright by the American Geophysical Union.

2.9 Effects of Transverse Dispersion

The ratio of longitudinal to transverse dispersivity (α_L/α_T) in an aquifer is an important control over the shape of a contaminant plume in two-dimensional mass transport. The lower the ratio, the broader the shape of the resulting plume will be. Figure 2.14 shows various two-dimensional shapes of a contaminant plume, where the only factor varied was the ratio of longitudinal to transverse dispersivity. This illustrates the fact that it is important to have some knowledge of the transverse dispersivity in addition to the

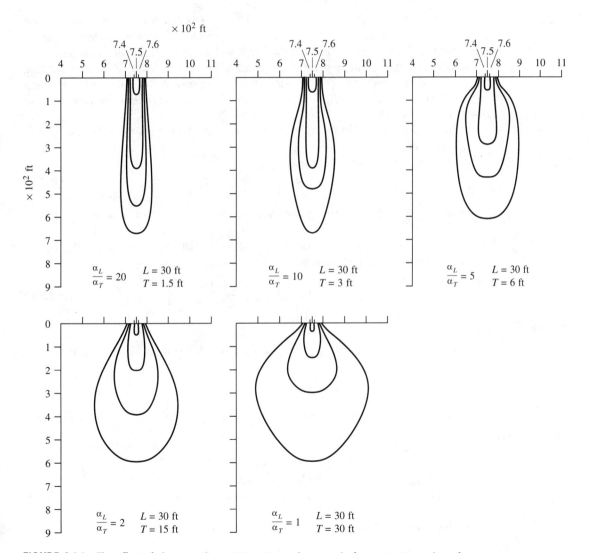

FIGURE 2.14 The effect of changing dispersivity ratio on the spread of a contaminant plume from a continuous source. *Source:* Robert L. Stollar.

longitudinal dispersivity. There is a paucity of data in the literature on the relationships of longitudinal to transverse dispersivities. From the few field studies available, α_L/α_T is in the range of 6 to 20 (Anderson 1979, Klotz et al. 1980). In addition, dispersivity ratios based on field studies are based on fitting the diffusional model of dispersion to cases where it might not be applicable.

2.10 Tests to Determine Dispersivity

2.10.1 Laboratory Tests

Diffusion and dispersivity can be determined in the laboratory using columns packed with the porous media under investigation. The results of column studies are often reported in terms of pore volumes of fluid that is eluted. One pore volume is the cross-sectional area of the column times the length times the porosity (ALn). The unit discharge rate from the column is the linear velocity times the porosity times the cross-sectional area ($v_x nA$). The total discharge over a period of time is the product of time and the discharge rate ($v_x nAt$).

The total number of pore volumes, U, is the total discharge divided by the volume of a single pore volume:

$$U = \frac{v_x nAt}{ALn} = \frac{v_x t}{L} = t_R \tag{2.33}$$

It can be seen that the number of pore volumes is equivalent to a dimensionless time, t_R.

With this equivalency Equation 2.25, the approximate one-dimensional dispersion equation, can be rearranged to yield (Brigham 1974)

$$\frac{C}{C_0} = 0.5\left[\operatorname{erfc}\left(\frac{1-U}{2(UD_L/v_x L)^{1/2}}\right)\right] \tag{2.34}$$

where

$U =$ the number of effluent pore volumes, where a pore volume is equal to the total column volume times the porosity

$L =$ the length of the column

Equation 2.34 can, through appropriate substitution, be made equivalent to Equation 2.26.

The concentration of the tracer in the effluent, C, is measured for various values of U, and then C/C_0 is plotted as a function of $[(U-1)/U^{1/2}]$ on linear probability paper. If the data plot as a straight line, they are normally distributed, the diffusive form of the advection-dispersion equation is valid, and the slope of the line is related to the longitudinal hydrodynamic dispersion.

The value of D_L can be found from

$$D_L = \left(\frac{v_x L}{8}\right)(J_{0.84} - J_{0.16})^2 \tag{2.35}$$

where

$$J_{0.84} = [(U-1)/U^{1/2}] \text{ when } C/C_0 \text{ is } 0.84$$
$$J_{0.16} = [(U-1)/U^{1/2}] \text{ when } C/C_0 \text{ is } 0.16$$

Since $D_L = \alpha_L v_x + D^*$, then

$$\alpha_L = \frac{D_L - D^*}{v_x} \qquad (2.36)$$

The average linear velocity in the column can be found from the quantity of water discharging per unit time divided by the product of the cross-sectional area and the porosity. The effective diffusion coefficient can either be measured in a column test or estimated.

EXAMPLE PROBLEM

Pickens and Grisak (1981) conducted a laboratory study of dispersion in sand columns with the following characteristics:

Tracer	Chloride
Column length	30 cm
Column diameter	4.45 cm
Mean grain size	0.20 mm
Uniformity coefficient of sand	2.3
Porosity	0.36
Flow rate	
Test R1	5.12×10^{-3} mL/sec
Test R2	1.40×10^{-2} mL/sec
Test R3	7.75×10^{-2} mL/sec
Average Linear Velocity	
Test R1	9.26×10^{-4} cm/sec
Test R2	2.53×10^{-3} cm/sec
Test R3	8.60×10^{-3} cm/sec

Test R1 was run using chloride at 200 mg/L, followed by test R2, in which the saline solution was flushed out of the column using deionized water, and then test R3, where the 200-mg/L chloride solution was again introduced into the column.

The results of the three tests are plotted in Figure 2.15. The results of test R2 have a reverse slope as deionized water replaced the saline solution. It can be seen that the results form a straight line.

For chloride in water at 25°C, the molecular diffusion coefficient is 2.03×10^{-5} cm^2/sec. Based on this, Pickens and Grisak estimated the effective diffusion coefficient to be 1.02×10^{-5} cm^2/sec. The hydrodynamic dispersion coefficients are based on the slope of the straight lines. The following values were obtained for the three tests:

Test	Hydrodynamic Dispersion	Dispersivity
R1	4.05×10^{-5} cm^2/sec	0.033 cm
R2	8.65×10^{-5} cm^2/sec	0.030 cm
R3	3.76×10^{-4} cm^2/sec	0.043 cm

The replicate values are not equal because of experimental error.

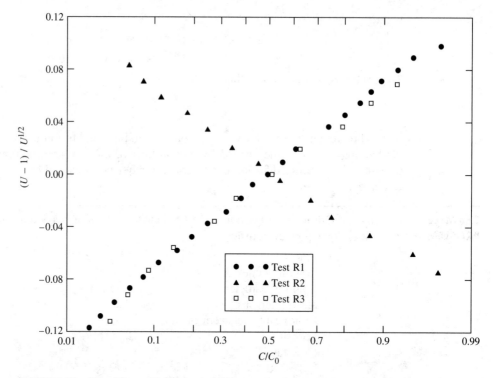

FIGURE 2.15 Plot of $(U - 1)/U^{1/2}$ versus C/C_0 on probability paper for determination of dispersion in a laboratory sand column. *Source:* S. F. Pickens and G. E. Grisak, *Water Resources Research* 17, no. 4 (1981):1191–1211. Copyright by the American Geophysical Union.

The computation of dispersivity for test R1 is illustrated here:

$$\alpha_L = \frac{D_L - D^*}{v_x}$$

$$= \frac{(4.05 \times 10^{-5} \text{ cm}^2/\text{sec} - 1.02 \times 10^{-5} \text{ cm}^2/\text{sec})}{9.26 \times 10^{-4} \text{ cm/sec}} = 0.033 \text{ cm}$$

2.10.2 Field Tests for Dispersivity

Dispersivity can be determined in the field by two means. If there is a contaminated aquifer, the plume of known contamination can be mapped and the advection-dispersion equation solved with dispersivity as the unknown. Pinder (1973) used this approach in a groundwater modeling study of a plume of dissolved chromium in a sand and gravel aquifer on Long Island, New York. He started with initial guesses of α_L and α_T and then varied them during successive model runs until the computer model yielded a reasonable reproduction of the observed contaminant plume. One of the difficulties of this approach is that the concentration and volume of the contaminant source are often not known.

Mass Transport in Saturated Media

A much more common approach is the use of a tracer that is injected into the ground via a well. There are a variety of variations to this approach. Natural gradient tests involve the injection of a tracer into an aquifer, followed by the measurement of the plume that developed under the prevailing water table gradient (e.g., Sudicky and Cherry 1979; Gillham et al. 1984; Mackay et al. 1986; LeBlanc et al. 1991; Garabedian et al. 1991). The plume is measured by means of small amounts of water withdrawn from down-gradient observation wells and multilevel piezometers. One- and two-well tests have also been used in which a tracer is pumped into the ground and then groundwater containing the tracer is pumped back out of the ground (e.g., Fried 1975; Grove and Beetem 1971; Sauty 1978; Pickens et al. 1981; Pickens and Grisak, 1981).

2.10.3 Single-Well Tracer Test

A single-well tracer test involves the injection of water containing a conservative tracer into an aquifer via an injection well and then the subsequent pumping of that well to recover the injected fluid. The fluid velocities of the water being pumped and injected are much greater than the natural ground-water gradients.

Equation 2.20 can be written (Hoopes and Harleman 1967) as

$$\frac{\partial C}{\partial t} + u\frac{\partial C}{\partial r} = \alpha_L u \frac{\partial^2 C}{\partial r^2} + \frac{D^*}{r}\frac{\partial}{\partial r}\left(r\frac{\partial C}{\partial r}\right) \quad (2.37)$$

Gelhar and Collins (1971) derived a solution to Equation 2.37 for the withdrawal phase of an injection-withdrawal well test in which the diffusion term is neglected because it is very much smaller than the dispersion term. The relative concentration of the water being withdrawn from the injection well is

$$\frac{C}{C_0} = \frac{1}{2}\operatorname{erfc}\left(\frac{(U_p - U_i) - 1}{\{\frac{16}{3}(\alpha_L/R_f)[2 - (1 - U_p/U_i)]^{1/2}[1 - (U_p/U_i)]\}^{1/2}}\right) \quad (2.38)$$

where

U_p = cumulative volume of water withdrawn during various times
U_i = total volume of water injected during the injection phase
R_f = average frontal position of the injected water at the end of the injection period, which is defined by

$$R_f = \left(\frac{Qt}{\pi bn}\right)^{1/2} \quad (2.39)$$

where

Q = rate of injection
t = total time of injection
b = aquifer thickness
n = porosity

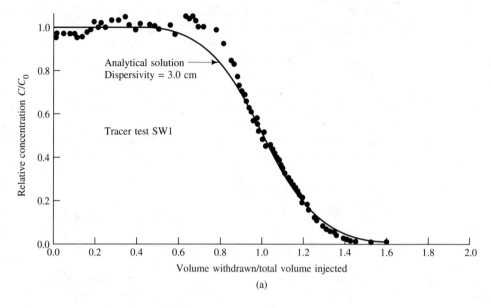

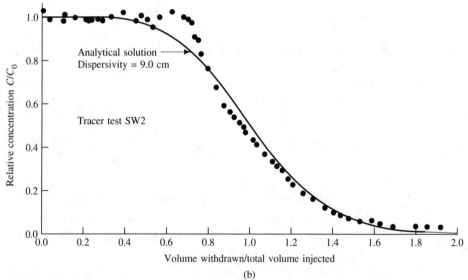

FIGURE 2.16 Comparison of measured C/C_0 values for a single-well injection-withdrawal test versus an analytical solution. *Source:* S. F. Pickens and G. E. Grisak, *Water Resources Research* 17, no. 4 (1981):1191–1211. Copyright by the American Geophysical Union.

EXAMPLE PROBLEM

Pickens and Grisak (1981) performed a single-well injection-withdrawal tracer test into a confined sand aquifer about 8.2 m thick with an average hydraulic conductivity of 1.4×10^{-2} cm/sec and a porosity of 0.38. The sediment tested in the column study described in the previous example problem came from this aquifer.

The injection well was 5.7 cm in diameter and the full thickness of the aquifer was screened. Clear water was injected at a constant rate for 24 hr prior to the start of the test to establish steady-state conditions. The tracer used during the tests was ^{131}I, a radioactive iodine, which was added to the injected water. All measurements were corrected for the radioactive decay that occurred during the test.

Two tests were performed on the well. The first test, SW1, had an injection rate of 0.886 L/sec and injection continued for 1.25 da. A total volume of 95.6 m³ of water was injected, and the injection front reached an average radial distance away from the well of 3.13 m. Water was then pumped for 2.0 da at the same rate, so that a total of 153 m³ of water was withdrawn. The second test, SW2, was longer. Water with the tracer was added at a rate of 0.719 L/sec for 3.93 da. A total of 244 m³ of water was added, and the average position of the injection front reached to 4.99 m from the well. During the withdrawal phase a total of 886 m³ of water was pumped over a period of 16.9 da at an average rate of 0.606 L/sec.

The results of the test are shown in Figure 2.16. Relative concentration, C/C_0, is plotted against U_p/U_i. The dots represent field values and the solid lines are curves, which were computed using Equation 2.38. Various curves were computed for different values of α_L, and the curves with the best fit to the field data were plotted on the graphs. In Figure 2.16(a) the calculated curve was based on a longitudinal dispersivity of 3.0 cm, whereas for curve 2.16(b) the best-fit curve was based on a longitudinal dispersivity of 9.0 cm. This test illustrates the scale-dependent nature of dispersivity. The second test, in which a larger volume of water was injected, tested a larger volume of the aquifer than the first test and yielded a higher dispersivity value.

2.11 Scale Effect of Dispersion

The two example problems derived from Pickens and Grisak (1981) illustrate what has been called the **scale effect of dispersion** (Fried 1975). At the laboratory scale the mean value of α_L was determined to be 0.035 cm when the flow length was 30 cm. With the single-well injection-withdrawal test, α_L was 3 cm when the solute front traveled 3.1 m and 9 cm when the solute front traveled 5.0 m. In a two-well recirculating withdrawal-injection tracer test with wells located 8 m apart, α_L was determined to be 50 cm. All these values were obtained from the same site. The greater the flow length, the larger the value of longitudinal dispersivity needed to fit the data to the advection-dispersion equation.

Lallemand-Barres and Peaudecerf (1978) published a graph on which dispersivity, as measured in the field, was plotted against flow length on log-log paper (Figure 2.17). This graph suggested that the longitudinal dispersivity could be estimated to be about

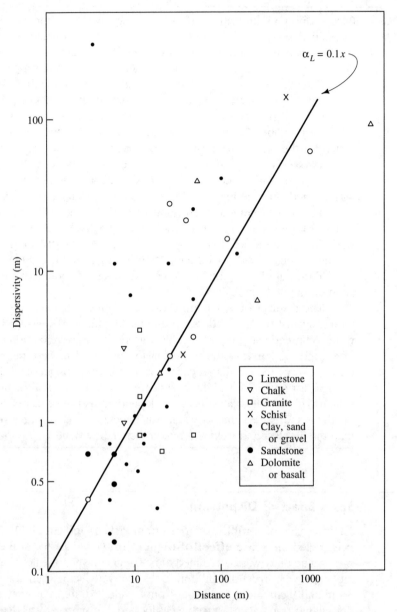

FIGURE 2.17 Field-measured values of longitudinal dispersivity as a function of the scale of measurement. *Source:* P. Lallemand-Barres and P. Peaudecerf, *Bulletin, Bureau de Recherches Géologiques et Miniéres,* Sec 3/4 (1978):277–84. Editions BRGM BP6009 45060 ORLEANS CEDEX 2.

Mass Transport in Saturated Media

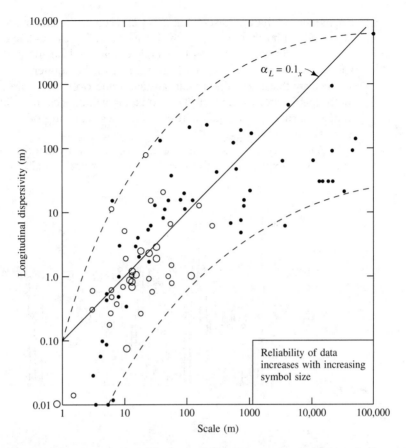

FIGURE 2.18 Field-measured values of longitudinal dispersivity as a function of the scale of measurement. The largest circles represent the most reliable data. *Source:* L. W. Gelhar, *Water Resources Research* 22, no. 9 (1986): 135S–145S. Copyright by the American Geophysical Union.

0.1 of the flow length. Gelhar (1986) published a similar graph (Figure 2.18), which contained more data points and was extended to flow lengths more than an order of magnitude greater than the Lallemand-Barres and Peaudecerf figure. The additional data on the Gelhar graph suggest that the relationship between α_L and flow length is more complex than a simple 1 to 10 ratio.

The longitudinal dispersivity that occurs at field-scale flow lengths can be called **macrodispersion.** In a flow domain that encompasses a few pore lengths, mechanical dispersion is caused by differences in the fluid velocities within a pore, between pores of slightly different size, and because different flow paths have slightly different lengths. However, at the field scale, even aquifers that are considered to be homogeneous will have layers and zones of somewhat different hydraulic conductivity. If mechanical dispersion can be caused by slight differences in the fluid velocity within a single pore,

74 Chapter Two

imagine the mechanical dispersion that will result as the fluid passes through regions of the aquifer with different conductivity values and corresponding different velocities.

Hydraulic conductivity is frequently determined on the basis of a pumping test, where water is removed from a large volume of the aquifer. As a result, the hydraulic conductivity that is obtained is an average value over the entire region of the aquifer contributing water to the well. This averaging will conceal real differences in hydraulic conductivity across the aquifer. These differences exist in both vertical and longitudinal sections.

Figure 2.19(a) shows the variation of laboratory-determined values of intrinsic permeability, expressed as the logarithm of the value in millidarcies, versus depth for

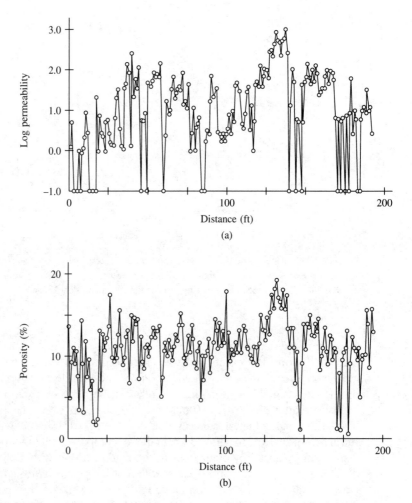

FIGURE 2.19 Permeability in millidarcies and porosity data from laboratory tests of cores from the Mt. Simon aquifer in Illinois. *Source:* L. W. Gelhar, *Water Resources Research.* 22, no. 9 (1986):135S–145S. Copyright by the American Geophysical Union.

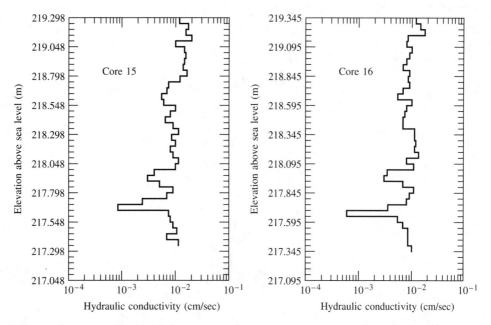

FIGURE 2.20 Hydraulic conductivity as determined by permeameter tests of remolded sediment samples from a glacial drift aquifer. The borings from which the cores were obtained are separated by one meter horizontally. *Source:* E. A. Sudicky, *Water Resources Research* 22, no. 13 (1986):2069–82. Copyright by the American Geophysical Union.

rock core samples from a well in the Mt. Simon aquifer in Illinois (Gelhar 1986, based on Bakr 1976). Figure 2.19(b) shows the variation of porosity with depth for the same core samples. Figure 2.20 shows profiles of the vertical variation in hydraulic conductivity based on permeameter tests of repacked core samples of sediment, from two borings located 1 m apart (Sudicky 1986). Figure 2.21 shows the distribution of the log of hydraulic conductivity of a cross section in a stratified sandy outwash aquifer with layers of primarily medium-grained, fine-grained, and silty, fine-grained sand. The cross section is 1.75 m deep by 19 m long (Sudicky 1986).

Figures 2.19 through 2.21 illustrate the natural variation of both hydraulic conductivity and porosity. Even aquifers that are usually considered to be homogeneous still have variations in porosity and hydraulic conductivity. Hydraulic conductivity of geologic materials varies over a very wide range of values, up to nine orders of magnitude. Porosity varies over a much, much smaller range: approximately from 1 to 60% or less than two orders of magnitude. From the standpoint of describing aquifers mathematically, it is sometimes useful to assume that hydraulic conductivity follows a lognormal distribution, which means that the logarithms of the conductivity values are normally distributed, whereas porosity is normally distributed (Freeze 1975). Since dispersion depends upon variations in the fluid velocity and from Darcy's law $[v = (K/n)(dh/dl)]$, it is obvious that variations in both hydraulic conductivity and porosity play a role. However, since hydraulic conductivity varies over a much larger range, it is the more important.

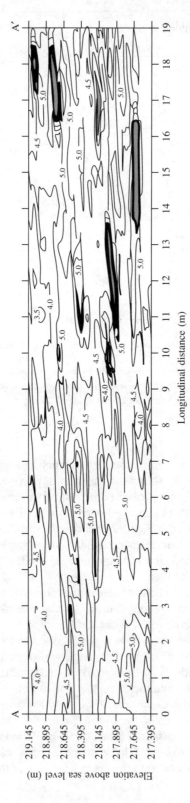

FIGURE 2.21 Distribution of the hydraulic conductivity along a cross section through a glacial drift aquifer. Hydraulic conductivity is expressed as a negative log value. (If $K = 5 \times 10^{-2}$ cm/sec, then $-\log K$ is 1.3.) Sample locations are every 5 cm vertically and every 1 m horizontally. Hydraulic conductivity was less than 10^{-3} cm/sec in the stippled zones. *Source:* E. A. Sudicky, *Water Resources Research* 22, no. 13 (1986):2069–82. Copyright by the American Geophysical Union.

This leads us to an explanation for the scale factor. As the flow path gets longer, ground water will have an opportunity to encounter greater and greater variations in hydraulic conductivity and porosity. Even if the average linear velocity remains the same, the deviations from the average will increase, and hence the mechanical dispersion will also increase. It is logical that the flow path will eventually become long enough that all possible variations in hydraulic conductivity will have been encountered and that the value of mechanical dispersion will reach a maximum. If one assumes that the distribution of hydraulic conductivity has some definable distribution, such as normal or lognormal, and that transverse dispersion is occurring, it can be shown that apparent macrodispersivity will approach an asymptotic limit at long travel distances and large travel times (Matheron and de Marsily 1980; Molz, Guven, and Melville 1983; Gelhar and Axness 1983; Dagan 1988). When the asymptotic limit is reached, the plume will continue to spread. In this region the variance of the plume will grow proportionally to the time or mean travel distance, as it does at the laboratory column scale. The advective-dispersion model is based on the assumption that dispersion follows Fick's law. Some authors contend that dispersion follows Fick's law only at the laboratory scale, where it is caused by local mechanical dispersion, and for very long flow paths, where the effects of advection through heterogeneous materials and local transverse dispersion create macroscale dispersion that follows Fick's law (e.g., Gelhar 1986; Dagan 1988). The contention that macroscale dispersion becomes Fickian (i.e., follows Fick's law) at long travel times and distances is somewhat controversial, especially if the flow is through geological formations that are heterogeneous at different scales (Anderson 1990).

2.12 Stochastic Models of Solute Transport

2.12.1 Introduction

The normal manner of determining a field-scale dispersion coefficient is to look for a natural tracer or inject a tracer into an aquifer and observe the resulting development of a plume. A solute-transport model is then constructed and the computed solute distribution is fitted to the observed field data by adjusting the dispersion coefficients. Dispersion coefficients obtained in this manner are fitted curve parameters and do not represent an intrinsic property of the aquifer. This is especially true when the aquifer is assumed to be homogeneous and is described by a single value for hydraulic conductivity and porosity. It is apparent that flow and transport modeling based on a single value for porosity and hydraulic conductivity is a gross simplification of the complexity of nature. For analytical solutions, we are constrained to use of a single value for average linear velocity, and for numerical models we often use a single value because that is all we have.

A **deterministic model** is one where a partial differential equation is solved, either numerically or analytically, for a given set of input values, aquifer parameters, and boundary conditions. The resulting output variable has a specific value at a given place in the aquifer. It is assumed that the distribution of aquifer parameters is known. The equations given earlier in this chapter are examples of deterministic models.

A **stochastic model** is a model in which there is a statistical uncertainty in the value of the output variables, such as solute distribution. The probabilistic nature of this outcome is due to the fact that there is uncertainty in the value and distribution of the underlying aquifer parameters, such as the distribution and value of hydraulic conductivity and porosity (Freeze 1975; Dagan 1988).

The idea behind stochastic modeling is very attractive. It is obvious that it takes a great effort to determine hydraulic conductivity and porosity at more than a few locations in an aquifer system. If we could determine the distribution of aquifer properties with a high degree of detail, then a numerical solution of a deterministic model would yield results with a high degree of reliability. However, with limited knowledge of aquifer parameters, a deterministic model makes only a prediction of the value of an output variable at a given point and time in the aquifer. The stochastic model is based on a probabilistic distribution of aquifer parameters. At the outset it is recognized in the stochastic model that the result will be only some range of possible outcomes. It cannot tell us what the concentration of a solute will be at a particular point in the aquifer at a given time. The stochastic model thus recognizes the probabilistic nature of the answer, whereas the deterministic model suggests that there is only one "correct" answer. Of course, the experienced hydrogeologist recognizes the uncertainty even in the deterministic answer. There have been literally hundreds of papers written since 1975 on various aspects of stochastic modeling of ground-water flow and solute transport (e.g., Freeze 1975; Gelhar, Gutjahr, and Naff 1979; Gelhar and Axness 1983; Gelhar 1986; Dagan 1982, 1984, 1986, 1988; Neuman, Winter, and Newman 1987). The stochastic transport models predict the movement of the center of mass of the solute plume (Sposito, Jury, and Gupta 1986) and the average moment of inertia or second spatial moment of the solute with respect to its center of mass (Dagan 1988).

Stochastic models have reached the stage of development where their accuracy has been tested by comparison of model-predicted results with the movement of a tracer in field tests (Sposito and Barry 1987; Barry, Coves and Sposito 1988).

2.12.2 Stochastic Descriptions of Heterogeneity

The greatest uncertainty in the input parameters of a model is the value of hydraulic conductivity, because it varies over such a wide range for geologic materials. If we make a measurement of hydraulic conductivity at a given location, the only uncertainty in its value at that location is due to errors in measuring its value. However, at all locations where hydraulic conductivity is not measured, additional uncertainty exists. If we make a number of measurements of the value of hydraulic conductivity, we can estimate this uncertainty using certain statistical techniques.

Let us define Y as the log of the hydraulic conductivity, K, and assume that the log value Y is normally distributed. We will assume a one-dimensional series of Y values $\{Y_1, Y_2, Y_3, Y_4, \ldots, Y_n\}$ (Freeze et al. 1990). Therefore,

$$Y_i = \log K_i \qquad (2.40)$$

The population that consists of all of the values of Y has a mean value, μ_Y, and a standard deviation, σ_Y. The only way to obtain precise values of μ_Y and σ_Y would be to sample the aquifer everywhere, clearly an impossible task, but we can find estimates of

Mass Transport in Saturated Media

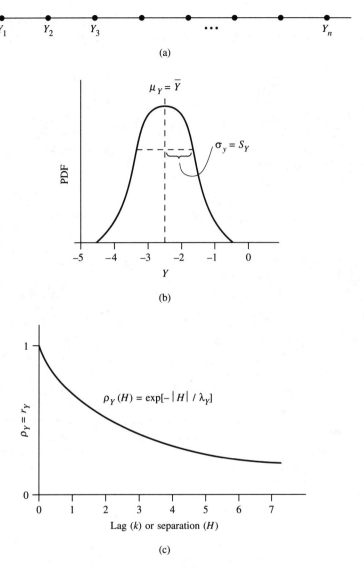

FIGURE 2.22 (a) One-dimensional sequence of log hydraulic conductivity values, Y; (b) probability distribution function for Y; (c) autocorrelation function for Y. *Source*: R. A. Freeze et al., *Ground Water* 28, no. 5 (1990):738–66. Used with permission. Copyright © 1990 Water Well Journal Publishing Co.

their values based on the locations where we have actually measured K. If we have a series of Y values $\{Y_1, Y_2, Y_3, \ldots, Y_n\}$, as in Figure 2.22(a), based on measured values of K, then our estimate of the population mean is obtained from the mean value of the sampled values, $\overline{Y}$, which can be found from

$$\overline{Y} = \frac{1}{n} \sum_{i=1}^{n} Y_i \qquad (2.41)$$

The estimate of the variance of the population is also obtained by the variance of the sampled values, S_Y^2, which is found from the following equation:

$$S_Y^2 = \frac{1}{n} \sum_{i=1}^{n} (Y_i - \overline{Y})(Y_i - \overline{Y}) \qquad (2.42)$$

For a normally distributed population, the probabilistic value is called a probability density function (PDF) and is described by the mean and the variance. The variance is a measure of the degree of heterogeneity of the aquifer. The greater the value, the more heterogeneous the aquifer. The PDF can be represented as a bell-shaped curve with the peak equal to the mean, as in Figure 2.22(b), and the spread of the bell can be defined by either the variance or the standard deviation, S_Y, which is the square root of the variance.

If we have measured the value of Y_i at a number of locations and wish to estimate the value Y_j at some other location j that is not close to any of the measured values, how can we estimate the value of Y_j? One approach is to say that the most likely estimate of Y_j is the mean of the measured values of Y_i, and the uncertainty in this value is normally distributed with a standard deviation equal to the standard deviation of the measured values, S_Y. In doing so we have accepted the **ergodic hypothesis.** This means that there is a 16% chance that the value of Y_i is greater than $\overline{Y} + S_Y$, a 50% chance that it is greater than $\overline{Y}$, and an 84% chance that it is greater than $\overline{Y} - S_Y$.

Hydraulic conductivity values measured at locations close to each other are likely to be somewhat similar. The farther apart the measurements, the less likely that the values will be similar. This is due to the fact that as distances become greater, the chance that there will be a change in geologic formation increases. The function that describes this is the **autocorrelation function,** ρ_Y. The value of the autocorrelation function decreases with the distance between two measurements. An estimate of the autocorrelation function, r_Y, can be obtained from the measured sample values by the following equation:

$$r_Y = \frac{1}{S_Y^2} \frac{1}{n} \sum_{i=1}^{n} (Y_i - \overline{Y})(Y_{i-k} - \overline{Y}) \qquad (2.43)$$

with k, the **lag,** being a whole number representing a position in the sequence away from the i position. Figure 2.22(c) shows an autocorrelation function plotted against the lag. If the lag is zero, then Equation 2.43 reduces to $r_Y = S_Y^2/S_Y^2 = 1$. This means that a Y value is perfectly correlated with itself.

The autocorrelation factor can be expressed in terms of either lag, ρ_{Yk}, or distance, $\rho_Y(H)$. When a measurement of Y_i is made at position X_i and a measurement of Y_{i-k} is made at position X_{i-k}, the absolute value of $X_i - X_{i-k}$ is called the **separation,** H.

If the autocorrelation function has an exponential form, then it can be expressed as

$$\rho_Y(H) = \exp[-|H|/\lambda_Y] \qquad (2.44)$$

where λ_Y, the **correlation length,** is representative of the length over which Y is correlated. It is the distance over which $\rho_Y(H)$ decays to a value of e^{-1}. The **integral scale,** ϵ_Y, is the area under the curve.

$$\epsilon_Y = \int_0^\infty \rho_Y(H)\, dH \qquad (2.45)$$

Integration of Equation 2.45 will show that $\epsilon_Y = 1/\lambda_Y$, so that the correlation structure can be described by either the correlation length or the integral scale.

The **autocovariance,** τ_{Yk} or $\tau_Y(H)$, is equal to the autocorrelation times the variance.

$$\tau_Y(H) = \sigma_Y^2 \rho_Y(H) \qquad (2.46)$$

We can describe the distribution of heterogeneity of Y by the use of three stochastic functions, μ_Y, σ_Y (or σ_Y^2), and λ_Y. If a stochastic process is said to be stationary, the values of μ_Y, σ_Y (or σ_Y^2), and λ_Y do not vary in space in the region being studied. If the hydraulic conductivity of an aquifer can be described as a stationary stochastic process, the aquifer is uniformly heterogeneous.

2.12.3 Stochastic Approach to Solute Transport

If we accept the idea that we don't know the value of the hydraulic conductivity and the porosity everywhere, then we must accept the idea that it is not possible to predict the actual concentration of a solute that has undergone transport through an aquifer. The best estimate of the concentration is the **ensemble mean concentration,** $\langle C \rangle$, or the mean of all the means of an ensemble of all possible random but equivalent populations, and the associated variance. The movement of a solute body may be described by the motion of the center of mass of the body and the second-order spatial moment, or the moment of inertia (Dagan 1988). However, it is important to note that the process of advective transport dominates macrodispersion. This means that whether one uses a deterministic model or a stochastic model, the large picture of solute transport will emerge, since both account primarily for advective transport, with the dispersion factor tending to smear the leading edge of the plume.

Dagan (1987, 1988) has derived a linear model of stochastic transport. This model unites the work of Dagan (1982, 1984), Gelhar and Axness (1983) and Neuman, Winter, and Newman (1987). In order to do so, he neglected all nonlinear terms, such as those arising from the deviation of solute particles from their mean trajectory. Neuman and Zhang (1990) and Zhang and Neuman (1990) have derived a quasilinear stochastic model that is more general than Dagan's linear model, which is applicable only for solute transport domains with a large Peclet number—that is, those with a long flow distance. However, as solute transport over long flow distances represents a practical problem and Dagan (1988) presents some useful closed-form analytical solutions, we will examine his results.

The case that is presented is for the movement of a slug of contamination such that at $t = 0$ a solute of total mass M is injected into the aquifer at concentration C_0. We wish to know the concentration $C(x, t)$ at some future time, t, and distance from the point of injection, x.

The velocity vector, **V**, can be determined from Darcy's law:

$$\mathbf{V} = -\left(\frac{K}{n}\right)\nabla h \tag{2.47}$$

Let $\mathbf{V} = \mathbf{U} + \mathbf{u}$, where $\mathbf{U} = \langle \mathbf{V} \rangle$, the ensemble average for **V**, and **u** is the fluctuation in **V**. **U** is assumed to be constant and its covariance $u_{jl} = \langle u_j(\mathbf{x}) u_l(\mathbf{x}') \rangle$ is a function of the separation vector $\mathbf{H} = \mathbf{x} - \mathbf{x}'$, where $j, l = 1, 2, 3$ and $\mathbf{x}$ (x_1, x_2, x_3) is the coordinate vector.

The position of the center of mass $\langle \mathbf{X} \rangle$ is defined as

$$\langle \mathbf{X} \rangle = \left(\frac{1}{M}\right) \int \mathbf{x} n \langle C \rangle \, d\mathbf{x} \tag{2.48}$$

The total mass M is defined as

$$M = \int nC \, d\mathbf{x} \tag{2.49}$$

The average moment of inertia or second spatial moment of the solute body is

$$X_{jl}(t) = \left(\frac{1}{M}\right) \int (x_j - \langle X_j \rangle)(x_l - \langle X_l \rangle) n \langle C \rangle \, d\mathbf{x} \tag{2.50}$$

For a conservative solute the position of the center of mass of the solute slug can be obtained from the advective equation.

$$\langle \mathbf{X} \rangle = \mathbf{U} t + \mathbf{X}_0 \tag{2.51}$$

where

$\langle \mathbf{X} \rangle$ = the ensemble mean average position of the center of mass
$\mathbf{U} = \langle \mathbf{V} \rangle$ the ensemble average of ground-water velocity
$\mathbf{X}_0$ = the starting position of the contamination
t = time

The value of **U** can be estimated from Darcy's law:

$$\mathbf{U} = -\left(\frac{K_G}{n_e}\right)(dh/dl) \tag{2.52}$$

where K_G is the geometric mean of the hydraulic conductivity values measured in the area over which **U** is being estimated. The geometric mean of K is found by taking the natural log of each value, finding the mean of the natural logs, and then finding the exponential of the mean of the natural logs.

The total particle displacement, $\mathbf{X}_t$, is defined as

$$\mathbf{X}_t = \mathbf{U} t + \mathbf{X}' + \mathbf{X}_d \tag{2.53}$$

with $d\mathbf{X}'/dt = \mathbf{u}(\mathbf{X}_t)$ and $d\mathbf{X}_d/dt = \mathbf{u}_d$, where **u** is the fluctuation in the convective velocity and $\mathbf{u}_d$ is velocity of a Brownian motion of zero mean. $\mathbf{X}'$ is the residual of the

Mass Transport in Saturated Media

displacement of a particle defined as $d\mathbf{X}'/dt = \mathbf{u}(\mathbf{X}_t)$, which is mechanical dispersion plus heterogeneous advection.

Dagan (1987, 1988) showed that

$$\frac{d^2 X_{jl}(t)}{dt^2} = \frac{2}{(2\pi)^3} \iint \langle \hat{u}_j(\xi) \hat{u}_l^*(\xi') \exp(i\xi' \cdot \mathbf{X}') \rangle$$
$$\cdot \exp(i\xi' \cdot \mathbf{U}t - D_{d,pq}\xi'_p\xi'_q t)\, d\xi\, d\xi' \quad (j, l = 1, 2, 3) \quad (2.54)$$

where ξ and ξ' are Fourier transform wave vector numbers, $\hat{u}_l^*$ is the complex conjugate of $\hat{u}_l$, and i stands for the imaginary unit. This equation has been solved for both longitudinal and transverse dispersion. The longitudinal spread of the solute body is characterized by $X_{11}(t)$ and the transverse spread, by $X_{22}(t)$.

Dimensionless forms of $X_{11}(t)$, $X_{22}(t)$, and t are given by $X_{11}' = X_{11}/\epsilon_h^2$, $X_{22}' = X_{22}/\epsilon_h^2$, $t' = tU/\epsilon_h$, where ϵ_h is the integral scale for the log conductivity for horizontal flow.

Solutions of 2.54 have been obtained for conditions of average flow parallel to the plane of anisotropy and for large Peclet numbers. Closed forms of these equations are available for two conditions: (1) isotropic conditions with three-dimensional heterogeneity and (2) anisotropic conditions where the log conductivity correlation scale for vertical flow is much less than the log conductivity correlation scale for horizontal flow. The anisotrophy ratio, ϵ_v/ϵ_h, is indicated by Ω.

Longitudinal dispersion with isotropic conditions:

$$\frac{X_{11}'(t')}{\sigma_Y^2} = 2t' - 2\left[\frac{8}{3} - \frac{4}{t'} + \frac{8}{t'^3} - \frac{8}{t'^2}\left(1 + \frac{1}{t'}\right)\Omega^{-t'}\right] \quad (2.55)$$

Longitudinal dispersion with two-dimensional anisotropic conditions:

$$\frac{X_{11}'(t')}{\sigma_Y^2} = 2t' - 3\ln t' + \frac{3}{2} - 3E$$
$$+ 3\left[Ei(-t') + \frac{\Omega^{-t'}(1 + t') - 1}{t'^2}\right] \quad (2.56)$$

Transverse dispersion with three-dimensional isotropic heterogeneity:

$$\frac{X_{22}'}{\sigma_Y^2} = \frac{X_{33}'}{\sigma_Y^2} = 2\left[\frac{1}{3} - \frac{1}{t'} + \frac{4}{t'^3} - \left(\frac{4}{t'^3} + \frac{4}{t'^2} + \frac{1}{t'}\right)\Omega^{-t'}\right] \quad (2.57)$$

Transverse dispersion with two-dimensional isotropic heterogeneity.

$$\frac{X_{22}'}{\sigma_Y^2} = \ln t' - \frac{3}{2} + E - Ei(-t') + 3\left[\frac{1 - (1 + t')\Omega^{-t'}}{t'^2}\right] \quad (2.58)$$

In Equations 2.56 and 2.58, E is the Euler number (0.577...) and Ei is the exponential integral, which is tabulated in Appendix D. Figure 2.23 shows the relationship

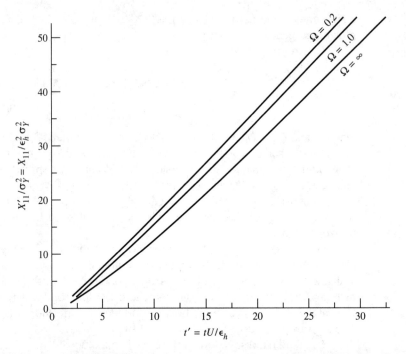

FIGURE 2.23 Dependence of longitudinal spatial moment, X_{11}, on travel time, t. Curve for $\Omega = 1$ is for isotropic three-dimensional heterogeneity and curve for $\Omega =$ infinity is for two-dimensional transport in the horizontal plane for isotropic heterogeneity. *Source:* G. Dagan, *Water Resources Research* 24, no. 9 (1988): 1491–1500. Copyright by the American Geophysical Union.

of the dimensionless longitudinal spatial moment, X_{11}', to dimensionless time, t'. Curves are given for various anisotropy ratios, Ω, where $\Omega = \epsilon_v/\epsilon_h$. The curve for $\Omega = 1.0$ is the solution to Equation 2.55, whereas the curve for $\Omega = \infty$ is the solution to Equation 2.56.

Dagan (1988) defined an apparent longitudinal macrodispersivity coefficient as "the value of the constant dispersivity that would lead to the solution of the convection-dispersion equation for the same X_{11} as the actual, time-dependent one." He indicated that this was equal to $X_{11}/2t$. Figure 2.24 shows the relationship of dimensionless apparent longitudinal macrodispersivity to dimensionless time for various anisotropy ratios. It can be seen that this parameter has limited dependence on the anisotropy ratio. The apparent macrodispersivity approaches an asymptotic value with increasing travel time, which corresponds to increasing travel distance. This figure demonstrates that at all times longitudinal macrodispersion is dominated by advective transport. The asymptotic value for D_L is $\sigma_Y^2 \epsilon_h U$ and is independent of Ω.

Since $\epsilon_h = 1/\lambda_h$, the asymptotic value of D_L is also equal to $(\sigma_Y^2 U)/\lambda_h$ and the asymptotic macrodispersivity, α_L, is equal to σ^2/λ_h.

Figure 2.25 shows the dependence of the horizontal, transverse spatial moment, X_{22}, on time and Figure 2.26 shows the same thing for the vertical, transverse spatial moment, X_{33}. It can be seen that the effect of the anisotropy ratio is significant for both these moments.

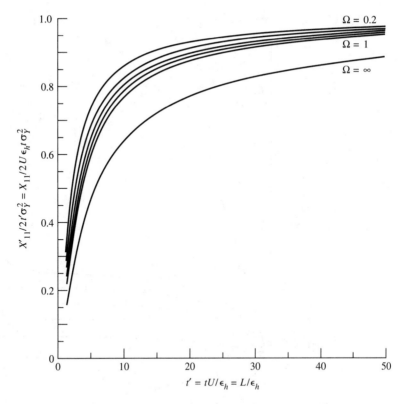

FIGURE 2.24 Dependence of the apparent longitudinal macrodispersivity upon travel time. Intermediate curves are for $\Omega = 0.4, 0.6,$ and 0.8. Source: G. Dagan, *Water Resources Research* 24, no. 9 (1988): 1491–1500. Copyright by the American Geophysical Union.

2.13 Fractal Geometry Approach to Field-scale Dispersion

2.13.1 Introduction

A recently developed field of mathematics is fractal geometry (Mandelbrot 1983). Fractal geometry is a way of looking at irregular objects, such as coastlines or aquifers. One of the precepts of fractal geometry is that irregular objects in nature tend to have patterns that repeat themselves at different scales, a phenomenon known as self-similarity. For example, in a sedimentary aquifer the relationship of individual pores to each other may be similar to the relationship of laminae to each other, which may be similar to the relationship of beds to each other, which may be similar to the relationship of geologic formations to each other.

2.13.2 Fractal Mathematics

In a classic paper Mandelbrot (1967) demonstrated that the measured length of an irregularly shaped object, in that case the coastline of Great Britain, depends upon the

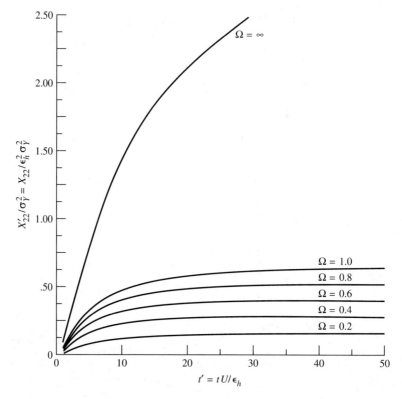

FIGURE 2.25 Dependence of the horizontal plane transverse spatial moment, X_{22}, on travel time. *Source:* G. Dagan, *Water Resources Research* 24, no. 9 (1988): 1491–1500. Copyright by the American Geophysical Union.

scale of measurement. The degree of irregularity of the coastline is independent of the scale at which the measurement takes place. A coastline has a similar irregular shape if viewed from the beach, an airplane, or a space station.

If we measure a straight line with a ruler, the length is constant and equal to the number of units times the unit length in which the measurement is made. If the unit length is halved, the number of units is doubled, but the overall length remains the same.

If we measure an irregular line, the accuracy of the measurement is a function of the scale of the measuring device. In the case of our coastline, if we used a ruler that had a minimum scale of 100 km, we would get a certain approximate length. If we then used a ruler that had a minimum scale of 1 km, we would get a different, more accurate measurement, which would be longer because we could more accurately trace the irregularities of the coast. If we made yet another measurement using a ruler with a 1-m scale, we would obtain a third, even more accurate length, that was longer still. Even so, we have yet to measure the curve of the coastline as it bends around individual rocks, much less the curve of the coastline around grains of sand. Using conventional geometry, the measured length is a function of the minimum scale of measurement.

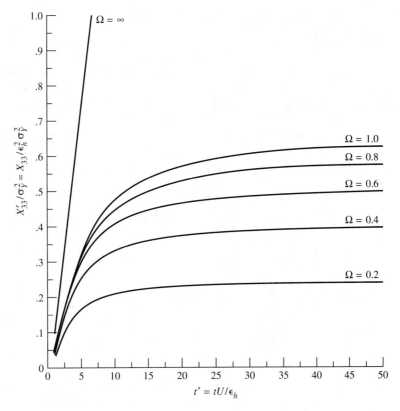

FIGURE 2.26 Dependence of the vertical plane transverse spatial moment, X_{33}, on travel time. *Source:* G. Dagan, *Water Resources Research* 24, no. 9 (1988): 1491–1500. Copyright by the American Geophysical Union.

Mandelbrot (1967) showed that for the case of a coastline, there was a constant length that was independent of the unit of measure. This was expressed as

$$J = N\eta^f = \text{constant} \tag{2.59}$$

where

$$J = \text{constant length}$$
$$N = \text{number of units}$$
$$\eta = \text{unit of measurement}$$
$$f = \text{fractal dimension}$$

The fractal dimension, f, is a number that, for a one-dimensional irregular object, lies between 1 and 2. If f is 1, then the line is straight, and if it is 2, the line is so irregular that it fills a plane.

The length measured at measurement scale η, $L(\eta)$ can be related to J by

$$L(\eta) = J\eta^{1-f} \tag{2.60}$$

2.13.3 Fractal Geometry and Dispersion

Wheatcraft and Tyler (1988) devised an approach to understanding dispersion based on the application of fractal geometry. The path taken by a particle of water traveling through a porous medium is not straight and thus is longer than the straight-line distance between the two endpoints. This phenomenon has already been discussed as tortuosity. If the tortuosity increases with increasing length of the flow path, the streamtube can be considered to be a fractal path.

One problem in using fractal mathematics in real-world problems is that, from Equation 2.60, as η tends toward zero, $L(\eta)$ grows toward infinity. There are no infinitely long flowpaths in an aquifer; therefore, there must be some lower bound to the length of η. This is on the order of the size of the average pore radius, as at a scale lower than this length the water particles can simply flow across the pore and don't necessarily follow a fractal path. This lower limit is called the **fractal cutoff limit, η_c**.

If we consider the fractal path taken by a particle of water to have a length L_f and the straight-line length from the start of the fractal streamtube to the end to be L_s, the following relationship holds:

$$L_f = \eta^{1-f} L_s^{f} \tag{2.61}$$

Here η cannot be less than the fractal cutoff limit, nor can it be longer than the straight-line length of the streamtube. In a more general sense, the length of any part of the streamtube, x_f, can be found from

$$x_f = \eta_c^{1-f} x_s^{f} \tag{2.62}$$

where x_s is the straight-line length from the origin of the streamtube to the coordinate of the end of x_f and η_c has a length less than x_s but greater than zero.

Fractal tortuosity, T_f, is defined as

$$T_f = \frac{x_f}{x_s} = \eta_c^{1-f} x_s^{f-1} \tag{2.63}$$

As the travel distance becomes greater, the fractal tortuosity also increases. This is a result of the water particle encountering more and more heterogeneity as the flowpath extends across heterogeneity boundaries.

The average linear velocity, v_x, is the average velocity determined by dividing the straight-line distance, x_s, by the travel time. The fractal velocity, v_f, is the fractal distance, x_f, divided by the travel time.

$$v_f = \frac{dx_f}{dt} \qquad v_x = \frac{dx_s}{dt}$$

Although we measure the value of v_x in the field, particles of water containing solute actually are traveling at v_f, the fractal velocity. This has an effect on dispersion.

If a tracer experiment is performed, a value of the coefficient of longitudinal dispersivity can be obtained. We will call this the field-measured dispersion coefficient, D_m. The concentration variance, σ_C^2, for the time when $C/C_0 = 0.5$ and for distance x_s is equal to

$$\sigma_C^2 = 2 D_m x_s \tag{2.64}$$

Mass Transport in Saturated Media

Since the tracer particles have moved along the fractal streamtube, they have actually gone a distance x_f, which is longer than x_s. If the experiment is continued over longer flow paths, the fractal streamtube length will grow faster than the straight-line distance. As a result the tracer will disperse more than is predicted by 2.64. The variance in the fractal streamtube is

$$\sigma_c^2 = 2D_L x_f \quad (2.65)$$

where D_L is the coefficient of pore scale longitudinal dispersion not affected by the scale of measurement. It is typically on the order of 0.001 to 1 cm. The measured coefficient of field dispersion, which is scale-dependent, can be related to the coefficient of pore scale dispersion by

$$D_m = D_L \eta^{1-f} x_s^{f-1} \quad (2.66)$$

Equation 2.66 suggests that the coefficient of field dispersion will increase as the scale of observation increases and that is it is proportional to x_s^{f-1}. However, for a nonfractal porous medium, $f = 1$ and $D_m = D_L$.

In a real aquifer a particle of water will follow one fractal streamtube, but the actual path it will follow is uncertain. There is an ensemble of fractal streamtubes, each with the same origin and each with the same straight-line distance, x_s, but each with a different fractal length, x_f. Therefore, there are many realizations of the fractal streamtube length, and we can compute a mean and a variance for each.

If we have an ensemble of fractal streamtubes, each with a different fractal length, x_f, then we can find a variance for the ensemble. From Equation 2.62 we have

$$\text{var}(x_f) = \text{var}(\eta_c^{1-f} x_s^f) \quad (2.67)$$

where we use the notation $\text{var}(x)$ as equivalent to σ_x^2.

If x_s, the straight-line distance, is fixed and the fractal dimension, f, is constant over the scale that we are studying, the only source of variation in x_f is variation in the fractal cutoff limit, η_c. Therefore, Equation 2.67 becomes

$$\text{var}(x_f) = x_s^{2f} \, \text{var}(\eta_c^{1-f}) \quad (2.68)$$

Equation 2.68 contains an expression for variation at the microscopic scale ($\text{var}(\eta_c^{1-f})$) based on the fractal cutoff limit. The term x_s^{2f} accounts for aquifer heterogeneities, because it provides for variation in the travel-path length as the scale of observation increases.

The variation in travel length on the basis of advection-dispersion (diffusional) theory can be found from

$$\text{var}(x_f) = 2D_m x_s \quad (2.69)$$

where D_m is the field-measured (calculated) dispersion coefficient and x_s is the straight-line travel distance, which is equivalent to the scale of measurement. Equations 2.68 and 2.69 can be combined to yield an expression for scale-dependent dispersion coefficient based on fractals.

$$D_m = 0.5 \, \text{var}(\eta_c^{1-f}) x_s^{2f-1} \quad (2.70)$$

2.13.4 Fractal Scaling of Hydraulic Conductivity

Neuman (1990) has presented a fractal analysis of the scale effect of dispersion, previously discussed in Section 2.11. Neuman refers to dispersivities that are measured in the field as apparent dispersivities because they are obtained by calculations that depend upon the theory that the observer was using. Figure 2.27 contains a logarithmic plot of apparent longitudinal dispersivity as measured in field and lab studies, α_m, as a function of the travel distance, L_s, or apparent length scale. All data with an apparent length scale greater than 3500 m were excluded for theoretical reasons. Regression analysis showed that although the data are widely scattered, a best-fit line with narrow 95% confidence bands

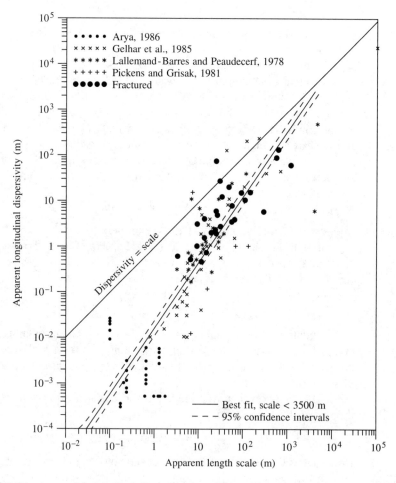

FIGURE 2.27 Apparent longitudinal dispersivity from field and laboratory studies as a function of the scale of the study. Results from the calibration of numerical models are not included. *Source*: S. Neuman, *Water Resources Research* 26, no. 8 (1990):1749–58. Copyright by the American Geophysical Union.

Mass Transport in Saturated Media

could be obtained. The equation for the line is

$$\alpha_m = 0.0175 L_s^{1.46} \qquad (2.71)$$

This line of best fit has a regression coefficient, r^2, of 0.74, which means that it accounts for 74% of the variation about the mean. The other 26% may be due to experimental and interpretive errors or may represent deviation of the real system from that described by Equation 2.71. The 95% confidence intervals about the coefficient of 0.0175 are 0.0113 and 0.0272 and the 95% confidence intervals about the exponent of 1.46 are 1.30 and 1.61.

The apparent longitudinal dispersivity can be calculated from the variance of the log hydraulic conductivity by the equation (Neuman and Zhang 1990)

$$\alpha_m = c_o L_s \sigma_Y^2 \qquad (2.72)$$

where c_o is a constant related to anisotropy that is approximately 0.5 for isotropic media (Neuman and Zhang 1990, Figure 6) and can be 1 or greater in anisotropic media (Zhang and Neuman, 1990, Figure 1).

Combining Equations 2.71 and 2.72, we obtain

$$\sigma_Y^2 = (0.0175/c_o) L_s^{0.46} \qquad (2.73)$$

The log hydraulic conductivity, $Y = \ln K$, has a semivariogram, γ_Y, at travel distance, s, given by

$$\gamma_Y(s) = Z_o s^{2\nu} \qquad (2.74)$$

where Z_o is a constant and ν is a coefficient known as the Hurst coefficient such that $0 < 2\nu < 1$. It can be shown (Neuman 1990) that such a semivariogram characterizes a self-similar random field with homogeneous increments that can be described by the fractal dimension

$$f = G + 1 - \nu \qquad (2.75)$$

where $G = 1$ for a one-dimensional field, 2 for a two-dimensional field, and 3 for a three-dimensional field.

The variance of Y has the semivariogram

$$\gamma_Y(L_s) = \left(\frac{0.0175}{c_o}\right) L_s^{0.46} \qquad (2.76)$$

From the exponential value of L_s, it can be seen that ν has a value of 0.23, indicating that the regression line of Figure 2.27 represents a log hydraulic conductivity field that is self-similar with a fractal dimension $f = G + 0.77$. Neuman (1990) performed a similar analysis on only the data representing flowpaths less than 100 m long. He obtained the following results on this data subset:

$$\alpha_m = 0.0169 L_s^{1.53} \qquad (2.77)$$

$$\gamma_Y(L_s) = \left(\frac{0.0169}{c_o}\right) L_s^{0.52} \qquad (2.78)$$

$$f = G + 0.74 \qquad (2.79)$$

On this basis Neuman has proposed that the log hydraulic conductivity values of the geological materials represented by Figure 2.27 can be characterized by the following, which form a universal scaling rule:

$$\gamma_Y(L_s) = \left(\frac{0.017}{C_o}\right) L_s^{0.5} \tag{2.80}$$

$$f = G + 0.75 \tag{2.81}$$

The preceding scaling rule can be considered to account for the self-similarity of log hydraulic conductivities from a wide variety of geologic materials over a large range of flowpath lengths and under diverse flow conditions. It would not necessarily describe any particular location.

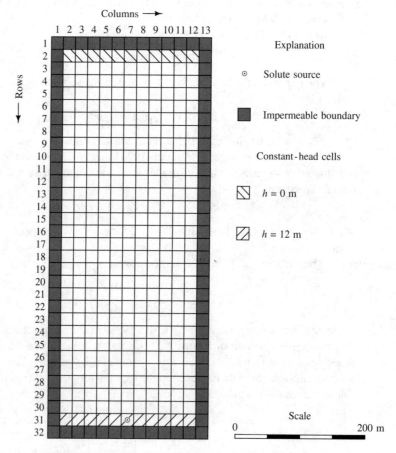

FIGURE 2.28 Finite-difference grid and boundary conditions for a deterministic model of solute transport. *Source*: A. D. Davis, *Ground Water* 24, no. 5 (1986):609–15. Used with permission. Copyright © 1986 Water Well Journal Publishing Co.

Mass Transport in Saturated Media

2.14 Deterministic Models of Solute Transport

Although workers in stochastic theory have asserted that the theoretical basis for the deterministic advective-dispersive solute transport equation is suspect except for long times and large distances (Anderson, 1984), it has been used with a great deal of success in many field and model applications. One of the problems raised with field scale application of the advective-dispersive equation is that it requires an unrealistically high value for the coefficient of longitudinal dispersion.

A model study by Davis (1986) demonstrates that deterministic models can be developed that incorporate heterogeneities. He modeled two aquifers with identical boundary conditions (Figure 2.28). One was uniform (Figure 2.29(a)) and one had

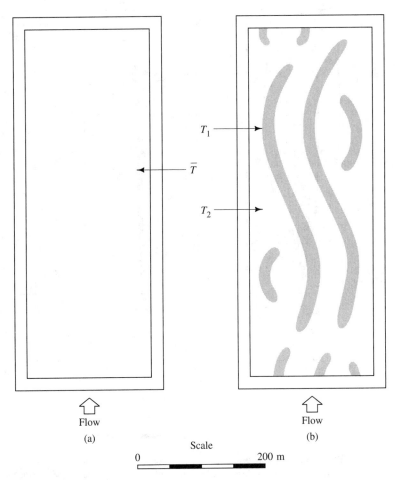

FIGURE 2.29 Model areas for finite difference solute transport model with (a) uniform transmissivity and (b) with heterogeneous transmissivity. *Source:* A. D. Davis, *Ground Water* 24, no. 5 (1986):609–15. Used with permission. Copyright © 1986 Water Well Journal Publishing Co.

variable transmissivity in the form of more permeable channels (Figure 2.29(b)). The deterministic model, based on the two-dimensional solute-transport equation, was used with small values of α_L and α_T, 0.0003 m, and 0.00009 m, respectively. The resulting solute plume in the uniform media is very long and narrow. See Figure 2.30(b). If larger values of α_L and α_T are used—3 m and 1 m, respectively—then a much broader plume results. See Figure 2.30(a). However, if the heterogeneous aquifer is used with the small values of dispersivity, the resulting plume, shown in Figure 2.30(c) has a size very similar to that created in the uniform media by using large values of dispersivity. This demonstrates that if deterministic models include the aquifer heterogeneities, then it may be possible to use dispersivity values that are more on the order of lab-scale values. Davis (1986) used the advective-dispersion equation in a model with varying transmissivities and with

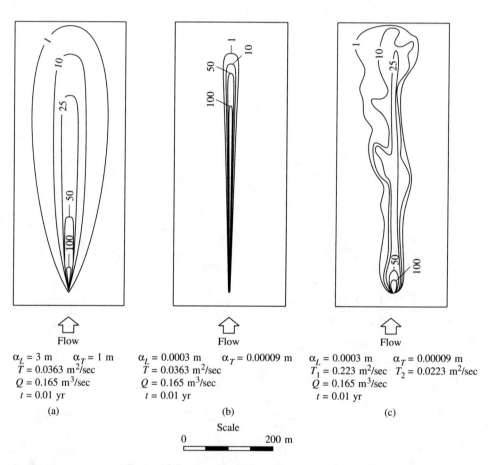

FIGURE 2.30 Model results for finite-difference solute-transport model. (a) Uniform media with large dispersivity values, (b) uniform media with small dispersivity values, and (c) heterogeneous media with small dispersivity values. *Source:* A. D. Davis, *Ground Water* 24, no. 5 (1986):609–15. Used with permission. Copyright © 1986 Water Well Journal Publishing Co.

Mass Transport in Saturated Media

a value of α_L of only 0.01 m was able to reproduce a solute plume that extended over a flow length of about 500 m. He found that a fine mesh for the finite-difference model grid was necessary for accurate results. Figure 2.31 compares the results of his model results with the field data.

From a practical standpoint, the advection-dispersion solute transport equation has been shown to give excellent results when applied to field situations. The Konikow-Bredehoeft solute-transport model (MOC) is one of the most widely used solute transport models and is based on the advection-dispersion equation (Konikow and Bredehoeft 1978). It was used with great success in modeling the movement of conservative groundwater contaminants at the Rocky Mountain Arsenal (Konikow 1977; Konikow and Thompson 1984). However, even if the advection-dispersion model can be fitted to data

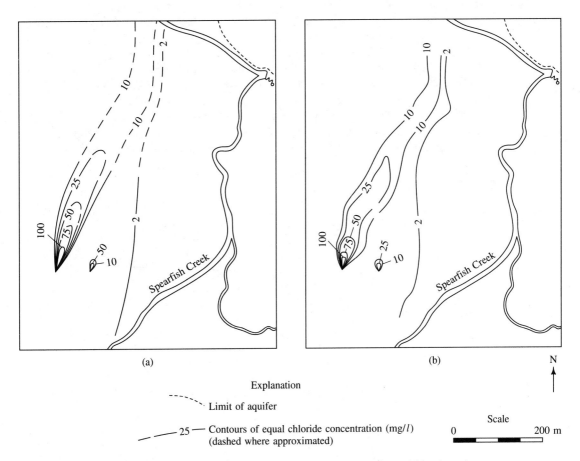

FIGURE 2.31 Comparison of (a) field observations of solute plume in an aquifer and (b) solute plume as computed by finite-difference solute-transport model for a heterogeneous aquifer. *Source:* A. D. Davis, *Ground Water* 24, no. 5 (1986):609–15. Used with permission. Copyright © 1986 Water Well Journal Publishing Co.

on the past movement of a solute plume, this does not necessarily mean the model will be accurate in predicting future movement of the plume.

Case Study: Borden Landfill Plume

An abandoned landfill in a shallow sand aquifer at Canadian Forces Base in Borden, Ontario, has been extensively studied (Cherry, 1983; MacFarlane et al., 1983). Frind and Hokkanen (1987) made a very interesting study of the plume based on a deterministic model.

The landfill was active from 1940 to 1976 and covers about 5.4 ha to a depth of 5 to 10 m. Figure 2.32 shows the location of water table wells and multilevel sampling devices. The multilevel sampling devices are concentrated along the long axis of the plume of ground-water contamination. The vertical location of the sampling points along cross section A-A' are shown in Figure 2.33. The aquifer is about 20 m thick beneath the landfill and thins to about 9.5 m in the direction of ground-water flow. The aquifer consists of laminated fine to medium sand. An average hydraulic conductivity of 1.16×10^{-2} cm/sec horizontally and 5.8×10^{-4} cm/sec vertically was used in the model with a porosity of 0.38. In 1979 a very extensive study of the water quality of the plume was conducted. Figure 2.34 shows the plume of chloride contamination along cross section A-A'. In 1979 the plume extended about 750 m from the landfill and had sunk to the bottom of the aquifer and then moved laterally with the flowing ground water. The sinking of the plume is believed to be caused by recharge concentrated in a sand pit to the north of the landfill, which is in the direction of flow.

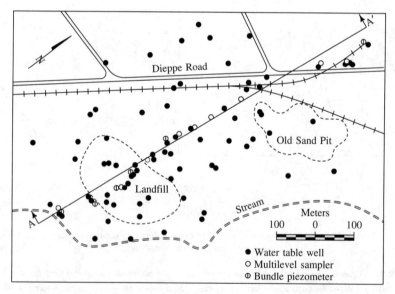

FIGURE 2.32 Location of landfill at Canadian Forces Base, Borden, Ontario, showing location of cross section and monitoring network. *Source:* E. O. Frind and G. E. Hokkanen, *Water Resources Research* 23, no. 5 (1987):918–30. Copyright by the American Geophysical Union.

Mass Transport in Saturated Media

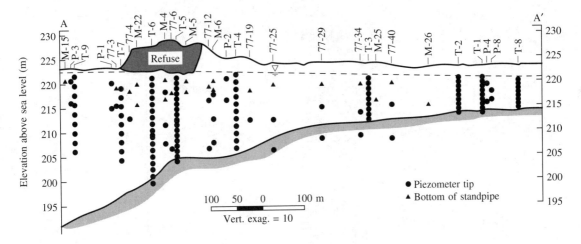

FIGURE 2.33 Cross section of aquifer at the Borden landfill showing the location of multilevel monitoring devices. *Source:* E. O. Frind and G. E. Hokkanen, *Water Resources Research* 23, no. 5 (1987):918–30. Copyright by the American Geophysical Union.

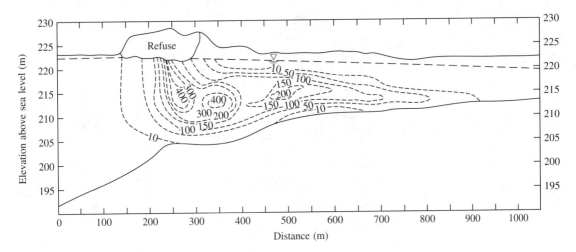

FIGURE 2.34 Chloride plume along the Borden landfill cross section in 1979. Values are in milligrams per liter. *Source:* E. O. Frind and G. E. Hokkanen, *Water Resources Research* 23, no. 5 (1987):918–30. Copyright by the American Geophysical Union.

The finite-difference grid system for the cross-sectional model is shown in Figure 2.35. Equipotential lines for observed conditions were essentially vertical (Figure 2.36). The model was calibrated against the water-table contours for steady-state conditions.

Sensitivity analyses were performed to determine the impact of varying α_L and α_T. Field tests had indicated that the value of α_L at the site is on the order of 5 to 10 m (Sudicky, Cherry, and Frind 1983). Figure 2.37 shows the sensitivity of the plume to the value

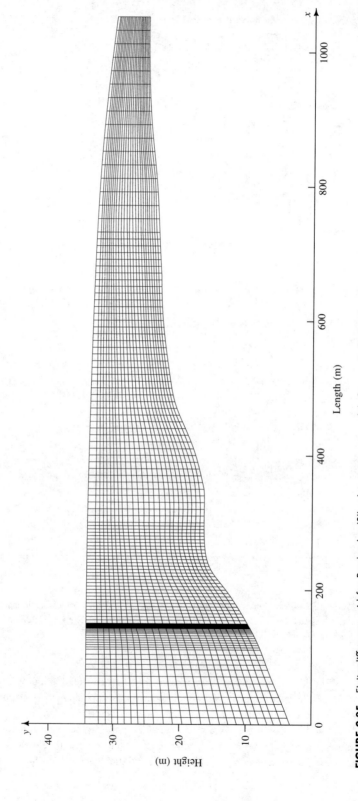

FIGURE 2.35 Finite-difference grid for Borden landfill solute transport model. *Source:* E. O. Frind and G. E. Hokkanen, *Water Resources Research* 23, no. 5 (1987):918–30. Copyright by the American Geophysical Union.

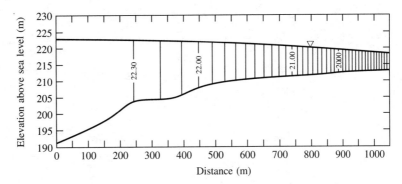

FIGURE 2.36 Equipotential lines from the calibration of the Borden landfill solute-transport model; values in meters above datum. *Source:* E. O. Frind and G. E. Hokkanen, *Water Resources Research* 23, no. 5 (1987):918–30. Copyright by the American Geophysical Union.

of α_T. The value of α_L was kept at 10 m and α_T was varied from 0.005 m to 1.0 m. It can be seen that the shape of the plume is very sensitive to the value of α_T. With a high value of α_T, the plume spread through the entire vertical thickness of the aquifer, whereas with a low value it tended to sink toward the bottom. Figure 2.38 illustrates the fact that the plume was not very sensitive to changes in the value of α_L over the range tested. The value of α_T was kept constant at 0.01 m, whereas α_L varied from 2.5 to 20 m. This figure is slightly misleading in that there is a 10:1 vertical exaggeration, so that the vertical spreading is more obvious than the horizontal. Also, the value of α_T was varied by a factor of 200, whereas α_L was varied only by a factor of 8.

Additional sensitivity analyses were conducted with respect to the water table boundary conditions and the concentration, size, and growth pattern of the source. The authors found that in order to reproduce the observed distribution, a source history that included multiple periods of high concentration was needed. Figure 2.39(a) shows the shape of the observed plume, Figure 2.39(b) illustrates the shape of a plume generated by a source with a history in which the concentration gradually increased (smooth source concentration), and Figure 2.39(c) contains the computed plume with the best match to the observed plume. It was generated by a run of the model in which the source concentration had two different periods of peak concentration. Although the solution was not unique—that is, several different combinations of model inputs might yield the same output—the shape of the plume could be reproduced with good accuracy. This was especially true at the leading edge of the plume, which is the most important part from the standpoint of predicting the movement of the plume into uncontaminated areas of the aquifer.

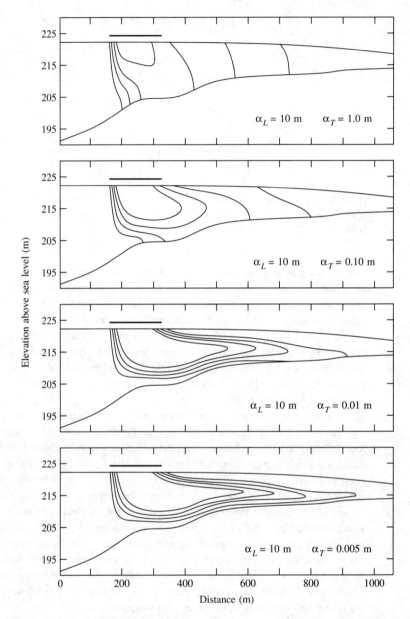

FIGURE 2.37 Sensitivity analysis of the Borden landfill solute-transport model with respect to transverse dispersivity. *Source:* E. O. Frind and G. E. Hokkanen, *Water Resources Research* 23, no. 5 (1987):918–30. Copyright by the American Geophysical Union.

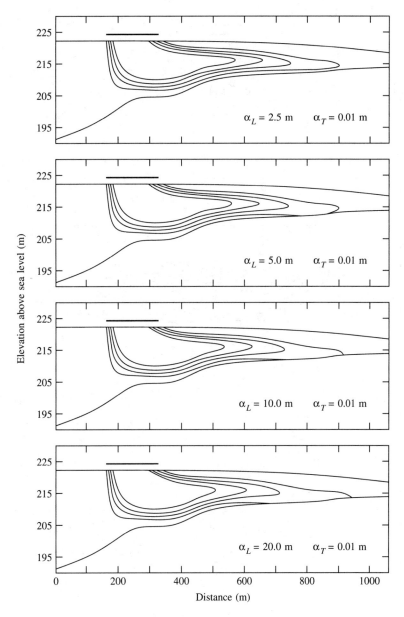

FIGURE 2.38 Sensitivity analysis of the Borden landfill solute-transport model with respect to longitudinal dispersivity. *Source*: E. O. Frind and G. E. Hokkanen, *Water Resources Research* 23, no. 5 (1987):918–30. Copyright by the American Geophysical Union.

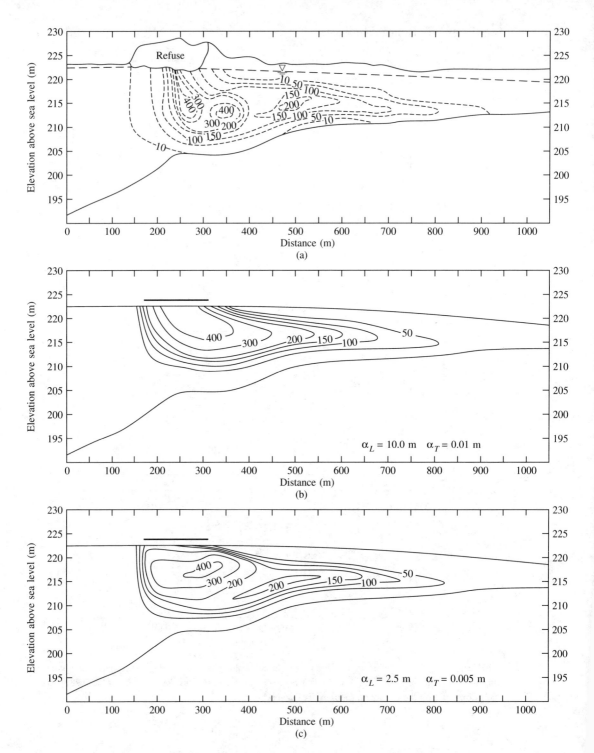

FIGURE 2.39 Comparison of (a) the observed chloride plume at the Borden landfill with (b) the chloride plume simulated by the solute transport model with a smooth source concentration and (c) the chloride plume simulated by the solute transport model with a doubly peaked source concentration. *Source*: E. O. Frind and G. E. Hokkanen, *Water Resources Research* 23, no. 5 (1987): 918–30. Copyright by the American Geophysical Union.

2.15 Transport in Fractured Media

Solute transport in fractured rock media is as important a process as transport in porous media. However, less research has been done on this topic than on transport in porous media. One reason may be the complexity of solute transport in fractured media. The rock in which fractures exist is porous. Hence, fluid moves in the fractures as well as in the rock matrix. Solutes in the fractures can diffuse into the fluid contained in the rock matrix and vice versa (Neretnieks 1980). The fractures themselves are not smooth channels but contain dead-end passages that hold nonmoving water into which solutes can diffuse (Raven, Novakowski, and Lapcevic 1988).

Berkowitz, Bear, and Braester (1988) suggested that solute transport in fractured media can be considered at a number of different scales. A very-near-field scale would be a single fracture near the source. A near-field scale would include a few fractures near the source. At a larger scale, the far field, the fracture network and the porous media matrix would have separate, discernable impacts on flow. At a very-far-field scale, which exists at considerable distance from the source, the entire flow domain can be considered as an equivalent porous medium in which the repeating fractures became large pores.

A number of different approaches to solute transport in fractured media have been attempted. These include analysis of transport in a single fracture in which effects of the transport in the fractures as well as interactions with a porous matrix are considered (e.g., Grisak and Pickens 1980, 1981; Tang, Frind, and Sudicky 1981; Rasmuson and Neretnieks 1981; Rasmussen 1984; Sudicky and Frind 1984). Sudicky and Frind (1982) and Barker (1982) examined transport in a media that consists of equally spaced fractures in a porous media. Endo and others (1984) made a deterministic study of flow in an irregular network of fractures contained in an impermeable host rock, whereas Schwartz et al. (1983) and Smith and Schwartz (1984) approached the same problem using a stochastic model. Berkowitz, Bear, and Braester (1988) and Schwartz and Smith (1988) examined the conditions under which the porous media matrix and the fractures can be considered to be a continuum that is representative of an equivalent porous media. Raven, Novakowski, and Lapcevic (1988) made a field study of flow through a single fracture to test a model that incorporates the effects of nonflowing water in the fractures. Tsang et al. (1988) and Moreno et al. (1988) examine fracture flow on the basis of the assumption that most of the flow is concentrated in a few channels.

One of the first considerations in dealing with fracture flow is deciding how to treat flow in a single fracture. Some authors (e.g., Tang, Frind, and Sudicky, 1981; Schwartz and Smith 1988) assume that the fluid in a fracture is all moving at a constant velocity. Conversely, Endo et al. (1984) treated flow in a fracture to be two-dimensional, with a parabolic velocity profile across the width of the fracture, as shown in Figure 2.40. Transport within a single fracture is due to advection, which occurs at different rates, depending upon the position between the parallel walls of the fracture, and molecular diffusion, both normal and parallel to the flow direction.

Hull, Miller, and Clemo (1987) examined the conditions whereby diffusion within the fracture needs to be considered. In a fracture with parallel sides, the solute transport within the fracture is described by

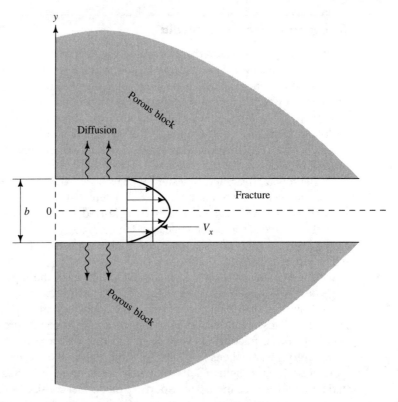

FIGURE 2.40 Horizontal distribution of flow in a vertical fracture and diffusion into the porous media matrix.

$$\frac{\partial C}{\partial t} = D^* \left(\frac{\partial^2 C}{\partial x^2} + \frac{\partial^2 C}{\partial y^2} \right) - 6V[(\tau) - (\tau)^2] \frac{\partial C}{\partial x} \qquad (2.82)$$

where

V = average fluid velocity in a fracture

τ = fractional transverse position in a fracture

At high flow rates, advection will dominate and the concentration will follow the velocity profile of Figure 2.40. At low velocities, diffusion will be important, since the concentration gradient at the solute front will be high and the distance will be short. Under these conditions, diffusion will homogenize the solute across the width of the fracture.

If L is the length of the fracture between cross fractures and β is the aperture of the fracture, the fracture residence time is L/V. This can be compared with $(\beta/2)^2/D$ to determine if diffusion needs to be considered (Crank 1956). If diffusion induces a change in the tracer concentration of less than 2% over a distance of 10% of the width of the fracture, the diffusion can be considered negligible, and the residence time in the fracture will be

$$\frac{L}{V} < 0.003 \frac{(\beta/2)^2}{D} \qquad (2.83)$$

If diffusion affects the tracer concentration to the extent that the tracer front is at 98% of the equilibrium value at all points across the fracture, the diffusion has homogenized the front, and the residence time in the fracture will be

$$\frac{L}{V} > 0.05 \frac{(\beta/2)^2}{D} \qquad (2.84)$$

Figure 2.41 indicates the circumstances under which fracture flow can be considered to be one- or two-dimensional. Fracture residence time (L/V) is plotted against fracture aperture on this figure, which is based on a diffusion coefficient of 1.7×10^{-9} m²/sec. The figure shows the conditions under which diffusion will homogenize the flow so that the transport within the fracture can be treated as one-dimensional (uniform conditions across the aperture). However, diffusion will still spread the tracer in advance of the advecting water. For even large fractures of 1 mm aperture, this will occur with a residence time of 1 min or more. This suggests that for most flow situations, one does not need to consider the velocity distribution across the fracture.

When the flow in a fracture is homogeneous, the mass transport can then be described by the one-dimensional advection-dispersion equation with the longitudinal dispersion coefficient equal to (Hull, Miller, and Clemo 1987)

$$D_L = \frac{(V\beta)^2}{210 D} \qquad (2.85)$$

One approach to solute transport modeling is to determine the flux of water through the fractures and then use a numerical technique known as a random walk model to simulate diffusion of the solute (Hull, Miller, and Clemo 1987). This ignores any diffusion into the porous media matrix. According to Witherspoon et al. (1980),

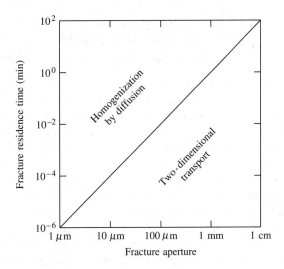

FIGURE 2.41 Fracture residence time necessary for homogenization of the tracer across the fracture width by molecular diffusion. *Source:* Modified from L. C. Hull, J. D. Miller, and T. M. Clemo, *Water Resources Research* 23, no. 8 (1987):1505–13. Copyright by the American Geophysical Union.

flow through a fracture can be described by Darcy's law using an equivalent hydraulic conductivity for a fracture, K_f, given by

$$K_f = \frac{\rho g}{12\mu} \beta^2 \qquad (2.86)$$

The quantity of flow, Q, can be found from the cubic law

$$Q = \frac{\rho g}{12\mu} I a \beta^3 \qquad (2.87)$$

where

g = acceleration of gravity

I = hydraulic gradient along the fracture

a = width of the fracture—that is, the third dimension after length and aperture

μ = viscosity of fluid

If the velocity in the channel needs to be described in two dimensions, this can be done with three equations: one for the maximum velocity in the center of the fracture, one for the flow velocity profile across the aperture, and one for the vertical velocity profile in the fracture.

The maximum velocity can be found from (Hull, Miller, and Clemo 1987):

$$V_x(\max) = \left[1.5 + 1.1664 \left(\frac{a}{\beta}\right)^{-1.0557}\right] V \qquad (2.88)$$

The velocity profile across the aperture is given by:

$$V_x(y) = 4(\tau - \tau^2) \qquad (2.89)$$

where τ = fractional transverse position in a fracture, y/β.

The vertical velocity profile is given by

$$V_x(\zeta) = 15.56\zeta - 97.72\zeta^2 + 308\zeta^3 - 513\zeta^4 + 431\zeta^5 - 143.7\zeta^6 \qquad (2.90)$$

where ζ = fractional vertical position in a fracture, z/a.

Raven, Novakowski, and Lapcevic (1988) pointed out that the fractures through which flow occurs are not smooth, parallel plates but have irregular walls that promote the formation of zones along the edge of the fracture where the water is immobile (Figure 2.42). The fluid moves through the mobile zone, but the solutes can diffuse into the immobile fluid zones. The solute would be stored in the immobile fluid during the early part of solute transport and would be released from storage if the solute concentration in the mobile fluid would decrease—for example, as might happen during the latter part of a slug injection test. They derived an advection-dispersion equation for mass transport in the fracture with "transient solute storage" in the immobile fluid zone (ADTS model). (An even more comprehensive model would have provision for transient solute storage in the immobile fluid zone as well as diffusion into a porous media matrix.)

A field test was performed on the flow through a single fracture that had been isolated by packers in the borehole. Water was injected into one borehole and withdrawn

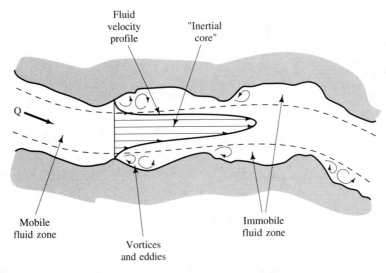

FIGURE 2.42 Zones of mobile and immobile water in a fracture. *Source:* K. G. Raven, K. S. Novakowski, and P. A. Lapcevic, *Water Resources Research* 24, no. 12 (1988):2019–32. Published by the American Geophysical Union.

from another. The water contained a tracer for the first few hours of the test, and then water without the tracer was again injected. Figure 2.43 contains circles representing the field data, in terms of relative concentration, plotted versus elapsed time. Also shown on this figure are the results of a conventional advection-dispersion (AD) model and an advection-dispersion transient storage (ADTS) model. Both models matched the observed data for the first few hours of the test. However, the ADTS model was far superior in matching the field data over the entire course of the test. The effect of transient storage was to reduce the peak concentration and to increase the concentrations above what would be produced by advection-dispersion alone during the later periods of the test.

2.16 Summary

Solutes dissolved in ground water are transported in two ways. Diffusion will cause solutes to move in the direction of the concentration gradient—that is, from areas of higher to lower concentration. This transport can occur even if the ground water is not flowing and may be the major factor in mass transport in geologic materials of very low permeability.

Solutes are also transported by the process of advection, which is also known as convection. This occurs as the flowing ground water carries the dissolved solutes with it. At the scale of a few pore diameters, ground water will move parallel to the flow path at different rates due to differences in pore size. This causes the solute plume to spread along the direction of the flow path, a process called longitudinal dispersion. The solute plume will also spread laterally as flow paths diverge around mineral grains, a process

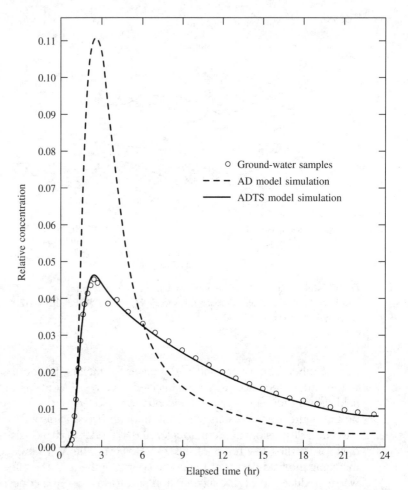

FIGURE 2.43 Comparison of field data from a tracer test in fractured rock with results of model simulation using an advection-diffusion (AD) model and an advection-diffusion transient storage (ADTS) model. *Source:* K. G. Raven, K. S. Novakowski, and P. A. Lapcevic, *Water Resources Research* 24, no. 12 (1988): 2019–32. Published by the American Geophysical Union.

known as transverse dispersion. At the laboratory column scale, the movement of a contaminant through a uniform porous media can be described by the advection-dispersion equation, which accounts for advection, diffusion, and pore-scale dispersion.

In field-scale studies it has been found that the coefficient of longitudinal dispersion obtained from the advection-dispersion equation increases with the length of the flow path. This is due to the heterogeneous nature of aquifer materials. As the length of the flow path increases, the range of permeability values that affect the rate of ground-water flow also increases. This causes the resulting solute plume to spread out more and more. This can be called macrodispersion. For flow paths less than 3500 m an apparent dispersion coefficient can be statistically correlated to the length of the flow path by the equation $\alpha_m = 0.0175 L_s^{1.46}$. Eventually the apparent dispersivity appears to reach a maximum value.

Stochastic methods of analysis have also been developed to analyze solute transport at the field scale. Stochastic methods are based on the variation in the hydraulic conductivity values because it is that variation that causes the solute plume to spread. The ground-water velocity depends upon the porosity as well as the hydraulic conductivity, but the hydraulic conductivity varies over a much greater range than porosity.

At the field scale the spreading due to hydraulic conductivity variation is much greater than that due to pore-scale dispersion. Both stochastic and advection-dispersion models demonstrate that the primary movement of the solute plume is due to advection. The stochastic model yields the movement of the center of mass of the solute plume from the average rate of movement of the ground water. The variance of the solute concentration about the mean position, or the second spatial moment, is also obtained from stochastic models.

If one has sufficient knowledge of the distribution of hydraulic conductivity in an aquifer, then a numerical advection-dispersion model of ground-water flow can be developed that uses a pore-scale dispersion value. This type of model has theoretical validity, because the necessary coefficient of longitudinal dispersion does not change with flow path length. It can be used to predict future solute concentrations at specific places and times. Naturally, such predictions will not be 100% accurate, because one can obviously never know the value of the hydraulic conductivity every place in the flow field.

Chapter Notation

A		Cross-sectional area
a		Width of a fracture
b		Aquifer thickness
B		$[(v_x x)^2/(2D_L)^2 + (v_x y)^2/(4D_L D_T)]^{1/2}$
C		Solute concentration
C_i		Concentration at some point x and time t
C_0		Concentration at time 0
C_R		Dimensionless solute concentration (C/C_0)
$\langle C \rangle$		Ensemble mean concentration
c_o		Constant related to anisotropy
d		Characteristic flow length for Peclet number, P
db/dl		Hydraulic gradient
D^*		Effective diffusion coefficient
D_d		Molecular diffusion coefficient
D_i		Coefficient of hydrodynamic dispersion in the i direction
D_L		Coefficient of longitudinal hydrodynamic dispersion
D_{LM}		Coefficient of longitudinal macrodispersivity at the asymptotic limit
D_m		Field-measured (calculated) coefficient of hydrodynamic dispersion
D_T		Coefficient of transverse hydrodynamic dispersion
E		Euler number (0.577...)
E_i		Exponential integral

f	Fractal dimension
F	Mass flux of solute per unit area per unit time
g	Acceleration of gravity
G	Topological dimension
h	Hydraulic head
H	Separation of autocorrelation function
i	Decay constant
I	Hydraulic gradient along a fracture
J	Constant length
k	Lag in autocorrelation function
K	Hydraulic conductivity
K_f	Equivalent hydraulic conductivity of a fracture
K_G	Geometric mean of hydraulic conductivity
K_0	Modified Bessel function of second kind and zero order
L	Straight-line distance between ends of a flowpath
L_e	Length of a tortuous flowpath
L_f	Length of a fractal flowpath
L_s	Straight-line length between ends of a fractal flowpath
M	Total mass of solute
N	Number of units
n	Porosity
n_e	Effective porosity
P	Peclet number ($v_x d/D_d$)
P_e	Peclet number ($v_x L/D_L$)
Q	Rate at which a tracer is being injected into an aquifer
r	Radial distance to a well
R	Length of well screen or open borehole
R_f	Average frontal position of water injected into a well
r_Y	Autocorrelation of sampled values of Y
S_Y	Standard deviation of sampled values of Y
S_Y^2	Variance of sampled values of Y
t	Time
t'	Dimensionless time (tU/ϵ_h)
t_R	Dimensionless time ($v_x t/L$)
T	Tortuosity
T_f	Fractal tortuosity
u	Average velocity of injection of water into a well
$\mathbf{u}$	Fluctuation in the velocity vector
u_{jl}	Covariance
$\mathbf{U}$	$\langle \mathbf{V} \rangle$ ensemble mean of velocity vectors
U	Pore volume
U_i	Total volume of water injected into a well
U_p	Cumulative volume of water withdrawn from a well
v_f	Velocity along a fractal flowpath
v_x	Average linear velocity in the x direction
V	Average fluid velocity in a fracture

$\mathbf{V}$	Velocity vector
$\langle \mathbf{V} \rangle$	Ensemble mean of the velocity vectors
$W[t, B]$	Hantush leaky well function
$\mathbf{x}$	Coordinate vector
x_f	Length of fractal flowpath
(x_0, y_0)	Origin of an xy field
x_s	Straight-line distance
$\mathbf{X}'$	Residual of the displacement of a particle
$\langle \mathbf{X} \rangle$	Ensemble mean of the center of mass
$X_{jl}(t)$	Second spatial moment of the solute mass at time t and location j, l
$\mathbf{X}_t$	Total particle displacement
$\overline{Y}$	Mean of sample values of Y
Y_i	$\log K_i$
Z_o	Constant related to a semivariogram
α	Dynamic dispersivity
α_L	Longitudinal dynamic dispersivity
α_T	Transverse dynamic dispersivity
α_m	Apparent dispersivity
β	Aperture of a fracture
ϵ_h	Correlation length for horizontal hydraulic conductivity
ϵ_v	Correlation length for vertical hydraulic conductivity
η	Fractal unit of measurement
η_c	Fractal cutoff limit
γ_Y	Semivariogram of Y
λ_Y	Correlation length of autocorrelation
μ	Viscosity of a fluid
μ_Y	Mean of population of Y
ν	Hurst coefficient for fractal dimensions
ω	Coefficient related to tortuosity
Ω	Anisotropy ratio (ϵ_v/ϵ_h)
ρ	Density of a fluid
ρ_Y	Autocorrelation of the population of Y
σ_Y	Standard deviation of population of Y
σ_Y^2	Variance of population of Y
τ	Fractional transverse position in a fracture
τ_Y	Autocovariance
θ	Angle in polar coordinate system
ξ	Fourier transform wave vector number
ζ	Fractional vertical position in a fracture

References

Anderson, Mary P. 1979. "Using models to simulate the movement of contaminants through groundwater flow systems." *Critical Reviews in Environmental Controls* 9, no. 2:97–156.

———. 1984, "Movement of contaminants in groundwater: Groundwater transport—advection and dispersion." In *Groundwater Contamination*, 37–45. Washington, D.C.: National Academy Press.

———. 1990. Aquifer heterogeneity—a geological perspective. *Proceedings, Fifth Canadian/American Conference on Hydrogeology.* National Water Well Association, pp. 3–22.

Ayra, A. 1986. Dispersion and reservoir heterogeneity. Ph.D. Diss. University of Texas, Austin.

Bakr, A. A. 1976. Effect of spatial variations of hydraulic conductivity on groundwater flow.", Ph.D. diss. New Mexico Institute of Mining and Technology, Socorro.

Barker, J. A. 1982. Laplace transform solutions for solute transport in fissured aquifers. *Advances in Water Resources* 5, no. 2:98–104.

Barry, D. A., J. Coves, and Garrison Sposito. 1988. On the Dagan model of solute transport in groundwater: application to the Borden site. *Water Resources Research* 24, no. 10:1805–17.

Bear, Jacob. 1961. Some experiments on dispersion. *Journal of Geophysical Research* 66, no. 8:2455–67.

———. 1972. *Dynamics of fluids in porous media.* New York: American Elsevier Publishing Company, 764 pp.

Bear, Jacob, and Arnold Verruijt. 1987. *Modeling groundwater flow and pollution,* Dordrecht, Netherlands: D. Reidel Publishing Company, 414 pp.

Berkowitz, Brian, Jacob Bear, and Carol Braester. 1988. Continuum models for contaminant transport in fractured porous formations. *Water Resources Research* 24, no. 8:1225–36.

Brigham, W. E. 1974. Mixing equations in short laboratory columns. *Society of Petroleum Engineers Journal,* 14:91–99.

Carman, P. C. 1937. Fluid flow through a granular bed. Transactions, *Institute of Chemical Engineers London.* 15:150–56.

Cherry, John. A. 1983. Migration of contaminants in groundwater at a landfill: a case study. *Journal of Hydrology* 63, no. 1/2.

Crank, J. 1956. *The mathematics of diffusion.* New York: Oxford University Press.

Dagan, Gedeon. 1982. Stochastic modeling of groundwater flow by unconditional and conditional probabilities, 2, The solute transport. *Water Resources Research* 18, no. 4:835–48.

———. 1984. Solute transport in heterogeneous porous formations. *Journal of Fluid Mechanics* 145:151–77.

———. 1986. Statistical theory of ground water flow and transport: Pore to laboratory, laboratory to formation, and formation to regional scale. *Water Resources Research,* 22, no. 9:120S–134S.

———. 1987. Theory of solute transport in water. *Annual Reviews of Fluid Mechanics,* 19:183–215.

———. 1988. Time-dependent macrodispersion for solute transport in anisotropic heterogeneous aquifers. *Water Resources Research* 24, no. 9: 1491–1500.

Davis, A. D. 1986. Deterministic modeling of dispersion in heterogeneous permeable media. *Ground Water* 24, no. 5:609–15.

De Josselin De Jong, G. 1958. Longitudinal and transverse diffusion in granular deposits. *Transactions, American Geophysical Union* 39, no. 1:67.

Endo, H. K., J. C. S. Long, C. R. Wilson, and P. A. Witherspoon. 1984. A model for investigating mechanical transport in fracture networks. *Water Resources Research* 20, no. 10:1390–1400.

Fetter, C. W. 1988. *Applied hydrogeology.* 2d ed. New York: Macmillan Publishing Company, 592 pp.

Freeze, R. Allen. 1975. A stochastic-conceptual analysis of one-dimensional groundwater flow in a nonuniform homogeneous media. *Water Resources Research* 11, no. 5:725–41.

Freeze, R. Allen, and John A. Cherry. 1979, *Groundwater.* Englewood Cliffs, N.J.: Prentice Hall, 604 pp.

Freeze, R. A., J. Massmann, L. Smith, T. Sperling, and B. James, 1990, Hydrogeological decision analysis: 1. A framework. *Ground Water* 28, no. 5:738–66.

Fried, J. J. 1975. *Groundwater Pollution.* Amsterdam: Elsevier Scientific Publishing Company.

Frind, E. O., and G. E. Hokkanen. 1987. Simulation of the Borden plume using the alternating direction Galerkin technique. *Water Resources Research* 23, no. 5:918–30.

Garabedian, S. P., D. R. LeBlanc, L. W. Gelhar, and M. A. Celia. 1991. Large-scale natural gradient tracer test in sand and gravel, Cape Cod, Massachusetts, 2. Analysis of spatial moments for a nonreactive tracer. *Water Resources Research* 27, no. 5:911–24.

Gelhar, L. W. 1986. Stochastic subsurface hydrology from theory to applications. *Water Resources Research* 22, no. 9:135S–145S.

Gelhar, L. W., and C. L. Axness. 1983. Three-dimensional stochastic analysis of macrodispersion in aquifers. *Water Resources Research* 19, no. 1:161–80.

Gelhar, L. W., and M. A. Collins. 1971. General analysis of longitudinal dispersion in nonuniform flow. *Water Resources Research* 7, no. 6:1511–21.

Gelhar, L. W., A. L. Gutjahr, and R. L. Naff. 1979. Stochastic analysis of macrodispersion in a stratified aquifer. *Water Resources Research* 15, no. 6:1387–91.

Gillham, R. W., E. A. Sudicky, J. A. Cherry, and E. O. Frind. 1984. An advection-diffusion concept for solute transport in heterogeneous unconsolidated geological deposits. *Water Resources Research* 20, no. 3:369–78.

Grisak, G. E., and J. F. Pickens. 1980. Solute transport through fractured media, 1. The effects of matrix diffusion. *Water Resources Research* 16, no. 4:719–30.

———. 1981. An analytical solution for solute transport through fractured media with matrix diffusion. *Journal of Hydrology* 52:47–57.

Grove, D. B., and W. A. Beetem. 1971. Porosity and dispersion constant calculations for a fractured carbonate aquifer using the two-well tracer method. *Water Resources Research* 7, no. 1:128–34.

Hoopes, J. A., and D. R. F. Harleman. 1967. Dispersion in radial flow from a recharge well. *Journal of Geophysical Research* 72, no. 14:3595–3607.

Hull, L. C., J. D. Miller, and T. M. Clemo. 1987. Laboratory and simulation studies of solute transport in fracture networks. *Water Resources Research* 23, no. 8:1505–13.

Klotz, D., K. P. Seiler, H. Moser, and F. Neumaier. 1980. Dispersivity and velocity relationship from laboratory and field relationships. *Journal of Hydrology* 45. no. 3:169–84.

Konikow, L. F. 1977. Modeling chloride movement in the alluvial aquifer at the Rocky Mountain Arsenal, Colorado. *U.S. Geological Survey Water Supply Paper 2044*.

Konikow, L. F., and J. D. Bredehoeft. 1978. *Computer model of two-dimensional solute transport and dispersion in ground water*. U.S. Geological Survey Techniques of Water Resources Investigations, Book 7, Chapter C2.

Konikow, L. F., and D. W. Thompson. 1984. "Groundwater contamination and aquifer restoration at the Rocky Mountain Arsenal, Colorado." In *Groundwater Contamination* 93–103. Washington D.C.: National Academy Press.

Lallemand-Barres, P., and P. Peaudecerf. 1978. Recherche des relations entre la valeur de la dispersivite macroscopique d'un milieu aquifere, ses autres caracteristiques et les conditions de mesure, etude bibliographique. *Bulletin, Bureau de Recherches Géologiques et Miniéres*. Sec. 3/4:277–87.

LeBlanc, D. R., S. P. Garabedian, K. M. Hess, L. W. Gelhard, R. D. Quadri, K. G. Stollenwerk, and W. W. Wood. 1991. Large-scale natural gradient tracer test in sand and gravel, Cape Cod, Massachusetts, 1. Experimental design and observed tracer movement. *Water Resources Research* 27, no. 5:895–910.

MacFarlane, D. S, J. A. Cherry, R. W. Gillham, and E. A. Sudicky. 1983. Migration of contaminants in groundwater at a landfill: A case study, 1. Groundwater flow and plume definition. *Journal of Hydrology* 63, no. 1/2:1–30.

Mackay, D. M., D. L. Freyberg, P. V. Roberts, and J. A. Cherry. 1986. A natural gradient experiment on solute transport in a sand aquifer, 1. Approach and overview of plume movement. *Water Resources Research* 22, no. 13:2017–29.

Mandelbrot, B. B. 1967. How long is the coastline of Great Britain? Statistical self-similarity and fractional dimension. *Science* 155:636–38.

———. 1983. *The fractal geometry of nature*. New York: W. H. Freeman.

Matheron, G., and G. de Marsily. 1980. Is transport in porous media always diffusive? A counterexample. *Water Resources Research* 16, no. 5:901–17.

Molz, Fred J., Oktay Guven, and Joel G. Melville. 1983. An examination of scale-dependent dispersion coefficients. *Ground Water* 21, no. 6:715–25.

Moreno, L., Y. W. Tsang, C. F. Tsang, F. V. Hale, and I. Neretnieks. 1988. Flow and tracer transport in a single fracture: A stochastic model and its relation to some field observations. *Water Resources Research* 24, no. 12:2033–48.

Neretnieks, I. 1980. Diffusion in the rock matrix: An important factor in radionuclide migration. *Journal of Geophysical Research* 85, no. B8:4379–97.

Neuman, S. P. 1990. Universal scaling of hydraulic conductivities and dispersivities in geologic media. *Water Resources Research* 26, no. 8:1749–58.

Neuman, S. P., C. L. Winter, and C. N. Newman. 1987. Stochastic theory of field-scale Fickian dispersion in anisotropic porous media. *Water Resources Research* 23, no. 3:453–66.

Neuman, S. P., and Y.-K. Zhang. 1990. A quasi-linear theory of non-Fickian and Fickian subsurface dispersion, 1. Theoretical analysis with application to isotropic media. *Water Resources Research* 26, no. 5:887–902.

Ogata, Akio. 1970. *Theory of dispersion in a granular medium*. U.S. Geological Survey Professional Paper 411-I.

Ogata, Akio, and R. B. Banks. 1961. *A solution of the differential equation of longitudinal dispersion in porous media*. U.S. Geological Survey Professional Paper 411-A.

Perkins, T. K., and O. C. Johnson. 1963. A review of diffusion and dispersion in porous media. *Society of Petroleum Engineers Journal* 3:70–84.

Pickens, J. F., R. E. Jackson, K. J. Inch, and W. F. Merritt. 1981. Measurement of distribution coefficients using a radial injection duel-tracer test. *Water Resources Research* 17, no. 3:529–44.

Pickens, John F., and Gerald E. Grisak. 1981. Scale-dependent dispersion in a stratified granular aquifer. *Water Resources Research* 17, no. 4:1191–1211.

Pinder, George. 1973. A Galerkin-finite element simulation of groundwater contamination on Long Island, New York, *Water Resources Research* 9, no. 6:1657–69.

Rasmuson, A. 1984. Migration of radionuclides in fissured rock: Analytical solutions for the case of constant source strength. *Water Resources Research* 20, no. 10:1435–42.

Rasmuson, A., and I. Neretnieks. 1981. Migration of radionuclides in fissured rock: The influence of micropore diffusion and longitudinal dispersion. *Journal of Geophysical Research* 86, no. B5:3749–58.

Raven, K. G., K. S. Novakowski, and P. A. Lapcevic. 1988. Interpretation of field tests of a single fracture using a transient solute storage model. *Water Resources Research* 24, no. 12:2019–32.

Robinson, R. A., and R. H. Stokes. 1965. *Electrolyte solutions*. London: Butterworth Press.

Sauty, Jean-Pierre. 1978. Identification des parametres du transport hydrodispersif das les aquiferes par interpretation de tracages en ecoulement cyclindrique convergent ou divergent. *Journal of Hydrology* 39, no. 3/4:69–103.

———. 1980. An analysis of hydrodispersive transfer in aquifers. *Water Resources Research* 16, no. 1:145–58.

Schwartz, F. W., and L. Smith. 1988. A continuum approach for modeling mass transport in fractured media. *Water Resources Research* 24, no. 8:1360–72.

Schwartz, F. W., L. Smith, and A. S. Crowe. 1983. A stochastic analysis of macroscopic dispersion in fractured media. *Water Resources Research* 19, no. 5:1253–65.

Smith, L., and F. W. Schwartz. 1984. An analysis of the influence of fracture geometry on mass transport in fractured media. *Water Resources Research* 20, no. 9:1241–52.

Sposito, Garrison, and D. A. Barry. 1987. On the Dagan model of solute transport in groundwater: foundational aspects. *Water Resources Research* 23, no. 10:1867–75.

Sposito, Garrison, W. A. Jury, and V. K. Gupta. 1986. Fundamental problems in the stochastic convection-dispersion model of solute transport in aquifers and field soils. *Water Resources Research* 22, no. 1:77–88.

Sudicky, E. A. 1986. A natural gradient experiment on solute transport in a sand aquifer: Spatial variability of hydraulic conductivity and its role in the dispersion process. *Water Resources Research* 22, no. 13:2069–82.

Sudicky, E. A., and J. A. Cherry. 1979. Field observations of tracer dispersion under natural flow conditions in an unconfined sandy aquifer. *Water Pollution Research, Canada* 14:1–17.

Sudicky, E. A., J. A. Cherry, and E. O. Frind. 1983. Migration of contaminants in groundwater at a landfill: A case study, 4, A natural gradient dispersion test. *Journal of Hydrology* 63, no. 1/2:81–108.

Sudicky, E. A., and E. O. Frind. 1982. Contaminant transport in fractured porous media: Analytical solution for a system of parallel fractures. *Water Resources Research* 18, no. 6:1634–42.

———.1984. Contaminant transport in fractured porous media: Analytical solution for a two-member decay chain in a single fracture. *Water Resources Research* 20, no. 7:1021–29.

Tang, D. H., E. O. Frind, and E. A. Sudicky. 1981. Contaminant transport in fractured porous media: Analytical solution for a single fracture. *Water Resources Research* 17, no. 3:555–64.

Tsang, Y. W., C. F. Tsanf, I. Neretnieks, and L. Moreno. 1988. Flow and tracer transport in fractured media: A variable aperture channel model and its properties. *Water Resources Research* 24, no. 12:2049–60.

Van Genuchten, M. Th. 1981. Analytical solutions for chemical transport with simultaneous adsorption, zero-order production, and first-order decay. *Journal of Hydrology* .49:213–233.

Wheatcraft, S. W., and S. W. Tyler. 1988. An explanation of scale-dependent dispersivity in heterogeneous aquifers using concepts of fractal geometry. *Water Resources Research* 24, no. 4:566–78.

Witherspoon, P. A., J. S. Y. Yang, K. Iwai, and J. E. Gale. 1980. Validity of the cubic law for fluid flow in a deformable rock fracture. *Water Resources Research* 16, no. 6:1016–24.

Zhang, Y.-K., and S. P. Neuman. 1990. A quasi-linear theory of non-Fickian and Fickian subsurface dispersion, 2. Application to anisotropic media and the Borden site. *Water Resources Research* 26, no. 5:903–13.

Chapter Three
Transformation, Retardation, and Attenuation of Solutes

3.1 Introduction

Solutes dissolved in ground water are subject to a number of different processes through which they can be removed from the ground water. They can be sorbed onto the surfaces of the mineral grains of the aquifer, sorbed by organic carbon that might be present in the aquifer, undergo chemical precipitation, be subjected to abiotic as well as biodegradation, and participate in oxidation-reduction reactions. Furthermore, radioactive compounds can decay. As a result of sorption processes, some solutes will move much more slowly through the aquifer than the ground water that is transporting them; this effect is called **retardation.** Biodegradation, radioactive decay, and precipitation will decrease the concentration of solute in the plume but may not necessarily slow the rate of plume movement.

Equation 2.18, the one-dimensional advection-dispersion equation, can be modified to include sorption and decay. This can be expressed as (Miller and Weber 1984):

$$\frac{\partial C}{\partial t} = D_L \frac{\partial^2 C}{\partial x^2} - v_x \frac{\partial C}{\partial x} - \frac{B_d}{\theta}\frac{\partial C^*}{\partial t} + \left(\frac{\partial C}{\partial t}\right)_{rxn} \quad (3.1)$$

(dispersion) (advection) (sorption) (reaction)

where

C = concentration of solute in liquid phase
t = time
D_L = longitudinal dispersion coefficient
v_x = average linear groundwater velocity
B_d = bulk density of aquifer
θ = volumetric moisture content or porosity for saturated media
C^* = amount of solute sorbed per unit weight of solid
rxn = subscript indicating a biological or chemical reaction of the solute (other than sorption)

The first term on the right side of Equation 3.1 represents the dispersion of the solute, the second term is the advection of the solute, the third term is the transfer of the solute from the liquid phase to the solid particles by sorption, and the last term simply indicates that there may be a change in concentration of the solute with time due to biological or chemical reactions or radioactive decay.

3.2 Classification of Chemical Reactions

Rubin (1983) listed six different classes of chemical reactions that can occur in solute transport (Figure 3.1). At the highest, or A, level reactions are either (1) "sufficiently fast" and reversible or (2) "insufficiently fast" and/or irreversible. Sufficiently fast reactions are reversible reactions that are fast relative to ground-water flow rates and are faster than any other reactions that act to change solute concentration. With these reactions one can assume that locally the solute is in chemical equilibrium with the surroundings (local equilibrium assumption, or LEA, system). If the reaction is not sufficiently fast for local equilibrium to develop or if it is irreversible, then it falls into the second major grouping.

At the second, or B, level reactions are either (1) homogeneous or (2) heterogeneous. Homogeneous reactions take place within a single phase, the dissolved phase, whereas heterogeneous reactions involve both the dissolved phase and the solid phase. Level C reactions, representing the greatest specification, apply only to heterogeneous

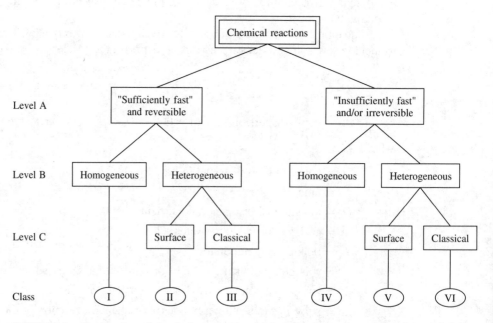

FIGURE 3.1 Classification of chemical reactions useful in solute transport analyses. *Source:* J. Rubin, *Water Resources Research* 19, no. 5 (1983): 1231–52. Published by the American Geophysical Union.

reactions. These can be either (1) surface reactions, such as hydrophobic adsorption of neutral organic compounds and ion exchange of charged ions, or (2) classical chemical reactions such as precipitation and dissolution.

3.3 Sorption Processes

Sorption processes include adsorption, chemisorption, absorption, and ion exchange. **Adsorption** includes the processes by which a solute clings to a solid surface. Cations may be attracted to the region close to a negatively charged clay-mineral surface and held there by electrostatic forces; this process is called **cation exchange.** Anion exchange can occur at positively charged sites on iron and aluminum oxides and the broken edges of clay minerals. **Chemisorption** occurs when the solute is incorporated on a sediment, soil, or rock surface by a chemical reaction. **Absorption** occurs when the aquifer particles are porous so that the solute can diffuse into the particle and be sorbed onto interior surfaces (Wood, Kramer, and Hern 1990).

In this chapter we will not attempt to separate these phenomena but will simply use the term sorption to indicate the overall result of the various processes. From a practical view the important aspect is the removal of the solute from solution, irrespective of the process. The process by which a contaminant, which was originally in solution, becomes distributed between the solution and the solid phase is called **partitioning.**

Sorption is determined experimentally by measuring how much of a solute can be sorbed by a particular sediment, soil, or rock type. Aliquots of the solute in varying concentrations are well mixed with the solid, and the amount of solute removed is determined. The capacity of a solid to remove a solute is a function of the concentration of the solute. The results of the experiment are plotted on a graph that shows the solute concentration versus the amount sorbed onto the solid. If the sorptive process is rapid compared with the flow velocity, the solute will reach an equilibrium condition with the sorbed phase. This process can be described by an **equilibrium sorption isotherm.** It is an example of a sufficiently fast, heterogeneous surface reaction. If the sorptive process is slow compared with the rate of fluid flow in the porous media, the solute may not come to equilibrium with the sorbed phase, and a **kinetic sorption model** will be needed to describe the process. These are insufficiently fast, heterogeneous surface reactions. Travis and Etnier (1981) give a comprehensive review of sorption isotherms and kinetic models.

3.4 Equilibrium Surface Reactions

3.4.1 Linear Sorption Isotherm

If there is a direct, linear relationship between the amount of a solute sorbed onto solid, C^*, and the concentration of the solute, C, the adsorption isotherm of C as a function of C^* will plot as a straight line on graph paper (Figure 3.2). The resulting linear sorption isotherm is described by the equation

$$C^* = K_d C \qquad (3.2)$$

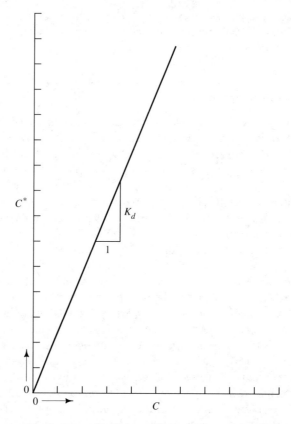

FIGURE 3.2 Linear sorption isotherm with C^* versus C plotting as a straight line.

where

C^* = mass of solute sorbed per dry unit weight of solid (mg/kg)

C = concentration of solute in solution in equilibrium with the mass of solute sorbed onto the solid (mg/L)

K_d = coefficient (L/kg)

The coefficient K_d is known as the **distribution coefficient.** It is equal to the slope of the linear sorption isotherm.

The linear sorption isotherm is very appealing from the standpoint of mathematical manipulation. If Equation 3.2 is substituted into Equation 3.1, the resulting advection-dispersion equation is

$$\frac{\partial C}{\partial t} = D_L \frac{\partial^2 C}{\partial x^2} - v_x \frac{\partial C}{\partial x} - \frac{B_d}{\theta} \frac{\partial (K_d C)}{\partial t} \qquad (3.3)$$

This can be reorganized as

$$\frac{\partial C}{\partial t}\left(1 + \frac{B_d}{\theta} K_d\right) = D_L \frac{\partial^2 C}{\partial x^2} - v_x \frac{\partial C}{\partial x} \qquad (3.4)$$

What has been termed the **retardation factor**, r_f, is given by

$$1 + \frac{B_d}{\theta} K_d = r_f \qquad (3.5)$$

If the average linear ground-water velocity is v_x, the average velocity of the solute front where the concentration is one-half of the original, v_c, is given by

$$v_c = \frac{v_x}{r_f} \qquad (3.6)$$

Equations 3.3 through 3.6 are convenient to solve mathematically and have been used in a number of studies to predict the rate of movement of a solute front (Anderson 1979; Faust and Mercer 1980; Prickett, Naymik, and Lonnquist 1981; Srinivasan and Mercer 1988).

There are two limitations of the linear-sorption isotherm model. One is that it does not limit the amount of solute that can be sorbed onto the solid. This is clearly not the case; there must be an upper limit to the mass of solute that can be sorbed. In addition, if there are only a few data points, what is actually a curvilinear experimental plot of C versus C^* might be misinterpreted to be a linear relationship. Figure 3.3 illustrates how it is important never to extrapolate from a limited data set to a range outside the data set and assume that a linear relationship exists in that region. The subset of the sorption data on Figure 3.3 marked by triangles can be used with the origin to form a linear relationship. The subset of the sorption data marked with squares can also be used with the origin to create a different linear relationship. However, if all the data are included, one can see that the sorption isotherm is not linear at all.

3.4.2 Freundlich Sorption Isotherm

A more general equilibrium isotherm is the **Freundlich sorption isotherm.** This is defined by the nonlinear relationship

$$C^* = KC^N \qquad (3.7)$$

where K and N are constants.

If the sorption characteristics can be described by a Freundlich sorption isotherm, when C is plotted as a function of C^* on graph paper the data will be curvilinear (Figure 3.4(a)). However, the data can be linearized by use of the following equation:

$$\log C^* = \log K + N \log C \qquad (3.8)$$

If $\log C$ is plotted against $\log C^*$ on graph paper, the result will be linear with a slope of N and an intercept of $\log K$. This is illustrated in Figure 3.4(b).

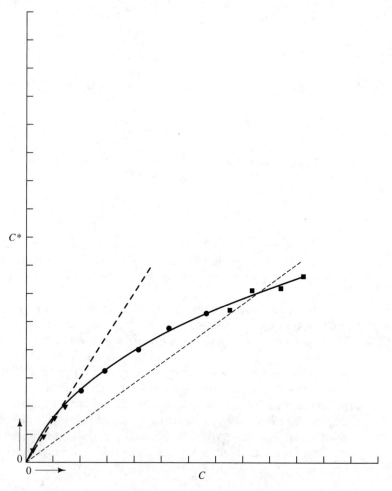

FIGURE 3.3 Nonlinear sorption isotherms can be misinterpreted as linear sorption isotherms if a small data set is extrapolated out of its range. The subset of the data represented by triangles can be interpreted as a linear sorption isotherm, as can the data subset consisting of squares. However, if the complete data set, which includes the triangles, circles, and squares, is used, it can be seen that the isotherm is nonlinear.

If Equation 3.7 is substituted into Equation 3.1, the result is

$$\frac{\partial C}{\partial t} = D_L \frac{\partial^2 C}{\partial x^2} - v_x \frac{\partial C}{\partial x} - \frac{B_d}{\theta} \frac{\partial (KC^N)}{\partial t} \tag{3.9}$$

After differentiation and reorganization, Equation 3.9 becomes

$$\frac{\partial C}{\partial t}\left(1 + \frac{B_d K N C^{N-1}}{\theta}\right) = D_L \frac{\partial^2 C}{\partial x^2} - v_x \frac{\partial C}{\partial x} \tag{3.10}$$

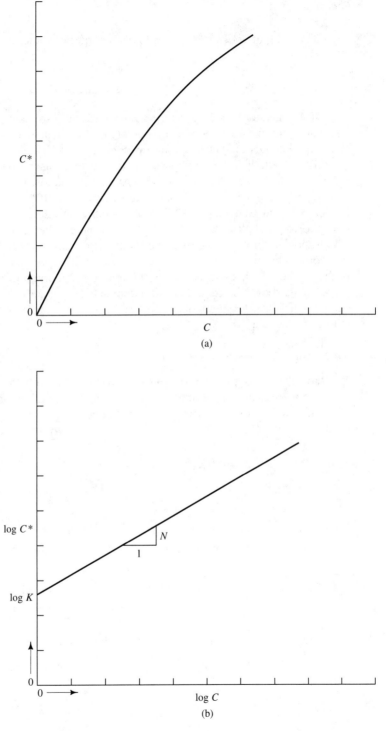

FIGURE 3.4 (a) Nonlinear Freundlich sorption isotherm with C^* versus C plotted on cross-section paper. (b) The Freundlich sorption isotherm can be made linear by plotting log C^* versus C.

The retardation factor for a Freundlich sorption isotherm, r_{ff}, is

$$1 + \frac{B_d K N C^{N-1}}{\theta} = r_{ff} \qquad (3.11)$$

If N is greater than 1, Equation 3.10 will lead to a spreading front, whereas if N is less than 1, the front will be self-sharpening. If N is equal to 1, the Freundlich sorption isotherm becomes the linear sorption isotherm.

The Freundlich sorption isotherm is one that has been widely applied to the sorption by soils of various metals and organic compounds such as sulfate (Bornemisza and Llanos 1967), cadmium (Street Lindsay, and Sabey 1977), copper and zinc (Sidle, Kardos, and van Genuchten 1977), molybdenum (Jarrell and Dawson 1978), organo-phosphorus pesticides (Yaron, 1978), *p*-chloroanaline residues (Van Bladel and Moreale 1977), and parathion and related compounds (Bowman and Sans 1977). The Freundlich sorption isotherm suffers from the same fundamental problem as the linear sorption isotherm; there is theoretically no upper limit to the amount of a solute that could be sorbed. One should be careful not to extrapolate the equation beyond the limits of the experimental data. The Freundlich sorption isotherm is usually obtained by an empirical fit to experimental data.

3.4.3 Langmuir Sorption Isotherm

The **Langmuir sorption isotherm** was developed with the concept that a solid surface possesses a finite number of sorption sites. When all the sorption sites are filled, the surface will no longer sorb solute from solution. The form of the Langmuir sorption isotherm is

$$\frac{C}{C^*} = \frac{1}{\alpha \beta} + \frac{C}{\beta} \qquad (3.12)$$

where

α = an absorption constant related to the binding energy (L/mg)

β = the maximum amount of solute that can be absorbed by the solid (mg/kg)

The Langmuir sorption isotherm can also be expressed as

$$C^* = \frac{\alpha \beta C}{1 + \alpha C} \qquad (3.13)$$

When Equation 3.13 is substituted into Equation 3.1, the following equation is obtained:

$$\frac{\partial C}{\partial t} = D_L \frac{\partial^2 C}{\partial x^2} - v_x \frac{\partial C}{\partial x} - \frac{B_d}{\theta} \frac{\partial \left(\frac{\alpha \beta C}{1 + \alpha C} \right)}{\partial t} \qquad (3.14)$$

Differentiation and reorganization of Equation 3.14 yields

$$\frac{\partial C}{\partial t}\left[1 + \frac{B_d}{\theta}\left(\frac{\alpha\beta}{(1+\alpha C)^2}\right)\right] = D_L \frac{\partial^2 C}{\partial x^2} - v_x \frac{\partial C}{\partial x} \qquad (3.15)$$

The retardation factor for the Langmuir sorption isotherm, r_{fl}, is

$$1 + \frac{B_d}{\theta}\left(\frac{\alpha\beta}{(1+\alpha C)^2}\right) = r_{fl} \qquad (3.16)$$

If the sorption of a solute onto a solid surface follows a Langmuir sorption isotherm, when experimental data of C^* versus C are plotted on graph paper they will have a curved shape that reaches a maximum value (Figure 3.5(a)). If C/C^* is plotted versus C on graph paper, the data will follow a straight line. The maximum ion sorption, β, is the reciprocal of the slope of the line, and the binding energy constant, α, is the slope of the line divided by the intercept (Figure 3.5(b)).

In studies of the sorption of phosphorous on soils, it has been found that a plot of C/C^* versus C will yield curves with two straight line segments (Fetter 1977; Munns and Fox 1976). This has been interpreted to mean that there are two types of sorption sites, which differ in their bonding energy. The **Langmuir two-surface sorption isotherm** is

$$\frac{C^*}{C} = \frac{\alpha_1 \beta_1}{1+\alpha_1 C} + \frac{\alpha_2 \beta_2}{1+\alpha_2 C} \qquad (3.17)$$

where

α_1 = the bonding strength at the type 1 sites
α_2 = the bonding strength at the type 2 sites
β_1 = the maximum amount of solute that can be sorbed at the type 1 sites
β_2 = the maximum amount of solute that can be sorbed at the type 2 sites

3.4.4 Effect of Equilibrium Retardation on Solute Transport

The effects of equilibrium retardation can be illustrated through use of a computer model, BIO1D. This model was developed by Srinivasan and Mercer (1988) and simulates both sorption processes and biodegradation in mass transport. It is very flexible and can simulate linear, Freundlich, and Langmuir adsorption as well as aerobic and anaerobic biodegradation.

The situation being modeled is one-dimensional mass transport through a saturated porous medium that is in a column 16 cm long. The pore water velocity is 0.1 cm/sec, the dispersion coefficient is 0.1 cm^2/sec, and the porosity is 0.37. The initial solute concentration is 0.0 mg/L. For 2 min a solute with a concentration of 0.05 mg/L is injected into the top of the soil column and allowed to drain from the bottom. After 2

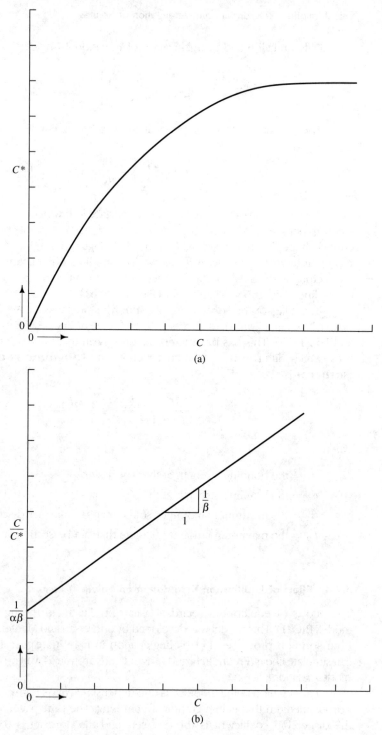

FIGURE 3.5 (a) Nonlinear Langmuir sorption isotherm will reach a maximum sorption value when C^* is plotted versus C. (b) The Langmuir sorption isotherm can be made linear by plotting C/C^* versus C.

Transformation, Retardation, and Attenuation of Solutes

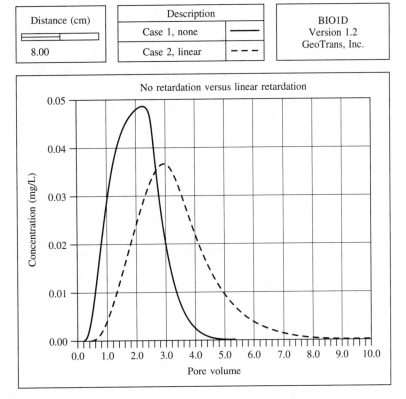

FIGURE 3.6 Illustration of the effect of retardation by comparing the breakthrough curve of a solute which isn't retarded with the breakthrough curve of a solute that undergoes linear-type retardation. Model simulation using BIO1D from Geotrans, Inc.

min the concentration of the solute in the water entering the column is set back to 0.00. The model yields the solute concentration in the water draining from the soil column as a function of the number of pore volumes that have been drained.

Figure 3.6 shows the general effect of retardation. One of the two curves, the solid one, is the solute breakthrough curve with no retardation (and no degradation). The dashed curve shows the breakthrough of a solute that is undergoing retardation, which follows a linear sorption isotherm, and has a K_d value of 0.476 $\mu g/g$. It can be seen that the retarded substance (dashed curve) has a lower peak value and that the peak comes later; i.e., it takes more pore volumes for it to occur than the unretarded peak (solid line).

Figure 3.7 illustrates the effect of different N values on the Freundlich sorption isotherm. The model is simulating exactly the same situation as before, except that there is Freundlich-type retardation. The same K_d value is used, with the solid line illustrating an N value of 1.3 and the dashed curve representing an N value of 0.7. (The linear

126   Chapter Three

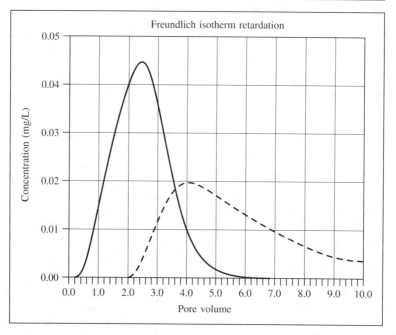

FIGURE 3.7 Illustration of the effect of the value of the constant N in the Freundlich sorption isotherm. The solid curve has an N greater than 1, whereas the dashed curve has an N less than 1. Model simulation using BIO1D from Geotrans, Inc.

sorption isotherm is a special case of the Freundlich sorption isotherm with an N value of 1.0.) Figure 3.7 shows that with an N value greater than 1, the breakthrough curve arrives earlier (i.e., takes fewer pore volumes) and has a greater peak value than the breakthrough curve with an N value less than 1.

Caution should be used if experimental absorption studies indicate an N value greater than 1.0 for a Freundlich sorption isotherm. There is no theoretical reason why the exponential constant should be greater than the linear value of 1.0. Some researchers believe that N values greater than 1.0 are a result of a combination of sorption and precipitation that is occurring because the experimental concentrations are exceeding the water solubility of the compound (Griffin 1991).

Figure 3.8 compares the linear sorption isotherm with a Langmuir sorption isotherm. The Langmuir sorption isotherm has a maximum binding energy of 0.345 $\mu g/g$ and a maximum sorption of 0.475 $\mu g/g$. The Langmuir sorption isotherm results in a higher peak value at breakthrough, which arrives at an earlier time than the linear sorption isotherm. In this particular case the Langmuir isotherm is not very different than the linear sorption isotherm.

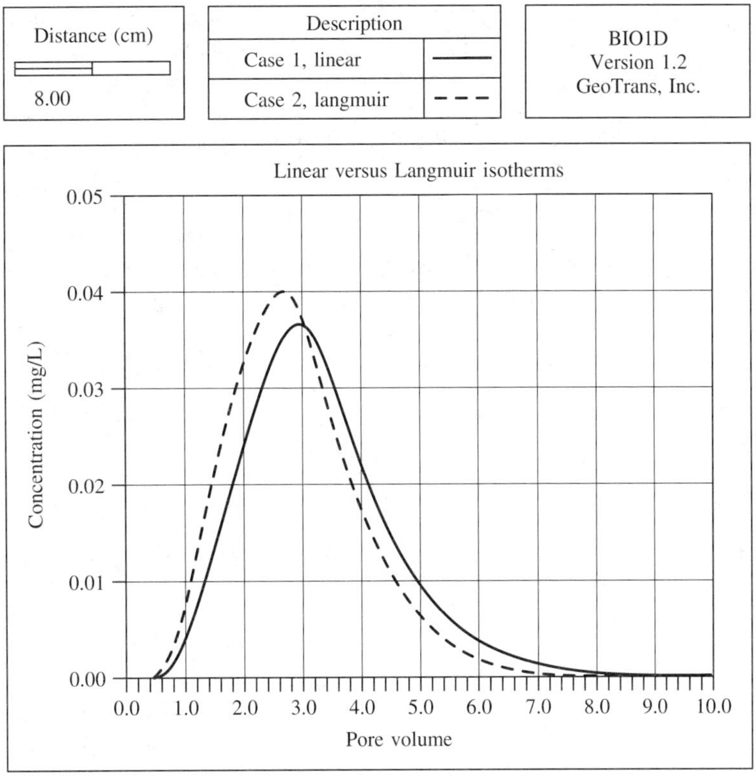

Transformation, Retardation, and Attenuation of Solutes

FIGURE 3.8 Illustration of the effect of different sorption isotherms in modeling solute transport. The solid curve is for a linear sorption isotherm while the dashed curve is for a Langmuir sorption isotherm. Model simulation using BIO1D from Geotrans, Inc.

EXAMPLE PROBLEM

Sorption of phosphorous by a calcareous glacial outwash was studied by means of a batch sorption test. The outwash was air-dried and then sieved to segregate the fraction that was finer than 2 mm. The coarser material was discarded. Ten-gram samples of the sediment were added to flasks containing 100 mL of 0.1 M NaCl and disodium phosphate in concentrations ranging from 0.53 to 12.1 mg/L. The flasks were shaken for 4 da on an autoshaker. The samples were then filtered and the filtrate analyzed for orthophosphate. The sediment was extracted with dilute HCl and the extract was analyzed to determine the amount of phosphorous sorbed to the sediment prior to the test. This amount was 0.016 mg/g.

The initial concentration of phosphorous in solution was known and the equilibrium concentration was determined by analysis. By knowing the volume of solution and the initial concentration, the mass of phosphorous could be computed. For example a 100-mL sample with a concentration of 3.85 mg/L has 0.385 mg of P. At equilibrium with the sediment, this aliquot had 2.45 mg/L of P, or 0.245 mg, still in solution. The amount sorbed was 0.14 mg (0.385 mg − 0.245 mg), or 0.014 mg/g of sediment. Prior to the sorption test the sediment had been extracted with dilute HCl and the extract tested for P. It was found to contain 0.016 mg/g of P. This amount already occupied some of the sorption sites and had to be added to the amount sorbed during the test. The following table lists the initial and equilibrium concentrations for P, the amount sorbed onto the soil, and the value of C/C^*. It is interesting to note that for the lowest initial concentration, the equilibrium concentration is greater than the initial concentration. This is due to P desorbing from the sediment.

Initial Concentration (mg/L)	Equilibrium Concentration C (mg/L)	Amount Sorbed per Gram during Test (mg/g)	Amount Sorbed in Test Plus 0.016 mg/g C^* (mg/g)	C/C^* (mg/L) (mg/g)
0.53	0.55	−0.002	0.014	39
1.95	1.25	0.007	0.023	54.5
3.85	2.45	0.014	0.030	81
6.05	3.85	0.022	0.038	103
8.0	5.00	0.030	0.046	108.5
12.1	7.70	0.044	0.060	127.5

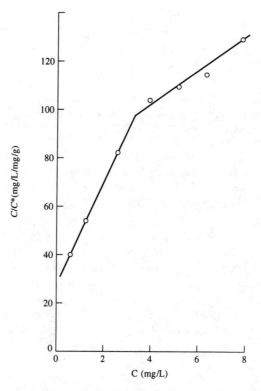

FIGURE 3.9 A linear Langmuir two-surface sorption isotherm for the sorption of phosphate on calcareous glacial outwash. *Source:* C. W. Fetter, *Ground Water* 15, no. 5 (1977):365–71. Used with permission. Copyright © 1977 Water Well Journal Publishing Co.

Figure 3.9 shows the plot of C/C^* versus C. This is clearly a Langmuir two-surface sorption isotherm. The sorption maxima for low concentrations is 0.05 mg P per gram of sediment, and for the higher concentrations it is 0.16 mg P per gram of sediment.

3.5 Nonequilibrium (Kinetic) Sorption Models

All the equilibrium models assume that the rate of change in concentration due to sorption is much greater than the change due to any other cause and that the flow rate is low enough that equilibrium can be reached. If this is not the case and equilibrium is not attained, a kinetic model is more appropriate. In a kinetic model the solute transport equation is linked to an appropriate equation to describe the rate that the solute is sorbed onto the solid surface and desorbed from the surface.

The most simple nonequilibrium condition is that the rate of sorption is a function of the concentration of the solute remaining in solution and that once sorbed onto the solid, the solute cannot be desorbed. This is an irreversible reaction and the process leads to attenuation of the solute (not retardation which by definition is reversible). The **irreversible first-order kinetic sorption model** that describes this consists of the following pair of equations:

$$\frac{\partial C^*}{\partial t} = k_1 C \tag{3.18}$$

$$\frac{\partial C}{\partial t} = D_L \frac{\partial^2 C}{\partial x^2} - v_x \frac{\partial C}{\partial x} - \frac{B_d}{\theta} \frac{\partial C^*}{\partial t} \tag{3.19}$$

where k_1 = a first-order decay rate constant.

If the rate of solute sorption is related to the amount that has already been sorbed and the reaction is reversible, then the **reversible linear kinetic sorption model** can be used. This consists of Equation 3.19 and the following expression for the rate of sorption:

$$\frac{\partial C^*}{\partial t} = k_2 C - k_3 C^* \tag{3.20}$$

where

k_2 = forward rate constant

k_3 = backward rate constant

If sufficient time is available for the system to reach equilibrium, then there is no further change in C^* with time and $dC^*/dt = 0$, so that $k_2 C = k_3 C^*$. This can be rearranged to $C^* = (k_2/k_3)C$, which is a linear equilibrium sorption isotherm.

Equation 3.20 is sometimes written in a slightly different form (Nielsen, van Genuchten, and Biggar 1986):

$$\frac{\partial C^*}{\partial t} = \gamma(k_4 C - C^*) \tag{3.21}$$

where

γ = a first-order rate coefficient

k_4 = a constant equivalent to K_d

Equation 3.21 describes a situation where reversible linear sorption is limited by a first-order diffusion process.

This model has been used to describe the sorption of pesticides (Leistra and Dekkers 1977; Hornsby and Davidson 1973) as well as some organics (Davidson and Chang 1972).

A third kinetic model is the **reversible nonlinear kinetic sorption model.** This couples Equation 3.19 with

$$\frac{\partial C^*}{\partial t} = k_5 C^N - k_6 C^* \qquad (3.22)$$

where k_5, k_6, and N are constants. This model describes a situation where the forward (sorption) reaction is nonlinear, whereas the backward (desorption) reaction is linear. This equation has been used, with a value of N less than 1, to describe the sorption of P (Fiskell et al. 1979) and herbicides (Enfield and Bledsoe 1975).

At the equilibrium condition for the reversible nonlinear model, $dC^*/dt = 0$ and $k_5 C^N = k_6 C^*$, which can be rearranged as $C^* = (k_5/k_6)C^N$, which is the Freundlich sorption isotherm.

The **bilinear adsorption model** is the kinetic version of the Langmuir sorption isotherm. This model has the form

$$\frac{\partial C^*}{\partial t} = k_7 C(\beta - C^*) - k_8 C^* \qquad (3.23)$$

where

β = the maximum amount of solute that can be sorbed
k_7 = the forward rate constant
k_8 = the backward rate constant

In some cases the sorption of ions may be controlled by the rate at which the ions are transported to the exchange sites by diffusion, even though the sorption may be instantaneous once the ions reach the sorption or exchange sites. In this situation a **diffusion-controlled rate law** must be employed (Nkedi-Kizza and et al. 1984). The liquid has a mobile phase, through which advective flow occurs, and an immobile phase near the solid surfaces. Transfer of solutes across the immobile water to the solid surfaces occurs by diffusion. The rate of solute transfer across the immobile water is assumed to be proportional to the difference in concentration between the two regions. The equations that are given are applicable to both saturated and unsaturated flow. For unsaturated flow, θ is the volumetric water content and for saturated flow, θ is the porosity. The equations can account for a system where some of the solid is in direct contact with the mobile phase and some is in direct contact with the immobile phase. This system requires a pair of equations:

$$\theta_m \frac{\partial C_m}{\partial t} = \theta_m D_m \frac{\partial^2 C_m}{\partial z^2} - \theta_m v_m \frac{\partial C_m}{\partial z} - fB_d \frac{\partial C_m^*}{\partial t}$$

$$- \theta_{im} \frac{\partial C_{im}}{\partial t} - (1-f)B_d \frac{\partial C_{im}^*}{\partial t} \qquad (3.24)$$

$$(1-f)B_d \frac{\partial C^*_{im}}{\partial t} = \tau(C_m - C_{im}) - \theta_{im}\frac{\partial C_{im}}{\partial t} \qquad (3.25)$$

where

θ_m = porosity occupied by mobile phase
θ_{im} = porosity occupied by immobile phase
C_m = solute concentration in the mobile phase
C_{im} = solute concentration in the immobile phase
C^*_m = absorbed concentration in contact with the mobile phase
C^*_{im} = absorbed concentration in contact with the immobile phase
v_m = velocity of the mobile phase
f = fraction of the solid surfaces in contact with the mobile phase
D_m = apparent diffusion constant for mobile phase
τ = first-order, mass-transfer coefficient

If the sorption of the solute is in equilibrium and reversible and follows a linear sorption isotherm, then

$$C^*_m = K_d C_m \quad \text{and} \quad C^*_{im} = K_d C_{im} \qquad (3.26)$$

The total sorption of solute from both the mobile and immobile regions is

$$C^* = fC^*_m + (1-f)C^*_{im} \qquad (3.27)$$

With these sorption conditions, Equations 3.24 and 3.25 can be written as

$$(\theta_m + B_d f K_d)\frac{\partial C_m}{\partial t} = \theta_m D_m \frac{\partial^2 C_m}{\partial z^2} - \theta_m v_m \frac{\partial C_m}{\partial z} - [\theta_{im} + (1-f)B_d K_d]\frac{\partial C_{im}}{\partial t} \qquad (3.28)$$

$$\frac{\partial C_{im}}{\partial t} = \frac{\tau(C_m - C_{im})}{[\theta_{im} + (1-f)B_d K_d]} \qquad (3.29)$$

An analytical solution to Equations 3.28 and 3.29 is available (van Genuchten and Wierenga 1976).

There are a number of additional nonlinear sorption models described in the literature (e.g., Travis and Etnier 1981). The sorption of solutes by solids is complex, and there does not appear to be a single universal model. The best approach is to make an experimental study of the sorptive capacity and rate of the particular solute and solid that is of concern. One can then search the literature for a model that adequately describes the experimental results. These models may be complex if two equations must be solved simultaneously. Some of the models, such as the bilinear adsorption model, do not have an analytical solution when coupled with the advection-dispersion equation. Most of the readily available computer programs, such as BIO1D, are not capable of addressing nonequilibrium sorption.

3.6 Sorption of Hydrophobic (Organic) Compounds

3.6.1 Introduction

Many organic compounds dissolved in ground water can be adsorbed onto solid surfaces by what is called the *hydrophobic effect* (Roy and Griffin 1985). These compounds exist as electrically-neutral species with differing degrees of polarity. The solubility of organic compounds in water is a function of the degree to which they are attracted by the polar water molecule. This attraction depends upon the polarity of the organic molecule itself. Hydrophobic compounds can be dissolved in many nonpolar organic solvents but have a low solubility in water. When dissolved in water, these molecules tend to be attracted to surfaces that are less polar than water. There is a small but limited amount of adsorption of organics on pure mineral surfaces (Ciccioli et al. 1980; Rogers, McFarlane, and Cross 1980; Griffin and Chian 1980). However, the primary adsorptive surface is the fraction of organic solids in the soil or aquifer (Karickhoff, Brown, and Scott 1979; Schwarzenbach and Westall 1981; Dzomback and Luthy 1984).

3.6.2 Partitioning onto Soil or Aquifer Organic Carbon

The partitioning of a solute onto mineral surface or organic carbon content of the soil or aquifer is almost exclusively onto the organic carbon fraction, f_{oc}, if it constitutes at least 1% of the soil or aquifer on a weight basis (Karickhoff, Brown, and Scott 1979). Under these circumstances a partition coefficient with respect to the organic fraction, K_{oc}, can be defined as

$$K_{oc} = \frac{K_d}{f_{oc}} \quad (3.30)$$

A partition coefficient based on soil or aquifer organic matter, K_{om}, is also used. Because the weight of the organic matter is greater than that of the organic carbon alone, K_{oc} will be larger than K_{om}. Based on lab studies K_{om} can be approximately related to K_{oc} by the equation (Olsen and Davis 1990)

$$K_{oc} = 1.724 K_{om} \quad (3.31)$$

If the organic fraction is less than 1%, then it is not automatic that the soil or aquifer organic carbon will be the primary surface onto which the organic compounds will partition. There is some critical level of soil or aquifer organic carbon at which the sorption onto the organic matter is equal to the sorption onto the mineral matter. Below this critical level, f_{oc}^*, the organic molecules will be primarily sorbed onto the mineral surfaces. McCarty, Reinhard, and Rittman (1981) have shown that this critical organic carbon level depends upon two variables, the surface area of the soil or aquifer, S_a, which is related to the clay content, and a property of the pure organic compound called the octanol-water partition coefficient.

The **octanol-water partition coefficient,** K_{ow}, is one measure of how hydrophobic a compound is. The organic compound is shaken with a mixture of *n*-octanol and water and the proportion dissolving into each phase is measured. The octanol-water

Transformation, Retardation, and Attenuation of Solutes

TABLE 3.1 Representative f_{oc}^* values for different organic compounds.

Chemical	K_{ow}	f_{oc}^*	Minimum Soil Organic Carbon (mg/kg)
Dichloroethane	62	0.002	2,000
Benzene	135	0.001	1,000
Trichloroethylene	195	0.0007	700
Perchloroethylene	760	0.0002	200
Naphthalene	2,350	0.00009	90
Pyrene	209,000	0.000002	2

partition coefficient is the ratio of the concentration in the octanol to the concentration in the water: $C_{octanol}/C_{water}$. It is usually expressed as a log value in reference books.

According to McCarty, Reinhard, and Rittman (1981) the value of f_{oc}^* can be found from

$$f_{oc}^* = \frac{S_a}{200(K_{ow})^{0.84}} \tag{3.32}$$

This equation suggests that soils or aquifers with low organic carbon content would retain organic compounds with high K_{ow} values but might not retain those with low K_{ow} values. Assuming a surface area of 12 m²/g, which would be found with a typical kaolinite clay soil, Table 3.1 contains f_{oc}^* values and the corresponding minimum soil organic carbon content necessary before organic compounds with different K_{ow} values will sorb primarily on the organic carbon.

3.6.3 Estimating K_{oc} from K_{ow} Data

A number of researchers have found that there is a relationship between the octanol-water partition coefficient and the K_{oc} value for various organic compounds. The use of such a relationship is predicated upon the following (Karickhoff 1984): (1) Sorption is primarily on the organic carbon in the soil or aquifer. (2) Sorption is primarily hydrophobic, as compared with polar group interactions, ionic bonding, or chemisorption. (3) There is a linear relationship between sorption and the concentration of the solute.

A number of different organic compounds have been studied, with the result that a number of different relationships have been developed. Olsen and Davis (1990) listed a total of nine different equations that have been developed. Karickhoff (1984) lists four equations that have published least squares regression correlation coefficients (r^2 values) that exceed 0.9. (A correlation coefficient of 1.00 would mean that there is a perfect correlation between K_{ow} and K_{oc}. A correlation coefficient of 0.9 means that 90% of the variation is accounted for by the equation.)

Table 3.2 lists a number of equations that relate K_{oc} to K_{ow}. Where known, correlation coefficients are listed below the equations.

The equations in Table 3.2 have been derived for many different organic compounds. Some have utilized related compounds, whereas others are based on a mixture of different organic molecules. The hydrogeologist or engineer who wishes to estimate a K_{oc} value is placed in the situation of deciding which equation to use. The best choice

TABLE 3.2 Equations for estimating K_{oc} from K_{ow}.

Equation Number	Equation	Chemicals Used	Reference
(T1)	$\log K_{om} = 0.52 \log K_{ow} + 0.62$	72 substituted benzene pesticides	Briggs, 1981
(T2)	$\log K_{oc} = 1.00 \log K_{ow} - 0.21$	10 polyaromatic hydrocarbons	Karickhoff, Brown, and Scott 1979
(T3)	$K_{oc} = 0.63 K_{ow}$	Miscellaneous organics	Karickhoff, Brown, and Scott 1979
(T4)	$\log K_{oc} = 0.544 \log K_{ow} + 1.377$	45 organics, mostly pesticides	Kenaga and Goring 1980
(T5)	$\log K_{oc} = 1.029 \log K_{ow} - 0.18$ $r^2 = 0.91; n = 13$	13 pesticides	Rao and Davidson 1980
(T6)	$\log K_{oc} = 0.94 \log K_{ow} + 0.22$	s-trizines and dinitroanalines	Rao and Davidson 1980
(T7)	$\log K_{oc} = 0.989 \log K_{ow} - 0.346$ $r^2 = 0.991; n = 5$	5 polyaromatic hydrocarbons	Karickhoff 1981
(T8)	$\log K_{oc} = 0.937 \log K_{ow} - 0.006$	Aromatics, polyaromatics, triazines	Lyman 1982
(T9)	$\ln K_{oc} = \ln K_{ow} - 0.7301$	DDT, tetrachlorobiphenyl, lindane, 2,4-D, and dichloropropane	McCall, Swann, and Laskowski 1983
(T10)	$\log K_{om} = 0.904 \log K_{ow} - 0.779$ $r^2 = 0.989; n = 12$	Benzene, chlorinated benzenes, PCBs	Chiou, Porter, and Schmedding 1983
(T11)	$\log K_{oc} = 0.72 \log K_{ow} + 0.49$ $r^2 = 0.95; n = 13$	Methylated and chlorinated benzenes	Schwarzenbach and Westall 1981
(T12)	$\log K_{oc} = 1.00 \log K_{ow} - 0.317$ $r^2 = 0.98; n = 22$	22 polynuclear aromatics	Hassett et al. 1980

is an equation that was derived on the basis of chemicals similar to the one under study. The various equations tend to yield similar results for many compounds. Table 3.3 shows log K_{oc} values computed from the equations in Table 3.2 for several different organic compounds that cover a wide range of K_{ow} values.

Although there are a number of equations, most of the computed log K_{oc} values for the example chemicals fall close to or within one standard deviation of the geometric means. There are some data in the literature on actual measured values of log K_{oc}. Table 3.4 gives some experimental K_{oc} values.

The values in Table 3.4 fall close to the means listed on Table 3.3. The equations that yield the maximum or minimum values for a particular compound may not be the most appropriate to use for that compound. It appears that there is no universal equation that relates K_{oc} to K_{ow} for all classes of organic compounds.

3.6.4 Estimating K_{oc} from Solubility Data

The value of K_{oc} can also be estimated from the solubility, S, of a particular compound. Several different equations describing this relationship have been published and are listed in Table 3.5.

TABLE 3.3 Estimated values of K_{oc} based on published K_{ow} values.

	Dichloroethane	Benzene	Trichloroethene	Ethyl Benzene	Tetrachloroethene	Napthalene	2,2'-Dichlorobiphenyl	Pyrene
Log K_{ow}	1.79	2.13	2.29	3.14	3.40	3.37	4.80	5.32
Equation Number[a]				Estimated K_{oc}				
(T1)	1.79	1.96	2.05	2.49	2.62	2.61	3.35	3.62
(T2)	1.58	1.92	2.10	2.93	3.19	3.16	4.59	5.11
(T3)	1.13	1.34	1.44	1.98	2.14	2.16	3.07	3.35
(T4)	2.35	2.54	2.62	3.09	3.23	3.21	3.99	4.27
(T5)	1.66	2.01	2.18	3.05	3.32	3.29	4.76	5.29
(T6)	1.90	2.22	2.37	3.17	3.42	3.39	4.73	5.22
(T7)	1.42	1.76	1.92	2.76	3.02	2.99	4.40	4.92
(T8)	1.67	1.99	2.14	2.94	3.18	3.15	4.49	4.98
(T9)	1.06	1.40	1.56	2.41	2.67	2.64	4.07	4.59
(T10)	1.08	1.39	1.53	2.06	2.30	2.51	3.80	4.27
(T11)	1.78	2.02	2.14	2.75	2.94	2.92	3.95	4.32
(T12)	1.47	1.81	1.97	2.82	3.08	3.05	4.48	5.00
Range	1.06–2.35	1.34–2.54	1.44–2.62	1.98–3.17	2.14–3.42	2.16–3.39	3.07–4.76	3.35–5.29
Mean	1.57	1.86	2.00	2.70	2.93	2.92	4.14	4.58
St. dev.	0.38	0.35	0.33	0.39	0.41	0.37	0.54	0.63
Coef. var.	0.24	0.19	0.17	0.14	0.15	0.13	0.13	0.14

[a] The equation numbers in this table refer to Table 3.2.

TABLE 3.4 Experimentally derived K_{oc} values.

Compound	K_{oc}	Reference
Benzene	1.50	Chiou, Porter, and Schmedding 1983
	1.92	Karickhoff, Brown, and Scott 1979
	1.98	Rogers, McFarlane, and Cross 1980
Ethylbenzene	2.22	Chiou, Porter, and Schmedding 1983
2,2'-Dichlorobiphenyl	3.92	Chiou, Porter, and Schmedding 1983
Tetrachloroethene	2.32	Chiou, Peters, and Freed 1979
Napthalene	3.11	Karickhoff, Brown, and Scott 1979
Pyrene	4.92	Karickhoff, Brown, and Scott 1979
	4.80	Means et al. 1980

TABLE 3.5 Empirical equations by which K_{oc} can be estimated from S.

Equation Number	Equation	Reference
(T13)	$\log K_{oc} = 0.44 - 0.54 \log S$ S in mole fraction, $r^2 = 0.94$	Karickhoff, Brown, and Scott 1979
(T14)	$\log K_{oc} = 3.64 - 0.55 \log S$ S in mg/L	Kenaga 1980
(T15)	$\log K_{oc} = 4.273 - 0.686 \log S$ S in mg/L	Means et al. 1980
(T16)	$\log K_{oc} = 3.95 - 0.62 \log S$ S in mg/L	Hassett et al. 1983
(T17)	$\log K_{om} = 0.001 - 0.729 \log S$ S in moles/L, $r^2 = 0.996$	Chiou, Porter, and Schmedding 1983

Aqueous solubility can be expressed in several ways. The most common is a **mass per volume** unit, such as milligram per liter. Equation T13 in Table 3.5 uses the concept of **mole fraction.** This is the ratio of the moles of a substance to the total number of moles of solution. A **mole** of a substance is equal to its formula weight in grams. In Equation T17 (Table 3.5) solubility is in terms of moles of solute per liter of solution, a unit known as **molarity.** For dilute solutions, to convert molarity to mole fraction, divide the molarity by 55.6, the number of moles of water in a liter.

Table 3.6 contains log K_{oc} estimated from the solubility for the same compounds that are listed in Table 3.3. A comparison of the results of Table 3.6 with the experimentally derived values for K_{oc} found in Table 3.4 shows that all the equations yield an estimate that is within an order of magnitude of the experimental result.

The use of solubility data is complicated by temperature and ionic strength effects on solubility. Published solubility data sometimes do not indicate the temperature at which the measurement was made. For a number of reasons, such as the purity of the organic chemical, the ionic strength of the water, the temperature, and the experimental procedure, there can be a range in the reported solubility data in the literature. For similar reasons there can be a range of octanol-water partition coefficient values reported for the same compound (Sabljić 1987). Because of this, one should recognize that K_{oc} values obtained from K_{ow} or solubility data are truly estimates.

TABLE 3.6 K_{oc} values estimated from the aqueous solubility.

Compound:	Dichloroethane	Benzene	Trichloroethene	Ethyl Benzene	Tetrachloroethene	Naphthalene	2,2'-Dichlorobiphenyl	Pyrene
Molecular weight:	98.96	78.12	131.38	165.82	106.18	128.18	223.10	202.26
				Solubility (mg/L)				
	5500	1780	1100	140	150	31	1.86	0.032
Log S:	3.74	3.25	3.04	2.15	2.18	1.49	0.269	−1.50
				Solubility (moles/L)				
	5.56×10^{-2}	2.28×10^{-2}	8.37×10^{-3}	8.44×10^{-4}	1.41×10^{-3}	2.42×10^{-4}	8.32×10^{-6}	1.58×10^{-7}
Log S:	−1.25	−1.64	−2.08	−3.07	−2.85	−3.62	−5.08	−6.80
				Solubility (Mole Fractions)				
	1.00×10^{-3}	4.10×10^{-4}	1.51×10^{-4}	1.52×10^{-5}	2.54×10^{-5}	4.35×10^{-6}	1.49×10^{-7}	2.84×10^{-9}
Log S:	−3.00	−3.39	−3.82	−4.82	−4.60	−5.36	−6.83	−8.55
Equation Number[a]				Estimated log K_{oc}				
(T13)	2.06	2.27	2.50	3.04	2.87	3.33	4.13	5.06
(T14)	1.58	1.85	1.97	2.46	2.44	2.82	3.79	4.47
(T15)	1.67	2.01	2.15	2.76	2.74	3.21	4.46	5.27
(T16)	1.63	1.94	2.07	2.62	2.60	3.03	4.12	4.88
(T17)	1.15	1.43	1.75	2.47	2.31	2.88	3.93	5.19
Range	1.15–2.06	1.43–2.27	1.75–2.50	1.80–3.04	2.31–2.87	2.82–3.33	3.79–4.46	4.47–5.27
Mean	1.62	1.90	2.09	2.67	2.59	3.05	4.09	4.97
St. dev.	0.32	0.31	0.27	0.24	0.22	0.22	0.25	0.32
Coef. var	0.08	0.07	0.06	0.05	0.04	0.04	0.05	0.08

[a] The equation numbers in this table refer to Table 3.5.

3.6.5 Estimating K_{oc} from Molecular Structure

The applicability of using octanol-water partition coefficient and aqueous solubility data to estimate K_{oc} has been questioned because some organic compounds have similar aqueous solubility but very different octanol solubility (Ellgehausen, D'Hondt, and Fuerer 1981; Mingelgrin and Gerstl 1983; Olsen and Davis 1990). A more fundamental approach has been suggested on the basis of molecular topology (Koch 1983; Sabljić 1984, 1987; Sabljić and Protic 1982).

Molecular topology refers to the shape of the organic molecule. The particular parameter of molecular structure that has been related to K_{oc} is the first-order molecular connectivity index, $^1\chi$. The first-order molecular connectivity index is calculated on the basis of the nonhydrogen part of the molecule. Each nonhydrogen atom has an atomic δ value, which is the number of adjacent nonhydrogen atoms. A connectivity index is then calculated for the molecule by the following formula:

$$^1\chi = \sum (\delta_i \delta_j)^{-0.5} \tag{3.33}$$

where δ_i and δ_j are the delta values for a pair of adjacent nonhydrogen atoms and the summation takes place over all the bonds between nonhydrogen atoms.

Sabljić (1987) made a regression analysis between the molecular connectivity and observed K_{om} values for 72 organic molecules, including chlorobenzenes, polyaromatic hydrocarbons, alkylbenzenes, chlorinated alkanes and alkenes, chlorophenols, and heterocyclic and substituted polyaromatic hydrocarbons. The relationship between K_{om} and $^1\chi$, which had an r^2 value of 0.95, is given by Equation 3.34. In order to convert K_{om} to K_{oc}, multiply by 1.724.

$$\log K_{om} = 0.53\,^1\chi + 0.54 \tag{3.34}$$

Equation 3.34 was derived on the basis of nonpolar organic compounds. Sabljić (1987) gave an empirical method of extending this equation to classes of polar and even ionic organic compounds. When Equation 3.34 is used for polar and ionic organic compounds, it predicts a K_{om} that is higher than observed values reported in the literature. The nonpolar organics from which Equation 3.34 was derived are more strongly sorbed to soil organic matter than polar organic compounds. Sabljić introduced a polarity correction value, P_f, for each of 17 different groups of polar organics (Table 3.7). Equation 3.35 can be used to predict the value of K_{om} for polar organics:

$$\log K_{om} = 0.53\,^1\chi + 0.54 - P_f \tag{3.35}$$

As these polarity correction factors are empirically determined, this method cannot be extended to other classes of polar organic compounds until additional experimental work is done. In addition, organic phosphates fall into two groups rather than one. However, this correction extends our ability to estimate K_{om}, and by extension K_{oc}, to a number of additional organic compounds. The molecular topology method has several advantages over the estimation of K_{oc} from octanol-water partition coefficients and aqueous solubility:

1. There is a theoretical basis to the molecular topology method for nonpolar organic compounds.

Transformation, Retardation, and Attenuation of Solutes

TABLE 3.7 Polarity correction factors for classes of polar organic compounds.

Class of Compounds	Polarity Correction Factor, P_f
Substituted benzenes and pyridines	1.00
Organic phosphates (group 1)	1.03
Carbamates	1.05
Anilines	1.08
Nitrobenzenes	1.16
Phenylureas	1.88
Triazines	1.88
Acetanilides	1.97
Uracils	1.99
Alkyl-N-phenylcarbamates	2.01
3-Phenyl-1-methylureas	2.07
3-Phenyl-1-methyl-1-methoxyureas	2.13
Dinitrobenzenes	2.28
3-Phenyl-1,1-dimethylureas	2.36
Organic acids	2.39
3-Phenyl-1-cycloalkylureas	2.76
Organic phosphates (group 2)	3.19

2. The literature contains a range of experimentally derived values for both the octanol-water partition coefficient and the aqueous solubility for a number of nonpolar organic compounds. One has no way of knowing which is the correct value.
3. Some compounds with similar aqueous solubility values have quite different octanol-water partition coefficients.
4. There are a number of competing equations that can be used for both the K_{oc} and solubility methods. One is never quite sure which to select.
5. The solubility and K_{ow} methods were devised strictly for nonpolar organic compounds. There is no way to extend them to polar organic compounds.

However, the molecular topology method does not account for the ionic strength, pH, and temperature of the solution.

EXAMPLE PROBLEM

Compute the value of $^1\chi$ for trichloroethylene. The structure of trichloroethylene is as shown:

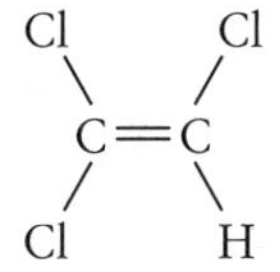

$$\begin{array}{c} \text{Cl} \qquad \text{Cl} \\ \diagdown \quad \diagup \\ \text{C}=\text{C} \\ \diagup \quad \diagdown \\ \text{Cl} \qquad \text{H} \end{array}$$

There are four nonhydrogen pairs: Carbon 1 is bonded to two chlorine atoms and carbon 2. Carbon 2 is bonded to one chlorine atom. The following table shows the δ_i and δ_j values as well as the computed $(\delta_i\delta_j)^{-0.5}$ value for each pair.

Nonhydrogen pair	δ_i	δ_j	$(\delta_i \delta_j)^{-0.5}$
Cl-C(1)	1	3	0.577
Cl-C(1)	1	3	0.577
C(1)-C(2)	3	2	0.408
C(2)-Cl	2	1	0.707
			$2.269 = {}^1\chi$

3.6.6 Multiple Solute Effects

Many hazardous waste sites contain more than one organic solvent in aqueous solution. It has been shown that the solubility of structurally similar hydrophobic organic liquids in aqueous solutions is dependent upon the mixture of solutes present (Banerjee 1984). These liquids behave in a nearly ideal fashion, which is described by the following equation:

$$\frac{C_i}{S_i} = (x_i)_{\text{org}} \tag{3.36}$$

where

C_i = the equilibrium molar concentration of the ith component in the mixture

S_i = the water solubility of the component in its pure form

$(x_i)_{\text{org}}$ = the mole fraction of the ith compound in the organic phase

The concentration of a solute in a solution that is saturated with several structurally similar compounds is less than it would be in water alone. Mixtures of dissimilar liquids and of organic solids are more complex (Banerjee 1984). In addition to having mutual effects on solubility, organic mixtures will also compete for sorption sites.

In one study the presence of organic solvents, such as methanol, ethanol, 2-propanol, and butanol, in concentrations of 5 to 10% reduced the soil-water partition coefficient for an organic molecule, kepone. This increased the rate at which kepone migrated through a soil column (Staples and Geiselmann, 1988).

If a chemical analysis indicates that an aqueous sample contains a hydrophobic organic compound in amounts in excess of its solubility, it is likely that part of the compound is present as a nonaqueous phase liquid. Under such conditions nonaqueous phase liquid transport theory must be considered (Hunt, Sitar, and Udell 1988).

3.7 Homogeneous Reactions

3.7.1 Introduction

According to the classification system of Rubin (1983) homogeneous reactions are ones that take place entirely within the liquid phase. If the reactions are revesible and proceed rapidly enough, the reaction can be described as being in local chemical equilibrium

(Walsh et al. 1984). If the reaction either does not reach equilibrium or is nonreversible, then it is treated as a homogeneous, nonequilibrium reaction.

3.7.2 Chemical Equilibrium

If two compounds in solution, A and B, react to form product C, which can dissociate into A and B, then the reaction is reversible and is expressed as:

$$a\text{A} + b\text{B} \rightleftharpoons c\text{C} \tag{3.37}$$

where a, b, and c represent the number of molecules of each compound that are needed to balance the reaction. When the reaction has progressed to the point that no further net production of C occurs, then it has reached equilibrium. The reaction continues, but the forward rate and the reverse rate have become equal. At that point, we can measure [A], [B], and [C], the concentrations of the reactants and the product. The relationship between them is expressed as an equilibrium constant, K_{eq}.

$$K_{eq} = \frac{[\text{C}]^c}{[\text{A}]^a[\text{B}]^b} \tag{3.38}$$

The reaction must have sufficient time to proceed to the point of equilibrium before Equation 3.38 is valid. This represents a Class I reaction according to Figure 3.1.

3.7.3 Chemical Kinetics

If reaction 3.37 were to proceed so slowly that it would not have time to come to equilibrium in the framework of the ground-water flow system, then local equilibrium cannot be assumed and we must consider it in the framework of chemical kinetics. We can look at the reaction from the standpoint of either the disappearance of the reactants, A and B, or the appearance of the product, C. We have three rates to consider, two that describe the disappearance of the reactants and one that describes the production of the product:

$$R_A = -\frac{d[\text{A}]}{dt} = \kappa[\text{A}]^p[\text{B}]^q \tag{3.39}$$

$$R_B = -\frac{d[\text{B}]}{dt} = \kappa'[\text{A}]^p[\text{B}]^q \tag{3.40}$$

$$R_C = \frac{d[\text{C}]}{dt} = \kappa''[\text{C}]^r \tag{3.41}$$

where

R_A and R_B = the reaction rates for the disappearance of A and B
R_C = the reaction rate for the appearance of C
[A], [B], and [C] = the measured concentrations of A, B, and C
κ, κ', and κ'' = reaction rate constants
p, q, and r = reaction order with respect to the indicated reactant or product

If one of p, q, or r in Equations 3.39, 3.40, and 3.41 is equal to zero, then the reaction rate is not a function of that product or reactant and the reaction is said to be zero order with respect to that reactant or product.

If one of p, q, or r is equal to 1, then there is a linear relationship with respect to that product or reactant, and the reaction rate is said to be first order.

If none of p, q, or r is 0 or 1, then there is a more complex relationship with respect to that product or reactant. Such a reaction is more difficult to analyze mathematically than zero- or first-order kinetics.

If a compound is present in a system in great excess, then the reaction rate may well be independent of the concentration. If we consider the depletion of reactant A in such a system, it could be described by the zero-order equation

$$[A] = [A]_0 - \kappa t \tag{3.42}$$

where

κ = the reaction rate constant
$[A]_0$ = the initial concentration
$[A]$ = the concentration at some time t

In a first order system, the rate at which the reactant disappears is described by

$$\ln \frac{[A]}{[A]_0} = -\kappa t \tag{3.43}$$

or

$$[A] = [A]_0 e^{-\kappa t} \tag{3.44}$$

Reaction-rate constants and the order of the reactions must be determined experimentally. These are not equal to the equilibrium constant of the reaction.

3.7.4 Tenads in Chemical Reactions

Rubin (1983) introduced the concept of *tenads* in chemical reactions. This classification scheme is useful as a means of formulating transport equations that incorporate chemical reactions. A **tenad** is a reacting or nonreacting chemical entity whose global mass in the system is reaction independent. A chemical entity may be either an uncharged atom or group of atoms, an ion, or an electron. Chemical species may be composed of one or more tenads. The global mass of a tenad is the sum of the mass of the tenad wherever it occurs. Reaction-independent means that the global mass does not depend upon the extent of progress of the reaction.

Reactions can have alternative sets of tenads. Consider a system with solid $CaSO_4$ in water. The reaction that occurs is

$$CaSO_4 \rightleftharpoons Ca^{2+} + SO_4^{2-} \tag{3.45}$$

The two alternative sets of tenads are (1) the atoms Ca, S, and O, and (2) the ions Ca^{2+} and SO_4^{2-}. $CaSO_4$ (solid) cannot be a tenad because the mass depends upon the reaction progress, but the Ca^{2+} in $CaSO_4$ (solid) is part of the tenad $\{Ca^{2+}\}$.

Transformation, Retardation, and Attenuation of Solutes

We will consider the chemical changes that can occur during one-dimensional flow of solute. The solute flux equation is

$$F_i = -n_e D_L \frac{\partial [C]_i}{\partial x} + n_e v_x [C]_i \quad (3.46)$$

where

F_i = the x-direction mass flux of a solute
$[C]_i$ = the concentration of the solute
D_L = the coefficient of longitudinal dispersion
n_e = the effective porosity

The one-dimensional continuity equation can be written as

$$\frac{\partial m^T}{\partial t} = -\frac{\partial F^T}{\partial x} + S^T \quad (3.47)$$

where

m^T = global mass of the given chemical entity stored in a unit volume of the porous media (solid and solute phases)
F^T = total mass flux of the entity into or out of the unit volume
S^T = total source/sink term for the entity

For a tenad that exists only in the form of solute i, m^T is equal to $n_e[C]_i$, and $F^T = F_i$. The total sink/source term, S^T, represents sources and sinks from external sources, S_{ext}, biological growth and decay, S_{bio}, and chemical reactions, S_{chem}.

$$S^T = \left(\frac{\partial m^T}{\partial t}\right)_{ext} + \left(\frac{\partial m^T}{\partial t}\right)_{bio} + \left(\frac{\partial m^T}{\partial t}\right)_{chem} \quad (3.48)$$

In order to illustrate the use of the tenad method, we will examine the class of sufficiently fast, reversible, homogeneous reactions, indicated by I on Figure 3.1. For the case of reversible homogeneous reactions, there is no source/sink term. In this case Equation 3.46 can be written as

$$n_e \frac{\partial [C]_i}{\partial t} = W[C]_i \quad (3.49)$$

where W is a linear operator defined as

$$W = n_e D_L \frac{\partial^2}{\partial x^2} - n_e V_x \frac{\partial}{\partial x} \quad (3.50)$$

For the case of flow in the vadose zone, the volumetric water content, θ, should be used in place of the effective porosity, n_e.

In the example we will have the ground water containing two solutes, E and F, which react to form a product, EF. The invading aqueous solution contains solutes E and G, which react to form product EG. The reactions are reversible and sufficiently fast for the local equilibrium assumption to apply.

The two reactions that occur are

$$EF \rightleftharpoons E + F \quad (3.51)$$

$$EG \rightleftharpoons E + G \quad (3.52)$$

The reaction has three tenads, E, F, and G. It could represent oxidation, reduction, or complex formation. There are a total of five participants in the reaction, the three tenads, EF, and EG. We need five equations in order to be able to solve for these five unknowns. These equations can be obtained by writing a convective transport equation for each of the three tenads, E, F, and G, and by writing the equilibrium equation for each of the two products, EF and EG.

$$n_e \frac{\partial [E]}{\partial t} + n_e \frac{\partial [EF]}{\partial t} + n_e \frac{\partial [EG]}{\partial t} = W[E] + W[EF] + W[EG] \quad (3.53)$$

$$n_e \frac{\partial [F]}{\partial t} + n_e \frac{\partial [EF]}{\partial t} = W[F] + W[EF] \quad (3.54)$$

$$n_e \frac{\partial [G]}{\partial t} + n_e \frac{\partial [EG]}{\partial t} = W[G] + W[EG] \quad (3.55)$$

$$K_{EFeq} = \frac{[E][F]}{[EF]} \quad (3.56)$$

$$K_{EGeq} = \frac{[E][G]}{[EG]} \quad (3.57)$$

Rubin (1983) presents methods of getting similar sets of equations for all six of the classes of chemical reactions listed in Figure 3.1.

3.8 Radioactive Decay

If radionuclides enter the ground-water system, those which are cations are subjected to retardation on soil surfaces. In addition they will undergo radioactive decay, which will reduce the concentration of radionuclides in both the dissolved and sorbed phases. A factor for radioactive decay can be substituted for the last term of Equation 3.1 in the following form:

$$\left(\frac{\partial C}{\partial t}\right)_{decay} = -\frac{\ln 2}{\lambda} C \quad (3.58)$$

where λ is the half-life of the radionuclide.

3.9 Biodegradation

The degradation of dissolved organic molecules in ground water is of great interest to practicing contaminant hydrogeologists. As was shown in Chapter 1, much of the ground-water contamination is due to organic chemicals, including hydrocarbons. The mechanism of biodegradation will be covered in detail in Chapter 7. In this section we will examine the transport and decay equations that can be used to describe the process. Although a wide range of organic molecules can be degraded, for the sake of this discussion we will refer to them simply as hydrocarbons. The hydrocarbons form a substrate for microbial growth—that is, they provide the energy source for the microbes which form a **biofilm** on the solid surfaces in the aquifer.

If the microbes require oxygen in their metabolism, then the process is called **aerobic biodegradation.** The removal of the hydrocarbons, the consumption of oxygen in the process, and the growth of the microbes in the aquifer, ignoring transport through the biofilm, can be described by the following equations, which are modifications of the **Monod function** (also known as Michaelis-Menten function) (Borden and Bedient 1986):

$$\frac{dH}{dt} = -M_t h_u \left(\frac{H}{K_h + H}\right)\left(\frac{O}{K_o + O}\right) \tag{3.59}$$

$$\frac{dO}{dt} = -M_t h_u G \left(\frac{H}{K_h + O}\right)\left(\frac{O}{K_o + O}\right) \tag{3.60}$$

$$\frac{dM_t}{dt} = M_t h_u Y \left(\frac{H}{K_h + H}\right)\left(\frac{O}{K_o + O}\right) + k_c Y C_{oc} - bM_t \tag{3.61}$$

where

H = hydrocarbon concentration in pore fluid (ML^{-3})

O = oxygen concentration in pore fluid (ML^{-3})

M_t = total aerobic microbial concentration (ML^{-3})

h_u = maximum hydrocarbon utilization rate per unit mass of aerobic microorganisms (T^{-1})

Y = microbial yield coefficient (g cells/g hydrocarbon)

K_h = hydrocarbon half-saturation constant (ML^{-3})

K_o = oxygen half-saturation constant (ML^{-3})

k_c = first-order decay rate of natural organic carbon

C_{oc} = natural organic carbon concentration (ML^{-3})

b = microbial decay rate (T^{-1})

G = ratio of oxygen to hydrocarbon consumed

The microorganisms will grow on both naturally occurring organic carbon as well as hydrocarbon contaminants. The microorganisms tend not to move in the aquifer because they generally adhere to aquifer materials (Harvey, Smith, and George 1984). Even if the microbes are free to move, the natural tendency of the aquifer matrix will be to filter them out. There will be some tendency for microbes to transfer from the solid surface to solution. As a first approximation this can be considered to be a linear function of the total mass of microorganisms.

We can combine Equations 3.59, 3.60, and 3.61 individually with Equation 3.1 to obtain solute transport equations for hydrocarbon, oxygen and microorganisms. The hydrocarbon is assumed to sorb onto the solid surfaces following a linear sorption isotherm. The resulting equations are (Borden and Bedient 1986)

$$\frac{\partial H}{\partial t} = \frac{1}{r_h}\left(D_L \frac{\partial^2 H}{\partial x^2} - v_x \frac{\partial H}{\partial x}\right) - \frac{h_u M_t}{r_h}\left(\frac{H}{K_h + H}\right)\left(\frac{O}{K_o + O}\right) \qquad (3.62)$$

$$\frac{\partial O}{\partial t} = D_L \frac{\partial^2 O}{\partial x^2} - v_x \frac{\partial O}{\partial x} - h_u M_t F\left(\frac{H}{K_h + H}\right)\left(\frac{O}{K_o + O}\right) \qquad (3.63)$$

$$\frac{\partial M_s}{\partial t} = \frac{1}{r_m}\left(D_L \frac{\partial^2 M_s}{\partial x^2} - v_x \frac{\partial M_s}{\partial x}\right) + h_u M_s Y\left(\frac{H}{K_h + H}\right)\left(\frac{O}{K_o + O}\right) + \frac{K_c Y C_{oc}}{r_m} - bM_s \qquad (3.64)$$

where

M_s = concentration of aerobic microbes in solution
r_h = retardation factor for hydrocarbon
r_m = microbial retardation factor
v_x = average linear ground-water velocity

Some microorganisms can degrade hydrocarbons in the absence of oxygen. These microbes use another electron acceptor, such as nitrate (Major, Mayfield, and Barker 1988). Anaerobic decomposition of hydrocarbons can be described by another variation of the Monod function, which describes two-step catalytic chemical reactions (Bouwer and McCarty 1984). This function is

$$\frac{dH}{dt} = -h_{ua} M_a \left(\frac{H}{K_a + H}\right) \qquad (3.65)$$

where

M_a = total mass of anaerobic microbes
h_{ua} = maximum hydrocarbon utilization rate per unit mass of anaerobic microbes
K_a = half-maximum rate concentration of the hydrocarbon for anaerobic decay

The solute transport and decay equation for anaerobic biodegradation in the aqueous phase is

$$\frac{\partial H}{\partial t} = \frac{1}{r_h}\left(D_L \frac{\partial^2 H}{\partial x^2} - v_x \frac{\partial h}{\partial x}\right) - \frac{h_{ua} M_a}{r_h}\left(\frac{H}{K_a + H}\right) \qquad (3.66)$$

If the concentration of the hydrocarbon, H, is much less than K_a, the half-maximum rate concentration, then Equation 3.65 can be simplified to a linear form by neglecting H in the denominator (Bouwer and McCarty 1984). This results in a first-order decay term:

$$\frac{dH}{dt} = -\left(\frac{h_{ua} M_a}{K_a}\right) H \qquad (3.67)$$

Under these conditions the solute-transport equation with anaerobic biodegradation becomes

$$\frac{\partial H}{\partial t} = \frac{1}{r_h}\left(D_L \frac{\partial^2 H}{\partial x^2} - v_x \frac{\partial h}{\partial x}\right) - \left(\frac{h_{ua} M_a}{r_h K_a}\right) H \qquad (3.68)$$

A single microbial growth substrate (that is, a hydrocarbon or other organic chemical that can serve as an energy source for the microbes) cannot be reduced for extended periods of time below the minimum concentration needed to maintain the microbial population. In other words, microbes cannot degrade a substance below the concentration that they need to continue to exist. This minimum concentration, H_{min}, is a function of the hydrocarbon, the electron acceptor, and the microorganism. It can be expressed as (Bouwer and McCarty 1984)

$$H_{min} = K_h \left(\frac{b}{Yb_u - b} \right) \tag{3.69}$$

In the case of aerobic microbial decay, there is also some minimum oxygen concentration below which aerobic decay will not occur. Although there is a minimum hydrocarbon concentration for each hydrocarbon that might be present, microbes may utilize more than one hydrocarbon as a growth substrate. As a result, if there is a mixture of hydrocarbons present, the microbes could degrade the separate hydrocarbons to concentrations that are lower than if the individual hydrocarbon were the only one present (McCarty, Reinhard, and Rittman, 1981). The primary substrate that is supporting the microbes can be a single substance or a mixture of substances. Although the primary substrate is supporting the growth of the microbes, substances that are present in trace amounts can be consumed by the microbial population through a process known as secondary utilization. The decay of a hydrocarbon that is undergoing secondary utilization is described by Equation 3.67.

Figure 3.10 shows the results of a laboratory experiment in microbial decay (Bouwer and McCarty 1984). An aerobic soil column was established with a microbial population. A solution with 10 mg/L of acetate as the primary substrate along with 10 μg/L of chlorobenzene and 10 μg/L of 1,4-dichlorobenzene as secondary substrates was introduced into the soil column as a constant flow. The utilization of the primary and the secondary substrates occurred simultaneously in the first 7 cm of the soil column. After that flow distance, all three compounds reached an irreducibly low concentration that would not support further microbial growth.

Srinivasan and Mercer (1988) developed a computer model, BIO1D, which simulates one-dimensional solute transport with biodegradation. In their model they relied upon the observations of Borden et al. (1986) that microbial growth reaches equilibrium rapidly with respect to the rate of ground-water flow and therefore the microbial population can be assumed to be constant. This means that Equation 3.64 is not needed, since $dM_s/dt = 0$. The model solves the equivalent of Equations 3.62 and 3.63 for aerobic biodegradation and Equation 3.66 for anaerobic decomposition. The computer code automatically switches from aerobic to anaerobic decomposition if the oxygen levels drop below the minimum to support aerobic decay.

If the substrate concentration, H, is less than $0.25 K_a$ and conditions are anaerobic, then the equation used is (3.68). If $H \ll K_a$, the Michaelis-Menten function and the first-order decay function will yield similar results. Figure 3.11 shows the modeling of the anaerobic biodegradation of trichloroethylene (Srinivasan and Mercer 1988). A pulse of contamination of 2780-μg/L trichloroethylene was injected into an aquifer for 150 hr, followed by continued injection at 0 μg/L. Figure 3.11 shows the concentration of trichloroethylene in the plume after 2500 hr and again after 3750 hr. Both Michaelis-Menten and first-order decay functions were used in the modeling, with very similar

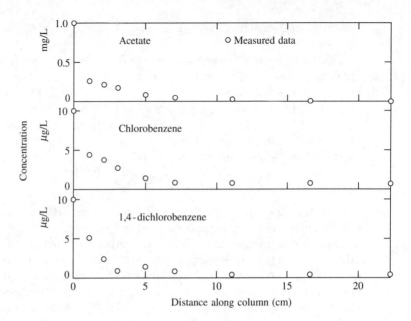

FIGURE 3.10 Measured steady-state profiles of the biodegradation of acetate, chlorobenzene and 1,4-dichlorobenzene in an aerobic biofilm reactor. *Source:* Modified from E. J. Bouwer and P. L. McCarty, *Ground Water* 22, no. 4 (1984):433–40. Used with permission. Copyright © 1984 Water Well Journal Publishing Co.

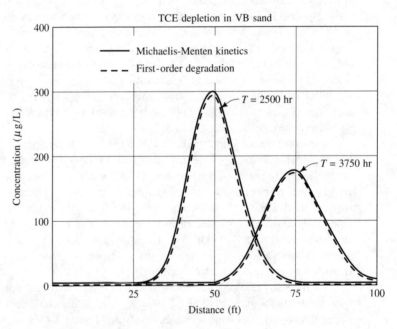

FIGURE 3.11 Modeling of the movement and anaerobic biodegradation of a trichloroethylene plume in a sand aquifer using both Michaelis-Menten and first-order decay functions. *Source:* P. Srinivasan and J. W. Mercer, *Ground Water* 26, no. 4 (1988):475–87. Used with permission. Copyright © 1988 Water Well Journal Publishing Co.

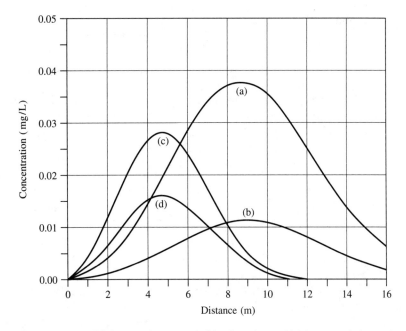

FIGURE 3.12 BIO1D model showing the position of a solute plume with (a) no retardation and no decay, (b) no retardation but biodegradation, (c) retardation that follows a linear sorption isotherm but no decay, and (d) retardation that follows a linear sorption isotherm and biodegradation.

results. The reduction in the peak concentration of the trichloroethylene plume with travel time can also be seen on this figure.

The BIO1D model can be used to illustrate the effects of retardation and biodegradation. Figure 3.12 shows the position of a solute plume with (a) no retardation and no decay, (b) no retardation but biodegradation, (c) retardation that follows a linear sorption isotherm but no decay and (d) retardation that follows a linear sorption isotherm and biodegradation. This model represents the movement of a slug of solute introduced into an aquifer over a 160-da period. The position of the solute front is shown after 800 da of travel. Biodegradation is modeled as first-order decay with a half-life of 400 da. Retardation delays the advance of the solute front, whereas biodegradation reduces the peak value.

3.10 Colloidal Transport

Colloids are particles with diameters less than 1 μm. Colloidal-size particles include dissolved organic macromolecules, such as humic substances, microorganisms, tiny droplets of insoluble organic liquids, and mineral matter (McCarthy and Zachara 1989). Some colloids may be small enough to flow through the pores of an aquifer. If dissolved solutes partition onto a colloid, this can create a second mobile phase. The solute can then be found in three regions: dissolved, sorbed onto mobile colloids, and sorbed onto immobile surfaces. The study of colloids in ground water is greatly complicated by the fact that the process of installing monitoring systems such as wells and piezometers may introduce colloids that were not originally present. Sampling processes may also create colloids, such as the precipitation of colloidal iron due to oxygenation of water and the dislocation of stable colloids due to too-rapid pumping.

In order for colloids to participate in contaminant transport, they must first be released to the ground water. This can occur due to chemical precipitation, biological activity, or disaggregation of stable aggregates. They can also be carried into the aquifer by infiltrating water, especially through cracks and macropores.

Colloids are mobile if their surface chemistry is such that the individual colloids are repulsed so that they remain disaggregated, rather than being attracted to form larger particles. In addition, the pore-size geometry must be such that the colloids are not filtered from the suspension.

There is ample evidence in the literature that colloids can migrate in aquifers. Keswick, Wang, and Gerwa (1982) report that bacteria have migrated up to 900 m in an aquifer and viruses have migrated up to 920 m. Layered silicate clays from surface soils have been found to travel up to several hundred meters to wells (Nightingale and Bianchi 1977). Asbestos fibers have been found in an aquifer recharged with surface water containing the mineral (Hayward 1984).

Colloids have been implicated in the unexpected movement of plutonium and americum, radioactive elements that are normally believed to be relatively immobile in the soil due to a high distribution coefficient (McCarthy and Zachara 1989).

An aqueous solution may contain dissolved organic macromolecules. Hydrophobic organic solvents in aqueous mixtures may partition onto these macromolecules rather than soil organic carbon (Enfield 1985). When this happens the mobility of the organic solvent is greatly enhanced, especially if it has a low mobility (high K_{oc} value). In fine-grained soils or aquifers the macromolecules may even have a velocity greater than the average linear ground-water velocity (Enfield and Bengtsson 1988). This is due to the **size-exclusion effect,** which occurs when molecules or ions are so large that they cannot travel through the smaller pores. As a result, they are restricted to the larger pores, in which the ground-water velocity is greater than average. Thus these molecules will travel at a rate greater than the average linear ground-water velocity. This effect is more prevalent in fine-grained soils and aquifers with some pores small enough to exclude some molecules. Organic macromolecules are likely to be produced in municipal landfill leachate, which has a high amount of dissolved organic carbon. This is one reason codisposal of toxic organic liquids and municipal refuse is not wise.

Case Study: Large-scale Field Experiment on the Transport of Reactive and Nonreactive Solutes in a Sand Aquifer Under Natural Ground-water Gradients—Borden, Ontario

In August, 1982, an experiment was begun by injecting about 12 m³ of water containing a number of solutes into a sand aquifer in which 275 multilevel ground-water samplers had previously been installed (Mackay et al. 1986; Freyberg 1986; Roberts, Goltz, and Mackay 1986). Each multilevel sampler had from 14 to 18 sampling ports vertically separated by about 0.2 to 0.3 m. Figure 3.13 shows the distribution of sampling points. The average porosity of the sand is 0.33, the geometric mean of hydraulic conductivity is 7.2×10^{-5} m/sec, and the mean annual horizontal gradient is 0.0043. The average linear ground-water velocity computed from these values is 29.6 m/yr. The direction of ground-water flow at the site is to the northeast in the direction indicated by line A-A' on Figure 3.13.

The injected water contained the following solutes:

Solute	Concentration (mg/L)	Mass (g)
Chloride ion	892	10,700
Bromide ion	324	3,870
Bromoform	0.032	0.38
Tetrachloroethylene	0.030	0.36
Carbon tetrachloride	0.031	0.37
1,2-dichlorobenzene	0.332	4.0
Hexachloroethane	0.020	0.23

From August 24, 1982, to June 2, 1984, synoptic monitoring was accomplished as water was withdrawn from a large number of the monitoring devices and analyzed for the ionic tracers and organics on 18 different occasions. This was done in order to assess the overall movement of the plume. In addition, at selected points along the flowpath time-series monitoring was done by sampling on a much more frequent basis than the synoptic monitoring. In all, 14,465 samples were analyzed during this time period for the synoptic monitoring program, and 1246 samples were analyzed for the time-series monitoring.

Figure 3.14 shows the breakthrough curves for chloride, carbon tetrachloride, and tetrachloroethylene at a monitoring point in the center of the plume located 5.0 m from the injection wells. Values are shown as relative concentration, which is the observed concentration for a parameter divided by the injected concentration. At 100 da the chloride slug has just about passed the observation point, the carbon tetrachloride plume has just about reached a peak, and the tetrachloroethylene plume has yet to reach it. By 200 da both the carbon tetrachloride and the chloride plumes have passed the monitoring point, whereas the tetrachloroethylene is near its peak value. The behavior of bromoform was very close to that of carbon tetrachloride and is not shown.

Figure 3.14 illustrates the chromatographic effect of retardation. The chloride ion is essentially unaffected by travel through the aquifer, whereas the carbon tetrachloride and the tetrachloroethylene are traveling at slower rates. The result is a separation of the components of the plume, a phenomenon known as the **chromatographic effect.**

Figure 3.15 shows the arrival at the monitoring point of chloride ion and two other organics, dichlorobenzene and hexachloroethane. The occurrences of both of these compounds are sporadic, and they have relative concentrations much less than carbon tetrachloride and tetrachloroethylene. Analysis of the area under the time-concentration curves shows that the relative areas of carbon tetrachloride, bromoform, and tetrachloroethylene are about the same size as the chloride curve. This indicates that the mass of these organics being measured is about the same as the mass that was introduced into the aquifer. However, for dichlorobenzene and hexachloroethane the relative areas are much less than that of chloride, indicating that the mass being measured is much less than the mass that was introduced. The

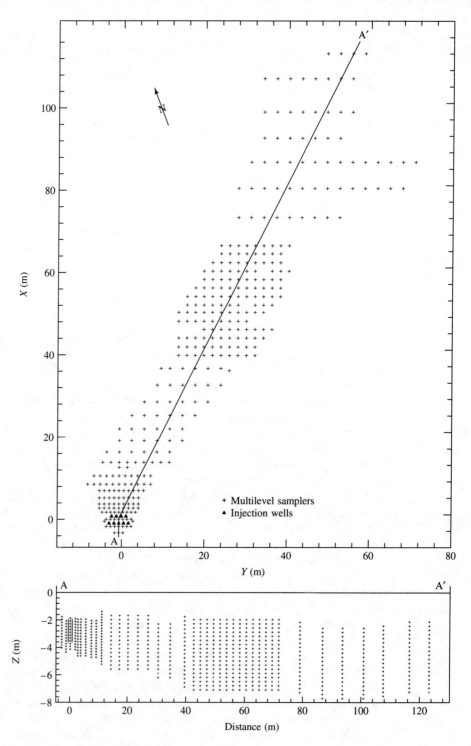

FIGURE 3.13 Location of multilevel sampling devices at the site of the Borden, Ontario, tracer test. *Source:* D. M. Mackay et al., *Water Resources Research* 22, no. 13 (1986):2017–30. Copyright by the American Geophysical Union.

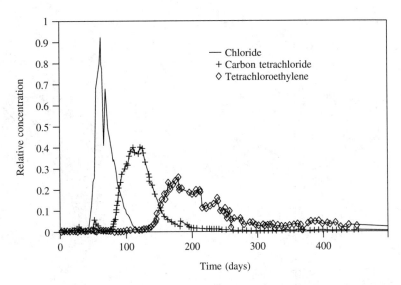

FIGURE 3.14 Arrival times of chloride, carbon tetrachloride, and tetrachloroethylene at a measuring point 5.0 m downgradient from the injection well at Borden, Ontario. *Source*: P. V. Roberts, M. N. Goltz, and D. M. Mackay, *Water Resources Research* 22, no. 13 (1986):2047–59. Copyright by the American Geophysical Union.

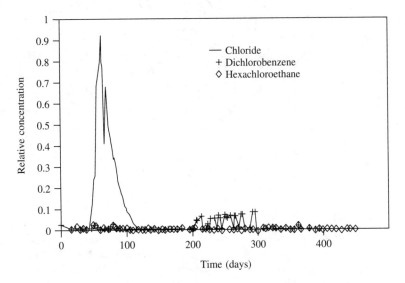

FIGURE 3.15 Arrival times of chloride, dichlorobenzene, and hexachloroethane at a measuring point 5.0 m downgradient from the injection well at Borden, Ontario. *Source*: P. V. Roberts, M. N. Goltz, and D. M. Mackay, *Water Resources Research* 22, no. 13 (1986):2047–59. Copyright by the American Geophysical Union.

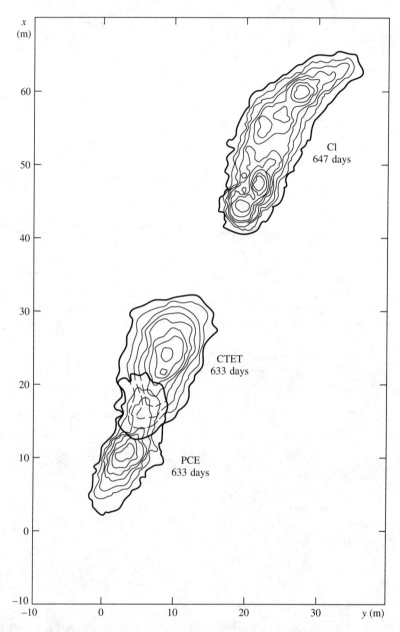

FIGURE 3.16 Plumes of chloride, carbon tetrachloride, and tetrachloroethylene at the end of the experimental period. The plumes are based on depth-averaged values. *Source:* P. V. Roberts, M. N. Goltz, and D. M. Mackay, *Water Resources Research* 22, no. 13 (1986):2047–59. Copyright by the American Geophysical Union.

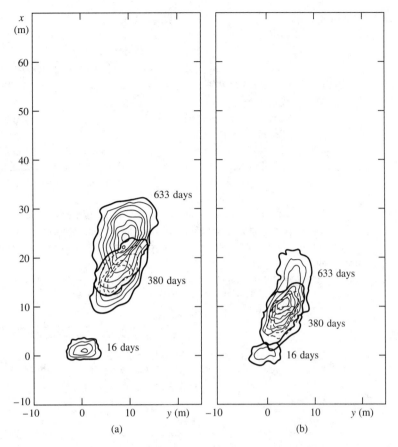

FIGURE 3.17 (a) Growth of carbon tetrachloride plume with time; (b) growth of tetrachloroethylene plume with time. *Source:* P. V. Roberts, M. N. Goltz, and D. M. Mackay, *Water Resources Research* 22, no. 13 (1986):2047–59. Copyright by the American Geophysical Union.

missing mass was presumably removed by biodegradation or abiotic pathways such as hydrolysis (Chapter 7).

Similar results were obtained during the synoptic sampling. Figure 3.16 shows the plumes of chloride (Cl), carbon tetrachloride (CTET) and tetrachloroethylene (PCE) at the end of the experimental period (after 633 days for the organics and 647 days for the chloride). The chloride can be seen to have moved significantly beyond the organics and carbon tetrachloride has moved farther than tetrachloroethylene. Figure 3.17(a) shows the growth of the carbon tetrachloride plume and 3.17(b) shows the growth of the tetrachloroethylene plume.

The relative velocities of the various solutes are indicated by the positions of the centers of mass of the plumes at the end of the experiment. The chloride plume was measured after 647 days of travel and the organic plumes were measured after 633 days. The distance from the center of the injection zone to the center of mass of the plume as well as the average velocities are given in the following table:

Compound	Distance to Center of Mass (m)	Average Velocity (m/day)
Chloride	58.21	0.0900
Carbon tetrachloride	24.82	0.0392
Bromoform	21.51	0.0340
Tetrachloroethylene	12.33	0.0195
Dichlorobenzene	8.09	0.0128
Hexachloroethane	None detected after 633 da	

The distances traveled by each of the compounds as a function of time are plotted in Figure 3.18. The distance traveled by chloride is linear with time, indicating a constant advective rate. On the other hand, the organic solutes indicate decreasing velocities with increasing time.

The organic compounds have the following aqueous solubilities, octanol-water partition coefficients, and first-order molecular connectivity indices:

	Solubility	log K_{ow}^a	log K_{ow}^b	$^1\chi$
Bromoform	3190 mg/L at 30°C[c]		2.30	2.00
Carbon tetrachloride	805 mg/L at 20°C[a]	2.83	2.70	2.00
Tetrachloroethylene	1503 mg/L at 25°C[a]	3.40	2.60	2.64
	150 mg/L at 25°C[c]			
1,2-dichlorobenzene	156 mg/L at 25°C[a]	3.38	3.40	3.83
Hexachloroethane	50 mg/L at 22.3°C[a]	3.82	3.60	3.25

[a.] Value from P. H. Howard, *Fate and Exposure Data for Organic Chemicals* (Chelsea, Mich.: Lewis Publishers, 1990).
[b.] Value from Mackay et al. (1986).
[c.] Value from Karel Verschueren, *Handbook of Environmental Data on Organic Chemicals*, 2d ed. (New York: Van Nostrand Reinhold Co., 1983).

This case history illustrates the problems with predicting the transport of organic compounds in ground water. The organic carbon content of the aquifer is low (0.02%), so absorption is limited. The observed order of relative velocity from greatest to least was carbon tetrachloride, bromoform, tetrachloroethylene, and dichlorobenzene. The published solubility data were collected at three different temperatures and thus cannot be directly compared. Two solubilities for tetrachloroethylene are published for the same temperature. Other than for 1,2-dichlorobenzene, the published K_{ow} values are not particularly close. Moreover, the relative order of the K_{ow} values are not consistent from one reference source to another nor does either correspond to the observed relative velocities. However, the computed first-order molecular connectivity indices have a rank order that corresponds to the observed movement of the different plumes. This suggests that molecular topology might be more useful than the other methods in predicting relative plume movements. There appears to be increased retardation with time, which indicates that nonequilibrium sorption may be occurring. Further research on physical properties of organic chemicals may resolve the inconsistencies in the published literature and make the use of K_{ow} values more reliable.

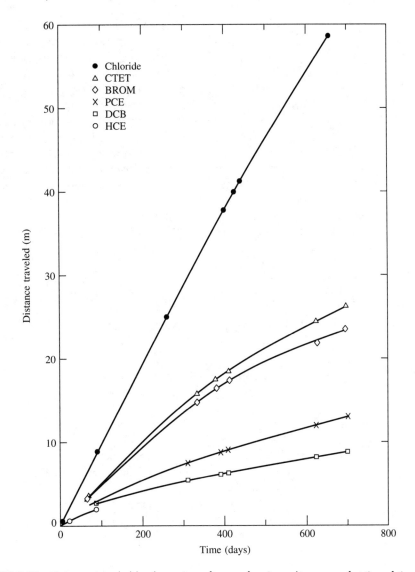

FIGURE 3.18 Distances traveled by the centers of mass of various plumes as a function of time since injection. *Source:* P. V. Roberts, M. N. Goltz, and D. M. Mackay, *Water Resources Research* 22, no. 13 (1986): 2047–59. Copyright by the American Geophysical Union.

3.11 SUMMARY

The advection-dispersion equation can be modified to reflect the effects of solutes that are removed from solution by sorption, chemical reaction, and biological and radioactive decay. Sorption may occur due to adsorption, chemisorption, ion exchange, and absorption. Sorption of inorganic solutes occurs primarily on mineral surfaces and is a function of the surface area available for sorption as well as the number of ion-exchange sites provided by clay and oxide minerals. The sorption of inorganic ions and organic chemical solutes can be quantified by experimentally derived adsorption isotherms.

Organic solutes may be sorbed on either soil mineral surfaces or by soil or aquifer organic matter. Nonpolar organic compounds are sorbed to a greater extent by soil organic matter than by mineral surfaces. The affinity of a particular organic molecule to be sorbed by soil or aquifer organic matter can be estimated from either the octanol-water partition coefficient, the aqueous solubility, or the molecular structure.

The advection-dispersion equation can also be modified to account for the disappearance of solutes or the appearance of a product due to other chemical reactions. Some of these reactions can be described by equilibrium reactions; others need to be dealt with on the basis of the kinetics of the reaction. A term to describe the disappearance of a solute due to radioactive decay can also be appended to the advection-dispersion equation. The biological degradation of organic compounds can occur under both aerobic and anaerobic conditions. Biodegradation terms can be joined to the advection-dispersion equation.

Chapter Notation

	$[A]$	Concentration of ion in solution
	B_d	Bulk density of soil
	b	Microbial decay rate
	C	Concentration of solute in liquid phase
	C^*	Amount of solute sorbed per unit weight of soil
	C_i	Equilibrium molar concentration of the ith component
	C_{im}	Solute concentration in immobile phase
	C_m	Solute concentration in mobile phase
	C_m^*	Absorbed concentration in contact with mobile phase
	C_{im}^*	Absorbed concentration in contact with immobile phase
	C_{oc}	Natural organic carbon concentration
	D_L	Longitudinal dispersion coefficient
	D_m	Apparent diffusion constant for mobile phase
	F_i	x-direction mass flux of a solute
	F^T	Total mass flux of chemical entity
	f	Fraction of solid surfaces in contact with mobile phase
	f_{oc}	Fraction of soil that consists of organic carbon
	f_{oc}^*	Critical level of soil organic carbon
	G	Ratio of oxygen to hydrocarbon consumed
	H	Hydrocarbon concentration in pore fluid
	H_{min}	Minimum hydrocarbon concentration to support microbial growth
	h_u	Maximum hydrocarbon utilization rate by aerobic microbes
	h_{ua}	Maximum hydrocarbon utilization rate by anaerobic microbes
	K	Coefficient in Freundlich sorption isotherm
	K_a	Half-maximum rate concentration for hydrocarbon for anaerobic decay

K_d	Distribution coefficient
K_h	Hydrocarbon half-saturation ratio
K_o	Oxygen half-saturation ratio
K_{oc}	Distribution coefficient for soil organic carbon
K_{om}	Distribution coefficient for soil organic matter
K_{ow}	Octanol-water partition coefficient
k_c	First-order decay rate for natural organic carbon
k_i	Kinetic rate constants ($i = 1$ to 8)
L	Liter
m^T	Global mass of chemical entity
M_a	Total mass of anaerobic microbes
M_s	Concentration of aerobic microbes in solution
M_t	Total aerobic microbial mass
N	Coefficient in Freundlich sorption isotherm
O	Oxygen concentration in pore fluid
P_f	Polarity correction factor
Q	x-direction flux of water
R_A	Reaction rate for disappearance of A
r_f	Retardation factor for linear sorption isotherm
r_{ff}	Retardation factor for Freundlich sorption isotherm
r_{fl}	Retardation factor for Langmuir sorption isotherm
r_h	Retardation factor for hydrocarbon
r_m	Retardation factor for microbes
S	Solubility of chemical in water
S_a	Surface area of soil
S_i	Water solubility of ith compound
S^T	Source/sink term for chemical entity
t	Time
v_c	Average linear velocity of solute front
v_m	Ground-water velocity in mobile phase
v_x	Average linear ground water velocity
W	Linear operator defined in Equation 3.50
$(x_i)_{org}$	Mole fraction of ith compound in organic phase
Y	Microbial yield coefficient
α	Absorption constant for Langmuir sorption isotherm
β	Maximum solute sorption from Langmuir sorption isotherm
γ	First-order rate coefficient
τ	First-order mass-transfer coefficient
λ	Half-life of radionuclide
$\delta_{i,j}$	Number of adjacent nonhydrogen atoms in an organic molecule
$^1\chi$	First-order molecular connectivity index
θ	Volumetric moisture content
θ_m	Porosity occupied by mobile phase
θ_{im}	Porosity occupied by immobile phase
$\kappa, \kappa', \kappa''$	Reaction-rate constants

References

Anderson, M. P. 1979. Using models to simulate the movement of contaminates through groundwater flow systems. *CRC Critical Reviews in Environmental Control* 9, no. 2:97–156.

Banerjee, Sujit. 1984. Solubility of organic mixtures in water. *Environmental Science and Technology* 18, no. 8:587–91.

Briggs, G. G. 1981. Theoretical and experimental relationships between soil adsorption, octanol-water partition coefficients, water solubilities, bioconcentration factors, and the parachor. *Journal of Agriculture and Food Chemistry.* 29:1050–59.

Bornemisza, E., and R. Llanos. 1967. Sulfate movement, sorption and desorption in three Costa Rica soils. *Soil Science Society of America Proceedings* 31:356–60.

Borden, R. C., and P. B. Bedient. 1986. Transport of dissolved hydrocarbons influenced by oxygen-limited biodegradation, 1. Theoretical development. *Water Resources Research* 22, no. 13:1973–82.

Borden, R. C., P. B. Bedient, M. D. Lee, C. H. Ward, and J. T. Wilson. 1986. Transport of dissolved hydrocarbons influenced by oxygen-limited biodegradation, 2. Field applications. *Water Resources Research* 22, no. 13:1983–90.

Bouwer, E. J., and P. L. McCarty. 1984. Modeling of trace organics biotransformation in the subsurface. *Ground Water* 22, no. 4:433–40.

Bowman, B. T., and W. W. Sans. 1977. Adsorption of parathion, fenitrothion, methylparathion, aminoparathion and paraoxon by Na^+, Ca^{2+}, and Fe^{3+} montmorillonite suspensions. *Soil Science Society of America Journal* 41:514–19.

Ciccioli, P., W. T. Cooper, P. M. Hammer, and J. M. Hayes. 1980. Organic solute-mineral surface interactions: A new method for the determination of groundwater velocities. *Water Resources Research* 16, no. 1:217–23.

Chiou, C. T., L. J. Peters, and V. H. Freed. 1979. A physical concept of soil-water equilibria for nonionic organic compounds. *Science* 206, no. 16:831–32.

Chiou, C. T., P. E. Porter, and D. W. Schmedding. 1983. Partition equilibrium of nonionic organic compounds between soil organic matter and water. *Environmental Science and Technology* 17, no. 4:227–31.

Davidson, J. M., and R. K. Chang. 1972. Transport of picloram in relation to soil-physical conditions and pore-water velocity. *Soil Science Society of America Proceedings* 36:257–61.

Dzombak, D. A., and R. G. Luthy. 1984. Estimating adsorption of polycyclic aromatic hydrocarbons on soils. *Soil Science* 137, no. 5:292–308.

Ellgehausen, H., C. D'Hondt, and R. Fuerer. 1981. Reversed-phase chromatography as a general method for determining octan-1-ol/water partition coefficients. *Pesticide Science* 12:219–27.

Enfield, C. G. 1985. Chemical transport facilitated by multiphase flow systems. *Water Science and Technology* 17:1–12.

Enfield, C. G., and B. E. Bledsoe. 1975. Fate of wastewater phosphorus in soil. *Journal of Irrigation and Drainage Division, American Society of Civil Engineers* 101, no. IR3:145–55.

Enfield, C. G., and G. Bengtsson. 1988. Macromolecular transport of hydrophobic contaminants in aqueous environments. *Ground Water* 26, no. 1:64–70.

Faust, C. R., and J. W. Mercer. 1980. Ground-water modeling: recent developments. *Ground Water* 18, no. 6:569–77.

Fetter, C. W. 1977. Attenuation of waste water elutriated through glacial outwash. *Ground Water* 15, no. 5:365–71.

Fiskell, J. G. A., R. S. Mansell, H. M. Selim, and F. G. Martin 1979. Kinetic behavior of phosphate sorption by acid, sandy soil. *Journal of Environmental Quality* 8:579–84.

Freyberg, D. L. 1986. A natural gradient experiment on solute transport in a sand aquifer, 2. Spatial movements and the advection and dispersion of nonreactive tracers. *Water Resources Research* 22, no. 13:2031–47.

Griffin, R. A. 1991. Personal communication.

Griffin, R. A., and E. S. K. Chian. 1980. Attenuation of water-soluble polychlorinated biphenyls by earth materials. *Environmental Geology Note 86*, Illinois State Geological Survey.

Harvey, R. W., R. L. Smith, and L. George. 1984. Effect of organic contamination upon microbial distribution and heterotrophic uptake in a Cape Cod, Mass., aquifer. *Applied and Environmental Microbiology* 48:1197–1202.

Hassett, J. J., J. C. Means, W. L. Banwart, and S. G. Wood. 1980. *Sorption properties of sediments and energy-related pollutants.* U.S. Environmental Protection Agency, EPA-600/3-80-041.

Hassett, J. J., W. L. Banwart, and R. A. Griffin. 1983. Correlation of compound properties with sorption characteristics of nonpolar compounds by soils and sediments: Concepts and limitations. In *Environment and Solid Wastes: Characterization, Treatment and Disposal*, ed. C. W. Francis and S. I. Auerback, 161–178. Boston: Butterworth Publishers.

Hayward, S. B. 1984. Field monitoring of chrysotile asbestos in California waters. *Journal, American Water Works Association* 76, no. 3:66–73.

Hornsby, A. G., and J. M. Davidson. 1973. Solution and adsorbed fluometuron concentration distribution in a water-saturated soil: Experimental and predicted evaluation. *Soil Science Society of America Proceedings.* 37:823–28.

Hunt, J. R., N. Sitar, and K. S. Udell. 1988. Nonaqueous phase liquid transport and cleanup, 1. Analysis of mechanisms. *Water Resources Research* 24, no. 8:1247–58.

Jarrell, W. M. and M. D. Dawson. 1978. Sorption and availability of molybdenum in soils of western Oregon. *Soil Science Society of America Journal* 42:412–15.

Karickhoff, S. W., D. S. Brown, and T. A. Scott. 1979. Sorption of hydrophobic pollutants on natural sediments.

Water Research 13:241–48.

Karickhoff, S. W. 1981. Semi-empirical estimation of sorption of hydrophobic pollutants on natural sediments and soils. *Chemosphere* 10:833–46.

Karickhoff, S. W. 1984. Organic pollutant sorption in aquatic systems. *Journal of Hydraulic Engineering* 110, no. 6:707–35.

Kenaga, E. E., and C. A. I. Goring. 1980. "Relationship between water solubility, soil sorption, octanol-water partitioning and bioconcentration of chemicals in biota." In *3rd Aquatic Toxicology Symposium, Proceedings of the American Society of Testing and Materials*, 78–115. No. STP 707.

Kenaga, E. E. 1980. Predicted bioconcentration factors and soil sorption coefficients of pesticides and other chemicals. *Ecotoxicology and Environmental Safety* 4:26–38.

Keswick, B. H., D. S. Wang, and C. P. Gerba. 1982. The use of microorganisms as ground-water tracers: A review. *Ground Water* 20, no. 2:142–49.

Koch, R. 1983. Molecular connectivity index for assessing ecotoxicological behavior of organic compounds. *Toxicological and Environmental Chemistry* 6:87–96.

Leistra, M., and W. A. Dekkers. 1977. Some models for the adsorption kinetics of pesticides in soils. *Journal of Environmental Science and Health* B12:85–103.

Lyman, W. J. 1982. "Adsorption coefficient for soils and sediment." In *Handbook of Chemical Property Estimation Methods*, ed. W. J. Lyman et al., 4.1–4.33. New York: McGraw-Hill.

Mackay, D. M., D. L. Freyberg, P. V. Roberts, and J. A. Cherry. 1986. A natural gradient experiment on solute transport in a sand aquifer, 1. Approach and overview of plume movement. *Water Resources Research* 22, no. 13:2017–30.

Major, D. W., C. I. Mayfield, and J. F. Barker. 1988. Biotransformation of benzene by denitrification in aquifer sand. *Ground Water* 26, no. 1:8–14.

McCall, P. J., R. L. Swann, and D. A. Laskowski. 1983. Partition Models for Equilibrium Distribution of Chemicals in Environmental Compartments. In *Fate of Chemicals in the Environment*, ed. R. L. Swann and A. Eschenroder, 105–23. American Chemical Society.

McCarthy, J. F., and J. M. Zachara. 1989. Subsurface transport of contaminants. *Environmental Science and Technology* 23, no. 5:496–502.

McCarty, P. L., M. Reinhard, and B. E. Rittman. 1981. Trace organics in groundwater. *Environmental Science and Technology* 15, no. 1:40–51.

Means, J. C., S. G. Wood, J. J. Hassett, and W. L. Banwart. 1980. Sorption of polynuclear aromatic hydrocarbons by sediments and soils. *Environmental Science and Technology* 14, no. 12:1524–28.

Miller, C. T., and W. J. Weber, Jr. 1984. Modeling organic contamination partitioning in ground-water systems. *Ground Water* 22, no. 5:584–92.

Mingelgrin, U., and Z. Gerstl. 1983. Reevaluation of partitioning as a mechanism of nonionic chemicals adsorption in soils. *Journal of Environmental Quality* 12:1–11.

Munns, D. N., and R. L. Fox. 1976. The slow reactions that continue after phosphate adsorption: kinetics and equilibrium in tropical soils. *Soil Science Society of America Proceedings* 40:46–51.

Nielsen, D. R., M. Th. van Genuchten, and J. W. Biggar. 1986. Water flow and solute transport in the unsaturated zone. *Water Resources Research* 22, no. 9:89S–108S.

Nightingale, H. I., and W. C. Bianchi. 1977. Ground-water turbidity resulting from artificial recharge. *Ground Water* 15, no. 2:146–52.

Nkedi-Kizza, P., J. W. Biggar, H. M. Selim, M. Th. van Genuchten, P. J. Wierenga, J. M. Davidson, and D. R. Nielsen. 1984. On the equivalence of two conceptual models for describing ion exchange during transport through an aggregated oxisol. *Water Resources Research* 20, no. 8:1123–30.

Olsen, R. L., and A. Davis. 1990. Predicting the Fate and Transport of Organic Compounds in Groundwater: Part 1. *Hazardous Materials Control* 3, no. 3:38–64.

Prickett, T. A., T. C. Naymik, and C. G. Lonnquist. 1981. *A "random walk" solute transport model for selected groundwater quality evaluations*. Illinois State Water Survey, Bulletin 65, 103pp.

Rao, P. S. C., and J. M. Davidson. 1980. Estimation of pesticide retention and transformation parameters required in nonpoint source pollution models. In *Environmental Impact of Nonpoint Source Pollution*, ed. M. R. Overcash and J. M. Davidson, 23–67. Ann Arbor, Mich.: Ann Arbor Science Publishers, Inc.

Roberts, P. V., M. N. Goltz, and D. A. Mackay. 1986. A natural gradient experiment on solute transport in a sand aquifer, 3. Retardation estimates and mass balances for organic solutes. *Water Resources Research* 22, no. 13:2047–59.

Rogers, R. D., J. C. McFarlane, and A. J. Cross. 1980. Adsorption and desorption of benzene in two soils and montmorillonite clay. *Environmental Science and Technology* 14, no. 4:457–60.

Roy, W. R., and R. A. Griffin. 1985. Mobility of organic solvents in water-saturated soil materials. *Environmental Geology and Water Sciences* 7:241–47.

Rubin, Jacob. 1983. Transport of reacting solutes in porous media: Relationship between mathematical nature of problem formation and chemical nature of reactions. *Water Resources Research* 19, no. 5:1231–52.

Sabljić, A., and M. Protic. 1982. Relationship between molecular connectivity indices and soil sorption coefficients of polycyclic aromatic hydrocarbons. *Bulletin of Environmental Contamination and Toxicology* 28:162–65.

Sabljić, A. 1984. Predictions of the nature and strength of soil sorption of organic pollutants by molecular topology. *Journal of Agricultural and Food Chemistry* 32:243–46.

Sabljić, A. 1987. On the prediction of soil sorption coefficients of organic pollutants from molecular structure: Application of molecular topology model. *Environmental Science and Technology* 21, no. 4:358–65.

Schwarzenbach, R. P., and J. Westall. 1981. Transport of nonpolar organic compounds from surface water to groundwater: Laboratory sorption studies. *Environmental Science and Technology* 15, no. 11:1360–67.

Sidle, R. C., L. T. Kardos, and M. Th. van Genuchten. 1977. Heavy metal transport model in a sludge-treated soil. *Journal of Environmental Quality* 6:438–43.

Srinivasan, P., and J. W. Mercer. 1988. Simulation of biodegradation and sorption processes in ground water. *Ground Water* 26, no. 4:475–87.

Staples, C. A., and S. J. Geiselmann. 1988. Cosolvent influences on organic solute retardation factors. *Ground Water* 26, no. 2:192–98.

Street, J. J., W. L. Lindsay, and B. R. Sabey. 1977. Solubility and plant uptake of cadmium in soils amended with cadmium and sewage sludge. *Journal of Environmental Quality* 6:72–77.

Travis, C. C., and E. L. Etnier. 1981. A survey of sorption relationships for reactive solutes in soil. *Journal of Environmental Quality* 10, no. 1:8–17.

Van Bladel, R. and A. Moreale. 1977. Adsorption of herbicide-derived *p*-chloroaniline residues in soils: A predictive equation. *Journal of Soil Science* 28:93–102.

Van Genuchten, M. Th., J. M. Davidson, and P. J. Wierenga. 1974. An evaluation of kinetic and equilibrium equations for the prediction of pesticide movement through porous media. *Soil Science Society of America Proceedings* 38:29–35.

Van Genuchten, M. Th. and P. J. Wierenga. 1976. Mass-transfer studies in sorbing porous media, 1. Analytical solutions. *Soil Science Society of America Journal* 40:473–80.

Walsh, M. P., S. L. Bryant, R. S. Schechter, and L. W. Lake. 1984. Precipitation and dissolution of solids attending flow through porous media. *American Institute of Chemical Engineers Journal* 30, no. 2:317–28.

Wood, W. W., T. P. Kramer, and P. P. Hern, Jr. 1990. Intergranular diffusion: An important mechanism influencing solute transport in clastic aquifers. *Science* 247 (March 30):1569–72.

Yaron, B. 1978. Some aspects of surface interactions of clays with organophosphorous pesticides. *Soil Science* 125:210–16.

Chapter Four
Flow and Mass Transport in the Vadose Zone

4.1 Introduction

Flow through the vadose zone is a topic that most hydrogeology texts tend to cover in a cursory manner. Classical hydrogeology is concerned primarily with obtaining water from wells. The **vadose,** or water-unsaturated zone, is seen as a somewhat mysterious realm through which recharge water must pass on the way to the water table—a watery purgatory. Soil scientists have been the primary force behind developing an understanding of unsaturated zone flow. Historically, soil scientists were concerned with such topics as the passage of water and solutes to roots of plants, water that flows primarily in the vadose zone. More recently soil scientists have studied the transport and fate of contaminants in the vadose zone. With the development of the science of contaminant hydrogeology, hydrogeologists have become much more interested in the mysteries of the vadose zone. Many releases of contaminants to the subsurface occur within or above the vadose zone. Contaminants are understood to include materials applied deliberately to the soil, such as fertilizers and pesticides, as well as those released accidentally. The hydrogeologist is suddenly faced with the daunting task of understanding the transport of dissolved contaminants through the vadose zone. To this end, much can be learned from the work of soil scientists. Transport in the vadose zone may also occur by flow of a pure nonaqueous phase liquid or gaseous phase.

The vadose zone extends from the land surface to the water table. It includes the capillary fringe, where pores may actually be saturated. The main distinguishing feature of the vadose zone is that the pore water pressures are generally negative. Areas of the vadose zone above the capillary fringe may temporarily be saturated due to surface ponding of water or because of the development of perched water tables above relatively low permeability soil layers.

4.2 Soil as a Porous Medium

The vadose zone includes the soil layers at the Earth's surface. It may also include sediment and/or consolidated rock. However, the presence of soil complicates the study of vadose zone hydrology, also known as soil physics.

Soil is a complex material. In physical form it consists mostly of mineral grains of varying size as well as varying amounts of organic matter. The mineral grains are arranged in such a fashion that the soil has structure; that is, there is a specific orientation and arrangement of the individual grains. The individual grains usually form larger units called **aggregates,** or **peds,** which are bound by organic matter (e.g., Hillel 1980). The porosity and permeability of the soil is a function of both soil texture and the soil structure. The structure is a function of the physical shape and size of the aggregates. Moreover, it may be vastly influenced by the soil chemistry, since soil minerals have an electrical charge on their surface. This surface charge, which is primarily due to the clay minerals, affects the stability of soil structural units. Soil contains mineral matter, organic matter, water containing dissolved solutes, and gases. The soil also has macropores, such as root casts and wormholes, and drying cracks in fine-textured soils. These form preferential channels for water movement.

The amount of moisture in a soil can be expressed as the gravimetric water content, w, which is the weight of the water as a ratio to the weight of the dry soil mass. The moisture state can also be expressed as the volumetric water content, θ, which is the volume of water as a ratio to the total volume of the soil mass. One must be careful in measuring volumetric water content, since in many soils (especially those with fine texture) the volume changes as water is imbibed or drained. This is due to the interactions between the charged soil particles and the polar water molecules.

4.3 Soil Colloids

The clay fraction of the soil consists of mineral particles that are less than 2 μm in diameter (Hillel 1980). Clay particles consist primarily of secondary minerals that have been formed by weathering. Clay minerals have an unbalanced negative electrical charge at the surface. Electrostatic attraction exists between the surface of the clay particles, the polar soil-water molecules, and solutes dissolved in the soil water. Fine-grained materials with an electrostatic surface charge are called **colloids.**

Clay minerals have a definite crystal structure, consisting primarily of aluminum, silica, and oxygen. **Kaolinite** (e.g., $Al_4Si_4O_{10}(OH)_8$) is a clay mineral with a low specific surface (surface area per unit mass); it ranges from 5 to 20 m^2/g. The low specific surface means that kaolinite is not particularly reactive. **Illite** (e.g., $Al_4Si_7AlO_{20}(OH)_4K_{0.8}$, with the potassium occurring between layers) has a larger specific surface area, ranging from 80 to 120 m^2/g. **Montmorillonite** (e.g., $Al_{3.5}Mg_{0.5}Si_8O_{20}(OH)_4$) is the most reactive clay with a specific surface area of 700 to 800 m^2/g. The reactive clays can absorb large amounts of water and ions between their sheetlike mineral grains. This property gives soils high in reactive clays the ability to swell as water is absorbed. It also means that they shrink and crack when dried. Montmorillonite has the largest shrink-swell behavior, and kaolinite has the least. **Chlorite** (e.g., $Mg_6Si_6Al_2O_{20}(OH)_4$, with $Mg_6(OH)_{12}$ occurring between the layers) is another common clay mineral with a behavior similar to illite.

Clay-size particles may also include **sesquioxides,** which are hydrated aluminum and iron oxides ($Al_2O_3 \cdot n\,H_2O$ and $Fe_2O_3 \cdot n\,H_2O$). Limonite, goethite, and gibbsite are examples. These substances are generally amorphous and have less electrostatic properties than the silicate clay minerals. Sesquioxides often act as cementing agents for soils.

Soil may also contain decomposed organic material, which is sometimes called **humus.** Humus is a complex mixture of organic molecules that are aggregated into colloidal-size particles. Humus particles are also negatively charged. Humus is found primarily in the A horizon of soils. The humus content of mineral soils can range from 0 to about 10% by weight. The organic content of organic soils, such as peat, can be up to 50% or more by weight.

4.4 The Electrostatic Double Layer

The colloidal particles of the soil have an unbalanced, negative surface charge. This negative charge is balanced by positively charged cations that are attracted to the surface of the colloid. These cations exist as solutes in water. When the colloid is dry, the layer of water held to the surface will be thin, and the neutralizing cations will be closely held to the particle surface. As the colloid becomes more hydrated, the cations will dissociate from the surface and form a swarm of ions near the negatively charged surface layer. The cations have a more or less fixed position near the negatively charged particle surface. The particle surface and the cation swarm form what is known as an **electrostatic double layer.**

The negatively charged particle surface will tend to repel anions. Hence, the region near the particle surface will have an abundance of cations and relatively few anions. Figure 4.1 shows the distribution of monovalent cations and anions near the surface of a montmorillonite particle. The effects of the electrostatic double layer extend to the distance at which the number of cations in solution equals the number of anions. Figure 4.1 shows that the electrostatic double-layer effect is a function of the solute concentration and extends farther in more dilute solutions. The extent of the electrostatic double layer is less for divalent cations as opposed to monovalent cations. The thickness of the electrostatic double layer can be computed from:

$$z_0 = \frac{1}{eV}\sqrt{\frac{\epsilon k_B T}{8\pi n_0}} \tag{4.1}$$

where

z_0 = the characteristic thickness of the double layer

e = the elementary charge of an ion, 4.77×10^{-10} esu

ϵ = the dielectric constant

k_B = the Boltzmann constant (the ratio of the gas constant to Avagadro's number)

V = the valence of the ions in solution

n_0 = the concentration of ions in bulk solution (ions/cm^3)

T = temperature in Kelvins

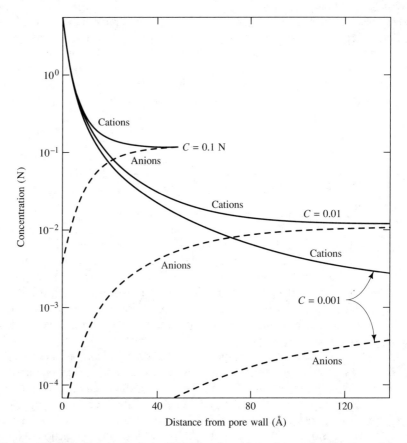

FIGURE 4.1 Distribution of monovalent cations near the surface of a montmorillonite particle. *Source:* L. Boersma et al., "Theoretical analysis." In D. R. Nielsen et al., *Soil Water*, 21–63. Madison, Wis.: American Society of Agronomy (1972).

The electrostatic double layer is very important in contaminant hydrogeology, because the cations associated with this layer can be replaced by other cations in solution. This replacement process is known as **cation exchange.** The total number of positive charges that can be exchanged in a soil is independent of the cation species and is expressed as the cation-exchange capacity. Cation exchange affects the transport of ions in solution. The type and concentration of cations involved in the exchange process can also affect the hydraulic conductivity of the soil.

Some colloids can also attract anions. Although the surface of clay particles carries a negative charge, the edges usually carry a net positive charge. Since the surface area far exceeds the edge area, the cation-exchange capacity for most soils far exceeds the anion-exchange capacity. Kaolinite and humus have the greater anion-exchange capacity.

Different cations are held with greater tenacity by the colloids. The larger the ionic radius and the greater the valence charge, the more tightly the cation is held. The ionic radius of the cation is affected by hydration, because polar water molecules are attracted to the ion. It is the radius of the hydrated ion that is important. The normal order of preference for cation exchange is

$$Al^{+3} \gg Ca^{+2} > Mg^{+2} \gg NH_4^+ > K^+ > H_3O^+ > Na^+ > Li^+$$

However, if a soil is flooded with a solution containing a large concentration of one cation, the normal cation-exchange order can be reversed.

4.5 Salinity Effects on Hydraulic Conductivity of Soils

The hydraulic conductivity of a soil can be affected by the strength and type of cations contained in the soil water (Nielsen, van Genuchten, and Biggar 1986). The impact of the solute increases with the amount of colloidal particles in the soil. Soil swelling caused by increased salinity can reduce hydraulic conductivity. As the electrostatic double layer grows thicker, the hydraulic conductivity decreases, because clay minerals tend to swell and expand into the pore space. Sodium is especially important in this process. The electrostatic double layer is thicker when it contains the monovalent sodium ions, and as such sodium tends to weaken the bonds between clay particles. The effect of swelling is reversible if the saline water is flushed from the pores. However, if smaller particles break loose from the soil structure, they can be transported by flowing water until they are carried into small pore throats, where they can lodge. This causes a more or less irreversible reduction in hydraulic conductivity (Dane and Klute 1977).

Examination of Equation 4.1 shows that the electrostatic double layer grows with decreasing concentration of cations in the soil water. It will also decrease with an increase in the ratio of the concentration of monovalent to divalent cations in the soil water. The principal cations in most natural waters are sodium, calcium, and magnesium. The sodium adsorption ratio (SAR) is a measure of the ratio of the concentration of monovalent sodium to divalent calcium and magnesium:

$$SAR = \frac{Na}{\sqrt{\frac{Ca + Mg}{2}}} \tag{4.2}$$

where

Na = concentration of sodium in milliequivalents per liter

Ca = concentration of calcium in milliequivalents per liter

Mg = concentration of magnesium in milliequivalents per liter

The greater the SAR of the soil water and the lower the total solute concentration, the lower the hydraulic conductivity of the soil, especially if it contains the expansive clays illite and montmorillonite. Figure 4.2 shows the effect for one soil of decreasing solute concentration on the unsaturated hydraulic conductivity of a soil. During these tests the SAR was kept at 40, so that the relative proportions of sodium, calcium, and magnesium did not change. An increase in the pH of soil water has also been demonstrated to result in a reduction of hydraulic conductivity for some soils (Suarez 1985).

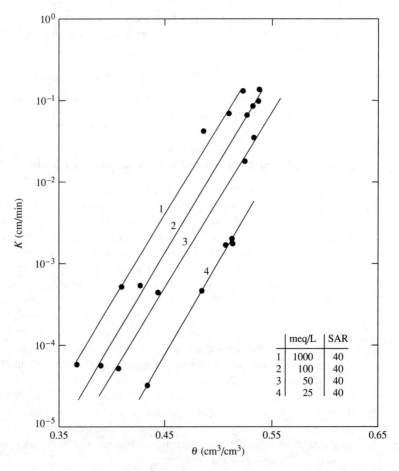

FIGURE 4.2 Hydraulic conductivity as a function of volumetric water content for solutions of varying solute concentration but with a constant SAR value of 40. *Source:* J. H. Dane and A. Klute, *Soil Science Society of America Journal* 41, no. 6 (1977): 1043–57.

4.6 Flow of Water in the Unsaturated Zone

Vadose zone hydrology is different from saturated zone hydrology because of the presence of air in the pore space. The relative proportion of air and water in the pores can vary, and with it can vary the hydraulic properties of the porous media.

4.6.1 Soil-Water Potential

In saturated flow the driving potential for groundwater flow is due to the pore-water pressure and elevation above a reference datum (Fetter 1988). However, in unsaturated flow the pore water is under a negative pressure caused by surface tension. Soil physicists call this the **capillary potential,** or **matric potential,** ψ; it is a function of the volumetric water content of the soil, θ. The lower the water content, the lower—i.e., more negative—the value of the matric potential.

The total soil-moisture potential, ϕ, is the sum of the matric potential, ψ, a pressure potential, the gravitational potential, Z, an osmotic potential, and an electrochemical potential. However, we will assume that the osmotic potential and the electrochemical potential do not vary within the soil and that the pressure is equal to atmospheric. Since we will eventually want to find the gradient of the potential, we can neglect osmotic pressure and electrochemical potentials, because their gradient will be zero. Total soil moisture potential is therefore reduced to the sum of the matric and gravitational potentials:

$$\phi = \psi(\theta) + Z \qquad (4.3)$$

Matric potential may be measured as a capillary pressure, P_c, which has the units of newtons per square meter, which are equivalent to joules per cubic meter or energy per unit volume ($LM^{-1}T^{-2}$). If the matric potential is measured on a pressure basis, then the gravitational potential, Z, is equal to $\rho_w g z$, where g is the acceleration of gravity, ρ_w is the density of water, and z is the elevation above a reference plane. The total soil moisture potential in terms of energy per unit volume can thus be found from

$$\phi_{EV} = P_c + \rho_w g z \qquad (4.4)$$

If Equation 4.4 is divided by $\rho_w g$, the result is the soil moisture potential expressed as energy per unit weight, which also has units of length (L). This is equivalent to head in saturated flow. The matric potential is also expressed in units of length, typically centimeters of water:

$$\phi_{EW} = \frac{P_c}{\rho_w g} + z = h + z \qquad (4.5)$$

where h, pressure potential, is the matric potential in units of length.

Dividing Equation 4.4 by ρ_w gives the soil-moisture potential expressed as energy per unit mass, with units of joules per kilogram (L^2T^{-2}):

$$\phi_{EM} = \frac{P_c}{\rho_w} + gz \qquad (4.6)$$

Common units for total potential and pressure potential include atmospheres of pressure and centimeters of water. One atmosphere is equivalent to about 1000 cm of water. Also, 10^5 pascals of pressure is equal to about 1 atmosphere (atm).

4.6.2 Soil-Water Characteristic Curves

The relationship between matric potential or pressure head and volumetric water content for a particular soil is known as a **soil-water characteristic curve,** or a soil-water retention curve. Figure 4.3 shows an idealized soil-water characteristic curve.

At atmospheric pressure the soil is saturated, with the water content equal to θ_s. The soil will remain saturated as the matric potential is gradually decreased. Eventually, the matric potential will become negative enough that water can begin to drain from the soil. This matric potential is known as the **bubbling pressure.** It is marked h_b on

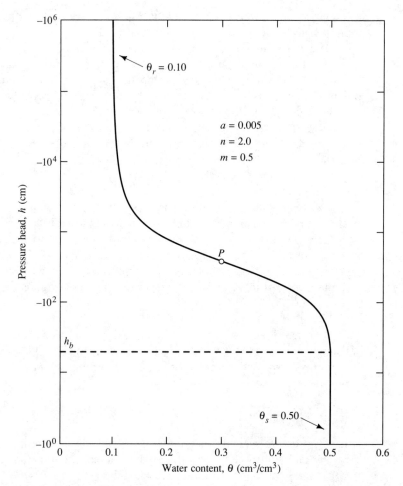

FIGURE 4.3 Typical soil-water retention curve. *Source:* M. Th. van Genuchten, *Soil Science Society of America Journal* 44 (1980):892–98.

Figure 4.3. The moisture content will continue to decline as the matric pressure is lowered, until it reaches some irreducible minimum water content, θ_r. Should the matric potential be further reduced, the soil would not lose any additional moisture.

The soil-moisture characteristic curve also shows the pore-size distribution of the soil. Figure 4.4 shows idealized soil-moisture characteristic curves for two soils, one of which is well sorted and one of which is poorly sorted.

The well-sorted soil has a narrower range of matric potential over which the water content changes than the poorly sorted soil. In a well-sorted soil, most of the grains are in a narrow size range, and hence the pores also have a narrow size range. The poorly sorted soil has a wider size range for both grains and pores. The well-sorted soil has a higher bubbling pressure because it has larger pores. However, once the well-sorted soil begins to desaturate, it does so rapidly, again because most of the pores are large.

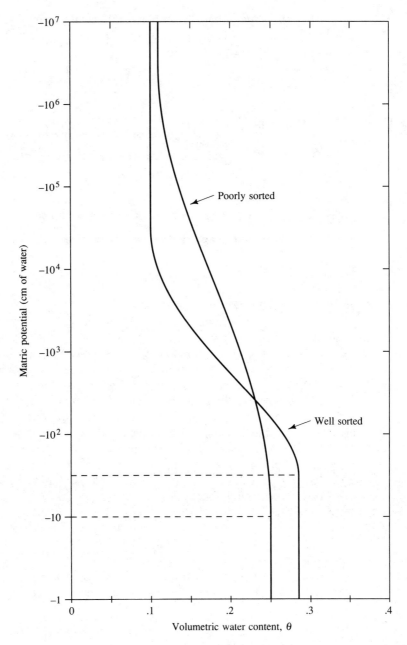

FIGURE 4.4 Typical soil-water retention curves showing the effect of grain-size sorting. The shape of the curves reflects the distribution of pore sizes in the soil.

There are some simple empirical expressions that can be used to relate the water content of a soil to the matric potential. Brooks and Corey (1966) used the relationship

$$\theta = \theta_r + (\theta_s - \theta_r)\left(\frac{\psi}{h_b}\right)^{-\lambda} \qquad (4.7)$$

where

θ = volumetric water content
θ_s = volumetric water content at saturation
θ_r = irreducible minimum water content
ψ = matric potential
h_b = bubbling pressure
λ = experimentally derived parameter

Brooks and Corey (1966) also defined an **effective saturation**, S_e, as

$$S_e = \left(\frac{S_w - \theta_r}{1 - \theta_r}\right) \qquad (4.8)$$

where S_w = the saturation ratio, θ/θ_s.

A graph of capillary pressure divided by the specific weight (P_c/γ) of the fluid versus the effective saturation (S_e) is shown for water in four different porous media in Figure 4.5(a). Figure 4.5(b) shows P_c/γ plotted versus S_e on log-log paper for the same data sets. Note that on Figure 4.5(b) the data sets plot mainly as straight lines, except close to the point where S_e is equal to 1.0. (At $S_e = 1.0$, the saturation ratio, S_w, is equal to 1.0). The negative slope of the line is called λ, the Brooks-Corey pore-size distribution index, and is one of two constants that characterize the media. The other constant is the intercept of the extension of the straight line with the $S_e = 1.0$ axis. This constant h_b, the bubbling pressure, has already been defined. The capillary behavior of a porous medium can thus be defined on the basis of these two constants, λ and h_b. In Figure 4.5(b) curve 1 for volcanic sand has a λ value of 2.29 and an h_b of about 15 cm.

The volcanic sand, fine sand, and glass beads shown in Figure 4.5 have a fairly narrow grain-size distribution, which results in a narrow range of pore-size distribution. They are not representative of most natural soils. The calculated λ values for these soils are also higher than one finds in most natural soils. The Touchet silt loam is more representative of a normal soil than the other three.

Van Genuchten (1980) also derived an empirical relationship between matric potential and volumetric water content. He defined the relationship by the expression

$$\theta = \theta_r + \frac{\theta_s - \theta_r}{[1 + (\alpha\psi)^n]^m} \qquad (4.9)$$

$$n = \frac{1}{1-m} \qquad (4.10)$$

$$\alpha = \frac{1}{h_b}(2^{1/m} - 1)^{1-m} \qquad (4.11)$$

where m is a parameter estimated from the soil-water retention curve.

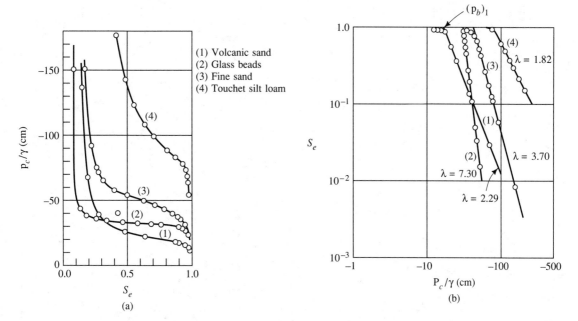

FIGURE 4.5 Capillary pressure head as a function of effective saturation for porous materials with various pore sizes. (a) Plotted on arithmetic paper and (b) plotted on log-log paper. *Source:* R. H. Brooks and A. T. Corey, Proceedings, American Society of Civil Engineers, Irrigation and Drainage Division 92, no. IR2 (1966):61–87.

To find the van Genuchten soil parameters, a soil-water retention curve ranging from a matric potential of 0 to a matric potential of $-15{,}000$ cm is constructed. The value of θ_s is found at a matric potential of 0 and the value of θ_r is that corresponding to a matric potential of $-15{,}000$ cm. Figure 4.3 shows such a plot. The point P on the curve corresponds to a water content θ_p, which is found from

$$\theta_p = \frac{\theta_s + \theta_r}{2} \qquad (4.12)$$

The slope, S, of the line at point P is determined graphically from the experimental soil-water retention curve. A dimensionless slope, S_p, is then found from the relationship:

$$S_p = \frac{S}{\theta_s - \theta_r} \qquad (4.13)$$

The parameter m can then be determined from the value of S_p using one of these formulas:

$$m = \begin{cases} 1 - \exp(-0.8 S_p) & (0 < S_p \leq 1) \\ 1 - \dfrac{0.5755}{S_p} + \dfrac{0.1}{S_p^{\,2}} + \dfrac{0.025}{S_p^{\,3}} & (S_p > 1) \end{cases} \qquad (4.14)$$

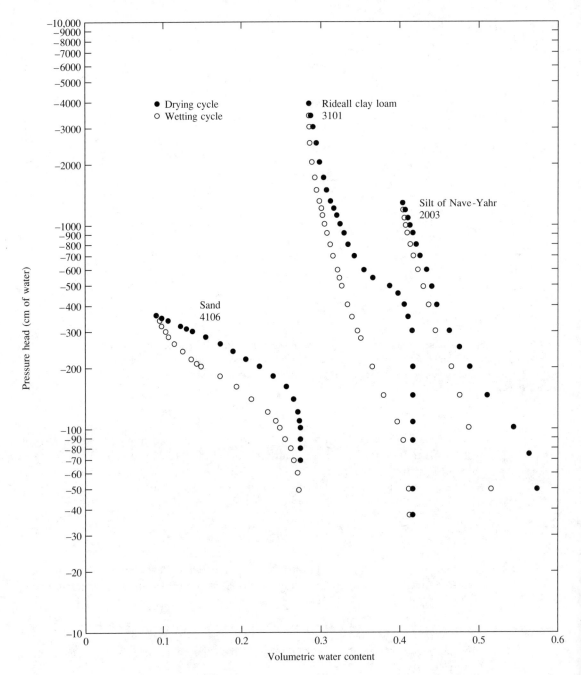

FIGURE 4.6 Soil-moisture-retention curves for three soils for both drying and wetting cycles. *Source:* Yechezkel Mualem, Catalogue of Hydraulic Properties of Unsaturated Soils (Haifa, Israel: Technion, 1976).

We can find the values of m and α from Equations 4.14 and 4.11, respectively, by using the bubbling pressure obtained from the soil-water retention curve.

From Figure 4.3 the slope of the curve at P is about 0.34. The dimensionless slope determined from Equation 4.13 is about 0.85. From Equation 4.14 m is determined to be about 0.5 and from Equation 4.10, n is 2.0. To estimate α from Equation 4.11, we need the value of h_b, the bubbling pressure. From Figure 4.3 the value of h_b is about -355, so $\log h_b \sim 2.55$ and $\alpha \sim 0.005$.

4.6.3 Hysteresis

If one constructs a soil-water retention curve by obtaining data from a sample that is initially saturated and then applying suction to desorb water, the curve is known as a **drying curve.** If the sample is then resaturated by decreasing the suction, it will follow a **wetting curve.** Figure 4.6 shows drying and wetting curves for three different soils. Typically, the drying curve and the wetting curve will not be the same. This phenomenon is called **hysteresis.** The causes of hysteresis include (Hillel 1980)

1. The geometric effects of the shape of single pores, which give rise to the so-called ink bottle effect. This effect is illustrated in Figure 4.7. The pore has a throat radius of r and a maximum radius of R. The matric potential when the air-water interface is at the pore throat, ψ_r, is equal to $2\sigma/r$, where σ is the interfacial tension between the pore water and the mineral surface (see Section 5.2.3). The pore will drain abruptly when ψ has a more negative pressure than ψ_r. The pore cannot then rewet until ψ falls below ψ_r. Since $R > r$, then $\psi_r > \psi_R$; that is, it takes a lower matric potential (more negative pressure) to drain a pore than to fill it.
2. The contact angle between the water and the mineral surface is greater during the period when a water front is advancing as opposed to when it is retreating. The advancing meniscus that forms during wetting will have a greater radius of curvature—and hence a lower matric potential—than that exhibited by a meniscus that forms during a drying cycle.

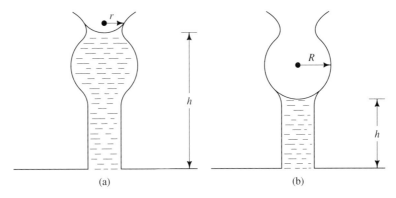

FIGURE 4.7 Pore geometry affects equilibrium height of capillary water during (a) drainage and (b) wetting.

3. Air that is trapped in pores during a wetting cycle will reduce the water content of soil as it is being wetted. Eventually that trapped air will dissolve.

The hysteresis effect may be augmented by the shrinking and swelling of the clays as the soil wets and dries and may also be affected by the rates of wetting and drying (Davidson, Nielsen, and Biggar 1966).

If the soil is not dried to the maximum extent possible (greatest negative pressure head), when it is rewet, the soil will follow an intermediate curve known as a wetting (or drying) scanning curve. There are many wetting and drying scanning curves, depending upon the point on the main wetting or drying curve where the scanning curve starts.

4.6.4 Construction of a Soil-Water-Retention Curve

Laboratory measurements of matric potential as a function of water content are made to construct a soil-water-retention curve. In the wet-soil range (-1 to -300 cm of water-pressure head), a tension plate assembly is used. A saturated soil sample of known water content is placed on a porous plate in a Buchner funnel. The porous plate is saturated and connected to a water column that ends in a burette (Figure 4.8). The position of the burette can be changed to decrease the pressure head. As the pressure head becomes more negative, water is drained from the soil sample and the amount is measured in the burette once equilibrium has been reached. Care must be taken to avoid evaporation

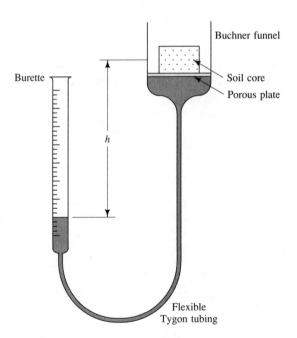

FIGURE 4.8 Equilibrating the water content of a soil sample with a known matric potential using a tension plate assembly.

Flow and Mass Transport in the Vadose Zone

of water from the soil sample and the burette. A number of measurements are made at progressively more negative pressure heads to determine the drying curve. Once the practical limit of the tension plate assembly is reached (-300 cm), the burette is raised in a number of steps to construct a wetting curve. This is a wetting scanning curve, as the soil is not fully drained at a pressure head of -300 cm of water.

For soils in the dry range (-300 to $-15{,}000$ cm of water), a pressure-plate assembly is used. The soil samples are placed on a saturated porous plate that is in a pressure chamber. The pressure below the porous plate is kept at atmospheric pressure and the pressure above the porous plate can be set between 0.3 atm (300 cm of water) and 15 atm (15,000 cm of water). The pressure across the soil sample and porous plate causes water to flow from the soil sample across the porous plate into a lower reservoir.

4.6.5 Measurement of Soil-Water Potential

Matric potential is measured in the field with a tensiometer. This apparatus consists of a porous ceramic cup attached to a tube, which is buried in the soil. The tube is filled with water and attached to a device such as a vacuum gauge, manometer, or pressure transducer, which can measure the tension. The matric potential of the soil tries to draw water from the water-filled porous cup, and the resulting tension is measured. Tensiometers can measure soil moisture tensions up to about 800 cm.

Figure 4.10 shows the operation of two tensiometers in determining the gradient of the matric potential. Tensiometer A has the porous ceramic cup at a depth of 100 cm,

---

EXAMPLE PROBLEM

The following data have been obtained for a soil sample using a tension-plate assembly. Construct a water-retention curve.

Pressure head (cm)	Volumetric Water Content	
	Wetting Cycle	Drying Cycle
0	0.447	0.448
-30	0.431	0.448
-60	0.411	0.443
-90	0.400	0.437
-120	0.392	0.424
-150	0.385	0.408
-180	0.379	0.391
-210	0.377	0.377
Wetting Scanning Curve		
-120	0.411	
-90	0.418	
-60	0.423	
-30	0.436	
0	0.477	

The data are plotted in Figure 4.9.

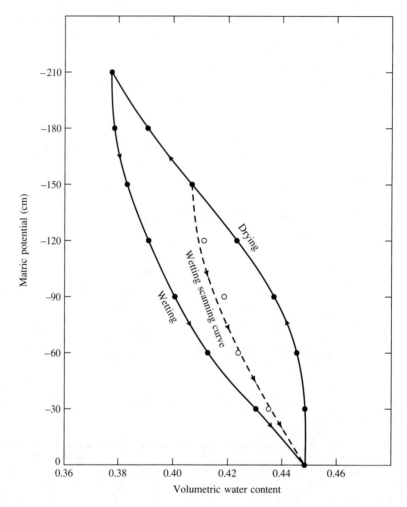

FIGURE 4.9 Matric potential as a function of volumetric water content for the example problem.

whereas tensiometer B is at a depth of 30 cm. The total potential measured for tensiometer A is −126 cm, whereas for tensiometer B it is −88 cm. Since total potential is the sum of pressure head and elevation head, one must subtract the elevation head from total head to get pressure head. Thus the pressure head measured in tensiometer A is −26 cm, whereas for tensiometer B it is −58 cm. However, the total potential gradient is downward, because the total head at A is more negative than that at B. The gradient is computed by finding the difference between the two total heads and dividing the difference by the distance between the two porous cups. In Figure 4.10 the gradient is

$$\frac{-126 - (-88)}{-70} = 0.54 \quad \text{(downward)}$$

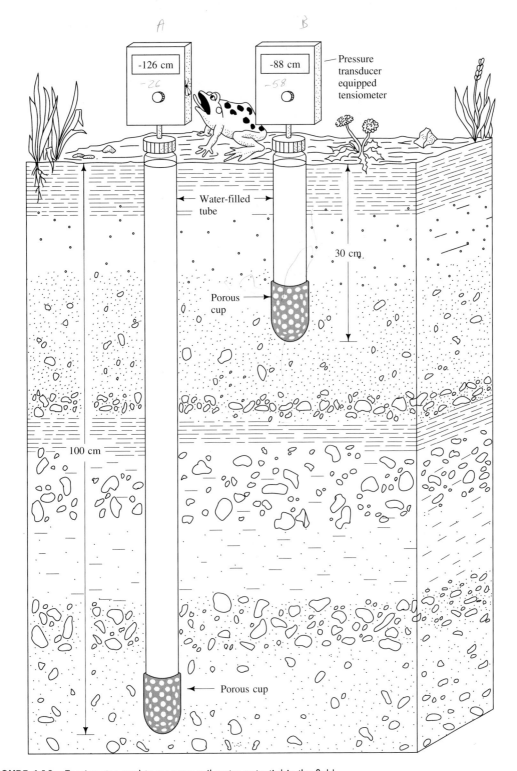

FIGURE 4.10 Tensiometer used to measure soil-water potential in the field.

4.6.6 Unsaturated Hydraulic Conductivity

When a rock or sediment is saturated, all the pores are filled with water, and most of them transmit water. Only "dead-end" pores do not participate in the transmittal of water. Unsaturated soils have a lower hydraulic conductivity because some of the pore space is filled with air and thus can't transmit water. Soil moisture in the vadose zone travels through only the wetted cross section of pore space. As a saturated soil drains, the larger pores empty first, especially in soils that are structured. Because these have the greatest pore-level hydraulic conductivity, there is an immediate large drop in the ability of the soil to transmit water. The unsaturated hydraulic conductivity is a function of the water content of the soil: $K = K(\theta)$. Unsaturated hydraulic conductivity can also be considered to be a function of the matric potential: $K = K(\psi)$. Figure 4.11 is a graph of unsaturated hydraulic conductivity as a function of matric potential. It can be seen that this relationship also exhibits hysteresis.

The flow of soil moisture is influenced by temperature. Figure 4.12 shows the influence of water temperature on curves of unsaturated hydraulic conductivity versus

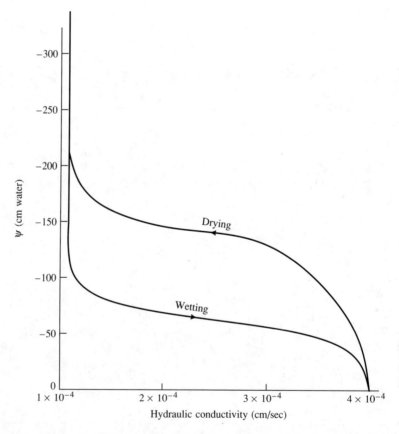

FIGURE 4.11 Relationship of hydraulic conductivity to matric potential for a wetting and drying cycle illustrating hysteresis.

Flow and Mass Transport in the Vadose Zone

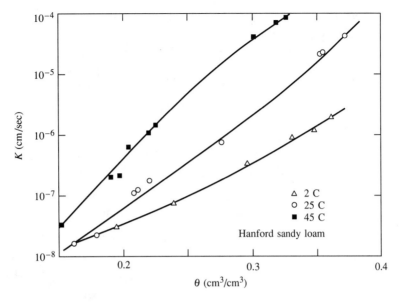

FIGURE 4.12 Unsaturated hydraulic conductivity as a function of water content for three temperatures. *Source:* J. Constantz, *Soil Science Society of America Journal* 46, no. 3 (1982):466–70.

water content. A change from 2°C to 25°C can cause unsaturated hydraulic conductivity to increase by as much as an order of magnitude. Constantz (1982) wrote the following expression for the relationship of unsaturated hydraulic conductivity to intrinsic permeability of the soil:

$$K(\theta) = \frac{k_r(\theta) k \rho_w g}{\mu_w} \quad (4.15)$$

where

$k_r(\theta)$ = the relative conductivity, which is a number from 0 to 1.0 that is the ratio of the unsaturated hydraulic conductivity at a given θ to the saturated hydraulic conductivity

k = the intrinsic permeability

ρ_w = the density of water at a given temperature

g = the acceleration of gravity

μ_w = the dynamic viscosity of soil water at a given temperature

Constantz (1982) reports that the effect of temperature on unsaturated hydraulic conductivity is primarily a function of the effect of temperature on dynamic viscosity.

Unsaturated hydraulic conductivity can be determined by both field methods (Green, Ahuja, and Chong 1986) and laboratory techniques (Klute and Dirksen 1986). However, both field and laboratory methods are time consuming and tedious, with numerous practical limitations (van Genuchten 1988). As a result, unsaturated hydraulic conductivity is often estimated from soil parameters obtained from soil-water retention curves.

Van Genuchten (1980) derived expressions that relate the unsaturated hydraulic conductivity to both the water content and the pressure head. The relationship between the unsaturated hydraulic conductivity and the water content is

$$K(\theta) = K_s S_e^{1/2}[1 - (1 - S_e^{1/m})^m]^2 \quad (4.16)$$

where

$$S_e = (\theta - \theta_r)/(\theta_s - \theta_r)$$
$K(\theta)$ = unsaturated hydraulic conductivity at water content θ
K_s = saturated hydraulic conductivity
m = van Genuchten soil parameter

The equivalent relationship between unsaturated hydraulic conductivity and pressure head is

$$K(h) = K_s \frac{\{1 - (\alpha h)^{n-1}[1 + (\alpha h)^n]^{-m}\}^2}{[1 + (\alpha h)^n]^{m/2}} \quad (4.17)$$

where

$K(h)$ = the unsaturated hydraulic conductivity at pressure head h
h = pressure head
m = van Genuchten soil parameter
n = van Genuchten soil parameter
α = van Genuchten soil parameter

Figure 4.13 shows observed values (open circles) and calculated curves (solid lines) based on Equation 4.17 for relative permeability ($K_r = K(h)/K_s$) as a function of pressure head for (a) a sandstone and (b) a silt loam. The predictive equation quite closely follows the observed values.

4.6.7 Buckingham Flux Law

The first to recognize the basic laws for the flow of water in soil was Buckingham (1907). He recognized that the matric potential, ψ, of unsaturated soils was a function of the water content, θ, temperature, and bulk density of the soil. He also realized that the flow of water across a unit cross-sectional area was proportional to the gradient of the soil water matric potential. The proportionality constant, $K(\theta)$, was recognized to be a function of the water content. Buckingham, a physicist, appears to have had no knowledge of Darcy's work on saturated flow some half-century before (Sposito 1986).

What is now known as the Buckingham flux law was formalized by Richards (1928), who extended the concept of the potential gradient to include the total soil moisture potential, ϕ. Written in vector notation, the Buckingham flux law is

$$\mathbf{q} = -K(\psi)\nabla(\phi) \quad (4.18)$$

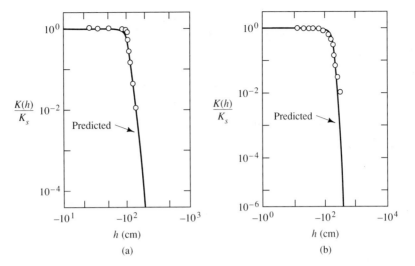

FIGURE 4.13 Observed values (open circles) and calculated curves (solid lines) for relative hydraulic conductivity of (a) Hygiene sandstone and (b) Touchet silt loam G.E.3. *Source:* M. Th. van Genuchten, *Soil Science Society of America Journal* 44 (1980):892–98.

where

$\mathbf{q}$ = the soil moisture flux ($L^3L^{-2}T^{-1}$)
$K(\psi)$ = the unsaturated hydraulic conductivity (LT^{-1}) at a given ψ
$\nabla(\phi)$ = the gradient of the total soil water potential, ϕ, where $\phi = \psi + Z$ (LL^{-1})

4.6.8 Richards Equation

The continuity equation for soil moisture through a representative elementary volume of the unsaturated zone can be stated as the change in total volumetric water content with time and is equal to the sum of any change in the flux of water into and out of the representative elemental volume. The continuity equation can be expressed as

$$\frac{\partial \theta}{\partial t} = -\left(\frac{\partial q_x}{\partial x} + \frac{\partial q_y}{\partial y} + \frac{\partial q_z}{\partial z}\right) \tag{4.19}$$

where q_x, q_y, and q_z are soil moisture fluxes.

In vector notation Equation 4.19 is

$$\frac{\partial \theta}{\partial t} = -\nabla \cdot \mathbf{q} \tag{4.20}$$

Combining Equations 4.20 with 4.18, we obtain the Richards equation (Richards 1931):

$$\frac{\partial \theta}{\partial t} = \nabla \cdot [K(\psi)\nabla \phi] \tag{4.21}$$

The value of ∇z is 0 in the x and y directions and is 1 in the z direction. In addition, ϕ is equal to $\psi + Z$. Therefore, if z is taken to be positive in a downward direction, Equation 4.21 can be rewritten as

$$\frac{\partial \theta}{\partial t} = \nabla \cdot [K(\psi) \nabla \phi] - \frac{\partial K(\psi)}{\partial z} \qquad (4.22)$$

For one-dimensional flow, Equation 4.22 reduces to

$$\frac{\partial \theta}{\partial t} = \frac{\partial}{\partial z}\left(K(\psi) \frac{\partial \psi}{\partial z}\right) - \frac{\partial K(\psi)}{\partial z} \qquad (4.23)$$

If the matric potential is much greater than the gravity gradient, then the last term of Equation 4.23 can be dropped; the resulting equation is

$$\frac{\partial \theta}{\partial t} = \frac{\partial}{\partial z}\left(K(\psi) \frac{\partial \psi}{\partial z}\right) \qquad (4.24)$$

The preceding equations assume a constant temperature and air pressure, a non-deformable soil matrix, incompressible water, and that soil water density is independent of solute concentration and does not vary throughout the flow domain. Furthermore, these equations assume that the presence of air can be ignored, except as it affects the value of K. Equation 4.23 is nonlinear and difficult to solve by analytical means. However, numerical methods of solution of this equation have been developed (Nielsen, van Genuchten, and Biggar 1986).

4.6.9 Vapor Phase Transport

Under special circumstances vapor can move through the soil under air pressure gradients as the atmospheric pressure fluctuates. However, water vapor normally moves by diffusion from areas where the vapor pressure in the unsaturated pores is higher to areas where it is lower.

Diffusion of soil moisture vapor is given by

$$q_v = -D_v \frac{\partial \rho_v}{\partial x} \qquad (4.25)$$

where

q_v = the vapor flux
ρ_v = the vapor concentration in the gaseous phase
D_v = the diffusion coefficient for water vapor

Assuming no convective transport, the nonsteady, one-dimensional transport of soil moisture in the vapor phase is given by (Jackson 1964) as

$$\frac{\partial \theta}{\partial t} = \frac{\partial}{\partial x}\left(D_v \frac{\partial \rho_v}{\partial \theta} \frac{\partial \theta}{\partial x}\right) \qquad (4.26)$$

Simultaneous transport of water in both the liquid and vapor phase is described by the following equation, which is obtained by combining Equations 4.24 and 4.26:

$$\frac{\partial \theta}{\partial t} = \frac{\partial}{\partial x}\left[\left(D_v \frac{\partial \rho_v}{\partial \theta} + K(\psi)\frac{\partial \psi}{\partial x}\right)\frac{\partial \theta}{\partial x}\right] \tag{4.27}$$

4.7 Mass Transport in the Unsaturated Zone

The steady-state diffusion of a solute in soil moisture is given by (Hillel, 1980)

$$J = -D_s^*(\theta)\, dC/dz \tag{4.28}$$

where

J = the mass flux of solute per unit area per unit time

$D_s^*(\theta)$ = the soil diffusion coefficient, which is a function of the water content, the tortuosity of the soil, and other factors related to the electrostatic double layer

dC/dz = the concentration gradient in the soil moisture

The second-order diffusion equation for transient diffusion of solutes in soil water is

$$\frac{\partial C}{\partial t} = \frac{\partial}{\partial z}\left[D_s^*(\theta)\frac{\partial C}{\partial z}\right] \tag{4.29}$$

Soil moisture traveling through the unsaturated zone moves at different velocities in different pores due to the fact that the saturated pores through which the moisture moves have different-sized pore throats. In addition, velocities within each saturated pore will vary across the width of the pore. As a result, soil water carrying a solute will mix with other soil moisture. This is analogous to the mechanical mixing of saturated flow. Mechanical mixing is found from the following equation (Nielsen, van Genuchten, and Biggar 1986):

$$\text{Mechanical mixing} = \zeta |v| \tag{4.30}$$

where

ζ = an empirical soil moisture dispersivity

v = the average linear soil moisture velocity

The soil moisture dispersion coefficient, D_s, is the sum of the diffusion and mechanical mixing:

$$D_s = D_s^* + \zeta|\lambda| \tag{4.31}$$

The total one-dimensional solute flux in the vadose zone is the result of advection, diffusion, and hydrodynamic dispersion. With diffusion and hydrodynamic dispersion combined as the soil-moisture dispersion coefficient, this result can be expressed as

$$J = v\theta C - D_s \theta\, dC/dz \tag{4.32}$$

where

> J = the total mass of solute across a unit cross-sectional area in a unit time
> v = the average soil-moisture velocity
> C = the solute concentration in the soil moisture
> θ = the volumetric water content
> dC/dz = the solute gradient
> D_s = the soil moisture diffusion coefficient, which is a function of both θ and v

The continuity equation for a solute flux requires that the rate of change of the total solute mass present in a representative elemental volume be equal to the difference between the solute flux going into the REV and that leaving the REV. The total solute mass is the sum of the dissolved solute mass and the mass of any solute associated with the solid phase of the soil. The dissolved solute mass is equal to the product of θ and C. The solute mass bound to the soil is the product of the soil bulk density, B_d, and the concentration of the solute phase bound to the soil, C^*. The continuity equation for the solute is:

$$\frac{\partial(B_d C^*)}{\partial t} + \frac{\partial(\theta C)}{\partial t} = -\frac{\partial J}{\partial z} \qquad (4.33)$$

By combining Equations 4.32 and 4.33, we obtain

$$\frac{\partial(B_d C^*)}{\partial t} + \frac{\partial(\theta C)}{\partial t} = \frac{\partial(v\theta C)}{\partial z} + \frac{\partial}{\partial z}\left(D_s \theta \frac{\partial C}{\partial z}\right) \qquad (4.34)$$

The convective soil moisture flux, q, is the product of the volumetric water content, θ, and the average soil moisture velocity, v. There may also be sources and sinks of the solute not accounted for by the absorbed concentration, C^*. For example, plants may remove nutrients from solution and solutes might be created by biological decay as well as microbial and chemical transformation and precipitations. These can be added to Equation 4.34 by means of a term for the summation of Υ_i, where Υ represents other sources and sinks. With a slight rearrangement Equation 4.34 becomes the fundamental mass transport equation for the vadose zone:

$$\frac{\partial(B_d C^*)}{\partial t} + \frac{\partial(\theta C)}{\partial t} = \frac{\partial}{\partial z}\left(D_s \theta \frac{\partial C}{\partial z} - qC\right) + \sum_i \Upsilon_i \qquad (4.35)$$

If B_d, D_s, θ, and q are assumed to be constant in time and space, Equation 4.35 reduces to Equation 3.1, the basic one-dimensional advection-dispersion equation.

4.8 Equilibrium Models of Mass Transport

In order to account for a solute, which can be in either a dissolved form or absorbed by the soil, we need to know the relationship between the concentration in solution, C, and the absorbed concentration, C^*. If the solute reaches equilibrium rapidly between the dissolved and absorbed phase, then the relationship can be described by an absorption isotherm. Sorption isotherms are discussed more thoroughly in Chapter 3. For illustration in this chapter, we will use the linear isotherm

$$C^* = K_d C \qquad (4.36)$$

where K_d, the distribution coefficient, is the slope of the plot of C^* as a function of C.

Under certain conditions the source and sink function, Υ_i, may be approximated by zero- and first-order decay and production terms (Nielsen, van Genuchten, and Biggar 1986). If η_l and η_s are rate constants for first-order decay in the liquid and solid phases, respectively, and ξ_l and ξ_s are zero-order rate terms for production in the liquid and solid phases, then

$$\Upsilon_i = -\eta_l \theta C - \eta_s B_d C^* + \xi_l \theta + \xi_s B_d \qquad (4.37)$$

By substitution of Equations 4.36 and 4.37 into 4.35 and simplification of terms, we can obtain the following equation:

$$R \frac{\partial C}{\partial t} = D_s \frac{\partial^2 C}{\partial z^2} - v \frac{\partial C}{\partial z} - \eta C + \xi \qquad (4.38)$$

where R is the retardation factor, which is given by

$$R = 1 + \frac{B_d K_d}{\theta} \qquad (4.39)$$

and η and ξ are consolidated rate factors, given by

$$\eta = \eta_l + \frac{\eta_s B_d K_d}{\theta} \qquad (4.40)$$

$$\xi = \xi_l + \frac{\xi_s B_d}{\theta} \qquad (4.41)$$

Equation 4.38 is based on steady-state flow, as the volumetric moisture content and the fluid velocity are taken to be constants.

Van Genuchten (1981) solved equation 4.38 for a number of different boundary conditions. In general, at time equals 0 and at some place in the soil column, z, the solute concentration is $C_i[C(z, 0) = C_i(z)]$. The concentration introduced into the top of the soil column where $z = 0$ at some time t is $C_0[C(0, t) = C_0(t)]$. The rate that solute is introduced into the top of the soil column by both advection and diffusion is equal to the pore water velocity, v, times C_0. (The usual boundary condition adopted for this is $(-D_s \, dC/dz + vC) = vC_0$.)

If a pulse of solute is introduced into the soil column for a time period of 0 to t_0, during that time the rate is vC_0. After time t_0, the pulse is ended and the rate at which solute is introduced is 0. (For $0 < t < t_0$, $(-D_s \, dC/dz + vC) = vC_0$; for $t > t_0$, $(-D_s \, dC/dz + vC) = 0$). The soil column is infinitely long ($dC/dz(\infty, t) = 0$).

In this case the solution to Equation 4.38 for times when the pulse is being injected $(0 < t < t_0)$ is

$$C(z, t) = \left(C_0 - \frac{\xi}{\eta}\right) A(z, t) + B(z, t) \qquad (4.42)$$

For times greater than t_0—that is, after injection of the pulse has stopped—the solution to Equation 4.38 is

$$C(z, t) = \left(C_0 - \frac{\xi}{\eta}\right) A(z, t) + B(z, t) - C_0 A(x, t - t_0) \qquad (4.43)$$

In both Equations 4.42 and 4.43, the following arguments are used:

$$A(z, t) = \frac{v}{(v + u)} \exp\left[\frac{(v - u)z}{2D_s}\right] \text{erfc}\left[\frac{Rz - ut}{2(D_s Rt)^{1/2}}\right]$$
$$+ \frac{v}{(v - u)} \exp\left[\frac{(v + u)z}{2D_s}\right] \text{erfc}\left[\frac{Rz + ut}{2(D_s Rt)^{1/2}}\right] \qquad (4.44)$$
$$+ \frac{v^2}{2\eta D_s} \exp\left(\frac{vz}{D_s} - \frac{\eta t}{R}\right) \text{erfc}\left[\frac{Rz + vt}{2(D_s Rt)^{1/2}}\right]$$

$$B(z, t) = \left(\frac{\xi}{\eta} - C_i\right) \exp\left(-\frac{\eta t}{R}\right) \left\{\frac{1}{2} \text{erfc}\left[\frac{Rz - vt}{2(D_s Rt)^{1/2}}\right] + \left(\frac{v^2 t}{\pi R D_s}\right)^{1/2}\right.$$
$$\times \exp\left[-\frac{(Rz - vt)^2}{4 D_s Rt}\right] - \frac{1}{2}\left(1 + \frac{vz}{D_s} + \frac{v^2 t}{D_s R}\right) \exp\left(\frac{vz}{D}\right)$$
$$\left.\times \text{erfc}\left[\frac{Rz + vt}{2(D_s Rt)^{1/2}}\right]\right\} + \frac{\xi}{\eta} + \left(C_i - \frac{\xi}{\eta}\right) \exp\left(-\frac{\eta t}{R}\right) \qquad (4.45)$$

and

$$u = v\left(1 + \frac{4\eta D_s}{v^2}\right)^{1/2} \qquad (4.46)$$

For the steady-state case, Equation 4.38 may be written as

$$D_s \frac{\partial^2 C}{\partial z^2} - v \frac{\partial C}{\partial z} - \eta C + \xi = 0 \qquad (4.47)$$

The solution to this is

$$C(z) = \frac{\xi}{\eta} + \left(C_0 - \frac{\xi}{\eta}\right)\left(\frac{2v}{v + u}\right) \exp\left[\frac{(v - u)z}{2D_s}\right] \qquad (4.48)$$

where u is defined in Equation 4.46.

Equations 4.42, 4.43, and 4.48 are only some of the analytical solutions that van Genuchten (1981) has obtained for Equation 4.38. Solutions for other boundary and initial conditions include variable initial solute concentration in the soil column and an exponentially decaying source term. The reader is directed to van Genuchten's original paper for additional solutions.

4.9 Nonequilibrium Models of Mass Transport

The soil moisture may move at such a quick rate that a solute may not be able to reach an equilibrium position with respect to chemical reactions that are occurring. One nonequilibrium formulation arises when the adsorption process can be described by a first-order linear rate equation. Under this condition, assuming steady-state flow and ignoring the source/sink term, Equation 4.34 can be written as a coupled system (Nielsen, van Genuchten, and Biggar 1986) as follows:

$$\frac{B_d}{\theta}\frac{\partial C^*}{\partial t} + \frac{\partial C}{\partial t} = D_s \frac{\partial^2 C}{\partial^2 z} - v\frac{\partial C}{\partial z} \quad (4.49a)$$

and

$$\frac{dC^*}{dt} = \zeta_r(K_d C - C^*) \quad (4.49b)$$

where ζ_r is a first-order rate coefficient. Equations 4.49a and b have been used by many to describe nonequilibrium transport in soils. As shown by van Genuchten, Davidson, and Wierenga (1974), the first-order rate model has not materially improved the description of nonequilibrium transport in soils. An alternative model of nonequilibrium transport arises when soil water is assumed to consist of a mobile phase and an immobile phase. The mobile phase occupies the center of saturated pores. Immobile water consists of thin coatings on soil particles, dead-end pores, and water trapped in small unsaturated pores (Coats and Smith 1964). Exchange of solute between the mobile and immobile phases occurs due to diffusion. In addition, solute in both the mobile and immobile phases can participate in adsorption-desorption reactions. For Freundlich-type linear equilibrium, this conceptualization can be described by the following equations (van Genuchten and Wierenga 1976):

$$\theta_m R_m \frac{\partial C_m}{\partial t} + \theta_{im} R_{im} \frac{\partial C_{im}}{\partial t} = \theta_m D_{sm} \frac{\partial^2 C_m}{\partial z^2} - \theta_m v_m \frac{\partial C_m}{\partial z} \quad (4.50a)$$

$$\theta_{im} R_{im} \frac{\partial C_{im}}{\partial t} = \beta(C_m - C_{im}) \quad (4.50b)$$

where

β = a mass transfer coefficient
R_m = a retardation factor for the mobile water
R_{im} = a retardation factor for the immobile water
C_m = solute concentration in the mobile water
C_{im} = solute concentration in the immobile water
θ_m = volumetric mobile water content
θ_{im} = volumetric immobile water content
D_{sm} = soil moisture dispersion coefficient for the mobile water

For saturated soils, the amount of immobile water is a function of the soil-water flux, the size of soil aggregates, and the concentration of the ionic solute. The mass transfer coefficient, β, is also a function of these same factors as well as the species being transferred (Nkedi-Kizza et al. 1983).

The preceding model with mobile and immobile water zones has been successfully used at the laboratory scale (Nkedi-Kizza, et. al. 1983); Nielsen, van Genuchten, and Biggar 1986). Figure 4.14 shows a breakthrough curve for a column study of the transport

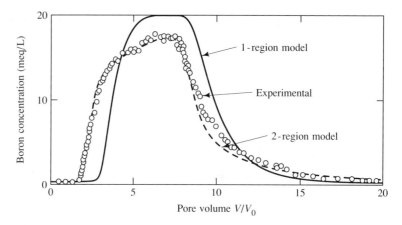

FIGURE 4.14 Observed and calculated breakthrough curves for a solution of boron passing through a column filled with an aggregated clay loam soil. *Source:* D. R. Nielsen, M. Th. van Genuchten, and J. W. Biggar, *Water Resources Research* 22, no. 9 (1986):89S–108S. Copyright by the American Geophysical Union.

of a solution of boron in an aggregated clay loam soil. The curve labeled "1-region model" is based on Equation 4.38, and the curve labeled "2-region model" is based on Equations 4.50a and b. In this case the 2-region model does a better job of matching the experimental data.

4.10 Anion Exclusion

For a nonreactive solute, Equation 4.35 can be written as

$$\theta \frac{\partial C}{\partial t} = D_s \theta \frac{\partial^2 C}{\partial z^2} - q \frac{\partial C}{\partial z} \quad (4.51)$$

Many solutes are considered to be nonreactive in the sense that they do not sorb onto particle surfaces. Included in this category are anions such as chloride. However, chloride carries a negative charge, and if there are many clay particles in the porous medium, the electrostatic double layer will repel anions. Consequently, there is a region around each colloidal particle from which the anions are repelled, with the resulting distribution shown on Figure 4.1. As a first approximation we can assume there is a two-phase distribution of the solute in the soil water. Within the exclusion volume the concentration is zero, with all of the solute thus being concentrated in the pore water outside of the exclusion volume (James and Rubin 1986). The volume of the exclusion zone is θ_{ex}.

The anion concentration in the pore water under conditions of anion exclusion is given by (Bresler 1973)

$$(\theta - \theta_{ex}) \frac{\partial C}{\partial t} = D_s \theta \frac{\partial^2 C}{\partial z^2} - q \frac{\partial C}{\partial z} \quad (4.52)$$

where C is the concentration of the solute in the bulk pore solution (that is, including the water in the exclusion zone).

The value of θ_{ex} can be found experimentally from a soil column test. Water with an initial concentration of a single anion of C_0 is introduced into a soil column that has a known water content. The water content is held constant, and sufficient solute of concentration C_0 is introduced so that the concentration of the water leaving the soil column, C_{out}, is equal to C_0. The mass of the anion contained in the soil column is determined, and a concentration is calculated based on the total water content in the soil column, C_{calc}. This calculated concentration of anionic solution is less than C_{out} because water in the exclusion volume, which contains no anions, was used in the calculation. The value of the exclusion volume can be found from (Bond, Gardiner, and Smiles 1982):

$$\theta_{ex} = \theta\left[1 - \left(\frac{C_{calc}}{C_{out}}\right)\right] \qquad (4.53)$$

In addition to decreasing the observed concentration of the anion in the soil column, anion exclusion causes the anions to travel faster than the average rate of moving pore water. The average rate of pore-water movement, v, is equal to the rate of the fluid flux, q, divided by the water content, θ. The anions cannot travel through the excluded part of the volumetric water content, which is close to the mineral surfaces and has a low or zero velocity. Therefore, they must move in the part of the pore water that is available to them—i.e., the center of the pores, where the fluid velocity is greater than average. As a result, excluded anions will travel further in a given period of time than they would in the absence of anion exclusion.

An approximate solution to Equation 4.52 for conditions of uniform water content and a steady flux of a dissolved anion of concentration C_0 into a semi-infinite soil column with an initial concentration of the dissolved anion, C_i, is

$$\frac{C - C_i}{C_0 - C_i} = 0.5 \, \text{erfc}\left\{\frac{(\theta - \theta_{ex})z - qt}{2[(\theta - \theta_{ex})D_s\theta t]^{0.5}}\right\} \qquad (4.54)$$

EXAMPLE PROBLEM

James and Rubin (1986) performed an experiment in which they measured the concentration profiles of chloride introduced into soil columns containing Delhi sand. The sand was 90% sand, 7% silt, and 3% clay by weight with a cation-exchange capacity of 0.05 mol/kg and an organic carbon content of 0.003 g of carbon per gram of soil. The soil column was constructed with a suction line on the bottom to create a soil-water tension on the entire column. The soil columns were initially leached with a nitrate solution to remove any chloride. Water content (θ) was kept uniform throughout the soil column and the volumetric water flux (q) was numerically equal to the unsaturated hydraulic conductivity evaluated at θ. A solution containing chloride was introduced into the soil column at the same constant volumetric water flux established during the leaching phase. The soil columns were constructed as a series of sections either 1.14 cm or 2.28 cm long so that at the end of the injection phase they could be disassembled and the chloride mass and water content of each section could be measured.

Several experiments were conducted in which the water content ranged from 0.167 to 0.225 and the volumetric water fluxes ranged from 0.0393 cm/hr to 0.397 cm/hr. Figure 4.15 shows the results of experiments A and B, which had water contents of

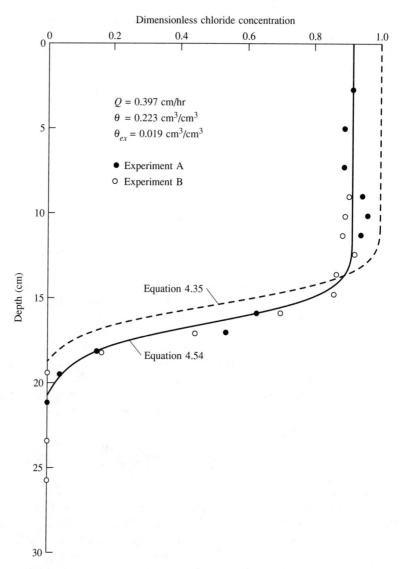

FIGURE 4.15 Chloride concentration profiles in a soil column affected by anion exclusion. *Source:* R. V. James and S. Rubin, *Soil Science Society of America Journal* 50 (1986): 1142–50.

0.221 and 0.225, respectively, while both had a water flux of 0.397 cm/hr. There are two curves on the figure, one computed using Equation 4.35 and one using Equation 4.54. Equation 4.35 is applied for a nonsorbing, nonreactive solute. The chloride concentrations are shown as dimensionless chloride, that is C_{calc}/C_0. No measured chloride values were equal to C_0, illustrating the results of anion exclusion. The experimental results can be seen to follow closely the curve computed from Equation 4.54.

The excluded water content was computed from the experimental results at the top of the soil columns using Equation 4.53. It was found to be 0.019 for both saturated and unsaturated conditions. The dispersion coefficient was determined by solving Equation 4.52, with appropriate boundary conditions as

$$\text{erfc}^{-1}\left[\frac{2(C - C_i)}{(C_0 - C_i)}\right] = \frac{(\theta - \theta_{ex})}{2\sqrt{(\theta - \theta_{ex})D_s t}}z - \frac{qt}{2\sqrt{(\theta - \theta_{ex})D_s \theta t}} \quad (4.55)$$

The left side of Equation 4.55 was plotted against the depth, z, and a straight line was determined by the method of least squares regression. The value of D_s was then calculated from the slope of the line. Inspection of Equation 4.55 shows that D_s is the only unknown. The curves on Figure 4.15 were calculated using the experimentally determined values of D_s.

Case Study: Relative Movement of Solute and Wetting Fronts

Pickens and Gillham (1980) studied the relative motion of the wetting front and the solute in an unsaturated soil column by use of a finite-element model. The model allows for the use of either nonhysteretic or hysteretic water-content–pressure-head relationships. Previous work

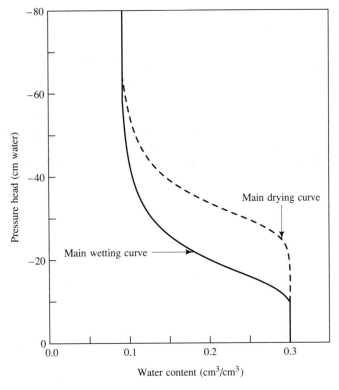

FIGURE 4.16 A water-content–pressure-head hysteresis loop used in a model study of solute infiltration in a soil column. *Source:* J. F. Pickens and R. W. Gillham, *Water Resources Research* 16, No. 6 (1980): 1071–78. Copyright by the American Geophysical Union.

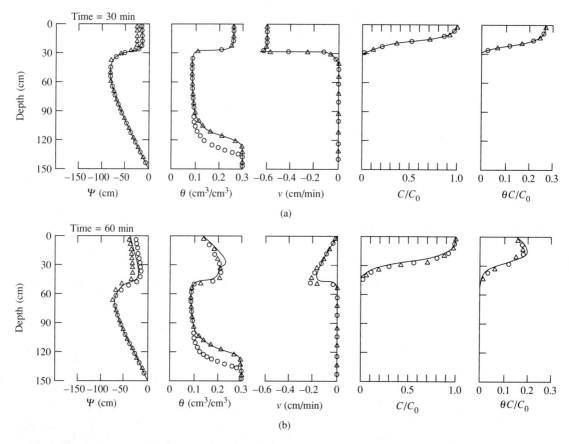

FIGURE 4.17 Model results of matric potential, (ψ), water content (θ), pore-water velocity (v), relative solute concentration (C/C_0), and volumetric solute concentration ($\theta C/C_0$). *Source*: J. F. Pickens and R. W. Gillham, *Water Resources Research* 16, no. 6 (1980):1071–78. Copyright by the American Geophysical Union.

(Gillham, Klute, and Heermann 1979) had shown that it was very important to include hysteresis in the model because the equation for computation of the pore-water velocity and the advection-dispersion equation contains a term for the water content.

The modeled soil column had a saturated hydraulic conductivity of 0.29 cm/min, a saturated water content of 0.301, and a longitudinal dispersivity of 0.5 cm. The initial soil-moisture condition in the model had a linear pressure head variation between 0 at the bottom to -150 cm at the top of the soil column. The corresponding initial water-content values were obtained from a water-content–pressure-head hysteresis loop (Figure 4.16) and followed the main drainage curve, so that the model represented conditions that would develop if a saturated-soil column were drained. The soil column did not contain any solute prior to infiltration. A slug of water containing a solute at concentration C_0 was allowed to infiltrate at a at a rate of 0.17 cm/min for a 30-min period, resulting in a total depth of infiltration of 5 cm. The model computed the pressure head, water content, vertical pore-water velocity, relative

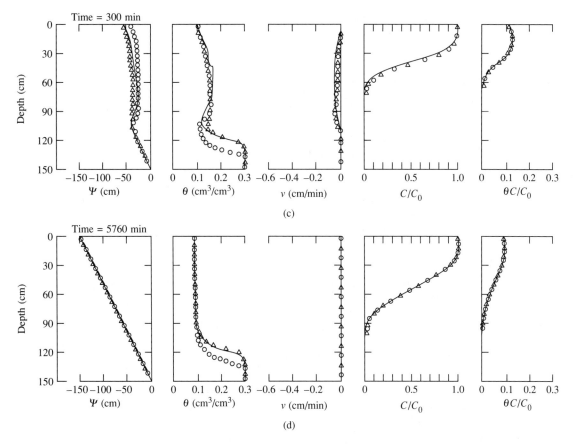

FIGURE 4.17 Continued

solute concentration (C/C_0), and volumetric solute concentration ($\theta C/C_0$) after 30 min, 60 min, 300 min, and 5760 min. Figure 4.17 shows the results of the model study. At 5760 min the head distribution had returned to the preinfiltration conditions, which indicates that equilibrium had been reached. Inspection of Figure 4.17 shows that although the water in the bottom of the soil column is above the irreducible water content at 5760 min, the solute has remained near the top of the soil column. This can occur because the water found at the bottom of the soil column was displaced downward from the top of the soil column by the infiltrating water containing the solute, which did not penetrate past the depth where $C/C_0 = 0$.

The model was run under three conditions. In the hysteretic mode the hysteretic relationship between water content and pressure head was used. As the slug of water infiltrated, the wetting relationships were used, followed by drying curves when the infiltrating slug had passed. This model run is represented as a solid line on Figure 4.17. In the nonhysteretic wetting-curve mode, the pressure-head–water-content relation was based on the main wetting curve, and in the nonhysteretic drying-curve mode, it was based on the main drying curve. The results of these two modes are represented by circles and triangles, respectively.

The importance of using the hysteretic mode appears to be greater for the pressure head and the water content than for the pore velocity and the solute front movement.

4.11 Preferential Flowpaths in the Vadose Zone

The preceding analyses all treat the unsaturated zone as a homogeneous, porous medium. However, this is certainly not the case. In the root zone there are numerous large pores and cracks formed by such agents as plant roots, shrinkage cracks, and animal burrows. These **macropores** can form preferential pathways for the movement of water and solute, both vertically and horizontally through the root zone (Beven and Germann 1982). This situation can lead to "short-circuiting" of the infiltrating water as it moves through the macropores at a rate much greater than would be expected from the hydraulic conductivity of the soil matrix; see Figure 4.18(a).

A second type of preferential flow is **fingering,** which occurs when a uniformly infiltrating solute front is split into downward-reaching "fingers" due to instability caused by pore-scale permeability variations. Instability often occurs when an advancing wetting front reaches a boundary where a finer sediment overlies a coarser sediment, see Figure 4.18(b) (Hillel and Baker 1988).

A third type of preferential flow is **funneling** (Kung 1990b). Funneling occurs in the vadose zone below the root zone and is associated with stratified soil or sediment profiles. Sloping coarse-sand layers embedded in fine-sand layers can impede the downward infiltration of water. The sloping layer will collect the water like the sides of a funnel and direct the flow to the end of the layer, where it can again percolate vertically, but in a concentrated volume, as shown in Figure 4.18(c). Field studies using water containing dye placed in furrows indicate that the water is moving in the fine-sand layer above the discontinuity of the coarse-sand layer (Kung 1990a). These same dye studies showed that because of funneling, the volume of the soil containing dye decreased with depth. The dyed soil region occupied about 50% of the soil volume at 1.5 to 2.0 m; from 3.0 to 3.5 m, it occupied only 10% of the soil volume, and by 5.6 to 6.6 m, it was found in about 1% of the soil volume. At this depth a single column of dyed soil was found, obviously formed by funneling of flow of dyed water from above (Kung 1990a).

These occurrences of preferential flow in particular and soil heterogeneity in general have disturbing implications for monitoring solute movement in the unsaturated zone. Some studies have recorded seemingly anomalous results, with deeper soil layers having greater concentrations of solute than more shallow layers (Kung 1990b). These anomalies can be explained by preferential flow patterns, with infiltrating solute being directed to certain regions of the vadose zone by short circuiting, fingering, and funneling. This suggests that a large number of sampling devices in the vadose zone might be needed if a reasonably accurate picture of the distribution of a contaminant is to be obtained. As contaminants reaching the water table may need to pass through the unsaturated zone, preferential flow paths in the unsaturated zone also have implications for groundwater monitoring. In the case of a contaminant that was evenly spread on the land surface—for example, an agricultural chemical—one would expect that there would be an evenly distributed solute load reaching the water table via the vadose zone. However, due to preferential flow paths the mass of solute may be concentrated in some locations,

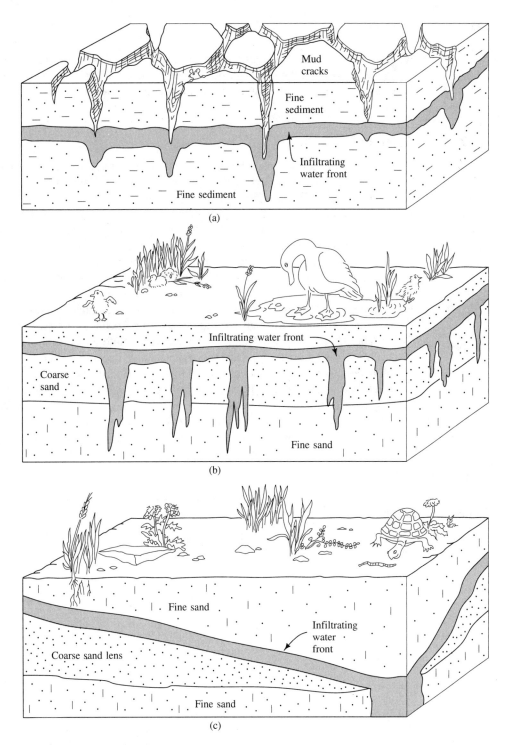

FIGURE 4.18 Preferential water movement in the vadose zone due to (a) short circuiting, (b) fingering, and (c) funneling.

resulting in an uneven distribution in the shallow ground water beneath the site. Monitoring wells beneath the site may show varying solute concentrations, depending upon how close they are to an up-gradient point of concentrated recharge.

4.12 Summary

The vadose zone extends from the land surface to the water table; moisture in the vadose zone is under tension. Soil particles can have charged surfaces, which can attract or repel anions and cations. Moisture moves through the vadose zone due to a potential that is the sum of the elevation potential and the matric potential. Matric potential is a function of the volumetric water content and depends upon whether the soil has previously undergone wetting or drying. The unsaturated hydraulic conductivity is also a function of the volumetric water content. The flux of moisture through the soil can be calculated from Darcy's law or the Buckingham flux law. The nonsteady flow of soil moisture is described by the Richards' equation. The nonsteady movement of soil vapor and soil moisture can also be described by a partial differential equation.

Solute movement through the vadose zone proceeds by both advection and diffusion. There is an advective-dispersive equation for solute transport in the vadose zone that accounts for retardation through sorption onto soil particles. Analytical solutions to this equation exist. Solute movement in the vadose zone is affected by regions of immobile water found in dead-end pores. It is also affected by an anion-exclusion zone when clays are present in the soil. Preferential pathways of solute movement may be present that create pathways for lateral movement of water in the vadose zone and concentrate infiltrating water into certain regions. Preferential pathways of water movement may make monitoring of the vadose zone and shallow water table difficult.

Chapter Notation

$A(z, t)$	Term defined by Equation 4.44
$B(z, t)$	Term defined by Equation 4.45
B_d	Bulk density of soil
C	Solute concentration
C^*	Solute phase bound to soil
C_i	Initial solute concentration in soil water
C_{im}	Solute concentration in immobile water
C_m	Solute concentration in mobile water
C_0	Solute concentration in injected water
C_{calc}	Concentration of solute calculated in soil if the water in the anion exclusion zone is included
D_v	Diffusion coefficient for water vapor
D^*_s	Soil-diffusion coefficient, which is a function of θ
D_s	Dispersion coefficient for soil moisture, which is a function of θ
D_{sm}	Dispersion coefficient for mobile soil moisture
e	Elementary charge of an ion

g	Acceleration of gravity
h	Pressure potential in units of length
J	Mass flux of solute
k_B	Boltzman constant
k	Intrinsic permeability
$K(\theta)$	Unsaturated hydraulic conductivity, which is a function of θ
$k_r(\theta)$	Relative permeability as a function of θ
K_d	Distribution coefficient
m	Van Genuchten soil parameter
n	Van Genuchten soil parameter
n_0	Concentration of ions in bulk solution
P_c	Capillary pressure
q	Volumetric soil water flux (specific discharge)
q_v	Vapor flux
u	Term defined by Equation 4.46
R	Retardation factor
R_m	Retardation factor for mobile water
R_{im}	Retardation factor for immobile water
S	Slope
S_p	Dimensionless slope
S_w	Saturation ratio (θ/θ_s)
T	Temperature in Kelvins
t	Time
t_0	Time when pulse of injection of solute ceases
V	Valence of ions in solution
v	Average linear soil moisture velocity (q/θ)
z	Elevation
Z	Gravitational potential
z_0	Characteristic length or extent of the double layer
α	Van Genuchten soil parameter
β	Mass transfer coefficient
ϵ	Dielectric constant
ζ	Empirical soil moisture dispersivity
ζ_r	First-order rate coefficient
η	Consolidated first-order decay term
η_l	First-order decay rate constant for liquid phase
η_s	First-order decay rate constant for solid phase
θ	Volumetric water content
θ_{ex}	Volume of excluded zone due to anion exclusion
θ_m	Volumetric water content of mobile water
θ_{im}	Volumetric water content of immobile water
λ	Brooks-Corey soil parameter
μ_w	Dynamic viscosity of water
ξ	Consolidated zero order source term
ξ_l	Zero-order source term for liquid phase
ξ_s	Zero-order source term for solid phase

ρ_v	Vapor concentration in vapor phase
ρ_w	Density of water
σ	Interfacial tension
Υ	Various sources and sinks for solute
ϕ_{EV}	Soil moisture potential in terms of energy per unit volume
ϕ_{EW}	Soil moisture potential in terms of energy per unit weight
ϕ_{EM}	Soil moisture potential in terms of energy per unit mass
ψ	Matric potential of soil

References

Beven, Keith, and Peter Germann. 1982. Macropores and water flow in soil. *Water Resources Research* 18, no. 5:1311–25.

Boersma, L., et al. 1972. "Theoretical research." In *Soil Water*, edited by Nielsen, D. R., R. D. Jackson, J. W. Cary, and D. D. Evans, 21–63. Madison, Wis. American Society of Agronomy, 175 pp.

Bond, W. J., B. N. Gardiner, and D. E. Smiles. 1982. Constant flux adsorption of a tritiated calcium chloride solution by a clay soil with anion exclusion. *Soil Science Society of America Journal* 46:1133–37.

Bresler, E. 1973. Anion exclusion and coupling effects in nonsteady transport through unsaturated soils: I. Theory. *Soil Science Society of America Proceedings* 37:663–69.

Brooks, R. H., and A. T. Corey. 1966. Properties of porous media affecting fluid flow. *Proceedings, American Society of Civil Engineers, Irrigation and Drainage Division* 92, no. IR2:61–87.

Buckingham, E. 1907. *Studies on the movement of soil moisture*. Bureau of Soils Bulletin 38. Washington, D.C.: U.S. Department of Agriculture.

Coats, K. H., and B. D. Smith. 1964. Dead-end pore volume and dispersion in porous media. *Society of Petroleum Engineer Journal* 4:73–84.

Constantz, J. 1982. Temperature dependence of unsaturated hydraulic conductivity of two soils. *Soil Science Society of America Journal* 46, no. 3:466–70.

Dane, J. H., and A. Klute. 1977. "Salt effects on the hydraulic properties of a swelling soil." *Soil Science Society of America Journal* 41, no. 6:1043–57.

Davidson, J. M., D. R. Nielsen, and J. W. Biggar. 1966. The dependency of soil water uptake and release on the applied pressure increment. *Soil Science Society of America Proceedings* 30, no. 3:298–304.

Fetter, C. W. 1988. *Applied Hyarogeology*. 2d ed. New York: Macmillan Publishing Company, 588 pp.

Gillham, R. W., A. Klute, and D. F. Heermann. 1979. Measurement and numerical simulation of hysteretic flow in a heterogeneous porous medium. *Soil Science Society of America Journal* 43:1061–67.

Green, R. E., L. R. Ahuja, and S. K. Chong. 1986. "Hydraulic conductivity, diffusivity, and sorptivity of unsaturated soils: Field methods." In *Methods of Soil Analysis, 1. Physical and Mineralogical Methods*, 2d ed., edited by A. Klute. Agronomy 9(1):771–98. Madison, Wis.: American Society of Agronomy.

Hillel, Daniel. 1980. *Fundamentals of Soil Physics*. New York: Academic Press, Inc., 413 pp.

Hillel, D., and R. S. Baker. 1988. A descriptive theory of fingering during infiltration into layered soils. *Soil Science* 146:51–56.

Jackson, R. D. 1964. Water vapor diffusion in relatively dry soil: 1. Theoretical considerations and sorption experiments. *Soil Science Society of America Proceedings* 28.:172–76.

James, R. V., and Jacob Rubin. 1986. Transport of chloride ion in a water-unsaturated soil exhibiting anion exclusion. *Soil Science Society of America Journal* 50:1142–50.

Javandel, I., C. Doughty, and C. F. Tsang. 1984. *Groundwater Transport: Handbook of Mathematical Models*. Water Resources Monograph 10. Washington, D. C.: American Geophysical Union. 228 pp.

Klute, A., and C. Dirksen. 1986. "Hydraulic conductivity and diffusivity: Laboratory methods." In *Methods of Soil Analysis, 1. Physical and Mineralogical Methods*, 2d ed., edited by A. Klute. Agronomy 9(1):687–734. Madison, Wis.: American Society of Agronomy.

Kung, K-J. S. 1990a. Preferential flow in a sandy vadose zone: 1. Field observation. *Geoderma* 46:51–8.

———. 1990b, "Preferential flow in a sandy vadose zone: 2. Mechanism and implications." *Geoderma* 46:59–71.

Nkedi-Kizza, P., J. W. Biggar, M. Th. van Genuchten, P. J. Wierenga, H. M. Selim, J. M. Davidson, and D. R. Nielsen. 1983. Modeling tritium and chloride-36 transport through an aggregated oxisol. *Water Resources Research* 19, no. 3:691–700.

Nielsen, D. R., M. Th. van Genuchten, and J. W. Biggar. 1986. Water flow and solute transport processes in the unsaturated zone. *Water Resources Research* 22, no. 9:89S–108S.

Pickens, J. F., and R. W. Gillham. 1980. Finite element analysis of solute transport under hysteretic unsaturated flow conditions. *Water Resources Research* 16, no. 6:1071–78.

Richards, L. A. 1928. The usefulness of capillary potential to soil-moisture and plant investigators. *Journal of Agricultural Research* 37:719–42.

———. 1931. Capillary conduction of liquids through porous mediums. *Physics* 1:318–33.

Sposito, Garrison. 1986. The physics of soil water physics. *Water Resources Research* 22, no. 9:83S–88S.

Suarez, D. L. 1985. Chemical effects on infiltration. In *Proceedings National Resources Modeling Symposium*. Edited by D. G. DeCoursey, 416–19. Washington, D. C. U. S. Department of Agriculture, Agricultural Research Service.

Suarez, D. L., J. D. Rhoades, R. Lavado, and C. M. Grieve. 1984. Effect of pH on saturated hydraulic conductivity and soil dispersion. *Soil Science Society of America Journal* 48, no. 1:50–55.

Van Genuchten, M. Th. 1980. A closed-form equation for predicting the hydraulic conductivity of unsaturated soils. *Soil Science Society of America Journal* 44:892–98.

———. 1981. Analytical solutions for chemical transport with simultaneous adsorption, zero-order production, and first-order decay. *Journal of Hydrology* 49:213–33.

Van Genuchten, M. Th., J. M. Davidson, and P. J. Wierenga. 1974. An evaluation of kinetic and equilibrium equations for the prediction of pesticide movement in porous soils. *Soil Science Society of America Proceedings* 38, no. 1:29–35.

Van Genuchten, M. Th., F. Kaveh, W. B. Russell, and S. R. Yates. 1988. "Direct and indirect methods of estimating the hydraulic properties of unsaturated soils." In *Land Qualities in Space and Time*, edited by J. Bouma and A. K. Bregt, 61–72. Wageningen, the Netherlands: International Society of Soil Science.

Van Genuchten, M. Th., and P. J. Wierenga. 1976. Mass transfer studies in sorbing porous media, I, Analytical solutions. *Soil Science of America Journal* 40, no. 4:473–80.

Chapter Five
Multiphase Flow

5.1 Introduction

The movement of liquids that are immiscible with water through the vadose zone as well as below the water table is an important facet of contaminant hydrogeology. Such liquids are often called **nonaqueous phase liquids.** They may have densities that are greater than water (dense nonaqueous phase liquids, or **DNAPLs**) or densities that are less than water (light nonaqueous phase liquids, or **LNAPLs**). They may be partially soluble in water, so that a dissolved phase as well as a nonaqueous phase may be present (Schwille 1981, 1984, 1988). Two-phase flow may occur below the water table with water and a DNAPL (McWhorter and Sunada 1990). Three-phase flow may occur in the vadose zone with air, water, and an NAPL (Abriola and Pinder 1985a, 1985b). In the vadose zone the NAPL may partition into the air as a vapor phase (Baehr 1987). There may be multiple compounds in the nonaqueous phase, each with different properties (Corapcioglu and Baehr 1987; Baehr and Corapcioglu 1987). Flow is dependent upon the densities, viscosities, and interfacial tensions of the liquids. In addition to dispersion and diffusion, compounds can undergo adsorption and chemical and biological degradation. These processes pose extremely complex hydrogeological challenges—so much so that Pinder and Abriola (1986) had to utilize 27 independent equations in the development of a comprehensive model for the flow of nonaqueous phase liquids.

The basic theory of nonaqueous phase liquid transport has been worked out by petroleum reservoir engineers, who are concerned with the movement of petroleum through reservoirs that also contain water (e.g., Buckley and Leverett 1942). This theory has been extended to the vadose zone, since most nonaqueous phase liquid problems start with a spill or leak that creates a release above the water table. In addition, soil-moisture flow in the vadose zone may be viewed as a subset of multiphase flow.

Common LNAPLs include gasoline and diesel fuel. DNAPLs include the chlorinated hydrocarbons, such as trichloroethylene and pentachlorophenol. Some industrial applications involve solutions of DNAPLs in mineral oils; the resulting solution is an LNAPL, even though the pure formulation of the compound of concern might be more dense than water. For example wood preservatives are sometimes formulated by dissolving pentachlorophenol in a carrier oil. Although pentachlorophenol has a specific gravity of 1.978, the commercial solution is less dense than water.

5.2 Basic Concepts

5.2.1 Saturation Ratio

The **saturation ratio** of a fluid is the fraction of the total pore space filled with that liquid. The total of the saturation ratios for all the fluids present, including air, add up to 1.0. Saturation ratio can also be expressed as a percent saturation.

5.2.2 Interfacial Tension and Wettability

A liquid in contact with another substance, which can be a solid, an immiscible liquid, or gas, possesses interfacial energy. This energy is the result of the difference in the degree of attraction for the molecules of the substance at the liquid surface to each other compared with their degree of attraction for molecules of the other substance. This phenomenon is called **interfacial tension.** It is defined as the amount of work necessary to separate a unit area of one substance from another. Units are dynes per centimeter. The interfacial tension between substances i and j is designated as $\sigma_{i,j}$.

Figure 5.1 shows the interfacial tension between two liquids, G and L, and a solid surface. The interface angle between the two liquids is indicated by θ. Equation 5.1 gives the relationship between θ and the interface tensions for the three interfaces: liquid G/solid, liquid L/solid, and liquid G/liquid L.

$$\cos \theta = \frac{\sigma_{SG} - \sigma_{SL}}{\sigma_{GL}} \quad (5.1)$$

By convention, θ is measured through the denser fluid. In general, one or the other of the fluids will preferentially spread over, or wet, the entire solid surface. If θ is less than 90°, then liquid L will preferentially wet the surface. If θ is more than 90°, then

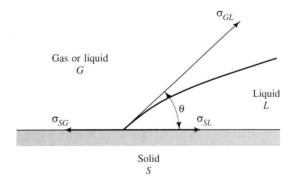

FIGURE 5.1 Interfacial tensions between a solid surface, a wetting liquid phase, L, and a nonwetting gas or liquid phase, G.

liquid G will preferentially wet the surface. That is, if we have two liquids competing for a surface, one dominates and coats the solid surface. In oil-water systems, water tends to preferentially wet the surface. However, if the surface is dry and first becomes coated with oil, then the system is oil-wet, as the water does not come into contact with the surface. An automobile that has been waxed is an example of an oil-wet surface. Water will form beads and flow in rivulets rather than move uniformly over the surface, as it does for an unwaxed car.

Aquifers are naturally water-wet because they contain water before any NAPLs are discharged to them. The vadose zone may be either water-wet or oil-wet, depending upon whether the soil is moist or dry when the oil is discharged. However, even soil in the vadose zone that appears to be dry will have water held to it by capillary pressures. At very low water content, water forms **pendular rings** around the grain contact points, with a thin film of water coating the rest of the grains. This water cannot flow, but it still coats the mineral grains, making the vadose zone water-wet.

5.2.3 Capillary Pressure

If two immiscible liquids are in contact, a curved surface will tend to develop at the interface. By measuring the pore pressure near the interface in each phase, one will find that the pressures are not the same. The difference is the **capillary pressure.**

In the vadose zone capillary pressure has a negative value. We can also refer to capillary pressure as a tension, in which case it would have a positive value. If P_w is the pressure of the wetting fluid and P_{nw} is the pressure of the nonwetting fluid, then P_c, the capillary pressure, is found from Equation 5.2.

$$P_c = P_w - P_{nw} \quad (5.2)$$

Figure 5.2 shows the radius of curvature, r', for a spherical air-water interface. Equation 5.3 gives the relationship between the capillary pressure, P_c, the interfacial tension, σ, and the radius of curvature.

$$P_c = -\frac{2\sigma}{r'} \quad (5.3)$$

Capillary pressure is directly proportional to the interfacial tension and inversely proportional to the radius of curvature. The radius of curvature is dependent upon the pore size and the amount of each fluid present. This means that the capillary pressure is a function of the properties of the two immiscible liquids present and is different for differing proportions of water and NAPL in the same porous media. Further, P_c is a property of the macroscopic geometry of the void spaces in the porous media, which cannot easily be described mathematically.

If we wish to know the capillary pressure in a real porous media, we must consider an analogous medium with known geometry. For example, we could consider the real porous media to be similar to a bundle of thin glass tubes of radius r. The capillary pressure in a thin tube, P_t, can be determined from Equation 5.4:

$$P_t = -\left(\frac{2\sigma}{r}\right)\cos\theta \quad (5.4)$$

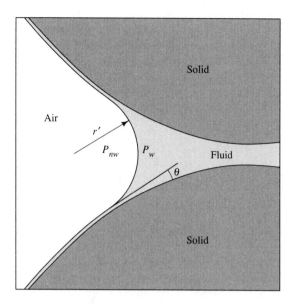

FIGURE 5.2 Radius of curvature for a spherical capillary interface.

For a given porous medium the relationship of the capillary pressure to the saturation ratio can be determined experimentally in the laboratory. Figure 5.3 shows capillary pressure curves. If the porous medium starts off saturated with the wetting fluid and the wetting fluid is slowly displaced by a nonwetting fluid, thus reducing the wetting-fluid saturation ratio, S_w, the result is a **drainage**, or **drying, curve**. As the wetting-fluid saturation ratio drops, the capillary pressure will become more negative. Eventually no more wetting fluid will be displaced by the nonwetting fluid, even with further decreases in capillary pressure. This saturation value is known as the **irreducible wetting-fluid saturation,** S_{wi}, also known as **residual wetting saturation.** Next, we will displace the nonwetting fluid by forcing the wetting fluid into the sample. The result is called an **imbibition,** or **wetting, curve.** Notice on Figure 5.3 that the imbibition curve does not follow the same pathway as the drainage curve. Recall that this phenomenon is called **hysteresis.** There is no unique relationship between capillary pressure and saturation ratio; the relationship depends upon which fluid is being displaced. When the imbibition curve reaches zero capillary pressure, some of the nonwetting fluid will remain in the porous media. This saturation value is known as the **irreducible,** or **residual, nonwetting fluid saturation,** S_{nwr}.

Notice that on Figure 5.3 the drainage curve starts off at a wetting fluid saturation ratio of 1.0 with a nonzero capillary pressure, P_d. This is the **threshold value,** also called the **displacement, imbibition, bubbling pressure,** or **air-entry, value.** In order for the nonwetting fluid to start to displace the wetting fluid, this pressure must be exceeded.

Figure 5.4(a) shows an experimentally derived capillary-pressure curve for air and water in a medium sand. Figure 5.4(b) shows a capillary-pressure curve for air and

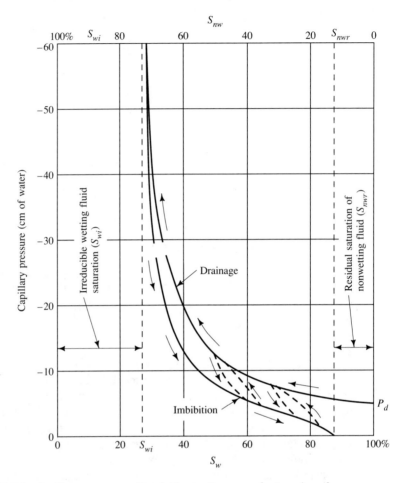

FIGURE 5.3 Capillary-pressure–wetting-fluid saturation curves for two-phase flow.

trichloroethylene in the same medium sand. Figure 5.4(c) contains an air and water capillary-pressure curve for fine sand and Figure 5.4(d) shows an air and trichloroethylene (TCE) curve for the same fine sand (Lin, Pinder, and Wood 1982). These diagrams illustrate that the capillary-pressure relationships are unique for a given porous media and depend as well on the specific immiscible fluid.

5.2.4 Relative Permeability

During simultaneous flow of two immiscible fluids, part of the available pore space will be filled with one fluid and the remainder will be filled with the other fluid. Figure 5.5 shows possible fluid-saturation states for water and oil with differing ratios of each and for both water-wet and oil-wet circumstances.

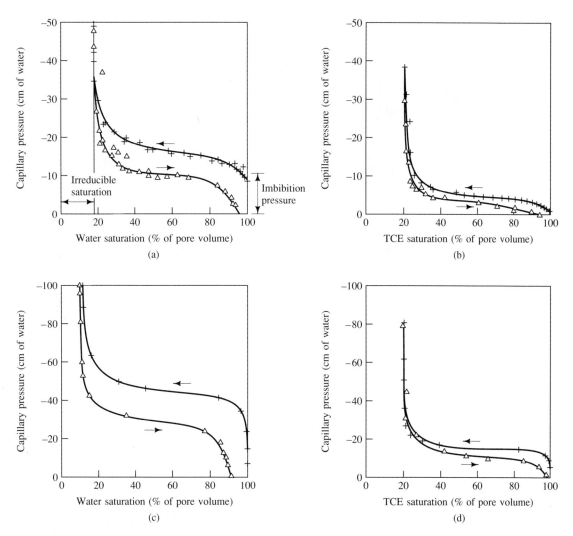

FIGURE 5.4 Experimentally derived capillary-pressure–wetting-fluid saturation curves for (a) air and water in medium sand, (b) air and trichloroethylene in medium sand, (c) air and water in fine sand, and (d) trichloroethylene and air in fine sand. *Source:* C. Lin, G. F. Pinder, and E. F. Wood, *Water Resources Program Report* 83-WR-2 (Princeton, N. J.: Princeton University, 1982).

Because the two fluids must compete for space in which to flow, the cross-sectional area of the pore space available for each fluid is less than the total pore space. This leads to the concept of relative permeability. Relative permeability is the ratio of the intrinsic permeability for the fluid at a given saturation ratio to the total intrinsic permeability of the rock. A relative permeability exists for both the wetting and the nonwetting

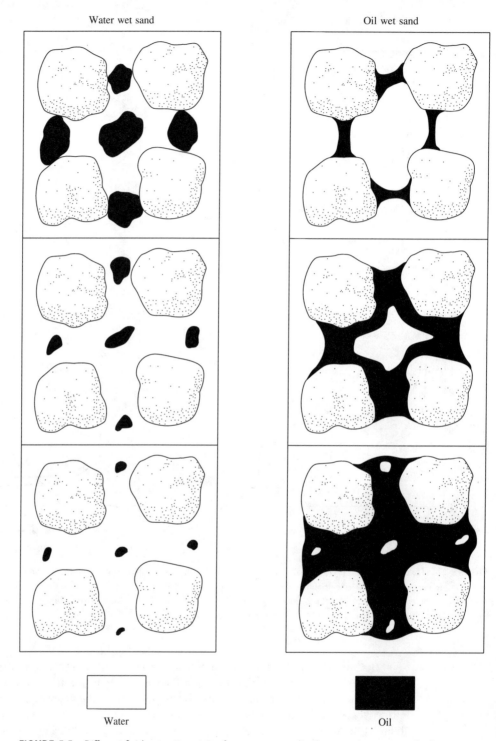

FIGURE 5.5 Different fluid saturation states for a porous media that contains water and oil.

Multiphase Flow

phase. Figure 5.6 shows two-phase relative-permeability curves for both wetting and nonwetting liquids.

The irreducible water saturation is the water content at which no additional water will flow. In Figure 5.6, water is the wetting fluid, and the irreducible water (wetting-fluid) saturation is shown on the left side. Thus water won't flow at all until the irreducible wetting-fluid saturation, S_{wi}, is exceeded. The nonwetting fluid won't begin to flow until the residual nonwetting-fluid saturation, S_{nwr}, is exceeded. This is shown on the right side of Figure 5.6. For a two-phase oil-water system, if the water content is less than the irreducible water saturation, oil can flow, but water will be held by capillary forces. Likewise, at an oil content less than the residual oil saturation, water can flow, but oil cannot, at least as a separate phase. Oil droplets dispersed in the water can still migrate.

Relative permeability is normally determined by laboratory tests of rock-core samples. There appears to be a hysteresis effect for relative permeability. This is not surprising, since such an effect exists for capillary pressure. Figure 5.7(a) shows experimentally derived relative-permeability curves in a fine sand for water in the presence of TCE (Lin, Pinder, and Wood 1982). The left curve was measured in a core that initially had a high saturation ratio of TCE and then had varying amounts of water injected into it to increase the water-saturation ratio. The right curve was measured for a core that was saturated with water and then was injected with TCE. Note that the two curves have greatly different shapes and that the relative permeability for water can vary substantially, depending upon whether the particular water-saturation value was reached by water displacing TCE or TCE displacing water. Figure 5.7(b) shows experimentally derived relative-permeability curves for TCE in the presence of water for the same fine sand as Figure 5.7(a) (Lin, Pinder, and Wood 1982). The two curves, representing water displacing TCE and TCE

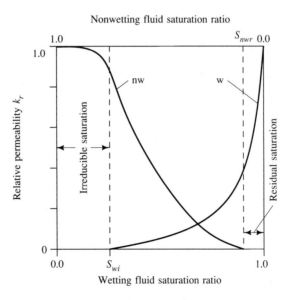

FIGURE 5.6 Typical relative permeability curves for a two-phase system.

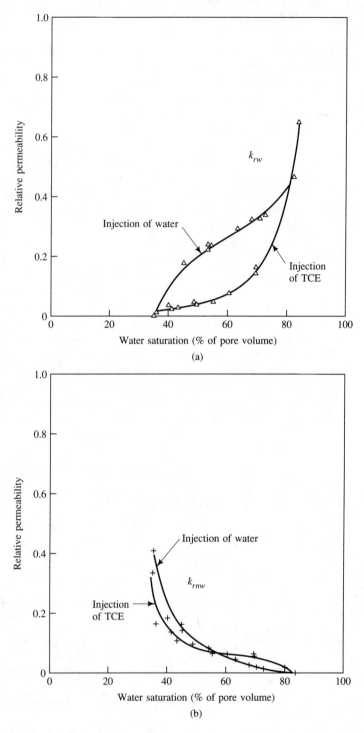

FIGURE 5.7 Experimentally derived relative permeability curves for (a) water with respect to TCE and (b) for TCE with respect to water. *Source:* C. Lin, G. F. Pinder, and E. F. Wood, *Water Resources Program Report* 83-WR-2 (Princeton, N. J.: Princeton University, 1982).

Multiphase Flow

displacing water, are much closer than the curves for relative permeability of water. However, they are not exactly the same. Hysteresis is an added complexity to analytical treatment of multiphase flow. Parker and Lenhard (1987) and Lenhard and Parker (1987a) have proposed models to describe the hysteretic relations of saturation-pressure ratios and permeability-saturation relations.

Relative permeabilities of three-phase systems of air-water–nonaqueous phase liquids are more complex. A method of estimating these values from data developed for two-phase systems was devised by Stone (1973). First the relative permeability of water, k_{rw}, as a function of the saturation ratio for water, S_w, is obtained for a water–nonaqueous phase system. Next, the relative permeability of air, k_{ra}, as a function of the saturation ratio for air, S_a, is determined for an air–nonaqueous phase system. Finally, the relative permeability of the nonaqueous phase liquid, k_{rn}, in the three-phase system is determined from Equation 5.5 (Faust 1985):

$$k_{rn} = k_{rnw}^* \left[\left(\frac{k_{rnw}}{k_{rnw}^*} + k_{rw} \right) \left(\frac{k_{rna}}{k_{rnw}^*} + k_{ra} \right) - (k_{rw} + k_{ra}) \right] \quad (5.5)$$

where

k_{rnw}^* = the relative permeability of the nonaqueous phase at the residual saturation of water in a water–nonaqueous phase system

k_{rwn} = the relative permeability of the nonaqueous phase system as a function of S_w

k_{rna} = the relative permeability of the nonaqueous phase in an air–nonaqueous phase system as a function of S_a.

Figure 5.8 is a ternary relative permeability diagram for the nonaqueous phase liquid in an air-water–nonaqueous phase system—i.e., the vadose zone—where the water is the wetting fluid (Faust 1985).

5.2.5 Darcy's Law For Two-Phase Flow

Darcy's law for the steady-state saturated flow of water in the presence of a nonaqueous phase liquid is given (Schwille 1984) as

$$Q_w = \frac{-k_{rw} k_i \rho_w}{\mu_w} A \frac{dh_w}{dl} \quad (5.6)$$

where

Q_w = volume of water flowing

k_{rw} = relative permeability of water in the presence of the nonwetting fluid

k_i = intrinsic permeability of the rock

ρ_w = density of the water

μ_w = dynamic viscosity of the water

A = cross-sectional area of flow

dh_w/dl = gradient of the head of the water

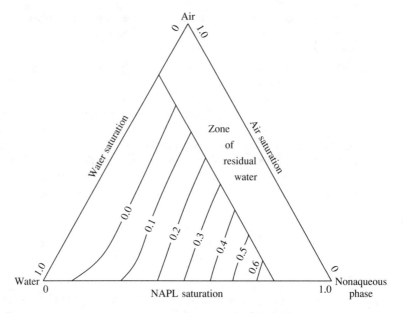

FIGURE 5.8 Ternary diagram showing the relative permeability of the NAPL phase in an air-water-NAPL system as a function of phase saturation. *Source:* C. R. Faust, *Water Resources Research* 21, no. 4(1985):587–96. Copyright by the American Geophysical Union.

A similar expression for the nonwetting fluid is

$$Q_{nw} = \frac{k_{rnw} k_i \rho_{nw}}{\mu_{nw}} A \frac{dh_{nw}}{dl} \tag{5.7}$$

5.2.6 Fluid Potential and Head

Hubbert (1953) defined fluid potential, Φ, as the amount of work needed to move a unit mass of fluid from some standard position and condition to a different position and condition. Position represents the potential energy of the fluid or elevation above the standard datum. The condition can be represented by the difference in pressure between the position under consideration and the standard pressure.

The fluid potential is thus defined as

$$\Phi = g(z - z_s) + (P - P_s)v_m \tag{5.8}$$

where

g = acceleration of gravity
z = elevation
z_s = standard elevation
P = pressure
P_s = standard pressure
v_m = volume per unit mass

Since volume per unit mass is the reciprocal of density, ρ, Equation 5.8 can be expressed as

$$\Phi = g(z - z_s) + \frac{P - P_s}{\rho} \qquad (5.9)$$

If the standard pressure is taken as atmospheric and z is defined as elevation above a convenient datum, such as sea level, then Equation 5.9 becomes

$$\Phi = gz + \frac{P}{\rho} \qquad (5.10)$$

If a pipe with an open bottom is inserted into an aquifer to a point at distance z above the datum, the fluid pressure at that location will cause the fluid in the aquifer to rise to a height h above the datum. The fluid pressure is equal to the weight of the fluid in the pipe per unit cross-sectional area:

$$P = \rho g(h - z) \qquad (5.11)$$

This can be substituted into Equation 5.10 to yield

$$\Phi = gz + \frac{\rho g(h - z)}{\rho} = gh \qquad (5.12)$$

where h is the total head.

Fluid will flow from an area of higher fluid potential, $\Phi + \Delta\Phi$, to an area of lower fluid potential, Φ. The force per unit mass exerted on the fluid by its environment is a vector, **E**. This force vector is perpendicular to the equipotential surfaces and in the direction of decreasing potential. It has a magnitude equal to the change in potential, $\Delta\Phi$, divided by the distance over which the change in potential is measured, Δn (Figure 5.9):

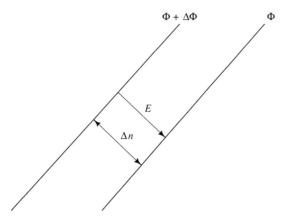

FIGURE 5.9 Relation between force field and potential gradient. *Source:* Modified from M. K. Hubbert, *American Association of Petroleum Geologists Bulletin* 37 (1953):1954–2026.

$$\mathbf{E} = -\frac{\Delta \Phi}{\Delta n} \tag{5.13}$$

The force vector **E** can be expressed several other ways:

$$\mathbf{E} = -\operatorname{grad} \Phi \tag{5.14}$$

$$\mathbf{E} = -g \operatorname{grad} h \tag{5.15}$$

$$\mathbf{E} = g - \frac{1}{\rho} \operatorname{grad} P \tag{5.16}$$

Equation 5.16 shows that at a point a unit mass of a fluid will be acted upon by a force **E**, which is the vector sum of gravity and the negative gradient of the pressure divided by the fluid density. Figure 5.10 shows the vector components of the force vector **E** for (a) the hydrostatic case where **E** = 0 and (b) the hydrodynamic case where **E** ≠ 0.

Equation 5.16 shows that the direction of the force vector is a function of the fluid density. Thus for the same point in the aquifer, different fluids will have different force vectors and, hence, different flow directions in the same potential field. Consider Figure 5.11, which has a force vector for water, $\mathbf{E}_w$, a force vector for an LNAPL, $\mathbf{E}_{LNAPL}$, and a force vector for a DNAPL, $\mathbf{E}_{DNAPL}$. For convenience, the water is shown to be flowing horizontally; that is, the force vector $\mathbf{E}_w$ is horizontal, although it could be going in any direction. Since the density of the LNAPL is less than the density of water, the vector $-\operatorname{grad} P/\rho_{LNAPL}$ is longer than the vector $-\operatorname{grad} P/\rho_w$ and the resulting vector $\mathbf{E}_{LNAPL}$ is angled upward compared with $\mathbf{E}_w$. The vector $\mathbf{E}_{DNAPL}$ is angled downward because the vector $-\operatorname{grad} P/\rho_{DNAPL}$ is shorter than $-\operatorname{grad} P/\rho_{DNAPL}$.

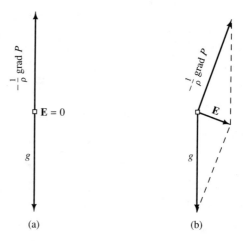

FIGURE 5.10 Vector components of the force vector for (a) hydrostatic case and (b) hydrodynamic case. *Source:* M. K. Hubbert, *American Association of Petroleum Geologists Bulletin* 37 (1953):1954–2026.

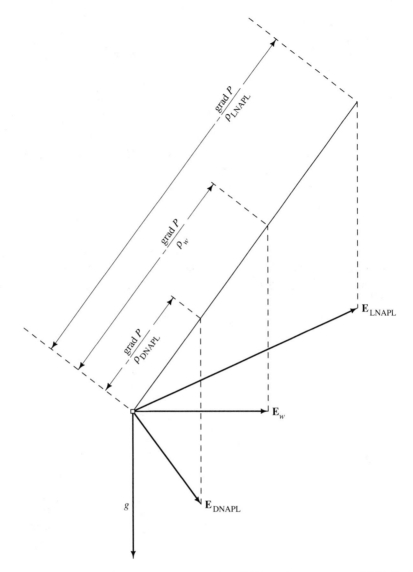

FIGURE 5.11 Force vectors for a DNAPL, water, and an LNAPL in the same potential field. The DNAPL sinks and the LNAPL rises with respect to the direction of ground-water flow. *Source:* Modified from M. K. Hubbert, *American Association of Petroleum Geologists Bulletin* 37 (1953):1954–2026.

This figure illustrates why a DNAPL will sink and an LNAPL will rise with respect to the direction of ground-water flow in the same potential field.

The fluid potential of a nonwetting fluid, either an LNAPL or a DNAPL, is given by

$$\Phi_{nw} = gz + \frac{P}{\rho_{nw}} \quad (5.17)$$

and the fluid potential for water is

$$\Phi_w = gz + \frac{P}{\rho_w} \tag{5.18}$$

If we solve Equation 5.18 for P and substitute it into 5.17, we obtain

$$\Phi_{nw} = \frac{\rho_w}{\rho_{nw}} \Phi_w - \frac{\rho_w - \rho_{nw}}{\rho_{nw}} gz \tag{5.19}$$

This expression relates the fluid potential of a nonwetting fluid to the fluid potential of water at the same location.

From Equation 5.12, $\Phi_{nw} = gh_{nw}$ and $\Phi_w = gh_w$, Equation 5.19 can be written as

$$h_{nw} = \frac{\rho_w}{\rho_{nw}} h_w - \frac{\rho_w - \rho_{nw}}{\rho_{nw}} z \tag{5.20}$$

In Equation 5.20, z is the elevation of the point in the aquifer, h_w is the height above the datum that water would stand in an open pipe terminating at the point, and h_{nw} is the height that a nonwetting fluid of density ρ_{nw} would stand. Figure 5.12 illustrates

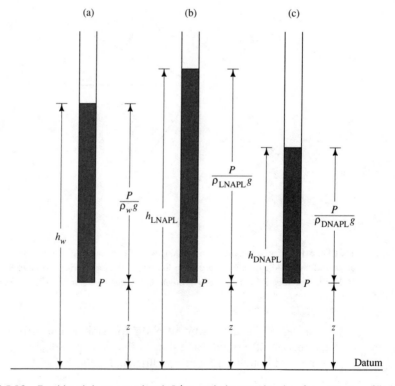

FIGURE 5.12 Total head, h, pressure head, $P/\rho g$, and elevation head, z, for open pipes filled with (a) water, (b) an LNAPL, and (c) an NAPL. All pipes have the same pressure at the open end.

Multiphase Flow

the relationships between h_w, h_{LNAPL}, and h_{DNAPL}. The fluid elevation in the pipe filled with LNAPL will be higher than the pipe filled with water, whereas the fluid elevation of the pipe filled with DNAPL will be lower.

5.3 Migration of Light Nonaqueous Phase Liquids (LNAPLs)

Light nonaqueous phase liquids are less dense that water. When spilled at the land surface, they migrate vertically in the vadose zone under the influence of gravity and capillary forces, just as water does. Unless the vadose zone is extremely dry, it will be water-wet, and the LNAPL will be the nonwetting phase.

Figure 5.13 shows the distribution of water in the vadose zone (Abdul 1988). Notice that at the top of the vadose zone the water is held at the irreducible water saturation. The water held here is called **pendular** water. Below that is a zone where the water content is above the irreducible saturation; this is sometimes called **funicular** water. When close to 100% water saturation is reached, we find the capillary fringe. The air-water relationship of the vadose zone behaves as a two-phase immiscible flow, so there is residual air saturation in the capillary zone. This is shown in Figure 5.13; however, we will ignore the residual air saturation in further analysis because it usually is a small value.

The capillary fringe is not a regular surface, such as the water table. The height of the capillary rise will be different in each vertical set of interconnected pores, depending

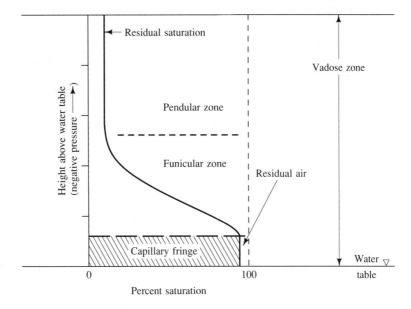

FIGURE 5.13 Vertical distribution of water in the vadose zone in the absence of nonaqueous phase liquids. *Source:* A. S. Abdul, *Ground Water Monitoring Review* 8, no. 4 (1988):73–81. Copyright © 1988 Water Well Journal Publishing Co.

upon the mean pore diameter of the set. Thus the capillary fringe has a ragged upper surface. However, for the sake of simplicity in diagrams of the capillary fringe, we will show it as a level surface. We can use the capillary tube model to estimate the average height of the capillary fringe. Equation 5.4 can give us the capillary pressure, P_t, based on a mean pore radius, r. This is equal to the weight of the water in the capillary tube, which is found by multiplying the height of the water in the tube, h_c, by the specific weight of water, γ.

$$P_t = -h_c\gamma = -\left(\frac{\sigma}{r}\right)\cos\theta \qquad (5.21)$$

$$h_c = \left(\frac{\sigma}{r\gamma}\right)\cos\theta \qquad (5.22)$$

For pure water in a clean glass tube, θ can be taken as zero, and cos 0 is 1.0. The value σ of for water at 20°C is 0.074 g/cm. With these values, Equation 5.22 becomes

$$h_c = \frac{0.15}{r} \quad \text{centimeter} \qquad (5.23)$$

Table 5.1 shows the heights of the capillary fringe that were observed experimentally in various materials. The visual capillary height is the level where the water saturation ratio is close to 1.0. The capillary water in the funicular zone is above this height, although it is not visible.

Figure 5.13 shows the capillary fringe as extending to the height where the water-saturation ratio begins to decline. This height is based on the larger pores, in which the capillary rise would be least. For the smaller pores, the capillary rise would be greater, extend upward into what is labeled the funicular zone. Water in that zone is not moving downward but is being held in place by capillary forces. This illustrates the irregularity of the capillary fringe. We use the phrase **capillary zone** to mean the part of the capillary fringe where the water saturation ratio is at or close to 1.0.

The LNAPL will travel vertically in the vadose zone. If a sufficient quantity is present so that the residual LNAPL saturation is exceeded, it will eventually reach the top of the capillary zone. However, much of the LNAPL may remain behind, trapped in the vadose zone. Eckberg and Sunada (1984) studied the distribution of oil in the vadose zone.

TABLE 5.1 Visual capillary rise in unconsolidated materials (porosity of all samples is about 41%).

Material	Grain Size (mm)	Capillary Rise (cm)
Fine gravel	2–5	2.5
Very coarse sand	1–2	6.5
Coarse sand	0.5–1	13.5
Medium sand	0.2–0.5	24.6
Fine sand	0.1–0.2	42.8
Silt	0.05–0.1	105.5
Fine silt	0.02–0.05	200+

Source: Lohman (1972).

Multiphase Flow

Figure 5.14 shows the changes in the distribution of water and oil in a sand column into which a quantity of oil was added. Note that much of the oil remains throughout the thickness of the vadose zone as a residual oil.

In moving downward, an LNAPL may displace some of the capillary water in the vadose zone, causing it to move ahead of the advancing LNAPL front. Once the capillary zone is reached, LNAPL will begin to accumulate. Initially, the LNAPL will be under tension, just as the water in the vadose zone is under tension. As additional LNAPL accumulates above the capillary zone, an "oil table" will develop, with some LNAPL having a positive pore pressure. The capillary zone will become thinner, and mobile, or "free," LNAPL will accumulate. Eventually, the capillary zone may disappear altogether and the oil table will rest directly on the water table. In the core of a thick zone of mobile LNAPL, the water table may be depressed by the weight of the LNAPL.

Abdul (1988) conducted an experiment to observe the development of an oil table. Vertical columns were packed with sand and partially filled with water. The columns had manometer/tensiometers installed at various heights to measure the pore-water

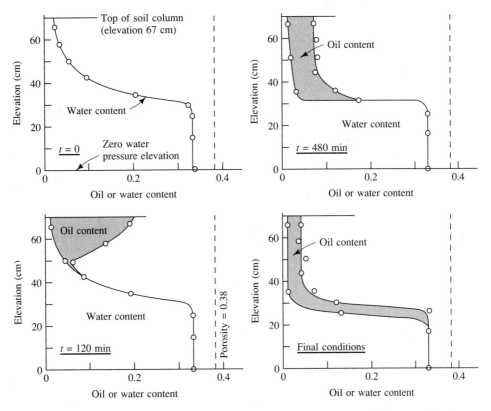

FIGURE 5.14 Changes in the vertical distribution of oil with time after a slug of oil is added to the top of a column of sand. Oil content and water content are expressed as a fraction of the total volume of the porous media. *Source:* D. K. Eckberg and D. K. Sunada, *Water Resources Research* 20, no. 12 (1984): 1891–97. Copyright by the American Geophysical Union.

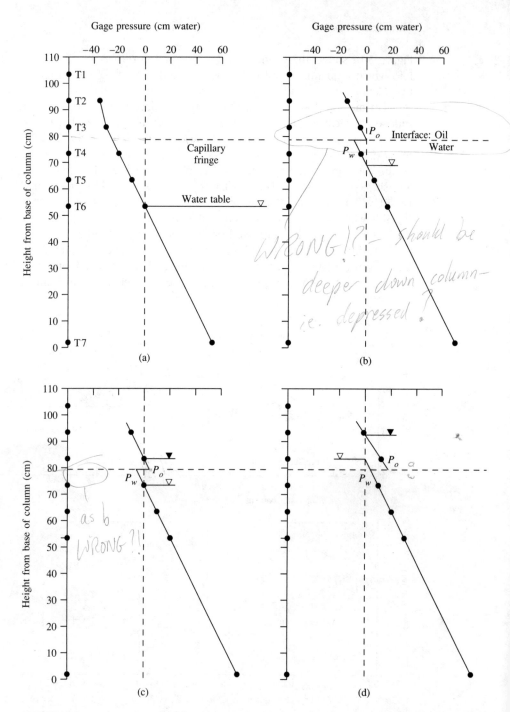

FIGURE 5.15 Hydrostatic pressure head/tension distribution in a sand column to which oil is being added to the top of the column. (a) Before the addition of the oil, (b) after addition of the oil showing the development of an oil fringe, (c) after addition of sufficient oil for an oil table to form, and (d) after sufficient mobile oil has accumulated to eliminate the water capillary fringe. *Source*: A. S. Abdul, *Ground Water Monitoring Review* 8, no. 4 (1988):73–81. Copyright © 1988 Water Well Journal Publishing Co. Used with permission.

pressure below the water table and tension above the water table. The elevation where the gauge pressure is zero is the water table. Figure 5.15(a) shows the distribution of pore pressures before any oil was added. Oil was then added to the top of the soil column, and the system was allowed to come into equilibrium. Figure 5.15(b) records the conditions after oil was added. The capillary fringe thinned, and the water table rose as the advancing oil displaced capillary water downward. Oil under tension accumulated above the capillary zone. Figure 5.15(c) shows that further addition of oil resulted in the formation of an oil table with positive pore pressure above the water capillary zone, which was still under tension. Eventually enough oil was added so the water capillary fringe disappeared and the oil table rested directly on the water table. An oil capillary fringe existed above the oil table (Figure 5.15(d)).

The mobile LNAPL can migrate in the vadose zone following the slope of the water table. Figure 5.16 shows the shape of a migrating spill of an LNAPL. Note that there is residual LNAPL in the unsaturated zone above the mobile LNAPL. In addition some of the LNAPL components can dissolve in the ground water and move by diffusion and advection with the ground water. For example, gasoline will release significant amounts of benzene, toluene, ethylbenzene, and xylene (BTEX) as soluble fractions.

The residual NAPL material in the vadose zone can partition into the vapor phase as well as a soluble phase in capillary water. The degree of the partitioning will depend upon the relative volatility of the material and its solubility in water.

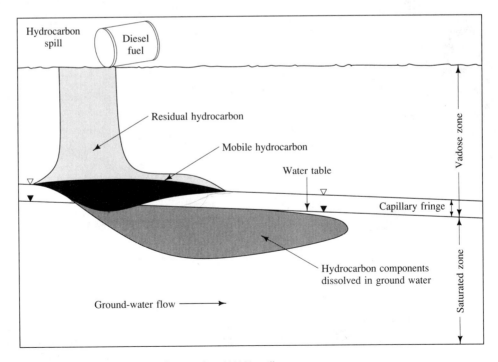

FIGURE 5.16 Subsurface distribution of an LNAPL spill.

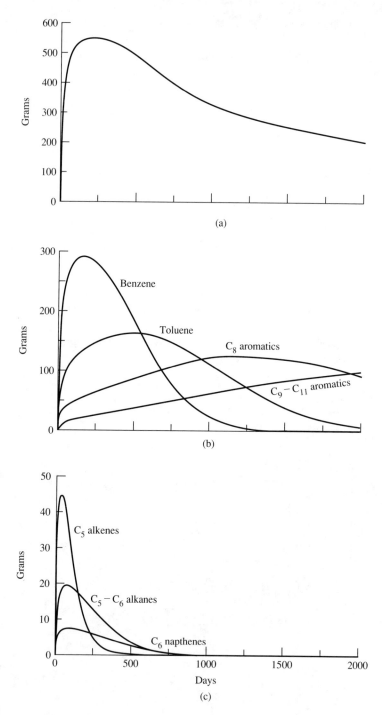

FIGURE 5.17 Mass of residual hydrocarbon in vadose zone partitioning into capillary water as a function of time, with (a) total hydrocarbons, (b) aromatic constituents, and (c) nonaromatic constituents. *Source:* A. L. Baehr, *Water Resources Research* 23, no. 10 (1987):1926–38. Copyright by the American Geophysical Union.

Henry's law states there is a linear relationship between the vapor pressure of a solute above its aqueous solution and the concentration in solution. The proportionality constant between the two is called a **Henry's law constant,** which can be expressed in units of atmospheres/(moles/cubic meter water) (Section 7.2). The proportionality constant from Henry's law has also been expressed as a water-air partition coefficient. This is the ratio of the aqueous solubility of a substance, expressed in milligrams per liter at a given temperature to the saturated vapor concentration of the pure phase of the substance, also expressed in milligrams per liter (Baehr 1987).

Those compounds with low water-air partition coefficients, such as the alkanes, favor the vapor phase, whereas those with high water-air partition coefficients, such as benzene, favor the aqueous phase. Hydrocarbons such as gasoline are a mixture of up to 200 different organic compounds; therefore, various water-air partition coefficients are needed to describe the behavior of the various constituent compounds. The diffusive properties of the soil are also very important in controlling vapor phase transport (Baehr and Corapcioglu 1987).

Baehr (1987) developed a model to describe the vapor phase and aqueous transport of residual hydrocarbons in the vadose zone. Figure 5.17 shows the partitioning of hydrocarbon mass from gasoline into vadose zone water as a function of time. This figure shows that the aromatic compounds, those based on the benzene ring, partition into water at a higher rate and for a longer time period than the nonaromatic compounds. This is to be expected, since the nonaromatic hydrocarbons studied, C_5 alkenes, C_5–C_6 alkanes, and C_6 napthenes, have much lower water-air partition coefficients than the aromatic constituents, benzene, toluene, ethylbenzene, xylene, etc. Table 5.2 gives water-air partition coefficients for selected gasoline constituents. The selective partitioning of benzene, toluene, ethylbenzene, and xylene in the aqueous capillary phase helps to explain why these compounds are so diagnostic of a gasoline spill. They can reach the water table via infiltration of capillary water through a zone of residual gasoline, even if no gasoline itself reaches the water table. Figure 5.18 illustrates this phenomenon.

TABLE 5.2 Water-air partition coefficients for selected organic compounds.

Compound	Formula	Molecular Weight	Water-Air Partition Coefficient
Aromatics			
Benzene	C_6H_6	78	5.88
Toluene	C_7H_8	92	3.85
o-Xylene	C_8H_{10}	106	4.68
Ethylbenzene	C_8H_{10}	106	3.80
Nonaromatics			
Cyclohexane	C_6H_{12}	84	0.15
1-Hexene	C_6H_{12}	84	0.067
n-Hexane	C_6H_{14}	86	0.015
n-Octane	C_8H_{18}	114	0.0079

Source: A. L. Baehr. *Water Resources Research* 23, no. 10:1928. Published 1987 by American Geophysical Union. Used with permission.

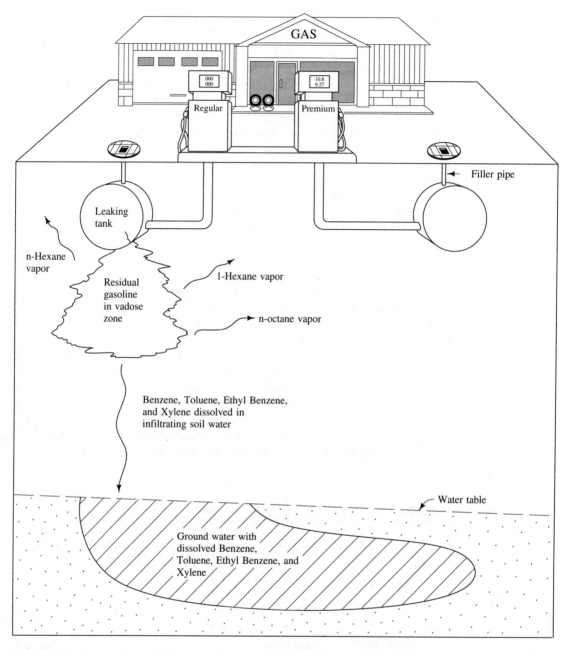

FIGURE 5.18 Process of ground water being contaminated by gasoline constituents from residual gasoline in the vadose zone.

5.4 Measurement of the Thickness of a Floating Product

The measurement of the amount of mobile LNAPL above the water table is not straightforward. Figure 5.19 shows the distribution of an LNAPL above the water table for the condition where a water capillary zone exists. This diagram shows that there is a zone of immobile LNAPL above the capillary zone where the LNAPL content is less than the residual LNAPL saturation. When the LNAPL content exceeds the residual LNAPL saturation and the sum of the water saturation and the LNAPL saturation is 100%, there will be positive pore pressures. In this zone LNAPL will be mobile and can flow laterally into a monitoring well. The screen zone of the monitoring well must thus extend above the top of the zone of free or mobile LNAPL. The water level in the monitoring well will initially be at the water table, which is below the level of the bottom of the mobile LNAPL zone. The LNAPL will flow down the monitoring well to the water table. The weight of the LNAPL will then depress the surface of the water in the monitoring well below that

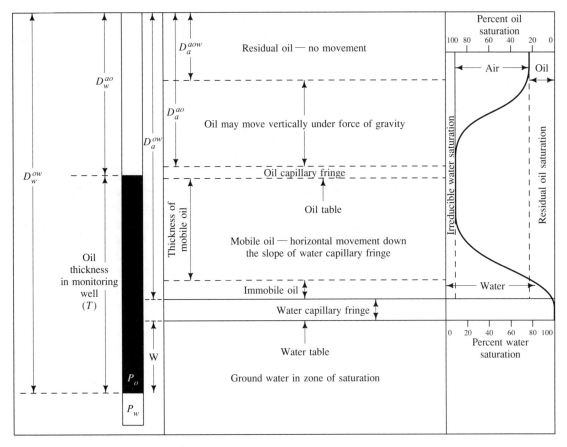

FIGURE 5.19 Comparison of distribution of mobile oil in an aquifer with the thickness of floating oil in a monitoring well for the case where a water capillary fringe exists below the zone of mobile oil.

of the water table. As a result the thickness of LNAPL measured in a monitoring well is greater than the actual thickness of the mobile LNAPL in the vadose zone. This effect is greater for thin zones of free LNAPL, where the capillary zone may be much thicker, than for a thick zone of free LNAPL, where the capillary zone may be thin or missing. It is also greater in fine-grained materials, where the capillary fringe may be thicker.

The depth below the water table at which the LNAPL will reach equilibrium in a monitoring well can be calculated. At equilibrium, the pressure in the monitoring well on the LNAPL side of the interface is P_o, and the pressure on the water side of the interface is P_w. The two pressures must be equal:

$$P_o = P_w \tag{5.24}$$

P_o is equal to the density of the oil, ρ_o, times the thickness of the oil layer, T.

$$P_o = \rho_o T \tag{5.25}$$

P_w is equal to the density of the water, ρ_w, times the distance from the water table to the interface, W.

$$P_w = \rho_w W \tag{5.26}$$

Since $P_o = P_w$, then

$$\rho_o T = \rho_w W \tag{5.27}$$

Solving for W yields

$$W = \left(\frac{\rho_o}{\rho_w}\right) T \tag{5.28}$$

Farr, Houghtalen, and McWhorter (1990) and Lenhard and Parker (1990) developed two methods to estimate the volume of recoverable LNAPL in an aquifer based on the thickness of the LNAPL floating in a monitoring well. These methods are based on the capillary soil properties. One of the two methods is based on the determination of soil properties as reported by Brooks and Corey (1966). We will look at this method in some detail using the derivation of Farr, Houghtalen, and McWhorter.

T as shown in Figure 5.19 is the difference between the depth to the water-oil interface in the well, D_w^{ow} and the depth to the oil-air interface, D_w^{ao}. The values of the depth to the oil table in the aquifer, D_a^{ow}, and the depth to the top of the capillary fringe, D_a^{ao}, can be computed.

$$D_a^{ao} = D_w^{ao} - \frac{P_d^{ao}}{\rho_o g} \tag{5.29}$$

$$D_a^{ow} = D_w^{ow} - \frac{P_d^{ow}}{(\rho_w - \rho_o)g} \tag{5.30}$$

where

P_d^{ao} = the Brooks-Corey air-organic displacement pressure
P_d^{ow} = the Brooks-Corey organic-water displacement pressure
g = the acceleration of gravity

Equation 5.30 may be rewritten as

$$D_a^{ow} = D_w^{ao} + T - \frac{P_d^{ow}}{(\rho_w - \rho_o)g} \quad (5.31)$$

If any of the organic liquid exists at a positive pore pressure, then D_a^{ow} will be greater than D_w^{ao} and from Equation 5.31,

$$T \geq \frac{P_d^{ow}}{(\rho_w - \rho_o)g} \quad (5.32)$$

If the organic liquid is all under tension in the capillary zone, then there will be no mobile organic layer and no organic liquid will collect in the monitoring well. Under these conditions, Equations 5.29, 5.30, 5.31, and 5.32 are not applicable. However, as soon as free organic liquid appears in the aquifer, it will collect to a depth of at least $P_d^{ow}/(\rho_w - \rho_o)g$.

The total volume of nonresidual organic liquid in the vadose zone is given by

$$V_o = n \left\{ \int_{D_a^{owa}}^{D_a^{ow}} (1 - S_w) \, dz - \int_{D_a^{owa}}^{D_a^{ao}} [1 - (S_w + S_o)] \, dz \right\} \quad (5.33)$$

where

V_o = the volume of organic liquid per unit area
n = the porosity
S_w = the water-saturation ratio
S_o = the organic liquid saturation ratio
z = the vertical coordinate measured positively downward
D_a^{ow} = a value determined from Equation 5.30
D_a^{ao} = a value determined from Equation 5.29
D_a^{owa} = the top of the zone where nonresidual oil occurs

Based on work by Lenhard and Parker (1987, 1988), the fluid-content relations are

$$S_o - S_w = (1 - S_{wi}) \left(\frac{P_c^{ao}}{P_d^{ao}} \right)^{-\lambda} + S_{wi}, \quad P_c^{ao} > P_d^{ao} \quad (5.34a)$$

$$S_o + S_w = 1, \quad P_c^{ao} < P_d^{ao} \quad (5.34b)$$

$$S_w = (1 - S_{wi}) \left(\frac{P_c^{ow}}{P_d^{ow}} \right)^{-\lambda} + S_{wi}, \quad P_c^{ow} > P_d^{ow} \quad (5.35a)$$

$$S_w = 1, \quad P_c^{ow} < P_d^{ow} \quad (5.35b)$$

where

S_{wi} = the irreducible water saturation
λ = the Brooks-Corey pore-size distribution index

In addition,

$$P_c^{ao} = \rho_o g(D_w^{ao} - (P_d^{ao}/\rho_o g) - z) + P_d^{ao} \quad (5.36)$$

$$P_c^{ow} = g(\rho_w - \rho_o) \left[D_w^{ow} - \frac{P_d^{ow}}{(\rho_w - \rho_o)g} - z \right] + P_d^{ow} \quad (5.37)$$

Integration of Equation 5.33 for $D_a^{aow} > 0$, using Equations 5.34, 5.35, 5.36, and 5.37, yields the following. For λ not equal to 1,

$$V_o = \frac{\phi(1 - S_{wi})D}{1 - \lambda}\left[\lambda + (1 - \lambda)\left(\frac{T}{D}\right) - \left(\frac{T}{D}\right)^{1-\lambda}\right] \quad \text{(5.38a)}$$

For λ equal to 1,

$$V_o = n(1 - S_{wi})[1 - D(1 + \ln T)] \quad \text{(5.38b)}$$

where

$$D = \frac{P_d^{ow}}{(\rho_w - \rho_o)g} - \frac{P_d^{ao}}{\rho_o g}$$

$$T = D_w^{ow} - D_w^{ow} \geq \frac{P_d^{ow}}{(\rho_w - \rho_o)g}$$

If organic liquid above the residual saturation exists all the way to the land surface, then D_a^{owa} does not exist. Under this condition integration of Equation 5.33 yields the following. For λ not equal to 1,

$$V_o = n(1 - S_{wi})\left\{(T - D) - \frac{P_d^{ao}}{\rho_o g(1 - \lambda)}\left[1 - \left(\frac{\rho_o g D_w^{ao}}{P_d^{ao}}\right)^{1-\lambda}\right]\right.$$
$$\left. + \frac{P_d^{ow}}{(\rho_w - \rho_o)g(1 - \lambda)}\left[1 - \left(\frac{(\rho_w - \rho_o)g D_w^{ow}}{P_d^{ow}}\right)^{1-\lambda}\right]\right\} \quad \text{(5.39a)}$$

For λ equal to 1,

$$V_o = n(1 - S_{wi})\left[(T - D) - \frac{P_d^{ow}}{(\rho_w - \rho_o)g}\ln D_w^{ow} + \frac{P_d^{oa}}{\rho_o g}\ln D_w^{ao}\right] \quad \text{(5.39b)}$$

The Brooks-Corey soil parameters can thus be used to estimate the volume of recoverable organic liquid in an aquifer based on the thickness of the organic liquid in the well, the measured depths in the well to the air-organic interface, and the organic-water interface, along with the densities of the organic liquid and the water. One must measure the Brooks-Corey soil parameters and the densities in the lab. The weakness in the Brooks-Corey approach is that it may not be accurate for very small volumes of mobile organic liquid in the soil. From Equation 5.32 the thickness of organic liquid in the well is at least $P_d^{ow}/[(\rho_w - \rho_o)g]$, even for very small volumes of mobile LNAPL.

Farr, Houghtalen, and McWhorter (1990) also presented an alternative method of analysis based on soil parameters developed by van Genuchten (1980). However, the equations based on the van Genuchten soil parameters are nonlinear and can't be solved analytically; they must be solved numerically. Thus this model is not as convenient as the one based on a Brooks-Corey soil. However, under certain conditions, such as thin layers of mobile organic liquid in the soil, the van Genuchten model may be more accurate than the Brooks-Corey model.

Figure 5.20(a) shows a graph of the volume of gasoline in a sandstone versus the measured thickness in a well computed by both the Brooks-Corey soil model and the van Genuchten soil model. Figure 5.20(b) shows the same thing for a different organic liquid, Soltrol (Farr, Houghtalen, and McWhorter 1990). It can be seen that for gasoline in sandstone, the van Genuchten and Brooks-Corey models yield quite similar results.

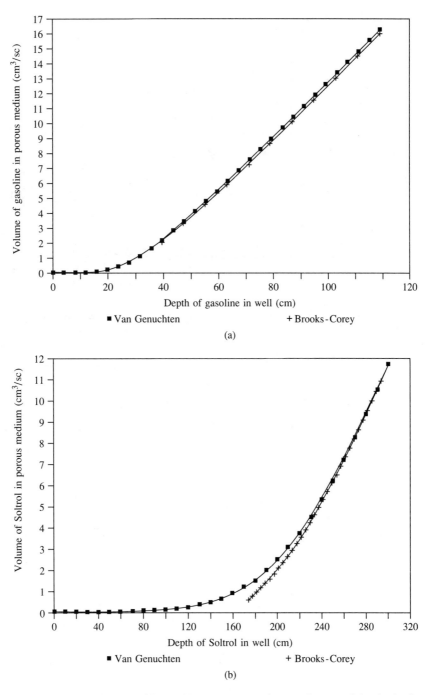

FIGURE 5.20 Volume of recoverable LNAPL in a porous media as a function of the depth of product floating in a monitoring well as computed by the method of Farr et al. (1990) based on both van Genuchten and Brooks-Corey soils models for (a) gasoline and (b) Soltrol. *Source*: A. M. Farr, R. J. Houghtalen, and D. B. McWhorter, *Ground Water*, 28, no. 1 (1990):48–56. Copyright © 1990 Water Well Journal Publishing Co. Used with permission.

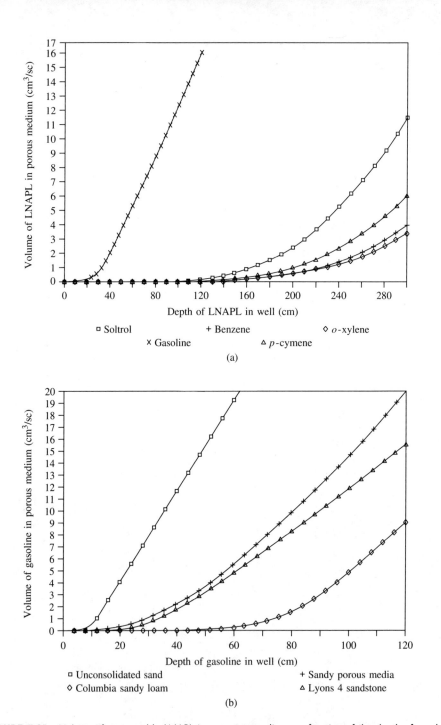

FIGURE 5.21 Volume of recoverable LNAPL in a porous media as a function of the depth of product floating in a monitoring well as computed by the method of Farr et al. (1990) based on a van Genuchten soil model (a) for various LNAPLs in a sandstone and (b) for gasoline in various porous media. *Source:* A. M. Farr, R. J. Houghtalen, and D. B. McWhorter, *Ground Water* 28, no. 1 (1990):48–56. Used with permission. Copyright © 1990 Water Well Journal Publishing Co.

Multiphase Flow

However, for Soltrol, they do not, because the Brooks-Corey model has a somewhat greater thickness of organic liquid in the well for very small volumes in the soil. Figure 5.21(a) shows the computed volume-thickness relationships, based on the van Genuchten soil model, for several LNAPLs in a sandstone. Figure 5.21(b) shows the computed volume-thickness relationships, again based on the van Genuchten soil model, for gasoline in a number of different porous media (Farr, Houghtalen, and McWhorter 1990). These diagrams illustrate the fact that there is no simple, constant relationship between the volume of an LNAPL in a soil and the thickness as measured in a monitoring well. It is a function of the properties of both the soil and the organic liquid.

5.5 Effect of the Rise and Fall of the Water Table on the Distribution of LNAPLs

The flow of LNAPLs is complicated by the rise and fall of the water table with the seasons. Figure 5.22(a) shows a layer of oil floating on the surface of the capillary zone. As the water table falls, the layer of mobile oil also falls. Residual oil is left in the vadose zone above the oil table as it falls. This is illustrated in Figure 5.22(b). When the water table rises, the oil table also rises. However, as Figure 5.22(c) illustrates, residual oil is left behind in the saturated zone. If the water table rises faster than the oil table can rise, "pockets" of free oil might become left below the water table. The flow of water and hydrocarbons is controlled by Darcy's law and depends upon the effects of density, viscosity, and relative permeability. Depending upon these factors, either the hydrocarbon or the water could have a greater velocity as the water table rises and falls.

In cleaning up LNAPL spills, the mobile LNAPL can be removed by skimming wells or trenches. However, a considerable amount of LNAPL will be left as a residual on the soil. Volatile LNAPLs can be removed by a soil-vapor extraction system. However, nonvolatile products will remain behind in the soil. The amount depends upon the properties of the LNAPL and the texture of the soil. The oil retention capacity of soil is estimated to range from 5 L/m^3 for gravel to 40 L/m^3 for silty sand (Testa and Paczkowski 1989). Many hydrocarbons can be degraded by soil bacteria, especially if the soil is aerobic. Systems that diffuse air into the soil have been effective in bioremediation of hydrocarbon spills. This problem is addressed in Chapter 9.

5.6 Migration of Dense Nonaqueous Phase Liquids

5.6.1 Vadose Zone Migration

Dense nonaqueous phase liquids (DNAPLs) have a specific gravity greater than 1. When spilled on the land surface or discharged to the subsurface, once the residual saturation value is exceeded they move vertically in the vadose zone under the influence of gravity. Since it is the wetting liquid, water occupies the smaller pores and capillary channels in the vadose zone. The DNAPL migrates through the larger pore openings, which initially have water coating the mineral grains, with air filling the remainder of the pore. The DNAPL displaces the air, so the pore becomes filled with the small amount of water wetting the mineral surface and the DNAPL. The vadose zone permeability for the DNAPL

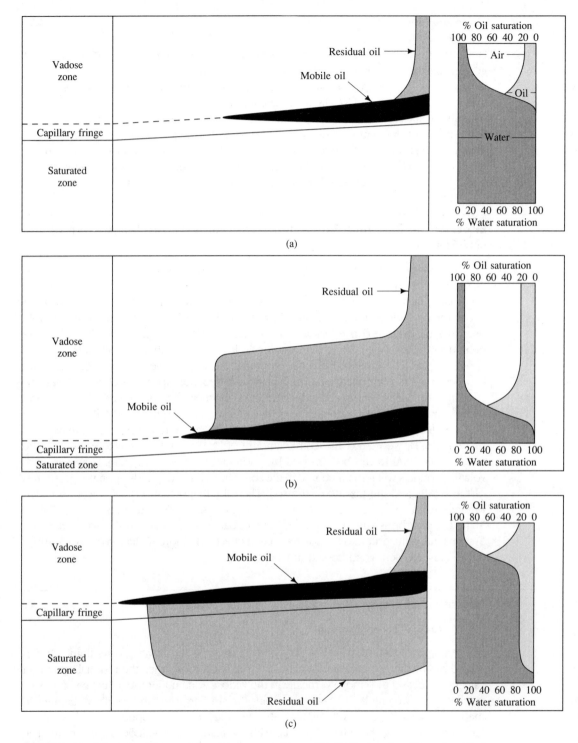

FIGURE 5.22 Effect of a falling and then rising water table on the distribution of mobile and residual phases of an LNAPL.

Multiphase Flow

is greater than for water because the pores through which the DNAPL, as the nonwetting fluid, is migrating are larger than the pores through which the water, as the wetting fluid, migrates.

When the DNAPL reaches the capillary zone, where the pores are all filled with water, it must start to displace water in order to migrate downward. This displacement of water with DNAPL also occurs below the water table.

5.6.2 Vertical Movement in the Saturated Zone

If water in the unsaturated zone is static, the DNAPL will continue to migrate downward under the force of gravity. In order to displace the water filling the pores, the DNAPL must have sufficient mass to overcome the capillary forces that hold the water in the pore. Vertical stringers of DNAPL can occupy vertically connected pores. When the vertical stringer has a sufficient height, its weight can displace the water in the pore. The critical height, h_o, can be determined from Hobson's formula (Berg, 1975):

$$h_o = \frac{2\sigma \cos\theta(1/r_t - 1/r_p)}{g(\rho_w - \rho_o)} \qquad (5.40)$$

where

σ = interfacial tension between the two liquids
θ = wetting angle
r_t = pore-throat radius
r_p = pore radius
g = acceleration of gravity
ρ_w = density of water
ρ_o = density of the DNAPL

For a well-rounded, well-sorted sediment of diameter d with rhombohedral packing, the pore-throat radius, r_t, and the pore radius, r_p, can be estimated from the following formulas:

$$r_p = 0.212d \qquad (5.41)$$
$$r_t = 0.077d \qquad (5.42)$$

The smaller the sediment diameter, the smaller both the pore-throat radius and pore radius. Equation 5.40 demonstrates that the value of h_o is inversely related to the pore diameter. For finer-grained materials, the DNAPL stringer must be longer than for coarser materials. Even thin, fine-grained layers could act as confining layers for a DNAPL.

If sufficient amounts of the DNAPL are present to overcome the capillary pressure and expel water from pores, DNAPL will continue to migrate downward under the force of gravity until it reaches an aquitard layer, where the pore openings are so small that the DNAPL cannot overcome the capillary forces binding the water in the pores. A layer of DNAPL then accumulates on the surface of the aquitard. This zone also has irreducible water present. Fluid pumped from this zone is exclusively DNAPL. Above the layer of DNAPL and irreducible water, there is a zone with DNAPL and water content above the irreducible saturation. Fluid pumped from this level includes both DNAPL and water.

From the top of this zone to the water table, the pores contain residual DNAPL and water. Fluid pumped from this zone is water. Since many DNAPLs are slightly to moderately soluble in water, water pumped from any of these zones can also contain dissolved organics. Figure 5.23 shows the DNAPL zones described previously.

Monitoring wells to detect DNAPLs should be placed at the bottom of the aquifer, just at the top of the confining layer. DNAPL from the zone of mobile DNAPL and irreducible water flows to the monitoring well, as will both water and DNAPL from the zone where both are mobile. The water and DNAPL from this zone will separate in the monitoring well, with the DNAPL sinking and the water rising. Well A in Figure 5.23 shows that the total depth of the DNAPL in the monitoring well is below the top of the zone where both water and DNAPL are present. The exact position depends upon the average DNAPL saturation in this zone. If a monitoring well extends below the top of the confining layer, a false thickness of the DNAPL will be measured, as it will fill the monitoring well below the level of the confining layer. Monitoring well B in Figure 5.23 shows this circumstance.

The relative thickness of the various zones described here depends upon the grain-size distribution, which is reflected in the permeability of the saturated zone. A low permeability aquifer (small pores) will have a thin layer of DNAPL collect on the bottom,

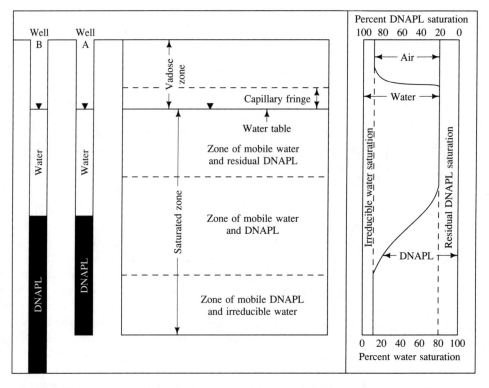

FIGURE 5.23 Zones of a DNAPL and the relationship of mobile DNAPL and nonmobile DNAPL to the DNAPL saturation; relationship of mobile DNAPL thickness to thickness of DNAPL is measured in a monitoring well.

Multiphase Flow

while a more permeable aquifer (large pores) will have a thicker zone of mobile NAPL on the bottom and a thinner zone where both DNAPL and water are mobile. Figure 5.24 illustrates this phenomenon (Villaume 1985).

If the DNAPL is not spilled in sufficient amounts to overcome the residual saturation in the vadose zone, ground-water contamination by the dissolved phase can still occur. The residual DNAPL forms a source that, although not mobile, can slowly partition into both the vapor phase and the aqueous phase of the water infiltrating through the vadose zone. If the DNAPL is volatile, as many are, the vapor can diffuse through the vadose zone as well. The vapor phase can partition into the pore water so that the area of soil water contamination can spread via the vapor phase.

5.6.3 Horizontal Movement in the Saturated Zone

If the DNAPL exists in a continuous phase, it will move below the water table according to the force vector $\mathbf{E}_{DNAPL}$ described in Equation 5.16. If the DNAPL is in the form of discontinuous stringers that are sinking in the aquifer, flowing ground water will tend to displace the DNAPL stringers in the direction of flow. Just as the sinking DNAPL stringers displace water by overcoming the capillary pressure holding the water in a pore, the laterally moving water must overcome the capillary pressure of the DNAPL stringer to displace it sideways.

Villaume (1985) indicates that the lateral pressure gradient needed to displace a DNAPL sideways, grad P, is given by

$$\text{grad } P = \frac{2\sigma}{L_o(1/r_t - 1/r_p)} \tag{5.43}$$

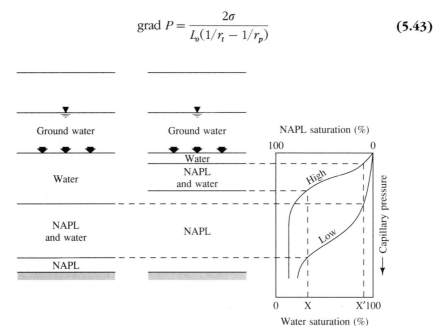

FIGURE 5.24 Effect of high and low permeability (and porosity) on the distribution of mobile DNAPL at the bottom of an aquifer; the arrows indicate level of original injection of the DNAPL. Source: J. F. Villaume, *Ground Water Monitoring Review* 5, no. 2 (1985):60–74. Copyright © 1985 Water Well Journal Publishing Co.

where

σ = interfacial tension

L_o = length of the continuous DNAPL phase in the direction of flow

r_t = pore-throat radius

r_p = pore radius

Once the percolating DNAPL reaches the aquitard layer, it can begin to move laterally, even in the absence of a hydraulic gradient on the water table. It migrates down the dip of the aquitard. DNAPLs can collect in low spots on the surface of an aquitard. It is possible for the DNAPL to migrate downdip, even if the hydraulic gradient and ground-water flow are in the opposite direction (Figure 5.25).

If a pocket of static DNAPL collects in a low spot on the surface of an aquitard and ground water is flowing in the aquifer above the DNAPL, the interface between the flowing ground water and the static DNAPL will be sloping. This angle can be found from (Hubbert 1953)

$$\tau = \frac{\rho_w}{\rho_w - \rho_{DNAPL}} \, dh/dl \tag{5.44}$$

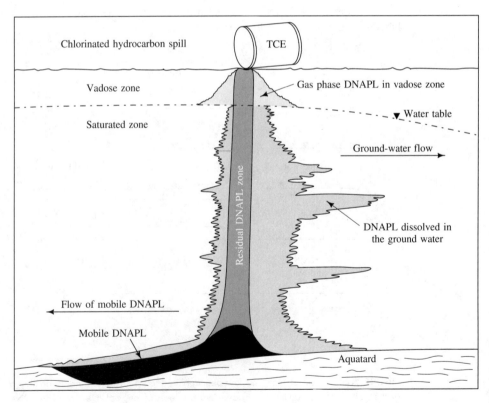

FIGURE 5.25 Distribution of a dense nonaqueous phase liquid in the vadose and saturated zone.

Multiphase Flow

where τ is the angle of the interface and dh/dl is the slope of the water table (Figure 5.26). A negative value for τ means that the angle has a slope opposite to the direction of the water table.

Buckley and Leverett (1942) derived the following equation for the one-dimensional flow of two immiscible, incompressible fluids:

$$\frac{\partial}{\partial x}\left(G\frac{\partial S_w}{\partial x}\right) - q_t \frac{df}{dS_w}\frac{\partial S_w}{\partial x} = n\frac{\partial S_w}{\partial t} \quad (5.45)$$

where

q_t = total volume flux

S_w = relative saturation of the wetting phase

n = porosity

f = a function that depends upon the value of S_w and is defined as:

$$f(S_w) = \left[1 + \frac{k_{nw}\mu_w}{k_w\mu_{nw}}\right]^{-1} \quad (5.46)$$

and G is a function that depends upon the value of S_w and is defined as

$$G(S_w) = -\frac{k_{nw}f}{\mu_{nw}}\frac{dP_c}{dS_w} \quad (5.47)$$

where

k_w = permeability to wetting fluid

k_{nw} = permeability to nonwetting fluid

μ_w = dynamic viscosity of wetting fluid

μ_{nw} = dynamic viscosity of nonwetting fluid

P_c = capillary pressure

The volume flux of the wetting phase, q_w, is given by

$$q_w = fq_t - G\frac{\partial S_w}{\partial x} \quad (5.48)$$

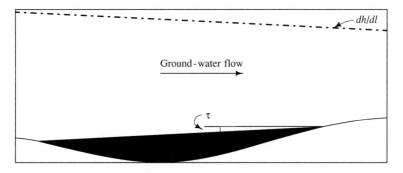

FIGURE 5.26 Sloping interface between a static layer of DNAPL and flowing ground water.

The volume flux of the nonwetting phase, q_{nw}, is given by

$$q_{nw} = q_t(1 - f) + G \frac{\partial S_w}{\partial x} \qquad (5.49)$$

Buckley and Leverett (1942) solved these equations by assuming that the force due to the rate at which the fluid is added at the boundary, q_t, is much greater than the capillary pressure force, P_c. Thus the second term in Equation 5.48 can be ignored. However, this may not be accurate. McWhorter and Sunada (1990) have developed an exact integral solution to this two-phase flow problem that accounts for both forces. They have solutions for either the displacement of the wetting fluid by the advance of a nonwetting fluid or the displacement of a nonwetting fluid by an invading wetting fluid. The former solution can be applied to the problem of the displacement of water in an aquifer by the lateral flow of a DNAPL.

5.7 Monitoring for LNAPLs and DNAPLs

Special consideration must be given to the design of monitoring wells and the collection of ground-water samples to test for the presence of LNAPLs and DNAPLs (floaters and sinkers). Naturally, different types of wells are used for each separate phase.

Because LNAPLs float on the capillary layer, a monitoring well for LNAPL detection should extend from above the capillary zone to below the water table. If an LNAPL is present, it will be floating at the surface of the liquid column in the monitoring well. Prior to any purging of the well, a top-loading bailer should be carefully lowered to just below the liquid surface so that the top layer of liquid drains into the bailer. If an LNAPL is present, the top-loading bailer should capture it. The LNAPL should be analyzed qualitatively. If there is a mixture of compounds present, it might be necessary to determine the proportion of each in the LNAPL. Special probes are available to measure the thickness of LNAPLs floating in a monitoring well.

To sample a DNAPL, a monitoring well should be constructed with a screen at the very bottom of the aquifer. It may be helpful to have a length of solid pipe as a sump at the bottom of the screen so that if even a thin layer of mobile DNAPL is present, it can collect in the sump in a sufficient thickness to sample. If a sump is used, the thickness of any measured product thickness must be reduced by the length of the sump. A bottom-loading bailer is used to collect the liquid from the bottom of the sump prior to any well purging. The bailer is slowly lowered all the way to the bottom of the monitoring well and then slowly raised. The collected sample should be placed into a glass jar to see if there is a dense layer on the bottom. In case the entire bailer is filled with a clear liquid, part of the contents should be placed in a jar partially filled with water to see if a separate phase forms.

If the chemical analysis of an aqueous water sample indicates that an organic compound is present in amounts greater than its published solubility value, the compound may be present as a nonaqueous phase that was emulsified in the sample collection process.

Monitoring for DNAPLs in areas of fractured bedrock geology is much more difficult than in sand aquifers. Figure 5.27 shows a buried barrel from which a DNAPL has leaked.

Multiphase Flow

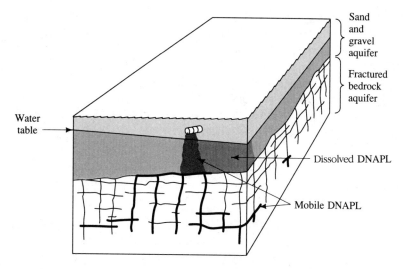

FIGURE 5.27 Movement of a DNAPL into a fractured bedrock aquifer that underlies a sand and gravel aquifer.

The DNAPL moves vertically through the vadose zone and the saturated sand and gravel aquifer that overlies the fractured bedrock. Some of the DNAPL dissolves in the flowing ground water and is transported by advection. When the DNAPL reaches the bedrock surface it flows downslope, in this case in a direction opposite to the flow of ground water. The DNAPL moves vertically into fractures in the bedrock below the pool that collects on the bedrock surface. The DNAPL can also migrate horizontally between vertical cracks. As a result, the DNAPL can spread in unexpected and unpredictable directions from the leaking barrel. Many bedrock monitoring wells would be needed even to find the portions of the DNAPL plume in the bedrock; it might be impossible to locate all of it (Mackay and Cherry 1989).

5.8 Summary

Many cases of ground-water contamination involve organic liquids that are either insoluble or only partially soluble in water. These liquids may be present both above and below the water table as separate nonaqueous phase liquids (NAPLs). A NAPL that is less dense than water (LNAPL) will float on the water table or the top of the capillary fringe. A NAPL that is more dense than water (DNAPL) may sink into the aquifer below the water table.

Water and a NAPL may both be present in the ground. Depending upon their relative proportions, only the water or only the NAPL or both may be mobile. Depending upon the surface tension of the liquids, one will be a wetting fluid and one a nonwetting fluid. The term relative permeability refers to the permeability of the soil for one fluid in the presence of a given volumetric content of a second fluid. Darcy's law can be written in terms of the relative permeability for both a nonwetting and a wetting fluid.

If a sufficient depth of an LNAPL collects on the surface of the capillary fringe, it can flow into a shallow monitoring well. The thickness of LNAPL measured in a monitoring well is greater than the thickness of the free LNAPL in the subsurface. If certain soil parameters are known, one can compute the thickness of the mobile LNAPL based on the thickness that accumulates in a monitoring well.

A DNAPL may sink in an aquifer until it reaches a fine-grained layer. DNAPL may accumulate in a mobile layer at the bottom of the aquifer. A monitoring well screened at the bottom of the aquifer may be used to detect the presence of a DNAPL.

Chapter Notation

A	Cross-sectional area of flow
d	Porous media grain diameter
dh_w/dl	Gradient of fluid potential for wetting fluid
dh_{nw}/dl	Gradient of fluid potential for nonwetting fluid
D	Argument $= (P_d^{ow}/(\rho_w - \rho_o)g) - (P_d^{ao} - \rho_o g)$
D_a^{ao}	Depth in the aquifer to the top of the capillary fringe
D_a^{ow}	Depth in the aquifer to the top of the oil table
D_a^{owa}	Depth in the aquifer to the top of the zone where nonresidual organic fluid occurs
D_w^{ao}	Depth in a well to the air-organic interface
D_w^{ow}	Depth in a well to the organic-water interface
E	Force vector
E_w	Force vector for water
E_{DNAPL}	Force vector for a DNAPL
E_{LNAPL}	Force vector for a LNAPL
f	A Buckley-Leverett function
G	A Buckley-Leverett function
g	Acceleration of gravity
grad P	Lateral pressure gradient
h_c	Height of capillary rise in a tube
h_w	Head (potential) of wetting fluid
h_{nw}	Head (potential) of nonwetting fluid
h_o	Critical height for a DNAPL stringer before it can displace water in the saturated zone
k_i	Intrinsic permeability
k_r	Relative permeability
k_{rw}	Relative permeability of water
k_{ra}	Relative permeability of air
k_{rn}	Relative permeability of a nonaqueous phase liquid
k_{rnw}^*	Relative permeability of nonaqueous phase liquid at the residual saturation of water in a water–nonaqueous phase system
k_{rnw}	Relative permeability of the nonaqueous phase system as a function of S_w

k_{rna}	Relative permeability of the nonaqueous phase in an air–nonaqueous phase system as a function of S_a
L_o	Vertical length of a DNAPL stringer
n	Porosity
P	Fluid pressure
P_d^{ao}	Brooks-Corey air-organic displacement pressure
P_d^{ow}	Brooks-Corey organic-water displacement pressure
P_w	Pore pressure for a wetting fluid
P_{nw}	Pore pressure for a nonwetting fluid
P_c	Capillary pressure
P_d	Threshold or displacement pressure
P_o	Pressure in an LNAPL at an LNAPL-water interface in a monitoring well
P_s	Reference pressure
P_t	Capillary pressure in a thin tube
P_w	Pressure in water at an LNAPL-water interface in a monitoring well
Q_w	Volume of wetting fluid flowing in a two-phase system
Q_{nw}	Volume of nonwetting fluid flowing in a two-phase system
q_t	Total volume flux
q_w	Volume flux of wetting fluid
q_{nw}	Volume flux of nonwetting fluid
r	Radius of a thin tube
r'	Radius of curvature for an air-water interface
r_p	Pore radius
r_t	Pore-throat radius
S_a	Saturation ratio for air
S_e	Effective saturation
S_o	Organic liquid saturation ratio
S_w	Wetting-fluid or water-saturation ratio
S_{wi}	Irreducible wetting-fluid saturation
S_{nwr}	Residual nonwetting-fluid saturation
T	Thickness of oil in a monitoring well
V_o	Volume of organic liquid per unit area
W	Depth from the water table to an organic fluid–water interface in a well
z	Vertical coordinate measured positively downward
z_s	Reference elevation
θ	Interfacial angle between two nonwetting fluids
$\sigma_{i,j}$	Interfacial tension between substances i and j
γ	Specific weight
λ	Brooks-Corey pore-size distribution index
ρ	Density
ρ_w	Density of water
ρ_o	Density of an organic fluid
ρ_{nw}	Density of a nonwetting fluid
ρ_{DNAPL}	Density of a DNAPL

ρ_{LNAPL} Density of a LNAPL
μ_w Dynamic viscosity of water
Φ Fluid potential
Φ_w Fluid potential for wetting fluid
Φ_{nw} Fluid potential for nonwetting fluid

References

Abdul, Abdul S. 1988. Migration of petroleum products through sandy hydrogeologic systems. *Ground Water Monitoring Review* 8, no. 4:73–81.

Abdul, Abdul S., Sheila F. Kau, and Thomas L. Gibson. 1989. Limitations of monitoring wells for the detection and quantification of petroleum products in soils and aquifers. *Ground Water Monitoring Review* 9, no. 2:90–99.

Abriola, Linda M., and George F. Pinder. 1985a. A multiphase approach to the modeling of porous media contamination by organic compounds: 1. Equation development. *Water Resources Research* 21, no. 1:11–18.

———. 1985b. A multiphase approach to the modeling of porous media contamination by organic compounds: 2. Numerical simulation. *Water Resources Research* 21, no. 1:19–26.

Baehr, Arthur L. 1987. Selective transport of hydrocarbons in the unsaturated zone due to aqueous and vapor phase partitioning. *Water Resources Research* 23, no. 10:1926–38.

Baehr, Arthur L., and M. Yavuz Corapcioglu. 1987. A compositional multiphase model for groundwater contamination by petroleum products: 2. Numerical solution. *Water Resources Research* 23, no. 1:201–13.

Bear, Jacob. 1972. *Dynamics of fluids in porous media*. New York: American Elsevier Publishing Company, 764 pp.

Berg, R., R. 1975. Capillary pressures in stratigraphic traps. *Bulletin, American Association of Petroleum Geologists* 59, no. 6:935–56.

Brooks, R. H., and A. T. Corey. 1966. Properties of porous media affecting fluid flow. *Proceedings, American Society of Civil Engineers, Irrigation and Drainage Division* 92, no. IR2:61–87.

Buckley, S. E., and M. C. Leverett. 1942. Mechanism of fluid displacement in sand. *Transactions, American Institute of Mining, Metallurgical, and Petroleum Engineering* 146:107–16.

Corapcioglu, Yavuz M., and Arthur L. Baehr. 1987. A compositional multiphase model for groundwater contamination by petroleum products: 1. Theoretical considerations. *Water Resources Research* 23, no. 1:191–200.

Eckberg, David K., and Daniel K. Sunada. 1984. Nonsteady three-phase immiscible fluid distribution in porous media. *Water Resources Research* 20, no. 12:1891–97.

Farr, A. M., R. J. Houghtalen, and D. B. McWhorter. 1990. Volume estimation of light nonaqueous phase liquids in porous media. *Ground Water* 28, no. 1:48–56.

Faust, Charles R. 1985. Transport of immiscible fluids within and below the unsaturated zone: a numerical model. *Water Resources Research* 21, no. 4:587–96.

Hall, Robert A., Steven B. Blake, and Stephen C. Champlin, Jr. 1984. Determination of hydrocarbon thicknesses in sediments using borehole data. *Proceedings of the Fourth National Symposium and Exposition on Aquifer Restoration and Ground Water Monitoring*. National Water Well Association, pp. 300–10.

Hochmuth, D. P., and David K. Sunada. 1985. Ground-water model of two phase immiscible flow in coarse material. *Ground Water* 23, no. 5:617–26.

Hubbert, M. King. 1953. Entrapment of petroleum under hydrodynamic conditions. *American Association of Petroleum Geologists Bulletin* 37, no. 8:1954–2026.

Lenhard, R. J., and J. C. Parker. 1987a. A model for hysteretic constructive relations governing multiphase flow: 2. Permeability-saturation relations. *Water Resources Research* 23, no. 12:2197–2206.

———. 1987b. Measurement and prediction of saturation-pressure relationships in three-phase porous media systems. *Journal of Contaminant Hydrogeology* 1:407–24.

———. 1988. Experimental validation of the theory of extending two-phase saturation-pressure relations to three-fluid phase systems for monotonic drainage paths. *Water Resources Research* 24, no. 3:373–80.

———. 1990. Estimation of free hydrocarbon volume from fluid levels in monitoring wells. *Ground Water* 28, no. 1:57–67.

Lohman, S. W. 1972. *Ground water hydraulics*. U.S. Geological Survey Professional Paper 708:70 pp.

Lin, C., George F. Pinder, and E. F. Wood. 1982. *Water resources program report 83-WR-2*. Water Resources Program. Princeton, N. J.: Princeton University.

Mackay, D. M., and J. A. Cherry. 1989. Groundwater contamination: pump and treat remediation. *Environmental Science and Engineering* 23, no. 6:630–37.

McWhorter, David B, and Daniel K. Sunada. 1990. Exact integral solutions for two-phase flow. *Water Resources Research* 26, no. 3:399–413.

Parker, J. C., and R. J. Lenhard. 1987. A model for hysteretic constructive relations governing multiphase flow: 2. Saturation-pressure relations. *Water Resources Research* 23, no. 12:2187–96.

Pinder, George F., and Linda M. Abriola. 1986. On the simulation of nonaqueous phase organic compounds in the subsurface. *Water Resources Research* 22, no. 9:109S–119S.

Schwille, F. 1981. "Groundwater pollution in porous media by fluids immiscible with water." *The science of the total environment* 21:173–85.

———. 1984. "Migration of organic fluids immiscible with water in the unsaturated zone." In *Pollutants in porous media*, edited by B. Yaron, G. Dagen and J. Goldshmid, 27–48. Berlin: Springer-Verlag.

Schwille, F. 1988. *Dense chlorinated solvents in porous and fractured media*. Translated by James F. Pankow. Chelsea, Mich.: Lewis Publishers, 146 pp.

Stone, H. L. 1973. Estimation of three-phase relative permeability and residual oil data. *Journal of Canadian Petroleum Technology* 12, no. 4:53–61.

Testa, Stephen M., and Michael T. Paczkowski. 1989. Volume determination and recovery of free hydrocarbon. *Ground Water Monitoring Review* 9, no. 1:120–27.

Van Genuchten, M. Th. 1980. A closed-form equation for predicting the hydraulic conductivity of unsaturated soils. *Soil Science Society of America Journal* 44:892–98.

Villaume, James F. 1985. Investigations at sites contaminated with dense nonaqueous phase liquids (NAPLS). *Ground Water Monitoring Review* 5, no. 2:60–74.

Chapter Six
Inorganic Chemicals in Ground Water

6.1 Introduction

Ground water is a solvent that is in contact with various earth materials. As a result, ground water naturally contains dissolved cations and anions as well as some nonionic inorganic material, such as silica (SiO_2). Naturally occurring ground water can contain dissolved solids that range in concentration from less than 100 mg/L to more than 500,000 mg/L (Hem 1985). The major ion constituents of natural water include calcium, magnesium, sodium, potassium, chloride, sulfate, and bicarbonate/carbonate. Dissolved gasses can include nitrogen, carbon dioxide, methane, oxygen, and hydrogen sulfide. There are a number of ions that can be naturally present in small amounts that can affect the water quality. In addition, inorganic ions that impact upon water quality can be released to the subsurface via human activity.

We have already seen in Chapter 3 that ions can be removed from solution by ion exchange and sorption. In this chapter we will examine other chemical processes that act to remove inorganic ions from solution. We will also examine the chemical properties of a number of inorganic materials frequently found in ground water. The geochemical zonation that can occur near landfills that have received municipal waste will be used to illustrate some basic principles.

6.2 Units of Measurement and Concentration

Chemical analyses are usually reported on the basis of weight of solute per volume of solvent. Common units are **milligrams per liter** (mg/L) and **micrograms per liter** (μg/L). Equivalent weight units are frequently used when the chemical behavior of a solute is being considered. The **equivalent weight** of an ion is the formula weight divided by the electrical charge. If the concentration of the ion in milligrams per liter is divided by the formula weight, the resulting concentration is expressed in terms of **milliequivalents per liter.** One **mole** of a substance is its formula weight in grams. A 1-**molal** solution has 1 mole of solute in 1000 g of solvent. A 1-**molar** (1-M) solution has 1 mole of solute in a liter of solvent.

If a solution is dilute and there is no need to make density corrections, the molality can be determined from the concentration by the following equation:

$$\text{Molality} = \frac{\text{milligrams per liter} \times 10^{-3}}{\text{formula weight in grams}} \quad (6.1)$$

6.3 Chemical Equilibrium and the Law of Mass Action

The law of mass action states that the rate of a chemical reaction will be proportional to the active masses of the participating substances (Hem 1985). If there are two substances, A and B, reacting to form two other substances, C and D, and if the process is reversible, then the reaction can be written as

$$a\text{A} + b\text{B} \rightleftharpoons c\text{C} + d\text{D} \quad (6.2)$$

The rate of the forward reaction, R_1, is

$$R_1 = k_1'[\text{A}]^a[\text{B}]^b \quad (6.3)$$

whereas the rate of the reverse reaction, R_2, is

$$R_2 = k_2'[\text{C}]^c[\text{D}]^d \quad (6.4)$$

where:

$[\text{A}]$ = active concentration of substance A
k_1' = proportionality constant for the forward reaction
k_2' = proportionality constant for the reverse reaction

If the reaction progresses to a point where the forward reaction rate is equal to the reverse reaction rate, then

$$k_1'[\text{A}]^a[\text{B}]^b \rightleftharpoons k_2'[\text{C}]^c[\text{D}]^d \quad (6.5)$$

Equation 6.5 can be rearranged to yield the following expression:

$$\frac{[\text{C}]^c[\text{D}]^d}{[\text{A}]^a[\text{B}]^b} = \frac{k_1'}{k_2'} = K_{eq} \quad (6.6)$$

where K_{eq} is the **equilibrium constant.**

If two or more ions react to form a solid precipitate and the reaction is reversible, then it can be represented as

$$a\text{A} + b\text{B} \rightleftharpoons c\text{AB} \quad (6.7)$$

The equilibrium relationship of this reaction is:

$$K_{sp} = \frac{[\text{A}]^a[\text{B}]^b}{[\text{AB}]^c} \quad (6.8)$$

where K_{sp} is called a **solubility product.** The activity of the solid together with the water is defined as unity. Solubility products can be used to compute the concentration of a solute in equilibrium with a solid phase, either via dissolution of the solid into an undersaturated solution or following precipitation of the solid from a saturated solution.

If one is dealing with a very dilute aqueous solution, then molal concentrations can be used to determine chemical equilibrium. However, for the general case, one must use **chemical activities** to employ the law of mass action.

The chemical activity of ion X, $[X]$, is equal to the molal concentration of X, m_x, times a factor known as an **activity coefficient,** γ_x:

$$[X] = m_x \gamma_x \tag{6.9}$$

The activity coefficient varies with the total amount of cations and anions in solution. The concentration and charge of the various ions in a solution determine its **ionic strength.** Ionic strength can be computed from the following formula:

$$I = \tfrac{1}{2} \sum m_i z_i^2 \tag{6.10}$$

where

$I =$ ionic strength

$m_i =$ molality of the ith ion

$z_i =$ charge of the ith ion

Once the ionic strength is determined, the activity coefficient can be calculated using the Debye-Hückel equation:

$$-\log \gamma_i = \frac{A z_i^2 \sqrt{I}}{1 + a_i B \sqrt{I}} \tag{6.11}$$

where

$\gamma_i =$ the activity coefficient for ionic species i

$z_i =$ the charge on ionic species i

$I =$ ionic strength of the solution

$A =$ constant equal to 0.5085 at 25°C

$B =$ constant equal to 0.3281 at 25°C

$a_i =$ the effective diameter of the ion from Table 6.1

The Debye-Hückel equation can be used with solutions that have an ionic strength of 0.1 or less (approximately 5000 mg/L). Figure 6.1 is a graph showing the relationship of activity coefficient to ionic strength for specific ions; it was calculated using the Debye-Hückel equation. Specific curves are for ions with the same effective diameter and charge as listed in Table 6.1. Not all the ions to which a curve applies are listed on the figure. For example, the curve labeled Ca^{2+} and Fe^{2+} can also be used for Cu^{2+}, Zn^{2+}, Sn^{2+}, Mn^{2+}, Ni^{2+}, and Co^{2+}, because all these ions have the same effective diameter and charge.

Chemical equilibrium is a useful concept in studies of contaminant hydrogeology. Ionic contaminants discharged into ground water may react with naturally occurring ions

Inorganic Chemicals in Ground Water

TABLE 6.1 Values of the parameter a_i in the Debye-Hückel equation.

a_i	Ion
11	Th^{4+}, Sn^{4+}
9	Al^{3+}, Fe^{3+}, Cr^{3+}, H^+
8	Mg^{2+}, Be^{2+}
6	Ca^{2+}, Cu^{2+}, Zn^{2+}, Sn^{2+}, Mn^{2+}, Fe^{2+}, Ni^{2+}, Co^{2+}, Li^+
5	$Fe(CN)_6^{4-}$, Sr^{2+}, Ba^{2+}, Cd^{2+}, Hg^{2+}, S^{2-}, Pb^{2+}, CO_3^{2-}, SO_3^{2-}, MoO_4^{2-}
4	PO_4^{3-}, $Fe(CN)_6^{3-}$, Hg_2^{2-}, SO_4^{2-}, SeO_4^{2-}, CrO_4^{3-}, HPO_4^{2-}, Na^+, HCO_3^-, $H_2PO_4^-$
3	OH^-, F^-, CNS^-, CNO^-, HS^-, ClO_4^-, K^+, Cl^-, Br^-, I^-, CN^-, NO_2^-, NO_3^-, Rb^+, Cs^+, NH_4^+, Ag^+

Source: Reprinted with permission from J. Kielland, "Individual Activity Coefficients of Ions in Aqueous Solutions," *American Chemical Society Journal* 59 (1937):1676–78. Published 1937 by the American Chemical Society.

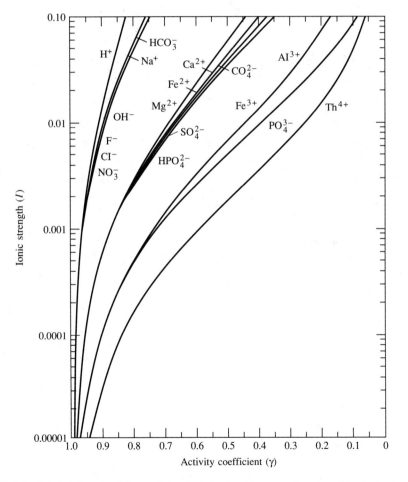

FIGURE 6.1 Relationship of activity coefficients of dissolved ions as a function of the ionic strength of a solution at 25°C. Source: J. D. Hem, "Study and interpretation of the chemical characteristics of natural waters," *Water Supply Paper* 2254, U.S. Geological Survey, 1985.

TABLE 6.2 Solubility products for selected minerals and compounds.

Compound	Solubility Product	Mineral Name
Chlorides		
$CuCl$	$10^{-6.7}$	
$PbCl_2$	$10^{-4.8}$	
Hg_2Cl_2	$10^{-17.9}$	
$AgCl$	$10^{-9.7}$	
Fluorides		
BaF_2	$10^{-5.8}$	
CaF_2	$10^{-10.4}$	Fluorite
MgF_2	$10^{-8.2}$	Sellaite
PbF_2	$10^{-7.5}$	
SrF_2	$10^{-8.5}$	
Sulfates		
$BaSO_4$	$10^{-10.0}$	Barite
$CaSO_4$	$10^{-4.5}$	Anhydrite
$CaSO_4 \cdot 2 H_2O$	$10^{-4.6}$	Gypsum
$PbSO_4$	$10^{-7.8}$	Anglesite
Ag_2SO_4	$10^{-4.8}$	
$SrSO_4$	$10^{-6.5}$	Celestite
Sulfides		
Cu_2S	$10^{-48.5}$	
CuS	$10^{-36.1}$	
FeS	$10^{-18.1}$	
PbS	$10^{-27.5}$	Galena
HgS	$10^{-53.3}$	Cinnebar
ZnS	$10^{-22.5}$	Wurtzite
ZnS	$10^{-24.7}$	Sphalerite
Carbonates		
$BaCO_3$	$10^{-8.3}$	Witherite
$CdCO_3$	$10^{-13.7}$	
$CaCO_3$	$10^{-8.35}$	Calcite
$CaCO_3$	$10^{-8.22}$	Aragonite
$CoCO_3$	$10^{-10.0}$	
$FeCO_3$	$10^{-10.7}$	Siderite
$PbCO_3$	$10^{-13.1}$	
$MgCO_3$	$10^{-7.5}$	Magnesite
$MnCO_3$	$10^{-9.3}$	Rhodochrosite
Phosphates		
$AlPO_4 \cdot 2 H_2O$	$10^{-22.1}$	Variscite
$CaHPO_4 \cdot 2 H_2O$	$10^{-6.6}$	
$Ca_3(PO_4)_2$	$10^{-28.7}$	
$Cu_3(PO_4)_2$	$10^{-36.9}$	
$FePO_4$	$10^{-21.6}$	
$FePO_4 \cdot 2 H_2O$	$10^{-26.4}$	

Source: K. B. Krauskopf, *Introduction to Geochemistry*, 2d ed. (New York: McGraw-Hill, 1979).

Inorganic Chemicals in Ground Water

in the ground water to form a precipitate or they may mobilize ions sorbed on solid surfaces. They may also undergo oxidation or reduction. Both these processes are reversible and can be described by chemical equilibrium. Many geochemical processes in ground water are not readily reversible, such as weathering of silicate minerals. These reactions must be treated using kinetics. However, as this type of reaction is not of significant interest in contaminant hydrogeology, we do not consider kinetic models.

Table 6.2 contains the solubility products for a large number of minerals, including many that can be formed from trace metals that can be ground-water contaminants.

6.4 Oxidation-Reduction Reactions

In some chemical reactions the participating elements change their valence state through the gain or loss of electron(s). If an electron is gained, there is a loss of positive valence called a reduction. A loss of negative valence is called an oxidation. Together, these are referred to as **oxidation-reduction**, or **redox**, reactions. In environmental systems they may be controlled by microorganisms that do not participate in the reaction but act as catalysts. The microbes occur as a biofilm on the surfaces of the aquifer materials. They obtain energy by oxidation of organic compounds or hydrogen or reduced inorganic forms of iron, nitrogen, and sulfur. Electron acceptors are necessary for these biologically mediated redox reactions to occur. Under aerobic conditions oxygen is the electron acceptor, whereas under anaerobic conditions nitrate, sulfate, and carbon dioxide are the electron acceptors (McCarty, Rittman, and Bouwer 1984).

An example of a reduction is

$$Fe^{2+} + 2\,e^- \rightleftharpoons Fe^0 \qquad (6.12)$$

In this example, ferrous iron is reduced to metallic iron by the addition of two electrons. This is a half-reaction, since the electrons must be supplied either by an electrical current or by a simultaneous reaction in which another element is oxidized and releases the requisite number of electrons. The standard electrical potential of a half-reaction is the voltage represented by the flow of electrons when the reaction is at equilibrium. Under standard conditions (25°C and 1 atm pressure) the standard potential is represented by the symbol E^0. The potential is in volts, with a negative value representing reducing conditions and a positive value representing oxidizing conditions (Hem 1985). By convention, the standard potential for the reduction of H^+ to hydrogen gas is 0:

$$2\,H^+ + 2\,e^- \rightleftharpoons H_2 \quad \text{(gas)} \qquad (6.13)$$

An example of oxidation occurs where ferrous iron loses an electron to form ferric iron:

$$Fe^{2+} \rightleftharpoons Fe^{3+} + e^- \qquad (6.14)$$

Oxidation-reduction reactions involve elements that can occur in more than one valence state. In Equations 6.12 and 6.14, iron occurred in the metallic (0) as well as the +2 and +3 states. Metals can usually occur in the metallic state with a zero valence and at least one other valence state. Some elements that are environmentally important can occur in several different valence states. Table 6.3 lists several elements that occur in different valence states and examples of compounds and ions formed from those elements.

In order for oxidation or reduction to occur in a chemical reaction, one element must be reduced while a second element is being oxidized. For example, the complete equation for the oxidation of ferrous iron to ferric iron is

$$4\,Fe^{2+} + O_2 + 4\,H^+ \rightleftharpoons 2\,H_2O + 4\,Fe^{3+} \qquad (6.15)$$

This complete reaction is composed of two half-reactions:

$$4\,Fe^{2+} \rightleftharpoons 4\,Fe^{3+} + 4\,e^- \qquad \text{[oxidation]} \qquad (6.16)$$
$$O_2 + 4\,H^+ + 4\,e^- \rightleftharpoons 2\,H_2O \qquad \text{[reduction]} \qquad (6.17)$$

An aqueous solution has an oxidation potential indicated by the symbol Eh. This can be calculated from the **Nernst equation:**

$$Eh = E^0 - \frac{RT}{nF} \ln \frac{[\text{products}]}{[\text{reactants}]} \qquad (6.18)$$

TABLE 6.3 Selected elements that can exist in more than one oxidation state.

Element	Valence State	Examples
Carbon	+4	HCO_3^-, CO_3^{2-}
	0	C
	−4	CH_4
Chromium	+6	CrO_4^{2-}, $Cr_2O_7^{2-}$
	+3	Cr^{3+}, $Cr(OH)_3$
Copper	+1	CuCl
	+2	CuS
Mercury	+1	Hg_2Cl_2
	+2	HgS
Iron	+2	Fe^{2+}, FeS
	+3	Fe^{3+}, $Fe(OH)_3$
Nitrogen	+5	NO_3^-
	+3	NO_2^-
	0	N
	−3	NH_4^+, NH_3
Oxygen	0	O
	−1	H_2O_2
	−2	H_2O, O^{2-}
Sulfur	−2	H_2S, S^{2-}, PbS
	+2	$S_2O_3^{2-}$
	+5	$S_2O_6^{2-}$
	+6	SO_4^{2-}

where:

        Eh = oxidation potential of the aqueous solution in volts

        E^0 = standard potential of redox reaction in volts

        R = gas constant, 0.00199 Kcal/(mole·K)

        T = temperature in Kelvins

        F = Faraday constant, 23.06 Kcal/V

        n = number of electrons in half-reaction

        [ ] = activity of products and reactants

The standard potential for a reaction can be determined from the relationship

$$E^0 = \frac{-\Delta G_R^0}{nF} \tag{6.19}$$

where ΔG_R^0 in volts is the **free energy,** or **Gibbs free energy,** of the reaction.

The free energy of a reaction is the sum of the free energies of the products minus the sum of the free energies of the reactants. For the reaction

$$a\text{A} + b\text{B} \rightleftharpoons c\text{C} + d\text{D}$$

the free energy can be found from:

$$\Delta G_R^0 = c\Delta G_c^0 + d\Delta G_d^0 - a\Delta G_a^0 - b\Delta G_b^0 \tag{6.20}$$

Values of free energy for many elements, ions, and compounds are found in standard reference works. Table 6.4 contains values for a number of species.

The equilibrium constant for a reaction is related to the free energy of the reaction by

$$\Delta G_R^0 = -RT \ln K_{eq} \tag{6.21}$$

At standard temperature and pressure and with ΔG_R^0 in kilocalories, Equation 6.21 can be rewritten as

$$\log K_{eq} = \frac{-\Delta G_R^0}{1.364} \tag{6.22}$$

The oxidation potential of an aqueous solution can be measured using a specific ion electrode. If the value is positive, the solution is oxidizing, and if it is negative, the solution is reducing. Oxidation potential is measured in volts relative to the hydrogen electrode, which is at zero. Commercially available Eh meters are available that can be attached to a ground-water sampling pump. The ground-water sample is pumped under positive pressure into the flowthrough cell where the electrode is located. The water sample is never subjected to a vacuum, which could cause degassing. Moreover, it is not exposed to the atmosphere, where it can come into contact with atmospheric oxygen. This has simplified the accurate and precise measurement of Eh in ground water.

TABLE 6.4 Standard Gibbs free energy of formation for selected species.

Species	$\Delta G°$ kcal/mole	Species	$\Delta G°$ kcal/mole	Species	$\Delta G°$ kcal/mole
Arsenic		**Manganese**		$UO_2(c)$ (uranite)	-246.61^e
$As_2O_5(c)$	-187.0^a	$MnO_2(c)$ (pyrolusite)	-111.18^b	UO_2^+ (aq)	-229.69^e
$As_4O_6(c)$	-275.46^a	Mn_2O_3 (bixbyite)	-210.6^b	UO_2^{2+} (aq)	-227.68^e
$As_2S_3(c)$	-40.3^a	Mn_3O_4 (hausmannite)	-306.7^b	$(UO_2)_2(OH)_2^{2+}$ (aq)	-560.99^e
$FeAsO_4(c)$	-185.13^a	$Mn(OH)_2$ (c) amorphous	-147.0^b	$(UO_2)_3(OH)_5^+$ (aq)	-945.16^e
H_3AsO_4(aq)	-183.1^a	$MnCO_3(c)$ (rhodochrosite)	-195.2^b	$(UO_2)_3(OH)_7^-$	-1037.5^e
$H_2AsO_4^-$ (aq)	-180.04^a	Mn^{2+} (aq)	-54.5^b	$UO_2CO_3^0(c)$	-367.07^e
$HAsO_4^{2-}$ (aq)	170.82^a	$MnOH^+$ (aq)	-96.8^b	$UO_2(CO_3)_2^{2-}$ (aq)	-503.2^e
AsO_4^{3-} (aq)	-155.0^a	**Molybdenum**		$UO_2(CO_3)_3^{4-}$ (aq)	-635.69^e
$HAsO_2$(aq)	-96.25^a	$MoO_3(c)$	-159.66^b	**Miscellaneous species**	
AsO_2^- (aq)	-83.66^a	$MoO_2(c)$	-127.40^b	$ZnFe_2O_4(c)$	-254.2^a
Chromium		$FeMoO_4(c)$	-233^b	$CuFeO_2(c)$	-114.7^b
$Cr_2O_3(c)$	-252.9^f	MoO_4^{2-} (aq)	-199.9^b	$CuFe_2O_4(c)$	-205.26^b
$HCrO_4^-$	-182.8^f	**Silver**		$NiFe_2O_4(c)$	-232.6^b
$Cr_2O_7^{2-}$ (aq)	-311.0^f	$Ag_2O(c)$	-2.68^b	$H_2O(l)$	-56.687^a
CrO_4^{2-} (aq)	-173.96^f	$AgCl(c)$	-26.24^b	OH^- (aq)	-37.594^a
Copper		$Ag_2S(c)$	-9.72^b	O_2(aq)	-3.9^a
$CuO(c)$	-31.0^b	$Ag_2CO_3(c)$	-104.4^b	HSO_4^- (aq)	-180.69^a
$CuSO_4 \cdot 3Cu(OH)_2(c)$	-434.5^b	Ag^+ (aq)	18.43^b	SO_4^{2-} (aq)	-177.97^a
(brochantite)		$AgOH$(aq)	-22.0^b	H_2S(aq)	-6.66^a
$Cu_2O(c)$	-34.9^b	$Ag(OH)_2^-$ (aq)	-62.2^b	HS^- (aq)	2.88^a
$Cu_2S(c)$	-20.6^b	$AgCl$(aq)	-17.4^b	S^{2-} (aq)	20.5^a
Cu^{2+} (aq)	15.67^b	$AgCl_2^-$ (aq)	-51.5^b	$CO_2(g)$	-94.254^a
$CuSO_4$(aq)	-165.45^b	**Vanadium**		CO_2(aq)	-92.26^a
$HCuO_2^-$ (aq)	-61.8^b	$H_4VO_4^+$	-253.67^k	H_2CO_3(aq)	-148.94^a
CuO_2^{2-} (aq)	-43.9^b	$H_3VO_4^0$	-249.2^k	HCO_3^- (aq)	-140.26^a
Cu^+ (aq)	11.95^b	$H_2VO_4^-$	-244^k	CO_3^{2-} (aq)	-126.17^a
Iron		HVO_4^{2-} (aq)	-233.0^i	Cl^- (aq)	-31.37^a
$Fe(OH)_3(c)$ ppt.	-166.0^h	VO_4^{3-} (aq)	-214.9^k	$CH_4(g)$	-12.13^a
$Fe(OH)_2(c)$ ppt.	-116.3^f	VO^{2+} (aq)	-106.7^i	CH_4(aq)	-8.22^a
$FeCO_3(c)$ (siderite)	-159.35^b	$V(OH)_3^0$(aq)	-212.9^k	H^+ (aq)	-0.00
$FeS_2(c)$ (pyrite)	-39.9^b	VOH^{2+} (aq)	-111.41^k	Cl^- (aq)	-31.38^c
Fe_2O_3 (hematite)	-177.4^b	$V(OH)_2^+$	-163.2^k	PO_4^{3-} (aq)	-243.5^a
Fe^{3+} (aq)	-1.1^b	$VOOH^+$	-155.65^k	HPO_4^{2-} (aq)	-260.34^a
$FeOH^{2+}$ (aq)	-54.83^b	V^{3+}	-57.8^k	$H_2PO_4^-$ (aq)	-270.14^k
$Fe(OH)_2^+$ (aq)	-106.7^i	**Uranium**		$H_3PO_4^0(c)$	-273.10^a
Fe^{2+} (aq)	-18.85^b	U^{4+} (aq)	-126.44^e	Na^+ (aq)	-62.59^c
$FeOH^+$ (aq)	-62.58^i	UOH^{3+} (aq)	-182.24^e	K^+ (aq)	-67.51^c
$Fe(OH)_3^-$ (aq)	-147.0^b	$U(OH)_4^0(c)$	-347.18^e	NH_4^+ (aq)	-18.99^c
$Fe(OH)_4^-$ (aq)	-198.4^i			Pb^{2+} (aq)	-5.83^a
$FeO(c)$	-60.03^g				
$Fe_2S(c)$ (pyrite)	-38.3^g				
$FeS(c)$	-24.22^g				

c = solid
aq = aqueous solution
g = gas
l = liquid

6.5 Relationship between pH and Eh

6.5.1 pH

Water undergoes a dissociation into two ionic species:

$$H_2O \rightleftharpoons H^+ + OH^- \tag{6.23}$$

The equilibrium constant for this reaction is

$$K_{eq} = \frac{[H^+][OH^-]}{[H_2O]} \tag{6.24}$$

The value of this equilibrium constant depends upon the temperature, but at 25°C it is 1×10^{-14}. Water that is neutral has the same number of H^+ and OH^- ions. If there are more H^+ ions, water is acidic, and if there are more OH^- ions, it is basic.

The pH of an aqueous solution is a measure of the number of hydrogen ions or protons present. The definition of pH is the negative logarithm of the hydrogen-ion activity. It ranges from 0 (most acidic) to 14 (most basic), and at 25°C a pH of 7 means that the solution is neutral. Because $[H_2O]$ is unity, from Equation 6.24 we have the relationship $[H^+][OH^-] = K_{eq} = 10^{-14}$. The pH of a solution is measured with a pH meter and an electrode. It should be measured in the field, preferably in a flowthrough cell so that dissolved gas isn't exchanged with the atmosphere prior to the measurement. The pH of a solution is especially sensitive to the amount of dissolved CO_2.

6.5.2 Relationship of Eh and pH

We thus have two ways to characterize a solution. The pH describes the number of protons present and the Eh is related to the number of electrons. Eh and pH can be related through the Nernst equation for a reaction that contains water and H^+ ions. Such a reaction can be written (Robertson 1975)

$$bB + mH^+ + ne^- \rightleftharpoons aA + wH_2O \tag{6.25}$$

Sources: for Table 6.4 (opposite)

[a] Wageman, D. D., W. H. Evans, V. B. Parker, I. Halow, S. M. Baily, and R. H. Schumm. 1968. *Selected values of chemical thermodynamic properties*. National Bureau of Standards Technical Note 270-3; 264 pp.

[b] Wageman, D. D., W. H. Evans, V. B. Parker, I. Halow, S. M. Baily, and R. H. Schumm. 1969. *Selected values of chemical thermodynamic properties*. National Bureau of Standards Technical Note 270-4; 141 pp.

[c] CODATA Task Group on Key Values for Thermodynamics. 1976. Recommended key values for thermodynamics 1975. *Journal of Chemical Thermodynamics* 8:603–5.

[d] CODATA Task Group on Key Values for Themodynamics. 1977. Recommended key values for thermodynamics 1976. *Journal of Chemical Thermodynamics* 9:705–6.

[e] Giridhar J., and Donald Langmuir. 1991. Determination of E° for the UO_2^{2+}/U^{4+} couple from measurement of the equilibrium: UO_2^{2+} + Cu(s) + 4 H^+ = U^{4+} + Cu^{2+} + 2 H_2O at 25°C and some geochemical implications. *Radiochemica Acta* 54:133–38.

[f] *Handbook of Chemistry and Physics*. Selected Values of Chemical Thermodynamic Properties. Boca Raton, Fla.: CRC Press.

[g] Robie, R. A., B. S. Hemingway, and J. R. Fisher. 1978. *Thermodynamic properties of minerals and related substances at 298.15 K and 1 bar (105 pascals) pressure and higher temperatures*. U. S. Geological Survey Bulletin 1452; 456 pp.

[h] Feitknecht, Walter, and P. W. Schindler. 1963. Solubility constants of metal oxides, metal hydroxides and metal salts in aqueous solution. *Pure and Applied Chemistry* 6:130–57.

[i] Baes, C. F., Jr., and R. E. Messmer. 1976. *The Hydrolysis of Cations*. New York: Wiley, 489 pp.

[j] Wageman, D. D., W. H. Evans, V. B. Parker, I. Halow, S. M. Baily, and R. H. Schumm. 1968. *Selected values of chemical thermodynamic properties*. National Bureau of Standards Technical Note 270-5.

[k] Langmuir, Donald. 1977. Uranium solution mineral equilibria at low temperatures. *Geochimica et Cosmochimica Acta* 42:547–69.

where

$$A = \text{reactant}$$
$$B = \text{product}$$
$$n = \text{number of electrons released}$$
$$a = \text{moles of reactant}$$
$$w = \text{moles of water}$$
$$b = \text{moles of product}$$
$$m = \text{moles of hydrogen ions}$$

The Nernst equation for Reaction 6.25 is

$$\text{Eh} = E^0 - \frac{RT}{nF} \ln \frac{[A]^a [H_2O]^w}{[B]^b [H^+]^m} \tag{6.26}$$

The activity of water is unity. For a particular reaction, E^0 is given and R, T, and F are constants. The significant variables are the Eh and the activities of the reactant, the product, and the hydrogen-ion activity, which can be expressed as a pH. Equation 6.26 can be rearranged and expressed in base 10 logs as either

$$\text{Eh} = E^0 - 2.303 \frac{RT}{nF} \log \frac{[A]^a}{[B]^b [H^+]^m} \tag{6.27}$$

or

$$\text{Eh} = E^0 - 2.303 \frac{RT}{nF} \log \frac{[A]^a}{[B]^b} + 2.303 \frac{RTm}{nF} \log[H^+] \tag{6.28}$$

At 25°C and 1 atm of pressure, Equation 6.28 can be expressed as

$$\text{Eh} = E^0 - \frac{0.0592}{n} \log \frac{[A]^a}{[B]^b} - 0.0592 \frac{m}{n} \text{pH} \tag{6.29}$$

6.5.3 Eh-pH Diagrams

The Eh-pH relationship is particularly useful when applied in the form of an Eh-pH diagram, with Eh the ordinate and pH the abscissa. If a solution has several ions present that can react to form different products or occur in different valence states, the stable product or valence state at a given concentration of reactants will be a function of the pH and Eh of the solution.

Figure 6.2 is a basic Eh-pH diagram. The range of pH is 0 to 14. For Eh, it is convenient to specify a range of about $+1.4$ to -1.0 V. In certain regions of the Eh-pH field, water will be oxidized to O_2, and in other regions water will be reduced to H_2. We will calculate these regions as an example problem.

EXAMPLE PROBLEM Calculate the stability field for water at standard conditions.
The oxidation of water is given by

$$O_2(g) + 4H^+ + 4e^- \rightleftharpoons 2H_2O(l)$$

Inorganic Chemicals in Ground Water

From Table 6.4,
$$\Delta G^0_{H_2O(l)} = -56.69 \text{ kcal}$$
$$\Delta G^0_{O_2(g)} = 0$$
$$\Delta G^0_{H^+} = 0$$

From Equation 6.20,
$$\Delta G^0_R = 2\Delta G^0_{H_2O(l)} - \Delta G^0_{O_2(g)} - 4\Delta G^0_{H^+}$$
$$\Delta G^0_R = 2(-56.69) - 0 - 4(0) = -113.38 \text{ kcal}$$

The value of ΔG^0_R in kilocalories is converted to a standard potential by use of Equation 6.19:
$$E^0 = \frac{-\Delta G^0_R}{nF} = \frac{-(-113.38)}{4 \cdot 23.06} = 1.229 \text{ V}$$

The Nernst equation (Equation 6.24) can be expressed as:
$$Eh = E^0 - \frac{RT}{nF} 2.303 \log \frac{[H_2O]}{[O_2][H^+]^4}$$

The activity of dissolved gaseous oxygen is expressed as a partial pressure, P_{O_2}. At standard conditions it has a value of 1 atm. The activity of water is unity. The Nernst equation thus reduces to
$$Eh = 1.229 - \frac{0.00199 \cdot 298}{4 \cdot 23.06} 2.303 \log[H^+]^{-4}$$

This expression can be reduced to
$$Eh_{(volts)} = 1.229 - 0.0592 pH$$

This equation defines the upper boundary of stability for water, above which oxidation would break apart the water molecule.

The reduction of hydrogen ions to form gaseous hydrogen is
$$2H^+ + 2e^- \rightleftharpoons H_{2(gas)}$$

From Table 6.4,
$$\Delta G^0_{H^+} = 0$$
$$\Delta G^0_{H_2(gas)} = 0$$

The value of ΔG^0_R for the formation of hydrogen gas is obviously zero. Therefore, the value of E^0 is also zero.

From the Nernst equation,
$$Eh = E^0 - \frac{0.00199 \cdot 298}{2 \cdot 23.06} 2.303 \log \frac{P_{H_2}}{[H^+]^2}$$

The value of P_{H_2} is 1 atm and the calculated value of E^0 is 0, hence the preceding expression can be reduced to
$$Eh_{(volts)} = 0.000 - 0.0592 pH$$

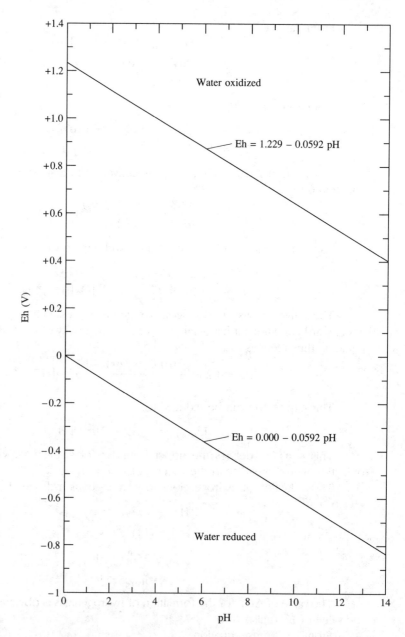

FIGURE 6.2 Eh-pH diagram showing the stability field for water.

This equation forms the lower boundary of the stability field for water. These boundaries are plotted in Figure 6.2.

6.5.4 Calculating Eh-pH Stability Fields

The stability fields within the Eh-pH diagram for various forms of an element can be computed using chemical thermodynamics. Basic sources of thermodynamic data include Wagman et al. (1968, 1969, 1971) and Robie, Hemingway, and Fisher (1978).

Boundaries for an element between dissolved species that have different valence states are computed using the Nernst equation (Equation 6.27, 6.28, or 6.29). If both ions are at the same valence state, then the equation for chemical equilibrium is used. If one is calculating the boundary between a solid species and a dissolved form, the chemical activity of the solid species is 1. For some of the boundaries of solid species, it will be necessary to assume an activity for the dissolved species.

EXAMPLE PROBLEM

Calculate an Eh-pH diagram for iron in which the solid species are $Fe(OH)_3$ and FeO and the activity of dissolved iron is 56 µg/L (10^{-6} M).

Soluble forms of the ferrous ion and the ferric ion include Fe^{2+}, Fe^{3+}, $FeOH^{2+}$, and $Fe(OH)_2^+$. Transformations between these ions are determined by redox equations:

$$FeOH^{2+} + H^+ + e^- \rightleftharpoons Fe^{2+} + H_2O \tag{6.30}$$

$$Fe(OH)_2^+ + 2H^+ + e^- \rightleftharpoons Fe^{2+} + 2H_2O \tag{6.31}$$

$$Fe^{3+} + e^- \rightleftharpoons Fe^{2+} \tag{6.32}$$

The free energy and the standard potential for these reactions can be determined from Equations 6.20 and 6.19, respectively.

From Table 6.4 free energies are as follows.

$FeOH^{2+} = -54.83$ kcal/mol $FeO = -60.03$ kcal/mol
$Fe(OH)_2^+ = -106.7$ kcal/mol $Fe(OH)_3 = -166$ kcal/mol
$Fe^{2+} = -18.85$ kcal/mol $Fe(OH)_4^- = -198.4$ kcal/mol
$Fe^{3+} = -1.1$ kcal/mol $H^+ = 0$
$H_2O = -56.69$ kcal/mol

For Reaction 6.30 ($FeOH^{2+} + H^+ + e^- \rightleftharpoons Fe^{2+} + H_2O$):

$$\Delta G_R^0 = [\Delta G_{H_2O}^0 + \Delta G_{Fe^{2+}}^0] - [\Delta G_{FeOH^{2+}} + \Delta G_{H^+}^0]$$

$$\Delta G_R^0 = -56.69 + (-18.85) - (-54.83) - 0$$

$$\Delta G_R^0 = -20.71 \text{ kcal/mol}$$

$$E^0 = \frac{-\Delta G_R^0}{nF} = \frac{-(-20.71)}{1 \cdot 23.06} \text{ V}$$

$$E^0 = +0.898 \text{ V}$$

For Reaction 6.31 ($Fe(OH)_2^+ + 2H^+ + e^- \rightleftharpoons Fe^{2+} + 2H_2O$):

$$\Delta G_R^0 = 2\Delta G_{H_2O}^0 + \Delta G_{Fe^{2+}}^0 - \Delta G_{Fe(OH)_2^+}^0 - 2\Delta G_{H^+}^0$$
$$\Delta G_R^0 = 2(-56.69) + (-18.85) - (-106.7) - 2(0)$$
$$\Delta G_R^0 = -25.53 \text{ kcal/mol}$$
$$E^0 = \frac{-\Delta G_R^0}{nF} = \frac{-(-25.53)}{1 \cdot 23.06} \text{ V}$$
$$E^0 = +1.107 \text{ V}$$

For Reaction 6.32 ($Fe^{3+} + e^- \rightleftharpoons Fe^{2+}$):

$$\Delta G_R^0 = \Delta G_{Fe^{2+}}^0 - \Delta G_{Fe^{3+}}^0$$
$$\Delta G_R^0 = -18.85 - (-1.1)$$
$$\Delta G_R^0 = -17.75$$
$$E^0 = \frac{-\Delta G_R^0}{nF} = \frac{-(-17.75)}{1 \cdot 23.06}$$
$$E^0 = +0.770$$

The boundaries between the stability fields are determined from the Nernst equation. At the boundary between two fields, the activities of the iron species on the left of the reaction is equal to the activity of the iron species on the right of the equation—i.e., the two species are at equilibrium.

For Reaction 6.30 ($FeOH^{2+} + H^+ + e^- \rightleftharpoons Fe^{2+} + H_2O$):

$$Eh = E^0 - \frac{0.0592}{n} \log \frac{[Fe^{2+}]}{[FeOH^{2+}]} - 0.0592 \frac{m}{n} pH$$

Since $[FeOH^{2+}] = [Fe^{2+}]$, m (the number of hydrogen ions) $= 1$, n (the number of electrons) $= 1$, $E^0 = +0.898$ V, and $\log 1 = 0$:

$$Eh_{(volts)} = 0.898 - 0.0592 \, pH \tag{6.33}$$

For Reaction 6.31 ($Fe(OH)_2^+ + 2H^+ + e^- \rightleftharpoons Fe^{2+} + 2H_2O$):

$$Eh = E^0 - \frac{0.0592}{n} \log \frac{[Fe^{2+}]}{[Fe(OH)_2^+]} - 0.0592 \frac{m}{n} pH$$

Since $[Fe(OH)_2^+] = [Fe^{2+}]$, $m = 2$, $n = 1$, and $E^0 = +1.107$ V:

$$Eh_{(volts)} = 1.107 - 0.1184 \, pH \tag{6.34}$$

For Reaction 6.32 ($Fe^{3+} + e^- \rightleftharpoons Fe^{2+}$):

$$Eh = E^0 - \frac{0.0592}{n} \log \frac{[Fe^{2+}]}{[Fe^{3+}]} - 0.0592 \frac{m}{n} pH$$

This reaction is independent of pH because neither $[H^+]$ nor $[OH^-]$ appears in the reaction. Hence the value of m is 0. Because as $[Fe^{3+}] = [Fe^{2+}]$ and $\log 1 = 0$, Eh is a constant equal to E^0, which is 0.770 V:

$$Eh = 0.770 \text{ V} \tag{6.35}$$

Inorganic Chemicals in Ground Water

The boundary between two dissolved species that are at the same valence state can be determined from chemical equilibrium.

For iron there are two boundaries between dissolved ions of ferric iron. These boundaries are represented by these reactions:

$$Fe^{3+} + H_2O \rightleftharpoons FeOH^{2+} + H^+ \quad (6.36)$$
$$FeOH^{2+} + H_2O \rightleftharpoons Fe(OH)_2^+ + H^+ \quad (6.37)$$

For Reaction 6.36 ($Fe^{3+} + H_2O \rightleftharpoons FeOH^{2+} + H^+$), the equilibrium constant can be obtained from the free energy of the reaction. The first step is to find the free energy of the reaction using Equation 6.20.

$$\Delta G_R^0 = \Delta G_{FeOH^{2+}}^0 + \Delta G_{H^+}^0 - \Delta G_{Fe^{3+}}^0 - \Delta G_{H_2O}^0$$
$$\Delta G_R^0 = -54.83 + 0 - (-1.1) - (-56.69)$$
$$\Delta G_R^0 = +2.96 \text{ kcal/mol}$$

The next step is to determine the equilibrium constant using Equation 6.22.

$$\log K_{eq} = -\frac{\Delta G_R^0}{1.364} = -\frac{2.96}{1.364} = -2.17$$
$$K_{eq} = 10^{-2.17}$$

From Equation 6.6,

$$K_{eq} = \frac{[FeOH^{2+}][H^+]}{[Fe^{3+}][H_2O]} = 10^{-2.17}$$

Since $[H_2O] = 1$ and at the boundary $[FeOH^{2+}] = [Fe^{3+}]$,

$$[H^+] = 10^{-2.17} \quad (6.38)$$

This means that a vertical line at a pH of 2.17 separates these two stability fields.

For Reaction 6.37 ($FeOH^{2+} + H_2O \rightleftharpoons Fe(OH)_2^+ + H^+$), find the free energy of the reaction:

$$\Delta G_R^0 = \Delta G_{Fe(OH)_2^+}^0 + \Delta G_{H^+}^0 - \Delta G_{FeOH^{2+}}^0 - \Delta G_{H_2O}^0$$
$$\Delta G_R^0 = -106.7 + 0 - (-54.83) - (56.69)$$
$$\Delta G_R^0 = +4.82 \text{ kcal/mol}$$

Next find the value of K_{eq}:

$$\log K_{eq} = \frac{-\Delta G_R^0}{1.364} = \frac{-4.82}{1.364} = -3.53$$
$$K_{eq} = 10^{-3.53}$$

Finally, from Equation 6.6,

$$K_{eq} = \frac{[Fe(OH)^{2+}][H^+]}{[FeOH^{2+}][H_2O]} = 10^{-3.53}$$

Since $[H_2O] = 1$ and $[Fe(OH)^{2+}] = [FeOH^{2+}]$,

$$[H^+] = 10^{-3.53} \tag{6.39}$$

Lines that delineate the stability field for solids can be obtained by similar reasoning. Remember that the activity of a solid in equilibrium with dissolved species is 1. The location of the boundaries of solid species is a function of the amount of dissolved iron present.

In this situation there are two stable iron precipitates, $Fe(OH)_3$ and FeO. The reactions at the boundaries include

$$Fe(OH)_3 + H^+ \rightleftharpoons Fe(OH)_2^+ + H_2O \tag{6.40}$$
$$Fe(OH)_3 + 3H^+ + e^- \rightleftharpoons Fe^{2+} + 3H_2O \tag{6.41}$$
$$Fe(OH)_4^- + H^+ \rightleftharpoons Fe(OH)_3 + H_2O \tag{6.42}$$
$$Fe(OH)_3 + e^- \rightleftharpoons FeO + H_2O + OH^- \tag{6.43}$$
$$Fe(OH)_4^- + 2H^+ + e^- \rightleftharpoons FeO + 3H_2O \tag{6.44}$$
$$FeO + 2H^+ \rightleftharpoons Fe^{2+} + H_2O \tag{6.45}$$

Reaction 6.40 ($Fe(OH)_3 + H^+ \rightleftharpoons Fe(OH)_2^+ + H_2O$) is solved using an equilibrium approach:

$$\Delta G_R^0 = \Delta G_{Fe(OH)_2^+}^0 + \Delta G_{H_2O}^0 - \Delta G_{Fe(OH)_3}^0 - \Delta G_{H^+}^0$$
$$\Delta G_R^0 = -106.7 + (-56.69) - (-166) - 0 = 2.61$$
$$\log K_{eq} = -\frac{\Delta G_R^0}{1.364} = -\frac{2.61}{1.364} = -1.91$$
$$K_{eq} = 10^{-1.91}$$
$$K_{eq} = \frac{[Fe(OH)_2^+][H_2O]}{[Fe(OH)_3][H^+]} = 10^{-1.91}$$

Since $[Fe(OH)_3] = 1$ and $[H_2O] = 1$,

$$[H^+] = \frac{[Fe(OH)_2^+]}{10^{-1.91}} \tag{6.46}$$

Reaction 6.41 ($Fe(OH)_3 + 3H^+ + e^- \rightleftharpoons Fe^{2+} + 3H_2O$) is solved using the Nernst equation:

$$\Delta G_R^0 = \Delta G_{Fe^{2+}}^0 + 3\Delta G_{H_2O}^0 - \Delta G_{Fe(OH)_3}^0 - 3\Delta G_{H^+}^0$$
$$\Delta G_R^0 = -18.85 + 3(-56.69) - (-166) - 0$$
$$\Delta G_R^0 = -22.95 \text{ kcal/mol}$$
$$E^0 = \frac{-\Delta G_R^0}{nF} = \frac{-(-22.95)}{1 \cdot 23.06} = +0.994 \text{ V}$$
$$Eh = E^0 - \frac{0.0592}{n} \log \frac{[Fe^{2+}]}{[Fe(OH)_3]} - 0.0592 \frac{m}{n} pH$$

Since there are three hydrogen ions ($m = 3$) and one electron ($n = 1$) and $[Fe(OH)_3] = 1$, then

$$Eh_{(volts)} = 0.994 - 0.0592 \log[Fe^{2+}] - 0.178 pH \qquad (6.47)$$

Reaction 6.42 ($Fe(OH)_4^- + H^+ \rightleftharpoons Fe(OH)_3 + H_2O$) is solved by using an equilibrium approach:

$$\Delta G_R^0 = \Delta G_{Fe(OH)_3}^0 + \Delta G_{H_2O}^0 - \Delta G_{Fe(OH)_4^-}^0 - \Delta G_{H^+}^0$$

$$\Delta G_R^0 = -166 - 56.69 - (-198.4) - 0$$

$$\Delta G_R^0 = -24.29$$

$$\log K_{eq} = -\frac{\Delta G_R^0}{1.364} = \frac{24.29}{1.364} = 17.8$$

$$K_{eq} = 10^{17.8}$$

$$K_{eq} = \frac{[Fe(OH)_3][H_2O]}{[Fe(OH)_4^-][H^+]} = 10^{17.8}$$

Since $[Fe(OH)_3] = 1$ and $[H_2O] = 1$,

$$[H^+] = \frac{10^{-17.8}}{[Fe(OH)_4^-]} \qquad (6.48)$$

Equation 6.43 ($Fe(OH)_3 + e^- \rightleftharpoons FeO + H_2O + OH^-$) is solved using the Nernst equation:

$$\Delta G_R^0 = \Delta G_{FeO}^0 + \Delta G_{H_2O}^0 + \Delta G_{OH^-}^0 - \Delta G_{Fe(OH)_3}^0$$

$$\Delta G_R^0 = -60.03 + (-56.69) + (-37.59) - (-166)$$

$$\Delta G_R^0 = +11.69$$

$$E^0 = \frac{-\Delta G_R^0}{nF} = \frac{-11.69}{1 \cdot 23.06} = -0.507 \text{ V}$$

$$Eh = E^0 - \frac{0.0592}{n} \log \frac{[FeO][H_2O][OH^-]}{[Fe(OH)_3]}$$

Since $[Fe(OH)_3] = 1$, $[FeO] = 1$, $[H_2O] = 1$, and $n = 1$,

$$Eh = -0.507 - 0.0592 \log[OH^-]$$

Because the diagram uses pH as a variable, $[OH^-]$ must be expressed in terms of pH. By definition, $[OH^-] = 10^{-14}/[H^+]$; therefore, $\log[OH^-] = \log 10^{-14} - \log[H^+]$, so that

$$Eh = -0.507 - 0.0592(pH - 14)$$

$$Eh_{(volts)} = 0.322 - 0.0592 pH \qquad (6.49)$$

Reaction 6.44 ($Fe(OH)_4^- + 2\,H^+ + e^- \rightleftharpoons FeO + 3\,H_2O$) is solved using the Nernst equation:

$$\Delta G_R^0 = \Delta G_{FeO}^0 + 3\,\Delta G_{H_2O}^0 - \Delta G_{Fe(OH)_4^-}^0 - 2\,\Delta G_{H^+}^0$$
$$\Delta G_R^0 = -60.03 + 3(-56.69) - (-198.4) - 0$$
$$\Delta G_R^0 = -31.7 \text{ kcal/mol}$$
$$E^0 = \frac{-\Delta G_R^0}{nF} = \frac{-(-31.7)}{1 \cdot 23.06} = +1.375$$
$$Eh = E^0 - \frac{0.0592}{n} \log \frac{[FeO][H_2O]^3}{[Fe(OH)_4^-][H^+]^2}$$

Since [FeO] and [H_2O] are 1 and $n = 1$,

$$Eh = +1.375 - 0.0592 \log \frac{1}{[Fe(OH)_4^-][H^+]^2}$$

This can be expressed in terms of pH as

$$Eh = 1.375 + 0.0592 \log[Fe(OH)_4^-] + 2(0.0592)\log[H^+]$$
$$Eh_{(volts)} = 1.375 + 0.0592 \log[Fe(OH)_4^-] - 0.118\,pH \qquad (6.50)$$

Reaction 6.45 ($FeO + 2\,H^+ \rightleftharpoons Fe^{2+} + H_2O$) is solved as an equilibrium reaction:

$$\Delta G_R^0 = \Delta G_{Fe^{2+}}^0 + \Delta G_{H_2O}^0 - \Delta G_{FeO}^0 - 2\,\Delta G_{H^+}^0$$
$$\Delta G_R^0 = -18.85 + (-56.69) - (-60.03) - 2(0)$$
$$\Delta G_R^0 = -15.51 \text{ kcal/mol}$$
$$\log K_{eq} = \frac{-\Delta G_R^0}{1.364} = 11.36$$
$$K_{eq} = 10^{11.36}$$
$$K_{eq} = \frac{[Fe^{2+}][H_2O]}{[FeO][H^+]^2} = 10^{11.36}$$

Since [H_2O] and [FeO] = 1,

$$[H^+]^2 = \frac{[Fe^{2+}]}{10^{11.36}}$$
$$[H^+] = \sqrt{\frac{[Fe^{2+}]}{10^{11.36}}} \qquad (6.51)$$

Several of the equations, including Equations 6.46, 6.47, 6.48, 6.50, and 6.51, depend upon the activity of the dissolved iron.

The following equations, which are independent of dissolved iron activity, have been derived.

Inorganic Chemicals in Ground Water

Boundary	Equation Number	Equation
$FeOH^{2+} - Fe^{2+}$	6.31	$Eh_{(volts)} = 0.898 - 0.0592pH$
$Fe(OH)_2^+ - Fe^{2+}$	6.32	$Eh_{(volts)} = 1.107 - 0.118pH$
$Fe^{3+} - Fe^{2+}$	6.33	$Eh_{(volts)} = 0.770$
$Fe^{3+} - FeOH^{2+}$	6.38	$pH = 2.17$
$FeOH^{2+} - Fe(OH)_2^+$	6.39	$pH = 3.53$
$Fe(OH)_3 - FeO$	6.49	$Eh_{(volts)} = 0.322 - 0.0592pH$

These equations need a dissolved iron activity, which is set at $10^{-6.00}$ mol:

Boundary	Equation Number	Equation
$Fe(OH)_3 - Fe(OH)_2^+$	6.46	$[H^+] = [10^{-6}]/10^{-1.91}$ $= 10^{-4.09}$ $pH = 4.09$
$Fe(OH)_3 - Fe^{2+}$	6.47	$Eh_{(volts)} = 0.994 - 0.0592 \log[10^{-6}]$ $\quad - 0.0178pH$ $Eh_{(volts)} = 1.349 - 0.178pH$
$Fe(OH)_4^- - Fe(OH)_3$	6.48	$[H^+] = 10^{-17.8}/[10^{-6}]$ $= 10^{-11.8}$ $pH = 11.8$
$Fe(OH)_4^- - FeO$	6.50	$Eh_{(volts)} = 1.375 + 0.0592 \log[10^{-6}]$ $\quad = -0.118pH$
$FeO - Fe^{2+}$	6.51	$Eh_{(volts)} = 1.0202 - 0.118pH$ $[H^+] = ([10^{-6}]/10^{11.36})^{0.5}$ $[H^+] = 10^{-8.68}$ $pH = 8.68$

Once the equations have been developed for the desired molar concentration of dissolved iron, the lines represented by the equations are plotted on an Eh-pH field. This has been done in Figure 6.3. The equation numbers are on the lines.

In order to finish the stability field diagram, we need to decide which segment of each line is needed. If we start with Fe^{3+}, it participates in two reactions, one characterized by Equation 6.33 and the other by Equation 6.38. Equation 6.33 is a horizontal line at 0.77 V and Equation 6.38 is a vertical line at pH = 2.17. These two lines define a corner at the upper left of the diagram in which Fe^{3+} can exist. Equation 6.38 divides the Fe^{3+} region from the $FeOH^{2+}$ region. This region is also bounded by Equations 6.31 and 6.39. The segments of these lines that intersect are used to define the $FeOH^{2+}$ region. In a similar manner of analysis, the line segments surrounding each of the stability fields are determined. The resulting Eh-pH diagram for the iron system is thus determined and is illustrated in Figure 6.4.

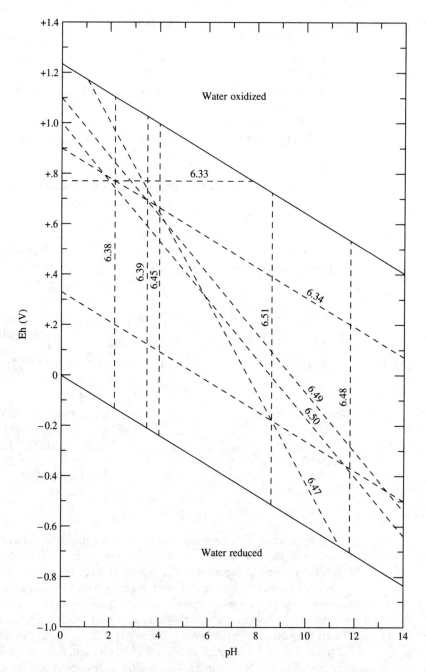

FIGURE 6.3 Equations for an Eh-pH diagram for dissolved iron with dissolved iron activity of 10^{-6} mol under standard conditions. (Numbers adjacent to dotted lines refer to equations discussed in the text.)

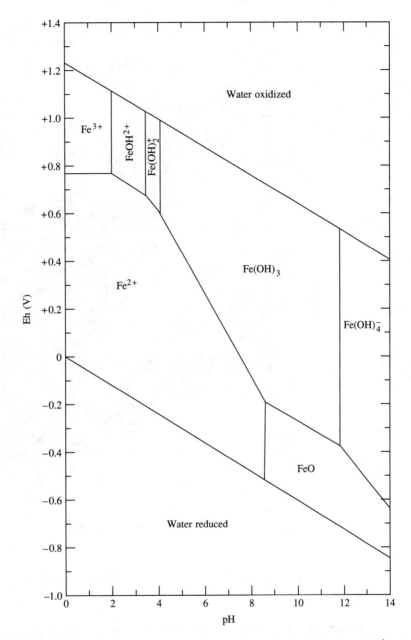

FIGURE 6.4 Final Eh-pH diagram for a dissolved iron system with dissolved iron at 10^{-6} moles under standard conditions.

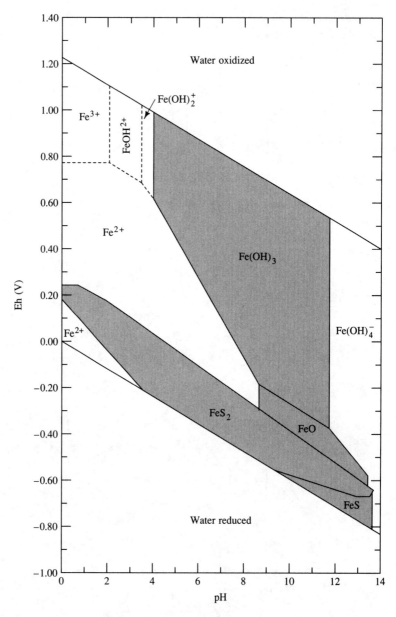

FIGURE 6.5 Eh-pH diagram showing fields of stability for dissolved iron under standard conditions. Activity of dissolved iron is 10^{-6} mol (56 μg/L), of sulfur species is 96 mg/L as SO_4^{2-}, and of carbon dioxide species is 61 mg/L as HCO_3^-. *Source:* J. D. Hem, *Study and interpretation of the chemical characteristics of natural waters,* U. S. Geological Survey Water Supply Paper 2254, 1985.

The iron Eh-pH diagram of Figure 6.4 is for a system that contains only dissolved iron. If other elements are present, such as sulfur, then additional iron compounds are possible. Figure 6.5 shows an Eh-pH diagram for a system with an iron activity of 56 µg/L (10^{-6} mol), sulfur of 96 mg/L as SO_4^{2-}, and carbon dioxide of 61 mg/L as HCO_3^-. Solids in the shaded area are thermodynamically stable. Under the conditions specified in this diagram, iron carbonate ($FeCO_3$) saturation was not reached and none is recorded as a solid phase.

The area of the region in which iron is precipitated rather than dissolved is a function of the concentration of dissolved iron. The more dissolved iron that is present, the greater the size of the stability field for the precipitates. This is illustrated in Figure 6.6. In this diagram the sulfur is 96 mg/L as SO_4^{2-} and the carbonate is 61 mg/L as HCO_3^-. Dissolved iron ranges from 5.6 µg/L to 56 mg/L.

Eh-pH diagrams have been used in the ground-water literature to explain such phenomena as the solubility of ferric oxyhydroxides (Whittemore and Langmuir 1975), hexavalent chromium (Robertson 1975), manganese (Hem 1985), iron, copper, silver, chromium, manganese, vanadium, molybdenum, and arsenic (Hem 1977), uranium (Langmuir 1978), thorium (Langmuir and Herman 1980), and arsenic (Matisoff et al. 1982).

6.6 Metal Complexes

6.6.1 Hydration of Cations

Although we consider that metallic ions exist in solution as an isolated ion, such as Cu^{2+}, in fact that is not the case. The Cu^{2+} ion is surrounded by polar water atoms that are chemically bound to the ion. Metallic ions, in general, have six water molecules surrounding them. The hydrated cupric ion is $Cu(H_2O)_6^{2+}$. Even outside the shell of chemically bound water molecules, there is a region where the polar water molecules are ordered by the electrostatic charge of the metallic ion. Anions in close association with a metal cation are called **ligands;** together they form a coordination compound. Water is considered to be a **ligand** that is bound to the metal ion. If other ligands bind to the metal, they must replace some of the water molecules acting as ligands. The stability of a complex relative to cation or ligand exchange can be described by equilibrium constants for the reaction.

6.6.2 Complexation

The following inorganic anions act as simple ligands in natural waters: OH^-, CO_3^{2-}, SO_4^{2-}, Cl^-, Br^-, F^-, NO_3^-, SiO_3^{2-}, S^{2-}, $S_2O_3^-$, PO_4^{3-}, $P_2O_7^{4-}$, $P_3O_{10}^{5-}$ and CN^-. Ammonia (NH_3) is a polar molecule that can also act as a ligand. Ligands can bond either covalently or electrostatically with a metal to form a complex ion or compound. We have already looked at the complex forms of the ferric ion and the hydroxyl ion. They form a series of complex ions: $FeOH^{2+}$, $Fe(OH)_2^+$, $Fe(OH)_3$ and $Fe(OH)_4^-$. Complex formation is involved with chemical equilibrium of ionic compounds and oxidation-reduction reactions.

In the case of monovalent ions, there is only one site where the ligand bonds to the metal ion. If the ligand has more than one site that can bond, then it forms what is known as a **chelating agent.** For example, the pyrophosphate ion, $P_2O_7^{4-}$, can bond

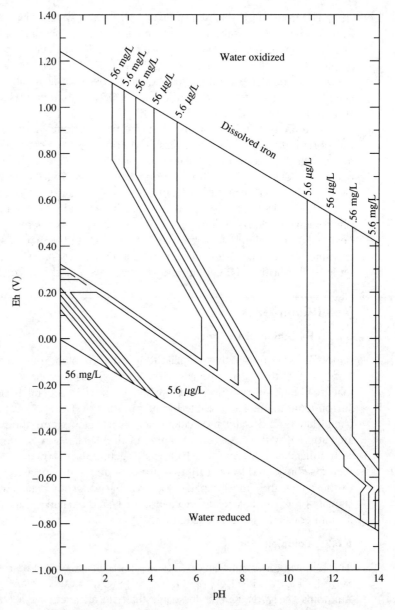

FIGURE 6.6 Equilibrium activity of dissolved iron as a function of Eh and pH under standard conditions, sulfur activity of 96 mg/L as SO_4^{2-} and activity of carbon dioxide of 61 mg/L as HCO_3^-. *Source:* J. D. Hem, *Study and interpretation of the chemical characteristics of natural waters,* U.S. Geological Survey Water Supply Paper 2254, 1985.

to a metal ion, such as cadmium, at two locations:

$$O=P(O)(O^-)-O-P(=O)(O^{2-})\cdots Cd$$

6.6.3 Organic Complexing Agents

Both natural waters and wastewaters contain a number of organic compounds that can act as chelating agents. In general, these organic compounds have a functional group that contains oxygen, nitrogen, phosphorous, or sulfur. If R symbolizes one or more carbon atoms with the appropriate number of hydrogens, then organic complexes can include functional groups such as

$$R-C(=O)-O^- \quad R-OH \quad R-NH_2 \quad R-O-P(=O)(OH)(O)$$

Carboxylate · Alcohol · Amine · Phosphate

There are a number of organic complexing agents that occur in nature. They are associated with **humic substances** that form from the decomposition of vegetation. These are complex organic molecules with molecular weights ranging upward into the tens of thousands. If a humic substance is extracted with a strong base and then acidified, there are three products. The nonextractable organic material is called **humin.** Substances called **fulvic acids** remain in the acidic solution, and other substances called **humic acids** precipitate from the acidified extract. These represent classes of compounds that contain many different individual organic molecules. Humic and fulvic acids contain many functional groups that can chelate to metals. Metals may be kept in solution by chelation with soluble fulvic acids or they may bind to the insoluble humic substances by cation exchange (Manahan, 1984).

Synthetic organic complexing agents are used in a number of industrial processes. They can be used as cleaning compounds, as constituents of detergent, in metal plating baths, and in water conditioning. These compounds include sodium tripolyphosphate, sodium ethylenediaminetetraacetate (EDTA), citric acid, and sodium nitrilotriacetate (NTA). The structures of some of these compounds are given in Figure 6.7. Synthetic chelating agents may keep metals in solution under conditions where the unchelated metal would precipitate or undergo cation exchange.

EDTA in wastewater can vastly increase the mobility of associated metals in the subsurface. Monitoring wells near radioactive waste disposal trenches at the Oak Ridge (Tennessee) National Laboratory contained significant levels of sodium EDTA, which was used as a cleaning agent. The same wells also contained radioactive ^{60}Co, a metal that is normally not expected to migrate very far due to cation exchange. The ^{60}Co had been chelated by the EDTA and hence had greatly increased mobility in the subsurface (Means, Crerar, and Duguid 1978).

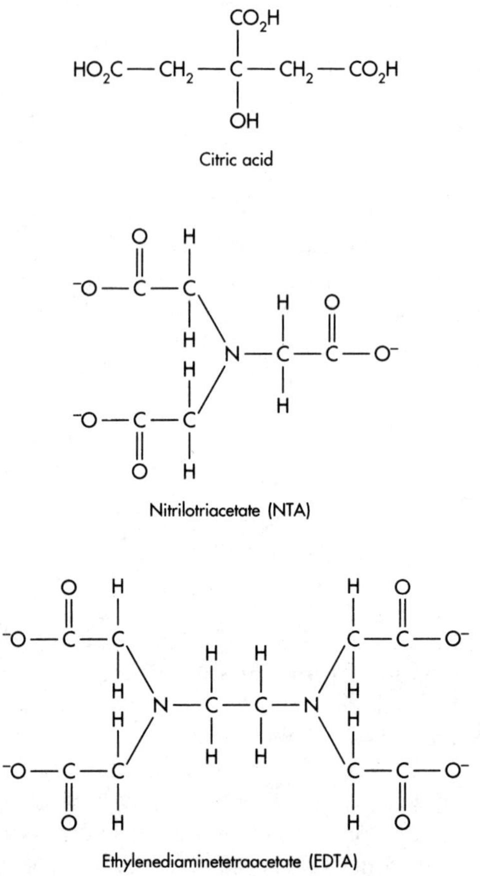

FIGURE 6.7 Structure of chelating agents including (a) citric acid, (b) nitrilotriacetate (NTA), and (c) ethylenediaminetetraacetate (EDTA).

6.7 Chemistry of Nonmetallic Inorganic Contaminants

6.7.1 Fluoride

Fluoride occurs in water as the F^- ion. In natural waters the amount of fluoride present is generally less than 1.0 mg/L, although concentrations as great as 67 mg/L have been reported (Hem 1985). Fluoride is present in minerals such as fluorite (CaF_2) and apatite ($Ca_5(Cl,F,OH)(PO_4)_3$). Weathering of these minerals may release fluoride. It may be released as a contaminant from industrial processes utilizing hydrofluoric acid. Cryolite (Na_3AlF_6) is used as a flux in the electrolytic production of aluminum. The manufacture of phosphate fertilizer from phosphate-rich rock may also release fluoride. Effluent from a Florida fertilizer plant had fluoride ranging from 2810 to 5150 mg/L (Cross and Ross 1970).

Fluoride can form complexes in water with a number of cations, including aluminum, beryllium and ferric iron (Hem, 1985). Dissolved fluoride can react with calcium to form fluorite. The solubility product for fluorite is $10^{-10.4}$. Precipitation of fluorite can act as a control on the amount of dissolved fluoride in solution if dissolved calcium is present. Table 6.5 shows the equilibrium amount of dissolved fluoride calculated for various activities of calcium. Actual activities of fluoride are likely to be somewhat higher due to the effect of the ionic strength of the solution as well as the effect of any complexes that might form with the fluoride ion.

Corbett and Manner (1984) have reported on the distribution of fluoride in water from both unconsolidated and bedrock aquifers of northeastern Ohio. They found that 239 out of 255 wells had fluoride concentrations less than 1 mg/L. However, 14 of the wells had fluoride ranging from 1 to 5.9 mg/L. All these high-fluoride wells were associated with a specific bedrock formation. Such information is useful from a public health standpoint because there is a 2.0-mg/L drinking-water criteria for fluoride. Some fluoride is needed to build strong teeth in growing children; however, fluoride in excess of 2.0 mg/L will cause teeth to discolor.

6.7.2 Chlorine and Bromine

The halides chlorine and bromine have similar chemistry, although chlorine is far more abundant in nature than bromine. Even though the elements can exist in a number of oxidation states, the chloride and bromide ions (Cl^-, Br^-) are the only ones of significance in natural waters. Chlorine gas is used an a disinfectant for purification of water and is a strong oxidizing agent when dissolved in water. The chloride ion occurs in natural waters in fairly low concentrations, usually less than 100 mg/L, unless the water is brackish or saline. Chloride is used by humans in many applications and can be added to the subsurface via industrial discharges, sewage, animal wastes, and road salting. Commercial fertilizers can contain chloride as KCl. Chlorine and bromine are components of halogenated organic compounds used for industrial solvents and pesticides. These compounds have been released to the environment both intentionally through the use of pesticides and accidentally through spills and leaks.

Chloride and bromide ions are not reactive. They don't participate in redox reactions, aren't sorbed onto mineral or organic surfaces, and don't form insoluble precipitates. Chloride is sometimes used as a tracer in ground-water studies because it is conservative.

TABLE 6.5 Equilibrium fluoride concentrations as a function of calcium activity.

Calcium		Fluoride	
Activity (mol)	Concentration (mg/L)	Activity (mol)	Concentration (mg/L)
2×10^{-2}	800	4.48×10^{-5}	0.85
10^{-2}	400	6.31×10^{-5}	1.20
5×10^{-3}	200	8.92×10^{-5}	1.70
10^{-3}	40	2.00×10^{-4}	3.79
5×10^{-4}	20	2.82×10^{-4}	5.36
10^{-4}	4	6.31×10^{-4}	11.99

6.7.3 Sulfur

Sulfur is released to the environment by the weathering of minerals containing the element. Rock containing pyrite can be oxidized to release sulfur, with microorganisms acting as a catalyst and mediating the oxidation. This is the source of the acidic water that drains from many areas that have been mined. Sulfuric acid is widely used in industrial processes. Sulfur can be released to the environment by the processing of sulfide ores and by the burning of fossil fuels, all of which contain sulfur to some degree.

Sulfur can exist in valence states ranging from S^{-2} to S^{+6}. Figure 6.8 is an Eh-pH diagram showing the stability of the two oxidized forms of sulfur, HSO_4^- and SO_4^{2-}, and the three reduced forms, S^{2-}, HS^-, and H_2S (aqueous). The field of stability for elemental sulfur is also shown. The total sulfur activity used in computing the diagram is 10^{-3} mol/L or 96 mg/L as SO_4^{2-}. If a greater total sulfur activity were used, the stability field for elemental sulfur would be larger. Although this is a very useful diagram for understanding the equilibrium conditions for dissolved sulfur, the redox reactions can be slow if microbes are not mediating the reactions. Hence, it may take a long time for the system to reach equilibrium.

Gypsum (calcium sulfate) is quite soluble in water ($K_{eq} = 10^{-4.6}$) and, except for waters with extremely high sulfate, would not be a sink for sulfate. Strontium sulfate is sparingly soluble ($K_{eq} = 10^{-6.5}$), whereas barium sulfate is nearly insoluble ($K_{eq} = 10^{-10.0}$). However, strontium and barium are not found in much abundance in natural waters. Sulfate could act as a sink for strontium and barium.

6.7.4 Nitrogen

Nitrogen is another element that can occur in both oxidized and reduced forms as well as the elemental state. The common forms of inorganic nitrogen include nitrate, NO_3^-, nitrite, NO_2^-, nitrogen gas, N_2, ammonium, NH_4^+ and cyanide, CN^-. Nitrogen is also a major constituent of organic matter in the form of amino acids. The majority (78%) of the Earth's atmosphere is nitrogen gas. Atmospheric nitrogen can be "fixed", or converted to nitrate, by cyanobacteria in lakes and the ocean and by bacteria living on the roots of plants such as legumes and lichens. Atmospheric nitrogen can also be converted to oxidized and reduced forms via fertilizer production and by heating it to high temperatures in internal combustion engines, power plants, lightening discharges and forest fires. Rainwater contains dissolved nitrate and ammonia. Nitrogen is released to the subsurface from sewage, animal wastes, and fertilizers.

In soil and ground water, oxidation and reduction of nitrogen species is accomplished by microorganisms. Under oxidizing conditions ammonia is converted to nitrite, which is converted to nitrate. Nitrite is a very reactive ion and is almost immediately converted to nitrate, so that little nitrite is normally found in the environment. Under reducing conditions nitrate is converted primarily to nitrogen gas, a process known as **denitrification.** Organic matter will decay to ammonia under reducing conditions. Septic tank effluent, for example, normally has high ammonia and very little nitrate. If the receiving ground water is reducing, the nitrogen will stay in the ammonia form. If it is oxidizing, bacteria will convert the ammonia to nitrate (Feth 1966).

Nitrate contamination of ground water has been documented in a number of areas (e.g. Hill 1982; Flipse et al. 1984; and Silver and Fielden 1980). Hill studied the distribution

Inorganic Chemicals in Ground Water

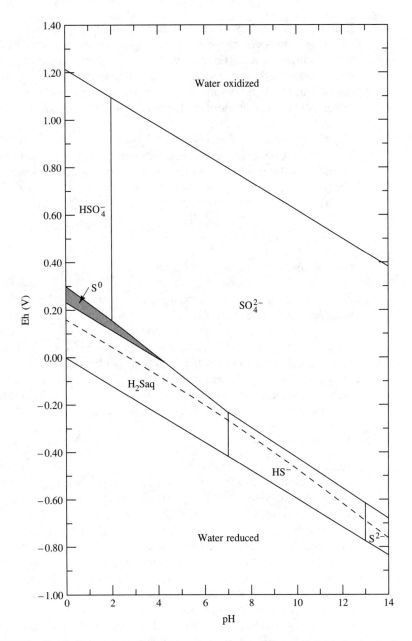

FIGURE 6.8 Eh-pH diagram for sulfur species at standard conditions with total dissolved sulfur activity of 96 mg/L. *Source:* J. D. Hem, *Study and interpretation of the chemical characteristics of natural waters,* U.S. Geological Survey Water Supply Paper 2254, 1985

of nitrate in ground water from a shallow unconsolidated sand aquifer. It was found that the ground water beneath areas of forest or permanent pasture has less than 1.0 mg/L of nitrate as nitrogen. The ground water beneath heavily fertilized potato fields typically contained in excess of 10 mg/L nitrate as nitrogen. Gray and Morgan-Jones (1980) found that the nitrate content of ground water in a study area increased over the past 40 yr and that the use of fertilizers in this catchment area also increased over the same time period.

Nitrogen occurs as two isotopes, ^{14}N and ^{15}N. Of the two, ^{14}N is by far the most abundant in the atmosphere. The relative abundance of ^{15}N—that is, the $^{15}N/^{14}N$ ratio—in nitrate may be used to distinguish nitrate that comes from animal and human waste from nitrate that comes from mineral fertilizers (Flipse et al. 1984).

The $^{15}N/^{14}N$ ratio is usually expressed as a $\delta^{15}N$ value, which is defined as

$$\delta^{15}N\ (\text{‰}) = \frac{(^{15}N/^{14}N)\ \text{sample} - (^{15}N/^{14}N)\ \text{standard}}{(^{15}N/^{14}N)\ \text{standard}} \times 1000$$

where ‰ stands for parts per thousand.

If the $\delta^{15}N$ is positive, then the nitrate of the sample has been enriched in ^{15}N with respect to the standard. For nitrogen, the standard is the atmospheric composition. Nitrate from animal and human waste typically has a $\delta^{15}N$ in excess of $+10$‰.

Flipse and Bonner (1985) found that mineral fertilizers used on Long Island had $\delta^{15}N$ values that averaged 0.2‰ at one site and -5.9‰ at another. However, the $\delta^{15}N$ of the ground water beneath the sites that had been fertilized was about $+6$‰. This increase in $\delta^{15}N$ from the mineral fertilizer was attributed to fractionation that occurred during infiltration of the nitrogen. However, the resulting $\delta^{15}N$ was still clearly lower than that expected from animal and human waste.

6.7.5 Arsenic

Arsenic can occur in valance states of $+5$, $+3$, $+1$, 0 and -3. However, the important states of dissolved arsenic in water are the arsenate $H_nAsO_4^{3-n}$, with a valance state of $+5$, and the arsenite $H_nAsO_3^{2-n}$, with a valance state of $+3$. An Eh-pH diagram for arsenic that shows the fields of stability for the arsenates and arsenites is given in Figure 6.9. Dissolved arsenic species can be absorbed by ferric hydroxides. Arsenic $(+5)$ is more strongly sorbed than arsenic $(+3)$. Ferric hydroxides are stable over a wide Eh-pH range, so this fact limits the mobility of arsenic. However, conditions that reduce Fe^{3+} to Fe^{2+} and As^{5+} to As^{3+} increase the mobility of arsenic in the environment, because the precipitated ferric hydroxides become soluble ferrous hydroxides (Matisoff, et al. 1982). In an oxidizing environment with a pH above 4.09, we will find colloidal ferric iron hydroxides, which will sorb arsenic and would thus expect to have little arsenic in solution. Under strongly reducing conditions, if both iron and hydrogen sulfide are present, arsenic sulfide coprecipitates with iron sulfide. Mildly reducing conditions that lack hydrogen sulfide present conditions under which one would expect to find the most mobile arsenic, as iron would be in the soluble ferrous state and arsenic would be in the arsenite form (Hounslow 1980).

Arsenic has been released to the environment through the burning of coal and the smelting of ores. In the past it was used in the formulation of insecticides and

Inorganic Chemicals in Ground Water

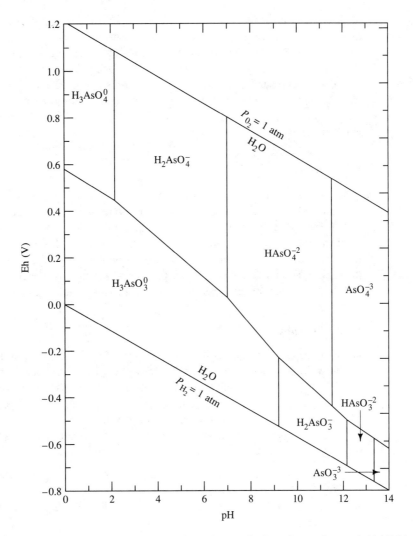

FIGURE 6.9 Eh-pH diagram for arsenic species under standard conditions. *Source*: A. H. Welch, M. S. Lico, and J. L. Hughes, *Ground Water* 26, no. 3 (1988): 333–47. Used with permission. Copyright © 1988 by Water Well Journal Publishing Co.

embalming corpses. Starting at the time of the Civil War in the United States (1860–1865), arsenic was an ingredient in a popular embalming fluid. As much as to 3 lb of arsenic could have used per corpse. The use of arsenic in embalming fluids was banned by the federal government in 1910 because its use interfered with the investigation of suspected arsenic poisonings. However, graveyards from the Civil War and the late nineteenth century may be a source of arsenic contamination (Konefes 1990). It has some modern industrial uses. Ground water has been found to have high (up to 96 μg/L) concentrations from natural sources in northeastern Ohio (Matisoff et al. 1982).

Elevated arsenic (up to 5 mg/L) in ground water in Nova Scotia, Canada, was reportedly due to the weathering of piles of mining waste that contained arsenopyrite (Grantham and Jones 1977). In the western United States high (>50 μg/L) concentrations of arsenic are common in ground water. These are associated with areas of sedimentary rocks derived from volcanic areas, geothermal systems and gold and uranium mining districts. Irrigation in some areas has liberated arsenic to the extent that concentrations of up to 1 mg/L are found in shallow ground water beneath irrigated fields (Welch, Lico, and Hughes 1988).

6.7.6 Selenium

Selenium occurs in oxidizing solutions as selenite, SeO_3^{2-}, with a $+4$ valance and as selenate, SeO_4^{2-}, with a $+6$ valence. It can be reduced to the insoluble elemental form, Se^0. It may also form a precipitate ferroselanite, $FeSe_2$, under reducing conditions. Selenate may be sorbed onto amorphous ferric hydroxides. Selenium has a number of industrial uses, such as the manufacture of pigments, stainless steel, and rubber compounds. It is contained in phosphate fertilizers. Selenium has been known to concentrate in irrigation return water draining from land that has soil high in selenium.

6.7.7 Phosphorus

Phosphorus can occur in a number of valance states, but in natural water it is really significant only in the $+5$ state. Dissolved phosphorus in water occurs as phosphoric acid (H_3PO_4) and its dissociation products, the orthophosphate ions: $H_2PO_4^-$, HPO_4^{2-} and PO_4^{3-}. The proportion of each present in an aqueous solution is a function of pH. Dissolved phosphorus is readily sorbed onto soil and has a very low mobility in ground water. In alkaline soils it can react with calcium carbonate to form a mineral precipitate, hydroxyapatite.

$$3\ HPO_4^{2-} + 5\ CaCO_3 + 2\ H_2O \longrightarrow Ca_5(PO_4)_3(OH) + 5\ HCO_3^- + OH^-$$

Phosphate is released to the environment from mineral fertilizers, animal wastes, sewage, and detergents.

6.8 Chemistry of Metals

Metals are cations, and most have fairly limited mobility in soil and ground water because of cation exchange or sorption on the surface of mineral grains. They can also form precipitates of varying solubility under specific Eh-pH conditions. Metals are mobile in ground water if the Eh-pH range is such that soluble ions exist and the soil has a low cation-exchange capacity (Dowdy and Volk 1983). They can also be mobile if they are chelated or if they are attached to a mobile colloid. Conditions that promote mobility include an acidic, sandy soil with low organic and clay content. Discharge of a metal in an acidic solution would keep the metal soluble and promote mobility.

6.8.1 Beryllium

Beryllium occurs only in the +2 valence state. In natural waters we can have Be^{2+}, $Be(OH)^+$, $Be(OH)_2$ and $Be(OH)_3^-$. Beryllium oxide and hydroxide have low solubilities and can act as a control on beryllium concentration. At equilibrium with $Be(OH)_2$, the dissolved form would have an activity of about 100 $\mu g/L$ at a pH of 6 (Hem 1985).

6.8.2 Strontium

Strontium also occurs in the +2 valence state and has a chemistry similar to that of calcium. The solubility product for strontium sulfate, $SrSO_4$, is $10^{-6.4}$. This suggests that there might be an equilibrium control on strontium concentration if sulfate is present in the water. Strontium carbonate, $SrCO_3$, has a solubility product of 10^{-10}. In general, strontium is present in ground waters in concentrations of less than 1 mg/L.

6.8.3 Barium

This alkaline earth element also has a valance of +2. Its distribution is controlled by the solubility of barite, $BaSO_4$. Barite has a solubility product of 10^{-10}. If the activity of sulfate is 96 mg/L (10^{-3} M), then the activity of barium is 10^{-7} M, or 0.014 mg/L.

6.8.4 Vanadium

This transition metal has oxidation states of +3, +4, and +5. In aqueous solutions it forms 10 different oxides and hydroxides. Dissolved iron can react with vanadium to form an insoluble ferrous vanadate, which can act as a control on vanadium in natural water (Hem 1977).

$$FeOH^+ + 2\,H_2VO_4^- + H^+ \rightleftharpoons Fe(VO_3)_2 + 3\,H_2O$$

6.8.5 Chromium

Chromium in natural waters occurs in a +3 and a +6 valance state. Stable ionic forms in aqueous systems include Cr^{3+}, $CrOH^{2+}$, $Cr(OH)_2^+$, $Cr_2O_7^{2+}$ and CrO_4^{2-}. Chromous hydroxide, $Cr(OH)_3$ is a possible precipitate under reducing conditions. Figure 6.10 is an Eh-pH diagram for the stability field for chromous hydroxide. Under some conditions chromate might react with ferrous iron to produce a chromous hydroxide precipitate (Robertson 1975).

$$CrO_4^{2-} + 3\,Fe^{2+} + 8\,H_2O \rightleftharpoons 3\,Fe(OH)_3 + Cr(OH)_3 + 4\,H^+$$

In general the hexavalent chromium in ground water is soluble and mobile and trivalent chromium will be insoluble and immobile. Industrial discharges of hexavalent chromium are common from metal-plating industries. This material may be quite mobile in ground water. A hexavalent chromium spill on Long Island, New York, traveled more than 3000 ft from a waste-discharge pond to a stream (Perlmutter, Lieber, and Frauenthal 1963). Hexavalent chromium from a natural source has been found in ground water in Paradise Valley, Arizona (Robertson 1975).

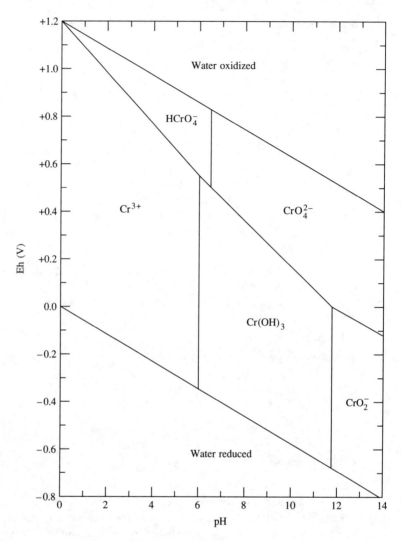

FIGURE 6.10 Eh-pH diagram for chromium under standard conditions. *Source:* Modified from F. N. Robertson, *Ground Water* 13, no. 6:516–27. Used with permission. Copyright © 1975 by Water Well Journal Publishing Co.

6.8.6 Cobalt

Cobalt occurs with valence states of $+2$ and $+3$. In the Eh and pH range of natural waters, only the $+2$ valence state is stable. It is thought that cobalt can coprecipitate or be absorbed by manganese and iron oxides. Cobalt carbonate has a solubility product of 10^{-10}. At a pH of 8.0 with 100 mg/L of carbonate, the equilibrium solubility of cobalt is 6 µg/L (Hem, 1985). The solubility product of cobalt sulfide is very low, $10^{-21.3}$. Virtually no cobalt would be in solution in a reducing environment. Radioactive cobalt

Inorganic Chemicals in Ground Water

is a waste product of certain defense activities (Means, Crerar, and Duguid 1978). Cobalt occurs in nature as smaltite ($CoAs_2$), and cobaltite (CoAsS).

6.8.7 Nickel

This metal occurs in aqueous solutions in the +2 valence state. Nickel ores include a variety of minerals, consisting of nickel, antimony, sulfur, and arsenic: NiSb, $NiAs_2$, NiAsS, and NiSbS. Nickel carbonate is more soluble that cobalt carbonate ($K_{sp} = 10^{-6.9}$), whereas the sulfide has a similar solubility ($K_{sp} = 10^{-19.4}$). Nickel is widely used in industry.

6.8.8 Molybdenum

Molybdenum occurs as the ore mineral molybdenite, MoS_2. The most common oxidation states are +4 and +6. Under oxidizing conditions the Mo^{6+} state dominates. Below pH 1.8 one finds H_2MoO_4 (aqueous). Between pH 1.8 and pH 5.3, $HMoO_4^-$ occurs, whereas above pH 5.3 the molybdate ion, MoO_4^{2-}, is stable. If ferrous iron is present, ferrous molybdate ($FeMoO_4$) presents a possible solubility control, since this has a solubility product of $10^{-10.45}$ at a pH range of 5.3 to 8.5 (Hem 1977). Molybdenum may also sorb onto amorphous ferric hydroxide (Kaback and Runnels 1980). The solubility product of calcium molybdate, $CaMoO_4$, is $10^{-8.7}$ (Hem, 1985). Molybdenum is used as an alloy in steel and as an additive to lubricants. Waste sources include mining and smelting of ore.

6.8.9 Copper

Copper occurs in either a +1 or a +2 valence state. Dissolved copper species in water include Cu^{2+}, $HCuO_2^-$, CuO_2^{2-}, and Cu^+. Cupric copper and ferrous iron can undergo an oxidation-reduction:

$$Cu^{2+} + 2\,Fe^{2+} + 7\,H_2O \rightleftharpoons Cu_2O + 2\,Fe(OH)_3 + 8\,H^+$$

Both cupric and cuprous sulfide have very low solubility products. Copper concentrations can be very high in acid mine drainage from metal mines, up to several hundred milligrams per liter. Copper can be leached from copper water-supply pipes and fixtures, especially by waters that have a pH of less than 7 (Hem, 1985).

6.8.10 Silver

Silver, a rare element, is widely used in industry, especially in photography. It occurs in the +1 valence state. Silver chloride, AgCl, has a solubility product of $10^{-9.7}$, which limits the solubility of silver in waters with chloride ion. Silver can also be naturally reduced to the metallic state by ferrous iron:

$$Ag^+ + Fe^{2+} + 3\,H_2O \rightleftharpoons Fe(OH)_3 + Ag + 3\,H^+$$

Silver sulfide has a low solubility. Thus, in water with chloride, iron, and sulfur present, stable solid forms of silver occur over the entire Eh-pH field (Hem, 1977). As a result, there is very little soluble silver in natural waters.

6.8.11 Zinc

Zinc is a fairly common metal and is extensively used in metallurgy and as a pigment, zinc oxide, which is often worn on the noses of lifeguards and other people in the sun. It occurs in the $+2$ valence state. Zinc carbonate has a rather low equilibrium constant, 10^{-10}, which would limit the solubility at pH ranges where the carbonate ion predominates. In a pH range of 8 to 11 and with 610 mg/L of HCO_3^-, there should be less than 100 µg/L of dissolved zinc (Hem 1985).

6.8.12 Cadmium

Cadmium has a very low maximum contaminant level (MCL) in drinking water—10 µg/L—due to its toxicity. It exists in aqueous solution in the $+2$ valence state. Cadmium carbonate has a very low solubility product, $10^{-13.7}$. Although this could serve as a control on solubility under some conditions, cadmium can be mobile in the environment. On Long Island, New York, a metal-plating waste containing cadmium and chromium traveled about 3000 ft in a shallow aquifer (Perlmutter, Lieber, and Frauenthal 1963). Cadmium has been implicated in an outbreak of a disease in Japan resulting in a softening of the bones of the victims that resulted in extreme bone pain. The cadmium was traced to rice and soybeans grown in soil contaminated by airborne cadmium that came from a nearby lead- and zinc-smelting operation (Emmerson 1970).

6.8.13 Mercury

Mercury has the lowest MCL for any inorganic chemical, 2 µg/L. It is considered to be very toxic. It has been known to concentrate in the food chain, especially in fish. Several outbreaks of mercury poisoning have been confirmed in Japan. Local discharges of mercury from industrial processes into surface-water bodies resulted in high mercury levels in fish. Inhabitants of fishing villages ate fish up to three times a day. Mortality of those affected was about 40%, and the poisoning was passed to unborn babies by apparently healthy mothers (Waldbott 1973). Mercury occurs as a metal and in the valance states $+1$ and $+2$. Most of the inorganic mercury compounds have a low solubility. The solubility product of Hg_2Cl_2 is $10^{-17.9}$, and for HgS it is about 10^{-50}. Under most natural conditions there is little soluble inorganic mercury. However, methane-generating bacteria can convert metallic mercury to organic forms such as methyl mercury, $HgCH_3^+$. The monomethyl mercury ion is soluble in water. Bacteria can also produce dimethyl mercury, $Hg(CH_3)_2$, which is volatile. Other organic forms of mercury, such as ethylmercuric chloride (C_2H_5HgCl), are manufactured and used as fungicides.

6.8.14 Lead

Lead occurs in aqueous solution as Pb^{2+} and in various hydroxides. Various lead compounds have solubility products that indicate that under the right Eh-pH conditions, lead solubility would be limited in natural waters: $PbCl_2$, $K_{sp} = 10^{-4.8}$; PbF_2, $K_{sp} = 10^{-7.5}$; $PbSO_4$, $K_{sp} = 10^{-7.8}$; $PbCO_3$, $K_{sp} = 10^{-13.1}$ and PbS, $K_{sp} = 10^{-27.5}$. Lead and the other metals are cations that can be expected to undergo cation exchange with clays. Hence, the mobility of lead in ground water is limited. This was born out by a study at a storage battery manufacturing facility at Medley, Florida. Soil near the facility was contaminated

Inorganic Chemicals in Ground Water

with lead in amounts of up to 98,600 mg/kg. However, shallow ground water immediately beneath the contaminated soil averaged less than 10 μg/L of lead, with the maximum being 31 μg/L. The soils at the site are high in carbonate, with some clay; however, the mechanism of lead removal by the soils is not known. Lead leached from lead pipes and solder used to join copper pipes is a potential threat for users of drinking water that is acidic or poorly buffered (Hem, 1985).

6.9 Radioactive Isotopes

6.9.1 Introduction

Certain isotopes of elements undergo spontaneous decay, resulting in the release of energy and energetic particles and consequent formation of different isotopes. Some of these radioactive isotopes are naturally occurring and others are created by the bombardment of the Earth by cosmic radiation. Humans have created nuclear isotopes through the detonation of nuclear weapons and the construction of nuclear reactors. Table 6.6 lists the sources of environmentally important isotopes.

Radionuclides emit ionizing radiation—alpha particles, beta particles, and gamma rays—when they decay. An alpha particle is a helium nucleus with atomic mass 4 and atomic number 2. A beta particle is either a negative electron or a positron (positive electron). Gamma radiation consists of electromagnetic radiation similar to X rays but more energetic (i.e., it has a shorter wave length). Gamma radiation is more destructive to tissue than X rays. The primary effect of these particles is to produce ions, hence the name ionizing radiation. Alpha particles do not penetrate very far into matter due to their large size, but they produce a lot of ions along their short path. Beta particles penetrate to a greater depth but produce fewer ions per unit path length.

Radionuclide concentrations can be reported in terms of their mass per volume concentration—e.g., milligrams per liter. However, they are more frequently reported in terms of a standard unit of radioactivity, the **curie** (Ci). A curie is 3.7×10^{10} disintegrations per second. In water we use the **picocurie** (pCi), which is 1×10^{-12} Ci, or

TABLE 6.6 Sources of environmentally important radioactive isotopes.

Source	Radionuclides
Naturally occurring	^{40}K, ^{222}Rn, ^{226}Ra, ^{230,232}Th, ^{235,238}U
Cosmic irradiation	^{3}H, ^{7}Be, ^{14}C, ^{22}Na
Nuclear weapons tests	^{3}H, ^{90}Sr, ^{137}Cs, ^{239,240}Pu
Mining waste—uranium, phosphate, coal	^{222}Rn, ^{226}Ra, ^{230,232}Th, ^{235,238}U
Industrial wastes—e.g., nuclear power plants, weapons manufacturing, research and medical waste	^{59,63}Ni, ^{60}Co, ^{90}Sr, ^{93,99}Zr, ^{99}Tc, ^{107}Pd, ^{129}I, ^{137}Cs, ^{144}Ce, ^{151}Sm, ^{152,154}Eu, ^{237}Np, ^{239,240,242}Pu, ^{241,243}Am

Source: G. W. Gee, Dhanpat Rai, and R. J. Serne, "Mobility of radionuclides in soil." In D. W. Nelson et al. (ed.) *Chemical Mobility and Reactivity in Soil Systems*, 203. (Madison, Wis.: Soil Science Society of America Spec. Publ. 11, 1983).

3.7 × 10^{-2} disintegrations per second. In the SI system the unit of radioactivity is the **becquerel** (Bq), which is 1 disintegration per second.

Radiation doses are measured in terms of **rads**, which are a measure of the absorption by the body of ionizing radiation of any type. A rad is equivalent to 100 ergs of energy from ionizing radiation absorbed per gram of soft tissue. In the SI system the unit of dose is a **gray** (Gy), which is equal to 100 rads.

The effect of ionizing radiation depends upon the type of particle and the body tissue with which it interacts. Therefore, the absolute measurement of dose must be converted to a **dose equivalent.** The unit of dose equivalent is the **rem.** Rads are converted to rems by multiplying by a factor that depends upon the type of ionizing radiation and its biological effect. For example, with gamma radiation the factor is 1 and a rad is equal to a rem. In the SI system the unit of dose equivalent is the **seivert** (Sv), and it is equal to a gray times the dose factor. A seivert is 100 rem.

6.9.2 Adsorption of Cationic Radionuclides

The cationic radionuclides may be subjected to ion exchange and other processes that sorb the radionuclide onto mineral or organic surfaces in the soil. The following transition metals and lanthanides have large distribution coefficients and hence low mobilities in waters that are in the neutral range: ^{60}Co, ^{59}Ni, ^{63}Ni, ^{65}Zn, ^{93}Zr, ^{107}Pd, ^{110}Ag, ^{114}Ce, ^{147}Pm, ^{151}Sm, ^{152}Eu, and ^{154}Eu. Many of them do not desorb significantly. The degree of sorption is strongly related to the pH of the solution. Insoluble metal hydroxides may also be formed. Technetium (Tc) solubility depends strongly upon the Eh of the solution, because under oxidizing conditions it forms the soluble pertechnetate ion (TcO_4^-) (Gee, Rai, and Serne 1983).

^{90}Sr, ^{137}Cs, and ^{226}Ra undergo cation exchange in a fashion similar to other exchangeable cations, such as Ca^{2+} and Mg^{2+}. Thorium and lead also have high distribution coefficients and limited mobility in neutral to alkaline soil. Lead is sorbed on hydrous oxides of iron, aluminum, and most likely manganese. Thorium hydroxides are of very limited solubility (Gee, Rai, and Serne 1983).

6.9.3 Uranium

Uranium occurs primarily as ^{238}U, with ^{235}U being much more rare. One of the disintegration products of ^{238}U is ^{226}Ra. This decay series ends with ^{206}Pb, a stable isotope. ^{235}U can decay to form ^{223}Ra. ^{238}U has a very long half-life, 4.5 × 10^9 years, which indicates that it is not very radioactive.

The chemistry of dissolved uranium is somewhat complex (Giridher and Langmuir 1991; Langmuir 1978). It has three valence states, +4, +5, and +6. Uranium can undergo oxidation-reduction reactions such as oxidation from the +4 to the +6 state:

$$U^{4+} + 2\,H_2O \rightleftharpoons UO_2^{2+} + 4\,H^+ + 2\,e^-$$

In a system with just uranium and water, stable species include (1) +4 valance: U^{4+}, UOH^{3+}, and $U(OH)_4^0$; and (2) +6 valance: UO_2^{2+}, UO_2H^+, $(UO_2)_3(OH)_5^+$ and $(UO_2)_3(OH)_7^-$. The stability fields for these ions and precipitates are shown in Figure 6.11 for an aqueous solution with a total uranium activity of 10^{-6} mol/L. The U(6+)

species has a tendency to form complexes with a wide variety of inorganic anions, including carbonate, hydroxide, phosphate, fluoride, and sulfate. This can be illustrated with an Eh-pH diagram for the same 10^{-6} mol/L solution of U but in contact with carbon dioxide at a partial pressure of 10^{-2} atm. This is represented by Figure 6.12. In Figure 6.11 the UO_2^{2+} formed a series of complexes with OH^-, starting with UO_2OH^+,

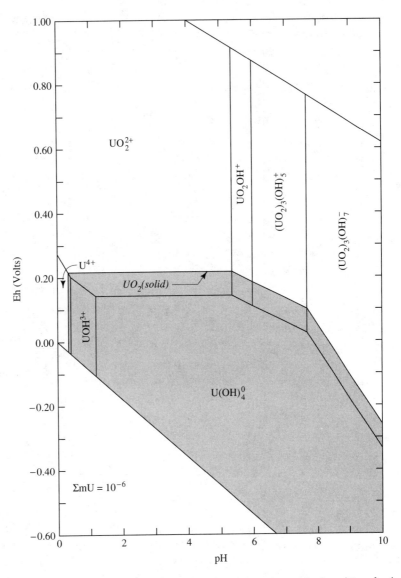

FIGURE 6.11 Eh-pH diagram for dissolved species of uranium under standard conditions for the system $U-O_2-H_2O$ with 10^{-6} mol/L U. The stability field for solid UO_2 (uranite) is shaded. *Source:* J. Giridhar and Donald Langmuir, *Radiochimica Acta* 54:133-38, 1991.

at a pH above about 5.2. With carbon present, as in Figure 6.12, UO_2^{2+} can form a series of carbonate complexes that replace the hydroxyl complexes.

The soluble complexes of oxidized uranium depend upon the pH of the water. Most natural water contains fluoride, phosphorus, carbon dioxide, and sulfur. Figure 6.13 shows the distribution of uranyl complexes for a ground water under standard conditions with $P_{CO_2} = 10^{-2.5}$ atm, $F^- = 0.3$ mg/L, $Cl^- = 10$ mg/L, $SO_4^{2-} =$

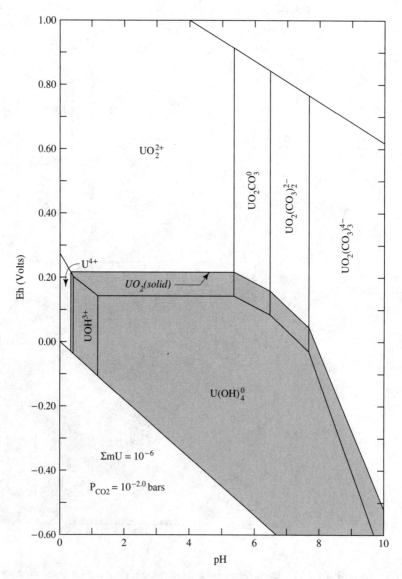

FIGURE 6.12 Eh-pH diagram for dissolved species of uranium under standard conditions for the system $U-CO_2-O_2-H_2O$ with 10^{-6} mol/L U. The stability field for solid UO_2 (uranite) is shaded. *Source:* J. Giridhar and Donald Langmuir, *Radiochimica Acta* 54:133–38, 1991.

Inorganic Chemicals in Ground Water

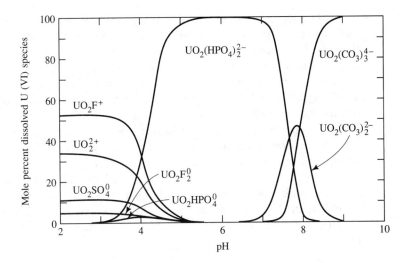

FIGURE 6.13 Distribution of uranyl complexes as a function of pH for ground water that contains 0.3 mg/L F^-, 10 mg/L Cl^-, 100 mg/L SO_4^{2-}, 0.1 mg/L PO_4^{2-} and P_{CO_2} of $10^{-2.5}$ atm. *Source:* Reprinted from *Geochimica et Cosmochimica Acta*, 42, D. Langmuir, "Uranium solution-mineral equilibrium at low temperatures with applications to sedimentary ore deposits." Copyright 1978, Pergamon Press plc.

100 mg/L, and $PO_4^- = 0.1$ mg/L. In different pH ranges the most prevalent stable species include UO_2F^+, $UO_2(HPO_4)_2^{2-}$, $UO_2(CO_3)_2^{2-}$, and $UO_2(CO_3)_3^{4-}$.

If reduced species of iron or sulfur are present, they could reduce U(6+) to U(4+) and precipitate the nearly insoluble mineral uranite, UO_2. This reduction could occur by oxidation of HS^- to SO_4^{2-}:

$$4\,UO_2(CO_3)_3^{4-} + HS^- + 15\,H^+ \rightleftharpoons 4\,UO_2\,(s) + SO_4^{2-} + 12\,CO_2\,(g) + 8\,H_2O$$

The same reduction could be accomplished by oxidation of ferrous iron to ferric hydroxide:

$$UO_2(CO_2)_3^{4-} + 2\,Fe^{2+} + 3\,H_2O \rightleftharpoons UO_2\,(s) + 2\,Fe(OH)_3 + 3\,CO_2$$

Because sulfur and iron are common in ground-water systems, under reducing conditions one could expect the formation of uranite, which would remove uranium from solution. Figures 6.11 and 6.12 show the stability field for uranite (UO_2). An MCL of 30 pCi/L for uranium has been proposed by the EPA (*Federal Register*, July 18, 1991).

6.9.4 Thorium

Thorium is a naturally occurring element with a principal isotope of ^{232}Th, which has a half-life of 1.39×10^{10} years. Daughter products of thorium decay include ^{224}Ra and ^{228}Ra.

The chemistry of thorium is much simpler than that of uranium. Thorium occurs only in a +4 valance, so it does not undergo oxidation-reduction. Thorium oxide, ThO_2, has a very low solubility. The primary thorium ore is monazite, which contains oxides of thorium, phosphorus, and the rare earths yttrium, lanthanum, and cerium.

The mobility of thorium is greatly enhanced if ligands are present to form complexes. Figure 6.14 shows the inorganic thorium complexes that form as a function of the pH of the solution. The solution contains 0.3 mg/L F^-, 10 mg/L Cl^-, 100mg/L SO_4^{2-}, and 0.1 mg/L PO_4^{2-}. It can be seen from this figure that the most abundant aqueous species in order of increasing pH are $Th(SO_4)_2^0$, ThF_2^{2+}, $Th(HPO_4)_2^0$, $Th(HPO_4)_3^{2-}$ and $Th(OH)_4^0$. However, the mobility of thorium complexes formed by organic ligands such as EDTA and citric acid are much greater than those formed by inorganic ligands (Langmuir and Herman, 1980).

Adsorption of dissolved thorium increases with increasing pH above pH 2. The sorption of thorium onto clays, oxides and soil organic matter is nearly total by a pH of 6.5. Strongly complexing organic ligands such as EDTA can retard sorption or even promote desorption (Langmuir and Herman 1980). Thorium in natural waters and soil should be nearly immobile due to the low solubility of the minerals and the strong tendency for dissolved forms to be sorbed only by clays, mineral oxides, and soil organic matter.

6.9.5 Radium

Radium occurs naturally in four isotopes: ^{223}Ra, ^{224}Ra, ^{226}Ra and ^{228}Ra. ^{232}Th decays into both ^{228}Ra and ^{224}Ra, whereas ^{235}U decays to ^{223}Ra and ^{238}U disintegrates to ^{230}Th, which in turn decays to ^{226}Ra. One isotope, ^{226}Ra, has a much longer half-life than any of the others, 1599 years. Because of their short half-lives, the radium isotopes are strongly radioactive (Hem 1985).

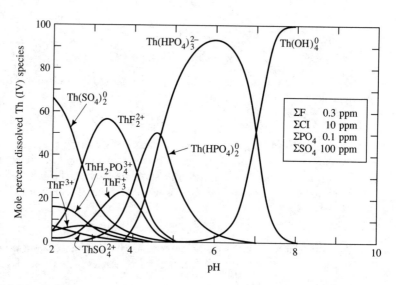

FIGURE 6.14 Distribution of thorium complexes as a function of pH under standard conditions for an aqueous solution containing 0.3 mg/L F^-, 10 mg/L Cl^-, 100 mg/L SO_4^{2-}, and 0.1 mg/L PO_4^{2-}. Source: Reprinted from *Geochemica et Cosmochemica Acta*, 44, D. Langmuir and J. S. Herman, "The mobility of thorium in natural waters at low temperatures." Copyright 1980, Pergamon Press plc.

Knowledge of the aqueous chemistry of radium is sketchy. It is reportedly similar in chemical behavior to barium (Hem, 1985) and calcium (Kathren, 1984). It is more soluble than uranium or thorium and can be bioconcentrated by plants (Brazil nuts have an especially high radium content). Radium can be strongly exchanged in the cation exchange series. According to Kathren (1984), the cation exchange sequence for soils is

$$Sr^{2+} < Ra^{2+} < Ca^{2+} < Mg^{2+} < Cs^{2+} < Rb^{2+} < K^+ < NH_4^+ < Na^+ < Li^+$$

^{228}Ra has a much shorter half-life, (5.8 yr) than ^{226}Ra. However, its parent, ^{232}Th, is more abundant in nature than ^{238}U, the parent of ^{226}Ra. As a result, both isotopes are found in ground water. The U.S. EPA has proposed MCLs of 20 pCi/L for both ^{226}Ra and ^{228}Ra (*Federal Register,* July 18, 1991). Wells with high radium levels in ground water have been discovered to be concentrated in two areas of the United States: the Piedmont and coastal plain of the Middle Atlantic states and the upper Midwestern states of Minnesota, Iowa, Illinois, Missouri, and Wisconsin (Hess et al. 1985). Table 6.7 summarizes the distribution of ^{226}Ra and ^{228}Ra in the Atlantic coastal plain and Piedmont region.

The radium content of ground water is a function of the rock type of the aquifer. Igneous rocks, such as granites, contain the highest proportion of uranium and thorium, the parent isotopes of radium. Granitic rock aquifers and sands and sandstones derived from the weathering of granites have the potential to have high radium. Phosphate rock is also very high in uranium. Radium is not only a problem that is naturally occurring, but there are localized areas of radium contamination from industrial operations. These are associated with uranium mill tailings as well as facilities where radioluminescent paints were prepared and used. For example, from World War I up until 1968 wrist watches with radium dials that glowed in the dark were sold in the United States.

6.9.6 Radon

There are several isotopes of radon, but ^{222}Rn is the only one that is important environmentally. The other isotopes have half-lives of less than 1 min. The half-life of ^{222}Rn

TABLE 6.7 Distribution of ^{2266}Ra and ^{228}Ra by aquifer type in the Atlantic coastal plain and Piedmont provinces.

Aquifer Type	Number of Samples	Ra-228 (pCi/L) Geometric Mean	Range	Ra-226 (pCi/L) Geometric Mean	Range
Igneous (acidic)	42	1.39	0.0–22.6	1.8	0.0–15.9
Metamorphic	75	0.33	0.0–3.9	0.37	0.0–7.4
Sand	143	1.05	0.0–17.6	1.36	0.0–25.9
Arkose	92	2.16	0.0–13.5	2.19	0.0–23.0
Quartzose	50	0.27	0.0–17.6	0.55	0.0–25.9
Limestone	16	0.06	0.0–0.2	0.12	0.0–0.3

Source: C. T. Hess, J. Michel, T. R. Horton, H. M. Prichard, and W. A. Coniglio, "The occurrence of radioactivity in public water supplies in the United States," *Health Physics* 48 (1985):553–86.

is 3.8 days. ^{222}Rn is produced by the decay of ^{226}Ra, so that it is associated with rocks that are high in uranium. Radon can be associated with water that is low in dissolved ^{226}Ra, because it comes primarily from the decay of the radium in the rock. Radon is a noble gas and does not undergo any chemical reactions, nor is it sorbed onto mineral matter. Radon is lost from water by diffusion into the atmosphere and by radioactive decay through a series of short-lived daughter products to ^{210}Pb, which has a half-life of 21.8 yr.

The EPA has proposed an MCL standard of 300 pCi/L for radon in drinking water (*Federal Register,* July 18, 1991). However, there is also a health concern for excessive radon accumulation in homes. Radon can enter homes through emanations from the soil as well as by diffusion from tap water with a high radon content.

Owners of private water systems are most at risk from radon in drinking water. Public water-supply systems normally have storage facilities to supply water during fires. The residence time for the water in these facilities allows the radon to both diffuse and decay. Private water systems rely upon wells and usually only have a very small storage facility used to maintain pressure.

Brutsaert et al. (1981) studied radon in ground water in Maine. They found radon levels in private wells of up to 122,000 pCi/L. The average ^{222}Rn content of wells obtaining water from granites was 22,100 pCi/L, from sillimanite-grade metasedimentary rocks, it was 13,600 pCi/L, and from chlorite-grade metasedimentary rocks, it was 1100 pCi/L. The high radon values in the high-grade metamorphic terrains were believed to be due to metamorphic pegmatites and associated uranium mineralization.

6.9.7 Tritium

Tritium, ^{3}H, is produced naturally by cosmic-ray bombardment of the atmosphere, by thermonuclear detonations and in nuclear reactors. As it is an isotope of hydrogen, it can form a water molecule or be incorporated into living tissue. Tritium has a half-life of 12.6 yr. Much of the radiation at low-level radioactive waste sites is due to tritium (Kathren 1984). Low-level radioactive waste is disposed by shallow burial on land. If waste packages at low-level radioactive waste sites leak and if the landfill is not secure, tritium is likely to escape. Tritium migration via ground water flow from low-level waste disposal has been detected at sites at the Savannah River facility, Los Alamos National Laboratory (Overcamp 1982), and the Sheffield, Illinois, commercial disposal site (Foster, Erickson, and Healy 1984), and the Hanford, Washington, site (Levi, 1992).

6.10 Geochemical Zonation

Landfills have proven to be a source of ground-water contamination in many different geologic terrains, climates, and hydrogeologic settings. Landfill **leachate** is the liquid that is the product of the liquid content of the waste, infiltrating precipitation, and ground water if the waste is below the water table. These liquids mix with the waste and dissolve both inorganic and organic constituents. Leachate is a complex mixture of dissolved and colloidal organic matter and inorganic compounds and ions. Many of the chemical processes are controlled by microbial activity. In this section we will examine the geochemical zonation that occurs in ground water affected by leachate from a landfill. This will serve to illustrate some of the reactions discussed in this chapter.

Organic matter is a major constituent of municipal landfills, where it is decomposed by bacteria and other microbes. The initial decomposition is under aerobic conditions. Once the oxygen in the landfill is consumed, anaerobic decomposition becomes prevalent. The complete aerobic decomposition of glucose, a sugar found in organic waste, yields carbon dioxide and water as products (Baedecker and Back 1979a):

$$C_6H_{12}O_6 + 6\,O_2 \longrightarrow 6\,CO_2 + 6\,H_2O$$

The carbon dioxide thus produced forms carbonic acid in the leachate and can also escape from the waste as a gas.

Anaerobic decomposition produces a variety of organic acids as intermediate steps in the formation of methane. This process can be represented by the fermentation of glucose:

$$C_6H_{12}O_6 \longrightarrow 2\,CH_3COCOOH\ (\text{pyruvic acid}) + 2\,H_2$$
$$CH_3COCOOH + H_2O \longrightarrow CH_3COOH\ (\text{acetic acid}) + HCOOH\ (\text{formic acid})$$
$$CH_3COOH \longrightarrow CH_4 + CO_2$$
$$HCOOH \longrightarrow CO_2 + H_2$$
$$CO_2 + 4\,H_2 \longrightarrow CH_4 + 2\,H_2O$$

Any hydrogen gas produced is utilized by the methane-forming bacteria in their reduction of organic compounds. In addition, oxidized forms of nitrogen and sulfur compounds are also reduced to form NH_3 and H_2S (Baedecker and Back 1979b). The methane formed can escape from the landfill as a gas and can also dissolve in the leachate.

The pH of landfill leachate is generally in the range of 6.5 to 7.0. This is due to the buffering offered by the generation of large amounts of CO_2, which dissociates to HCO_3^-, to a lesser extent the formation of NH_3, which forms NH_4^+ in water, and the reduction of SO_4^{2-} to H_2S. As has been shown earlier in this chapter, the solubility of mineral species is a function of the Eh and pH of the environment. The Eh of landfill leachate is controlled by the reduction of organic compounds and oxidized forms of nitrogen, sulfur, iron, and manganese. When reduced leachate mixes with oxygenated ground water, the Eh of the resulting solution can become more oxidizing.

Baedecker and Back (1979b) have compared the redox zonation of a landfill with that of marine sediments, a geochemical environment that has been extensively studied. The oxygenated water at the bottom of the sea represents an aerobic environment. The sediments at the bottom are anaerobic. In the transition from aerobic to anaerobic conditions, there is a sulfate-reducing zone, where decomposition of organic matter reduces sulfate to sulfide. Ferric iron will also be reduced to ferrous iron in this zone. Ferrous iron reacts with the hydrogen sulfide form insoluble sulfide precipitates. The most reduced zone in marine sediments is an anaerobic carbonate-reducing zone that has become depleted in sulfur. In this zone we find the production of CH_4 and NH_4^+.

In the study of leachate plumes at landfills, a three-part zonation has been found. The landfill itself represents an anaerobic zone. This anaerobic zone extends along the plume of leachate mixing with ground water. In this zone we find the production of methane and ammonia as microorganisms decompose organic matter and obtain oxygen from the reduction of sulfate and nitrate. Reduction will liberate soluble ferrous iron. Manganese may also become soluble due to dissolution of native minerals. Ferrous sulfide may precipitate in this zone.

As the leachate mixes with oxygenated ground water, it becomes less reducing and forms a transition zone. Most of the soluble organic matter has already been decomposed in this zone. In the transition zone there is coprecipitation of trace metals with iron and manganese hydroxides. Ahead of the transition zone, there is an aerobic zone, where the leading edge of the leachate has changed the native ground water quality, but not enough to deplete the oxygen. The aerobic zone contains nitrate and sulfate at the background levels of the aquifer.

Baedecker and Back (1979a, 1979b) have defined the boundaries of these zones at a landfill on the basis of (1) dissolved oxygen content, (2) the ratio of reduced nitrogen to nitrate, (3) the presence of methane gas, and (4) the ratio of dissolved manganese to iron.

Total reduced nitrogen, the sum of organic nitrogen and ammonia, is measured by a test called Kjeldahl nitrogen (Kjl N). Baedecker and Back (1979a, 1979b) examined the ratio Kjl N/NO_3^- and found that it was greatest in the anaerobic zone, where ammonia was present, decreased in the transition zone, and was quite low in the aerobic zone, where the nitrogen was primarily in the form of nitrate.

If there is limited sulfur present, there will be soluble iron and manganese in the anaerobic zone. As the iron traverses to the transition zone, it will begin to be oxidized and precipitate as ferric hydroxide. Manganese is soluble over a much larger Eh-pH range than iron and will remain in solution longer as the plume moves into oxygenated ground water. Because iron is much more abundant than manganese, there will be a large ratio of dissolved iron to dissolved manganese in the anaerobic zone. This ratio will gradually decrease in the transition zone as the iron is preferentially precipitated. Eventually both dissolved iron and manganese will disappear as the aerated zone is reached.

Methane gas and ammonia are found only in the anaerobic zone and dissolved oxygen is found only in the aerobic zone. Naturally, the preceding model of geochemical zonation is valid only for circumstances where the receiving ground water contains dissolved oxygen. Figure 6.15 shows an idealized leachate plume and the geochemical zonation.

If landfill leachate is discharged into a carbonate rock aquifer, or an unconsolidated aquifer with large amounts of carbonate, reactions between the acids of the leachate and the carbonate can occur (Kehew and Passero 1990). Both organic acids produced by the reduction of organic compounds and carbonic acid from the production of carbon dioxide are present in leachate. It will react with the carbonate rock, releasing calcium and magnesium:

$$CaCO_3 \rightleftharpoons Ca^{2+} + CO_3^{2-}$$
$$CO_3^{2-} + H^+ \rightleftharpoons HCO_3^-$$
$$CaMg(CO_3)_2 \rightleftharpoons Ca^{2+} + Mg^{2+} + 2\,CO_3^{2-}$$
$$2\,CO_3^{2-} + 2\,H^+ \rightleftharpoons 2\,HCO_3^-$$

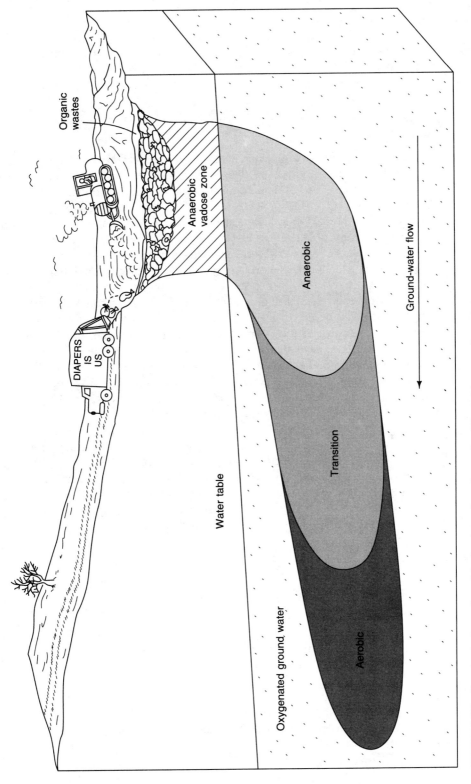

FIGURE 6.15 Geochemical zonation of the leachate plume from a landfill receiving organic waste.

These reactions liberate calcium and magnesium, and the resulting leachate has a high hardness. The increase in bicarbonate buffers the leachate to nearly neutral and also promotes the precipitation of carbonate minerals such as siderite ($FeCO_3$).

6.11 Summary

The inorganic chemistry of contaminated ground water can be studied from the standpoint of chemical thermodynamics. The first step is to obtain a representative chemical analysis of the water. This should contain the major anions and cations: Ca^{2+}, Mg^{2+}, Na^+, K^+, HCO_3^-, Cl^-, PO_4^{2-}, SO_4^{2-}, NO_3^-, and F^-. In addition, pH and Eh should be measured in the field in a flowthrough cell so that the sample does not come into contact with the atmosphere. If the Eh is positive, dissolved oxygen should be measured; if it is negative, hydrogen sulfide and ammonia should be measured. These data should be obtained to aid in interpretation of species analyses for suspected inorganic contaminants.

Inorganic contaminants in ground water can be removed by precipitation. The law of mass action can be used to study chemical equilibrium. Thermodynamically stable species may precipitate as the inorganic contaminant mixes with native ground water.

Many inorganic elements can exist in different valence states. The elements can participate in oxidation-reduction reactions. Some of the redox reactions are mediated by microorganisms. The Nernst equation relates the Eh and pH of an aqueous solution. The Nernst equation and chemical equilibrium can be used to construct Eh-pH diagrams, which show the fields of stability for various chemical spills.

Organic matter can be decomposed by microbes in a landfill. Under oxidizing conditions oxygen is the electron acceptor, whereas under reducing conditions the electron acceptor can be sulfate, nitrate, ferric iron, or carbon dioxide. Three redox zones have been identified in the leachate plumes at landfills: an anaerobic zone, where the waste is being decomposed, methane is being formed, and iron, sulfur, and nitrogen exist in reduced forms; a transition zone; and an anaerobic zone, with oxidizing conditions and oxidized forms of iron, sulfur, nitrogen, and carbon.

Chapter Notation

A		Constant equal to 0.5085 at 25°C
$[A]$		Activity of reactant A
a		Moles of reactant
a_i		Effective diameter of an ion from (Table 6.1)
B		Constant equal to 0.3281 at 25°C
$[B]$		Activity of product B
b		Moles of product
Eh		Oxidation potential of the aqueous solution in volts
E^0		Standard potential of redox reaction in volts
F		Faraday constant
I		Ionic strength

K_{eq}	Equilibrium constant
K_{sp}	Solubility product
m	Moles of hydrogen ions
m_i	Molality of the ith ion
m_x	Molal concentration
n	Number of electrons in half-reaction
R	Gas constant
T	Temperature in Kelvins
w	Moles of water
z_i	Charge on ionic species i
ΔG_R^0	Free energy of a reaction
γ_i	Activity coefficient for ionic species i
γ_x	Activity coefficient

References

Baedecker, Mary Jo, and William Back. 1979a. Hydrogeological processes and chemical reactions at a landfill. *Ground Water* 17, no. 5:429–37.

———. 1979b. Modern marine sediments as a natural analog to the chemically stressed environment of a landfill. *Journal of Hydrology* 43:393–414.

Brutsaert, W. F., S. A. Norton, C. T. Hess, and J. S. Williams. 1981. Geologic and hydrologic factors controlling radon-222 in ground water in Maine. *Ground Water* 19, no. 4:407–17.

Corbett, R. G., and B. M. Manner. 1984. Fluoride in the ground water of Northeastern Ohio. *Ground Water* 22, no. 1:13–17.

Cross, F. L., and R. W. Ross. 1970. Fluoride uptake from gypsum ponds. *Fluoride* 3:97-101.

Dowdy, R. H., and V. V. Volk. 1983. Movement of heavy metals in soil. In *Chemical Mobility and Reactivity in Soil Systems*, 229–39. Madison, Wis.: Soil Science Society of America.

Emmerson, B. T. 1970. "Ouch-ouch" disease: The osteomalacia of cadmium nephropathy. *Annals of Internal Medicine* 73; 854.

Feth, J. H. 1966. Nitrogen compounds in natural water—a review. *Water Resources Research* 2, no. 1:41–58.

Flipse, W. J., Jr., B. G. Katz, J. B. Linder, and R. Markel. 1984. Sources of nitrate in ground water in a sewered housing development, central Long Island, New York. *Ground Water* 22, no. 4:418–25.

Flipse, W. J., Jr., and F. T. Bonner. 1985. "Nitrogen-isotope ratios of nitrate in ground water under fertilized fields, Long Island, New York. *Ground Water* 23, no. 1:59–67.

Foster, J. B., J. R. Erickson, and R. W. Healy. 1984. Hydrogeology of a low-level radioactive-waste disposal site near Sheffield, Illinois. U. S. Geological Survey, *Water-Resources Investigations Report* 83-4125. 87 pp.

Gee, G. W., Dhanpat Rai, and R. J. Serne. 1983. "Mobility of radionuclides in soil." In *Chemical Mobility and Reactivity in Soil Systems*, 203–27. Madison, Wis.: Soil Science Society of America.

Giridhar, J., and D. Langmuir. 1991. Determination of E^0 for the UO_2^{2+}/U^{4+} couple from measurement of the equilibrium: $UO_2^{2+} + Cu(s) + 4H^+ = U^{4+} + Cu^{2+} + 2H_2O$ at 25°C and some geochemical implications. *Radiochimica Acta* 54: 133–38.

Grantham, D. A., and J. F. Jones. 1977. Arsenic contamination of water wells in Nova Scotia. *Journal American Water Works Association* 69:653–57.

Gray, E. M., and M. Morgan-Jones. 1980. A comparative study of nitrate levels at three adjacent ground-water sources in a chalk catchment area west of London. *Ground Water* 18, no. 2:159–67.

Hem, John D. 1977. Reactions of metal ions at surfaces of hydrous iron oxide. *Geochimica et Cosmochimica Acta* 41, no. 4:527–38.

———. 1985. *Study and interpretation of the chemical characteristics of natural water*. U. S. Geological Survey Water Supply Paper 2254, 263 pp.

Hess, C. T., J. Michel, T. R. Horton, H. M. Prichard, and W. A. Coniglio. 1985. The occurrence of radioactivity in public water supplies in the United States. *Health Physics* 48:553–86.

Hill, A. R. 1982. Nitrate distribution in the ground water of the Alliston region of Ontario, Canada. *Ground Water* 20, no. 6:696–702.

Hounslow, A. W. 1980. Ground-water geochemistry: Arsenic in landfills. *Ground Water* 18, no. 4:331–33.

Kaback, D. S., and D. D. Runnels. 1980. Geochemistry of molybdenum in some stream sediments and waters. *Geochimica et Cosmochimica Acta* 44:447–56.

Kathren, R. L. 1984. *Radioactivity in the environment*. Chur, Switzerland: Harwood Academic Publishers, 397 pp.

Kehew, A. E., and R. N. Passero. 1990. pH and redox buffering mechanisms in a glacial drift aquifer contaminated by landfill leachate. *Ground Water* 28, no. 5:728–37.

Konefes, John. 1990. Cited in Ground Water in the News. *Ground Water* 28, no. 6:997.

Langmuir, Donald. 1978. Uranium solution-mineral equilibria at low temperatures with applications to sedimentary ore deposits. *Geochimica et Cosmochimica Acta* 42, no. 6:547–70.

Langmuir, Donald, and J. S. Herman. 1980. The mobility of thorium in natural waters at low temperatures. *Geochimica et Cosmochimica Acta* 44:1753–66.

Levi, B. G. 1992. Hanford seeks short-and long-term solutions to its legacy of waste. *Physics Today* 45: no. 3 (March): 17–21.

Manahan, S. E. 1984. *Environmental chemistry*, 4th ed. Boston: PWS Publishers, 612 pp.

Matisoff, Gerald, C. J. Khourey, J. F. Hall, A. W. Varnes, and W. H. Strain. 1982. The nature and source of arsenic in Northeastern Ohio ground water. *Ground Water* 20, no. 4:446–56.

McCarty, P. L., B. E. Rittman, and E. J. Bouwer. 1984. "Microbiological Processes Affecting Chemical Transformations in Groundwater." In *Groundwater pollution microbiology*. Edited by Gabriel Bitton and C. P. Gerba, 89–115. New York: John Wiley and Sons.

Means, J. L., D. A. Crerar, and J. O. Duguid. 1978. Migration of radioactive wastes: radionuclide mobilization by complexing agents. *Science* 200:1477–81.

Overcamp, T. J. 1982. "Low-level radioactive waste disposal by shallow land burial." In *Handbook of environmental radiation*. Edited by A. W. Klement, Jr., 207–67. Boca Raton, Fla.: CRC Press.

Perlmutter, N. M., Maxim Lieber, and H. L. Frauenthal. 1963. *Movement of waterborne cadmium and hexavalent chromium wastes in South Farmingdale, Nassau County, Long Island*. U. S. Geological Survey Professional Paper 475C, pp. C170–C184.

Robertson, F. N. 1975. Hexavalent chromium in the ground water in Paradise Valley, Arizona. *Ground Water* 13, no. 6:516–27.

Robie, R. A., B. S. Hemingway, and J. R. Fisher. 1978. *Thermodynamic properties of minerals and related substances at* 298.15 K *and* 1 *bar* (10^5 *pascals*) *pressure and at higher temperatures*. U. S. Geological Survey Bulletin 1452, 456 pp.

Silver, B. A., and J. R. Fielden. 1980. Distribution and probable source of nitrate in ground water of Paradise Valley, Arizona. *Ground Water* 18, no. 3:244–51.

Wagman, D. D., W. H. Evans, V. B. Parker, I. Halow, S. M. Bailey, and R. H. Schumm. 1968. *Selected values of chemical thermodynamic properties (tables for elements* 1–34 *in standard order of arrangement)*. National Bureau of Standards Technical Note 270-3, 264 pp.

———. 1969. *Selected values of chemical thermodynamic properties (tables for elements* 35–53 *in standard order of arrangement)*. National Bureau of Standards Technical Note 270-4, 141 pp.

Wagman, D. D., W. H. Evans, V. B. Parker, I. Halow, S. M. Bailey, R. H. Schumm, and K. L. Churney. 1971. *Selected values of chemical thermodynamic properties (tables for elements* 54–61 *in standard order of arrangement)*. National Bureau of Standards Technical Note 270-5, 41 pp.

Waldbott, G. L. 1973. *Health Effects of Environmental Pollutants*, St. Louis, Mo.: C. V. Mosby Co., 316 pp.

Welch, A. H., M. S. Lico, and J. L. Hughes. 1988. Arsenic in ground water of the western United States. *Ground Water* 26, no. 3:333–47.

Whittemore, D. O. and Donald Langmuir. 1975. The solubility of ferric oxyhydroxides in natural waters. *Ground Water* 13, no. 4:360–65.

Chapter Seven
Organic Compounds in Ground Water

7.1   Introduction

Organic compounds can occur in the ground either pure as compounds, a mixture of compounds, or dissolved in water. In this chapter we will first examine the physical properties of organic compounds and the way that physical properties affect the behavior of organic chemicals in the subsurface. We will then learn the structure of common organic compounds, their nomenclature, and their possible sources in the subsurface. Finally, we will see how organic chemicals can undergo transformations by both chemical reactions and microbial degradation.

7.2   Physical Properties of Organic Compounds

There are several important physical properties of organic compounds that help us to understand how they will behave. These compounds can exist as gases, liquids, and vapors. We can use the **melting points** and **boiling points** to evaluate if a particular compound will be a gas, liquid, or vapor at a certain temperature. If the specified temperature is below the melting point, the compound will be a solid. If the temperature falls between the melting point and the boiling point, the compound will be a liquid, and if the temperature is above the boiling point, the compound will be a gas. Boiling points are usually given at a pressure of 760 mm mercury (1 atm). Boiling points for a homologous series of compounds will increase with increasing molecular weight. (A **homologous series** is one with the same basic structure but increasing numbers of carbon atoms.)

The **specific gravity** of a substance (liquid or solid) is the ratio of the weight of a given volume of that substance to the weight of the same volume of water. The water weight is usually measured at 4°C, whereas the organic liquid weight may be measured at some other temperature, which is typically 20°C. If the specific gravity of the pure substance is less than 1.0, the substance will float on water, whereas if the specific gravity is greater than 1.0, the substance will sink in water.

Water solubility is an important property of organic substances. For a gas this must be measured at a given vapor pressure. For a liquid it is a function of the temperature of the water and the nature of the substance. Solubilities of organic materials can range from completely miscible with water to nearly insoluble. More soluble materials have a greater potential mobility in the environment. Laboratory solubility is measured using distilled water, which may not have the same effect as natural waters.

The **octanol-water partition coefficient** is a measure of the degree to which an organic substance will preferentially dissolve in water or an organic solvent. The substance is mixed with equal amounts of two immiscible fluids, water and octanol (an eight-carbon chain alcohol). The coefficient is the ratio of the equilibrium concentration of the substance in octanol to the equilibrium concentration in water:

$$P_{octanol} = \frac{C_{octanol}}{C_{water}} \qquad (7.1)$$

This is usually given as a logarithm. The greater the value, the greater the tendency to dissolve in the organic liquid rather than the water. The greater the octanol-water partition coefficient, the less mobile the compound tends to be in the environment.

Vapor pressure is a measure of the tendency of a substance to pass from a solid or a liquid to a vapor state. It is the pressure of the gas in equilibrium with the liquid or the solid at a given temperature. The greater the vapor pressure, the more volatile the substance.

The **vapor density** of a gas indicates if it will rise or sink in the atmosphere. If the gas is lighter than air, it will rise; if it is denser than air, it will sink. The vapor density, V_d, is related to the equilibrium vapor pressure, the gram molecular weight of the gas, and the temperature by Equation 7.2.

$$V_d = \frac{PM}{RT} \qquad (7.2)$$

where

P = equilibrium vapor pressures in atms

M = gram molecular weight

R = gas constant (0.082 L·atms/mol/K)

T = temperature in K

Henry's law states that there is a linear relationship between the partial pressure of a gas above a liquid and the mole fraction of the gas dissolved in the liquid. It is given as Equation 7.3.

$$H_L = \frac{P_x}{C_x} \qquad (7.3)$$

where

P_x = partial pressure of gas (atms at a given temperature)

C_x = equilibrium concentration of the gas in solution (mol/m³ water)

H_L = Henry's law constant in atm/(mol/m³ water)

Henry's law is valid if the gas is sparingly soluble, the gas phase is reasonably ideal, and the gas will not react with the solute. It can also be applied to organic compounds

7.3 Organic Structure and Nomenclature

7.3.1 Hydrocarbon Classes

Organic chemistry is based on the behavior of carbon atoms. The most simple organic compounds are **hydrocarbons,** which consist solely of carbon and hydrogen. Carbon has four bonding locations and can bond with such elements as oxygen, nitrogen, sulfur, phosphorus, chlorine, bromine, and fluorine as well as hydrogen. Carbon can form single, double, and triple bonds.

Hydrocarbons can be divided into two classes, **aromatic** hydrocarbons, which contain a **benzene** ring, and **aliphatic** hydrocarbons, which don't contain a benzene ring. The benzene ring contains six carbon atoms joined in a ring structure with alternating single and double bonds. The single and double bonds change positions on the ring so that they are considered to be equal. A benzene ring is represented by the symbol in Figure 7.1(a).

Aliphatic hydrocarbons

The carbons of aliphatic hydrocarbons with more than one carbon atom can be joined by single bonds **(alkanes),** double bonds **(alkenes),** or triple bonds **(alkynes).** If there is a combination of single and multiple bonds, the compound is classified on the basis of the multiple bond.

Alkanes

Alkanes are also known as **saturated hydrocarbons,** or **paraffins.** The general formula is C_nH_{2n+2}. The first six straight-chain alkanes are as follows:

Name	Formula	Structure
Methane	CH_4	$H-CH_3$ (H−C−H with H above and below)
Ethane	CH_3CH_3	$H-CH_2-CH_2-H$
Propane	$CH_3CH_2CH_3$	$H-CH_2-CH_2-CH_2-H$

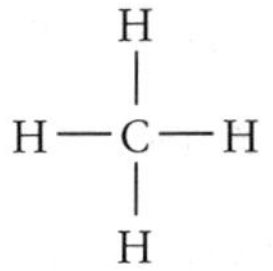

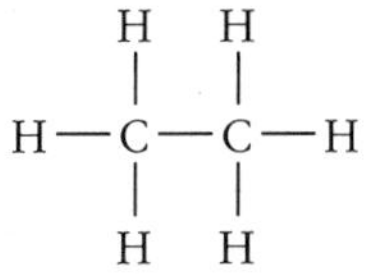

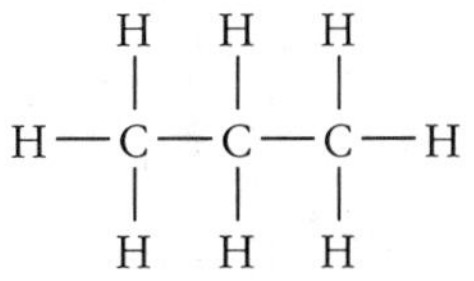

Butane $CH_3CH_2CH_2CH_3$

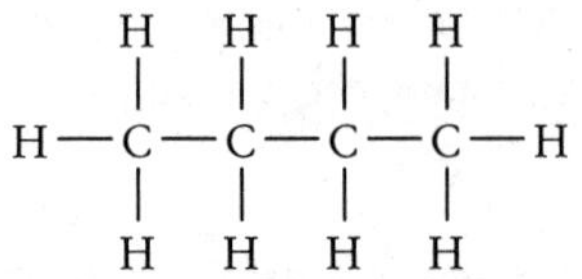

```
    H   H   H   H
    |   |   |   |
H — C — C — C — C — H
    |   |   |   |
    H   H   H   H
```

Pentane $CH_3CH_2CH_2CH_2CH_3$

```
    H   H   H   H   H
    |   |   |   |   |
H — C — C — C — C — C — H
    |   |   |   |   |
    H   H   H   H   H
```

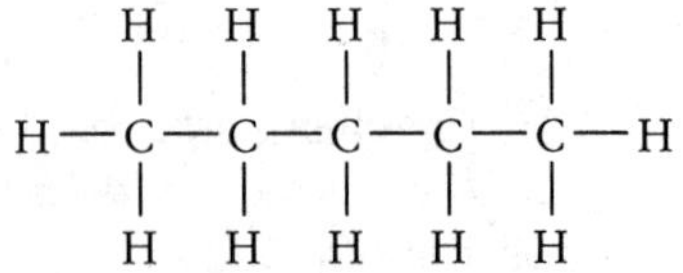

Hexane $CH_3CH_2CH_2CH_2CH_2CH_3$

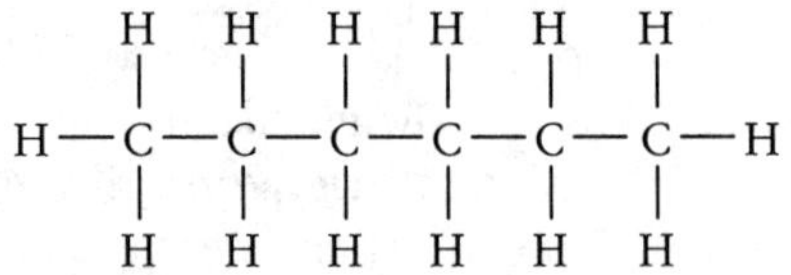

```
    H   H   H   H   H   H
    |   |   |   |   |   |
H — C — C — C — C — C — C — H
    |   |   |   |   |   |
    H   H   H   H   H   H
```

Organic compounds with the same formula may have different structural relationships. This is because alkanes can also have **branched chains.** These compounds are **structural isomers,** and although they have the same formula, they are different compounds with different properties. For example the formula C_5H_{12} represents three isomers:

Pentane

$CH_3CH_2CH_2CH_2CH_3$ or

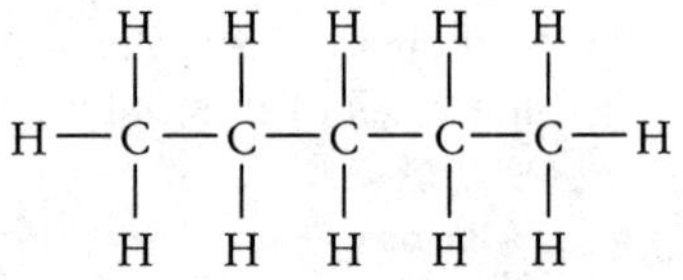

```
    H   H   H   H   H
    |   |   |   |   |
H — C — C — C — C — C — H
    |   |   |   |   |
    H   H   H   H   H
```

2-Methylbutane

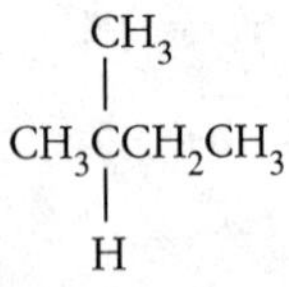

```
      CH_3
      |
CH_3CCH_2CH_3      or
      |
      H
```

```
              H
              |
          H — C — H
    H         |         H   H
    |         |         |   |
H — C ————— C ————— C — C — H
    |         |         |   |
    H         H         H   H
```

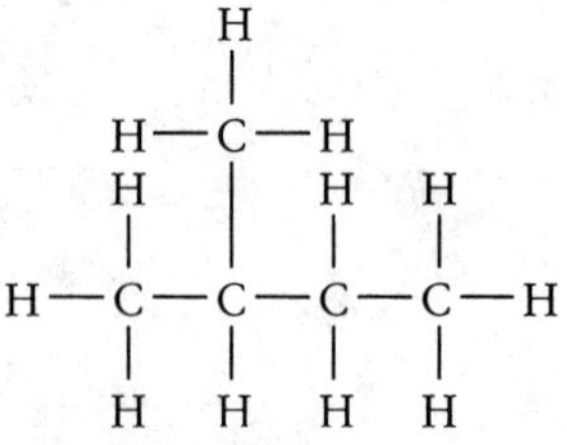

2,2-Dimethylpropane

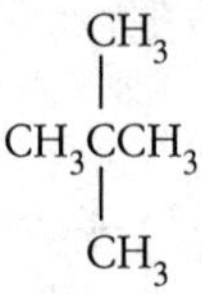

```
      CH_3
      |
CH_3CCH_3      or
      |
      CH_3
```

```
              H
              |
          H — C — H
    H         |         H
    |         |         |
H — C ————— C ————— C — H
    |         |         |
    H         |         H
          H — C — H
              |
              H
```

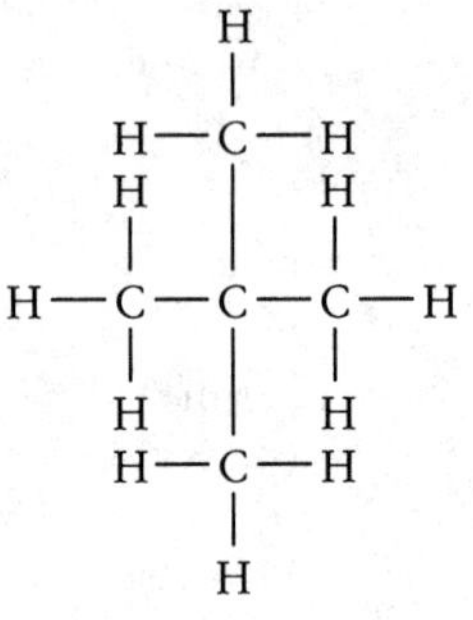

Note that there are two ways of illustrating the structural formulas. On the left is a condensed form, where only some of the bonds are shown, whereas on the right all the bonds between atoms are shown. In most cases we will use the condensed form as it is much more convenient.

These isomers have been named using the rules of the International Union of Pure and Applied Chemistry (IUPAC). For alkanes these rules are as follows:

1. The base name of the compound is the name of the longest straight-chain alkane that is present.
2. Any chain that branches from the straight chain or any functional group that is attached to the straight chain is named. Carbon chains are named as alkyl groups. The following are important alkyl groups:

Methyl	CH_3-	Butyl	$CH_3CH_2CH_2CH_2-$
Ethyl	CH_3CH_2-		
Propyl	$CH_3CH_2CH_2-$	t-Butyl	$CH_3\underset{\mid}{\overset{\overset{\displaystyle CH_3}{\mid}}{C}}CH_3$
Isopropyl	$CH_3\underset{\mid}{CH}CH_3$		

Other functional groups include such things as chloride ions, indicated by the prefix *chloro*-.

3. The location of the functional group is indicated by numbering the carbon atoms in the longest straight chain, with the end carbon closest to the position of the first functional group being carbon 1.
4. For more than one of the same functional group, the prefixes *di-*, *tri-*, and *tetra-* are used.

Consider the following branched-chain alkane:

$$\underset{1\quad 2\quad 3\quad 4\quad 5\quad 6\quad 7}{CH_3\underset{\underset{\displaystyle CH_3}{\mid}}{CH}CH_2\underset{\underset{\displaystyle CH_2CH_3}{\mid}}{CH}CH_2CH_2CH_3}$$

The longest straight chain has seven carbons, so this is a heptane. The carbons are numbered starting with the end closest to an attached functional group. A methyl group is attached to the second carbon and an ethyl group is attached to the fourth carbon. The name of the compound is 2-methyl-4-ethylheptane.

Alkanes can also have a cyclical structure—that is, the two ends of the carbon chain can be joined together. The shortest chain that can form a **cycloalkane** is propane. Cycloalkanes are also known as *cycloparaffins*, or *naphthenes*, and have a formula of C_nH_{2n}. The structure of cyclobutane is

```
      H   H
      |   |
　H—C — C—H
      |   |
　H—C — C—H
      |   |
      H   H
```

Alkenes

Alkenes have a carbon-carbon double bond, with the general formula C_nH_{2n}. Alkenes are also known as **unsaturated hydrocarbons**, or **olefins**. They are named by finding the longest chain that contains the double bond. The base name ends in *-ene* (or *-ylene*) rather than *-ane*. The carbon atoms in the straight chain are numbered starting at the end nearest the double bond. The simplest alkene is ethene (ethylene); it has the formula $CH_2=CH_2$. If the double bond can occur in more than one position, that position is indicated by a numerical prefix, which is the number of the first carbon atom containing the double bond. For example, butene can have the double bond in two positions:

$$CH_2=CHCH_2CH_3 \qquad CH_3CH=CHCH_3$$
$$\text{1-Butene} \qquad\qquad \text{2-Butene}$$

If there are functional groups present, then their position is indicated by the carbon atom to which they are bonded. Some of the alkenes can exist as structural isomers because of the double bond. Carbon molecules in alkanes can rotate around the single bonds so that a number of structural forms are equivalent. With double bonds, such rotation is not possible. For example there are 2 isomers of 1,2-dichloroethene:

$$\underset{\text{cis-1,2-Dichloroethene}}{\begin{array}{c}Cl \quad\quad Cl\\ \diagdown \quad \diagup \\ C=C \\ \diagup \quad \diagdown \\ H \quad\quad H\end{array}} \qquad \underset{\text{trans-1,2-Dichloroethene}}{\begin{array}{c}Cl \quad\quad H\\ \diagdown \quad \diagup \\ C=C \\ \diagup \quad \diagdown \\ H \quad\quad Cl\end{array}}$$

In addition to the formal names derived from the IUPAC system, many organic compounds have common names as well. Common names will be given as synonyms.

7.3.2 Aromatic Hydrocarbons

Aromatic hydrocarbons are based on the benzene ring. Figure 7.1(a) shows the condensed form of the benzene ring. Other molecules can be formed by joining functional groups to the benzene ring. Figure 7.1(b) shows methylbenzene, which has the common name toluene. If only one functional group is attached, then a position is not specified, since all six carbon atoms are equivalent. If there are two functional groups attached, then there are three isomers. If the two functional groups are the same, they may be distinguished by the prefixes *ortho-* (*o*), *meta-* (*m*) and *para-* (*p*), as shown in Figure 7.1(c) for dimethylbenzene, which is commonly called xylene. A numbering system for the six carbon atoms may also be used, as illustrated in Figure 7.1(d) for 1,3,5-trimethylbenzene.

Two or more benzene rings may be joined together. The most simple **polycyclic aromatic hydrocarbon** (PAH) is naphthalene, which consists of two benzene rings. Three or more benzene rings can also join. The structure and some properties of a number of PAH compounds are shown in Figure 7.2. In these compounds the benzene rings share carbon atoms. PAHs are found in heavy fractions of petroleum distillation,

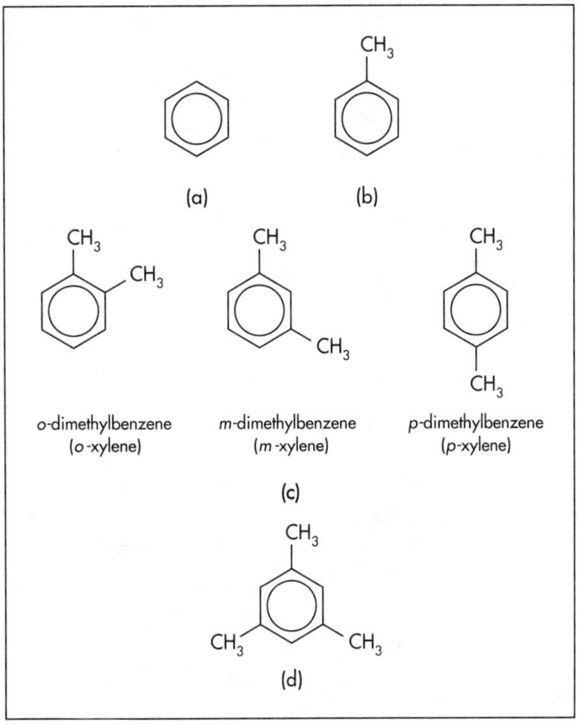

Organic Compounds in Ground Water

FIGURE 7.1 Structure and nomenclature of the benzene ring. (a) Benzene ring; (b) methyl benzene (toluene); (c) o-dimethyl benzene, m-dimethyl benzene, and p-dimethyl benzene; (d) 1,3,5,-trimethyl benzene.

asphalt, coal tar, and creosote. They also form from the incomplete combustion of fossil fuels.

If the benzene ring is joined to another group, it may be named as a functional group, *phenyl-*. For example, Figure 7.3 shows the structure of biphenyl and diphenylmethane. If biphenyl is chlorinated, it yields a mixture of isomers containing 1 to 10 chloride ions. These compounds are called **polychlorinated biphenyls, or PCBs,** and always contain a mixture of isomers. They are quite resistant to chemical, thermal, or biological degradation and tend to persist in the environment.

7.4 Petroleum Distillates

Crude oil consists of a mixture of hydrocarbons of varying molecular weight and on the average contains about 84.5% carbon, 13% hydrogen, 1.5% sulfur, 0.5% nitrogen, and 0.5% oxygen. A typical crude oil might consist of about 25% alkanes (paraffins), 50% cycloalkanes (naphthenes), 17% aromatics, including polycyclic aromatics, and 8% asphaltics, which are molecules of very high molecular weight with more than 40 carbon

Name	Structure	Molecular Weight	Solubility in Water	Soil-Water Partition Coefficient
Benzene		78.11	1780 mg/L	97
Toluene	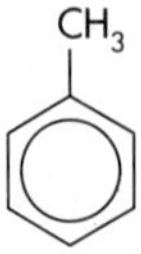	92.1	500 mg/L	242
Xylene, ortho	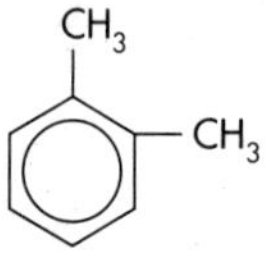	106.17	170 mg/L	363
Ethyl benzene	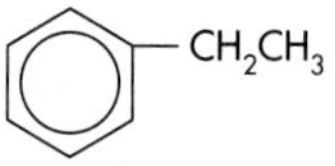	106.17	150 mg/L	622
Naphthalene	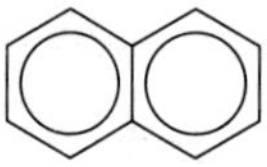	128.16	31.7 mg/L	1,300
Acenaphthene	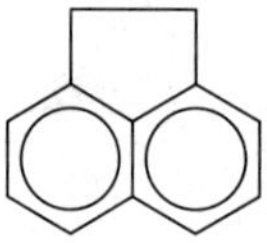	154.21	7.4 mg/L	2,580
Acenaphthylene	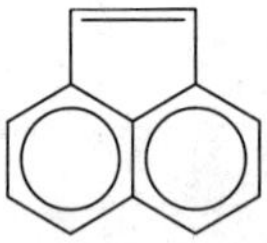	152.2	3.93 mg/L	3,814
Fluorene	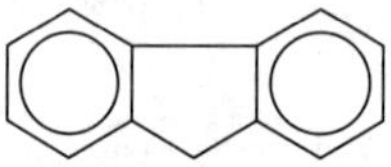	166.2	1.98 mg/L	5,835
Fluoranthene	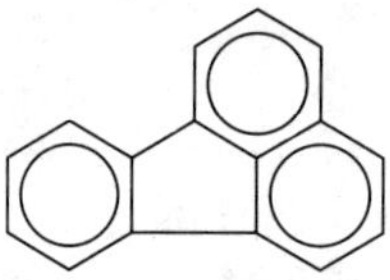	202	0.275 mg/L	19,000
Phenanthrene	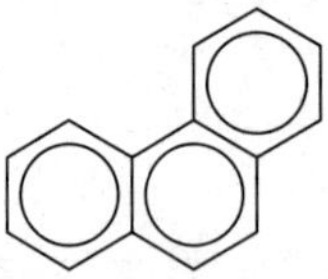	178.23	1.29 mg/L	23,000
Anthracene	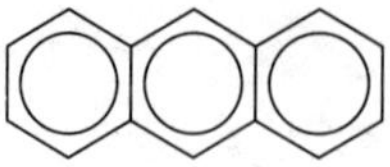	178.23	0.073 mg/L	26,000

FIGURE 7.2 Structure and properties of some aromatic and polycyclic aromatic hydrocarbons.

Name	Structure	Molecular Weight	Solubility in Water	Soil-Water Partition Coefficient
Pyrene	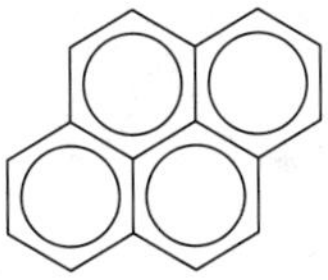	202.26	0.135 mg/L	63,000
Benzo[a]anthracene	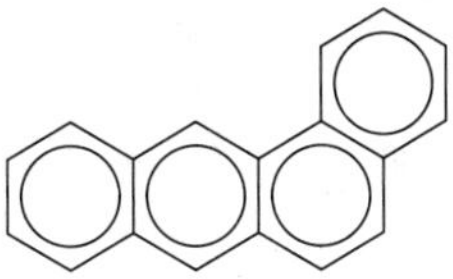	228	0.014 mg/L	125,719
Benzo[a]pyrene	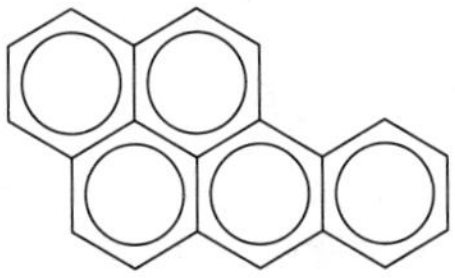	252.3	0.0038 mg/L	282,185
Chrysene	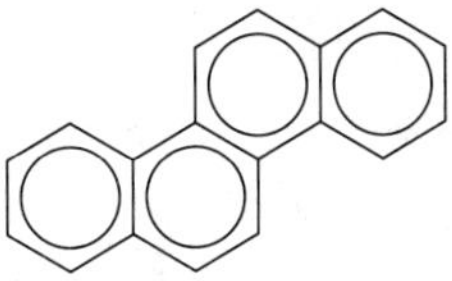	228.2	0.006 mg/L	420,108
Benzo[b]fluoranthene	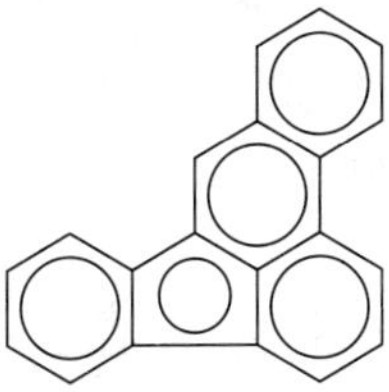	252	0.0012 mg/L	1,148,497
Benzo[g,h,i]perylene	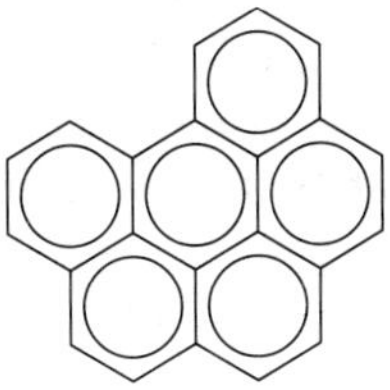	276	0.00026 mg/L	1,488,389
Dibenz[a,h]anthracene	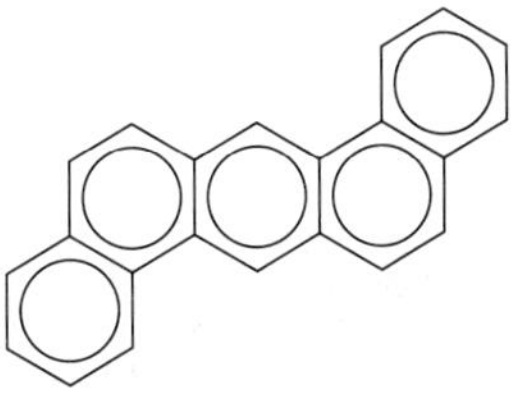	278.35	0.00249 mg/L	1,668,800
Benzo[k]fluoranthene	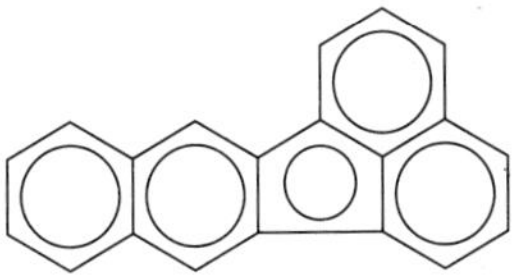	252	0.00055 mg/L	2,020,971

FIGURE 7.2 Continued

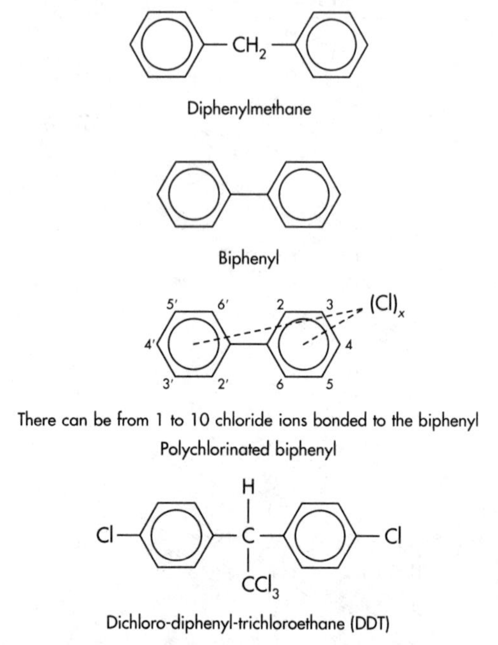

FIGURE 7.3 Structure of some phenyl compounds.

atoms. There have been more than 600 hydrocarbon compounds identified in petroleum (Hunt, 1979).

Petroleum is separated into fractions by distillation. The boiling point of hydrocarbons is correlated to the number of carbon atoms. For example, Table 7.1 lists several hydrocarbons with six carbon atoms and their boiling points. Although all have different structures, the boiling points are similar.

In a homologous series, the boiling point of the hydrocarbon will rise with the number of carbon atoms. Table 7.2 shows the boiling points for some of the normal alkanes. Because of these two characteristics, during distillation petroleum is separated

TABLE 7.1 Boiling points of several hydrocarbons with six carbon atoms.

Hydrocarbon	Boiling Point (°C)
Benzene	80.1
Hexane	68.7
Cyclohexane	81.0
2-Methylpentane	60
Methylcyclopentane	72

Organic Compounds in Ground Water

TABLE 7.2 Boiling points of normal alkanes.

Alkane	Carbon Atoms	Boiling Point (°C)
Butane	4	0
Pentane	5	36
Hexane	6	69
Heptane	7	98
Octane	8	126
Nonane	9	151
Decane	10	174

into the hydrocarbon fractions with similar numbers of carbon atoms according to Table 7.3.

Although the composition of each fraction is complex and is subject to a great deal of variation depending upon the crude oil and the refinery, the hydrocarbons in each fraction will have similar numbers of carbon atoms and boiling points. Gasoline, for example, contains over 100 separate organic compounds. Table 7.4 gives an analysis of two different brands of gasoline. In addition to the listed chemicals, there are many compounds present that constitute less than 0.01% of the mixture.

7.5 Functional Groups

7.5.1 Organic Halides

We have already seen that chlorine atoms can be substituted for hydrogen atoms at various places on organic molecules. This is also true for bromine and fluorine atoms. Such compounds are named by prefixing the name of the basic molecule with the term *chloro-*, *bromo-*, or *fluoro-* and specifying the number and position. Figure 7.4 shows

TABLE 7.3 Hydrocarbon fractions separated by distillation.

Fraction	Range of C Atoms per Molecule	Boiling Point Range	Uses
Gas	1 to 4	20°C	Cooking, home heating, chemical feed stock
Gasoline	5 to 10	20–190°C	Fuel, benzene for chemical feed stock
Kerosene	11 to 13	190–260°C	Fuel, jet fuel
Diesel	14 to 18	260–360°C	Diesel fuel and fuel oil
Heavy gas and lubricating oils	19 to 40	360–530°C	Lubricating oil, greases, waxes
Residuum	>40	>560°C	Asphalt

TABLE 7.4 Chemical composition of 87.1 octane Union 76 unleaded regular and 92.9 octane Amoco premium unleaded gasolines.

Hydrocarbon[a]	Percent Composition		Hydrocarbon[a]	Percent Composition	
	Union 76 No-lead Regular	Amoco Premium No Lead		Union 76 No-lead Regular	Amoco Premium No-Lead
Isobutane	1.86	1.40	3-Methylheptane	0.70	0.23
n-Butane/1,3-butadiene	7.75	3.52	2,2,5-Trimethylhexane	0.81	0.76
trans-2-Butene/2,2-dimethylpropane	0.25	0.13	n-Octane	0.76	0.20
cis-2-Butene/1-butyne	0.25	0.13	2,3,5-Trimethylhexane	0.18	0.13
3-Methyl-1-butene	0.10	0.07	2,4-Dimethylheptane	0.14	0.08
Isopentane	6.16	7.12	2,5- and 3,5-Dimethylheptane	0.24	0.09
1-Pentane/2-butyne	0.32	0.18			
n-Pentane	3.06	2.37	Ethylbenzene/2,3-Dimethylheptane	1.17	0.94
trans-2-Pentene	0.89	0.73			
cis-2-Pentene	0.51	0.41	p- and m-Xylene	4.58	2.60
2-Methyl-2-butene	1.22	1.50	2,4,5-Trimethylheptane	0.37	0.10
2,2-Dimethylbutane	0.41	0.08	o-Xylene/unknown C_9 paraffin	2.46	1.61
Cyclopentene	0.37	0.31			
Cyclopentane/3- and 4-methyl-1-pentene	0.48	0.42	2,4-Dimethyloctane	0.14	0.05
			n-Nonane	0.27	0.18
2,3-Dimethylbutane	0.86	0.78	C_{10} paraffin	0.16	0.32
2-Methylpentane/2,3-dimethyl-1-butene	2.76	2.76	n-Propylbenzene	0.70	0.90
			1,3,5-Trimethylbenzene	2.74	3.35
3-Methylpentane	1.76	1.47	3,4-Dimethyloctane	1.12	1.42
1-Hexene/2-ethyl-butene	0.64	0.64	1-Methyl-3-ethylbenzene	1.52	1.53
n-Hexane/cis-3-hexene	1.32	0.83	1-Methyl-2-ethylbenzene	0.28	0.07
trans-3-Hexene	0.80	0.73	1,2,4-Trimethylbenzene	3.75	4.59
2-Methyl-2-pentene	0.61	0.65	sec-Butylbenzene	0.25	0.17
2-Hexene (cis and trans)	0.33	0.27	1,2,3-Trimethylbenzene	1.21	1.26
Methylcyclopentane/3-methyl-trans-2-pentene	1.17	0.77	Indane	0.62	0.66
			Isobutylbenzene	0.42	0.48
2,4-Dimethylpentane	1.15	0.86	1-Methyl-3-n-propylbenzene	1.15	1.32
Benzene/cyclohexane	1.76	1.96	1,3-Diethylbenzene	0.67	0.78
Cyclohexene/2,3-dimethylpentane/2-methylhexane	2.73	1.31	1-Methyl-3-isopropylbenzene	0.82	0.99
			1,2-Diethylbenzene	0.57	0.68
3-Methylhexane	1.91	1.04	2-Methyldecane	1.83	1.38
2,2,4-Trimethylpentane	3.75	2.07	C_{10} aromatic	1.53	1.62
n-Heptane	1.23	0.42	C_{10} aromatic	0.51	0.67
Methylcyclohexane	1.57	0.33	n-Undecane	0.75	0.69
Dimethylhexene	0.28	0.25	C_{10} aromatic	0.77	1.03
2,2-Dimethylhexane	0.12	0.18	Unknowns	7.90	10.17
2,4- and 2,5-Dimethylhexane	1.14	0.84			
2,3,3-Trimethylpentane	2.26	1.82	Percent aromatics	31.23	44.20
Toluene/2,3-dimethylhexane	5.54	20.25	Percent olefins	10.54	9.33
2-Methylheptane/1-methylcyclohexene	0.37	0.10	Percent paraffins	58.23	46.47
4-Methylheptane	1.20	0.25			

Source: Reprinted with permission from J. E. Sigsby, Jr., Silvestre Tejada, and William Ray, "Volatile Organic Compound Emissions from 46 In-Use Passenger Cars," *Environmental Science and Technology* 21, no. 5 (1987):467. Copyright 1987 American Chemical Society.

Name	Structure	Uses and Other Sources
Trichloromethane (chloroform)	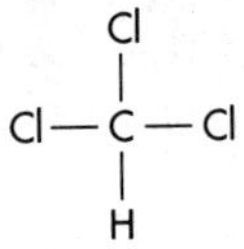 $Cl-CHCl_2$ (Cl, Cl, Cl, H on central C)	Liquid used in manufacture of anesthetics, pharmaceuticals, fluorocarbon refrigerants and plastics. Used as solvent and insecticide. Formed from methane when chlorinating drinking water.
Vinyl chloride (chloroethene)	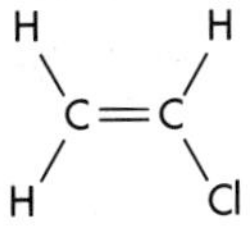 $H_2C=CHCl$	Gas used in the manufacture of polyvinyl chloride. End product of microbial degradation of chlorinated ethenes.
Chloroethane	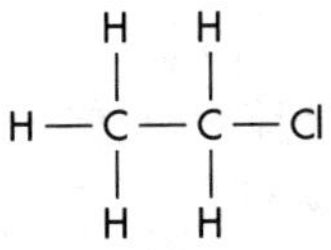 H_3C-CH_2Cl	Liquid used to manufacture tetraethyl lead. Degradation product of chlorinated ethanes.
1,2-Dichloroethane	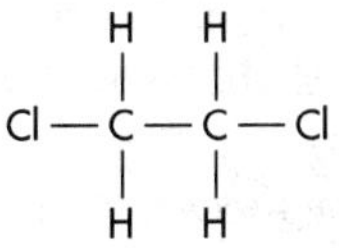 ClH_2C-CH_2Cl	Liquid used to manufacture vinyl chloride. Degradation product of trichloroethane.
Trichloroethene (Trichloroethylene)	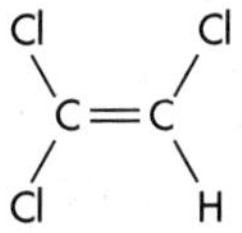 $Cl_2C=CHCl$	Solvent used in dry cleaning and metal degreasing. Organic synthesis. Degradation product of tetrachloroethene.
Tetrachloroethene (perchloroethene) (perchloroethylene)	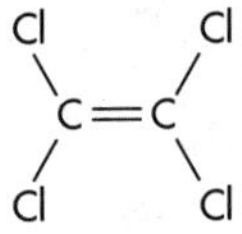 $Cl_2C=CCl_2$	Solvent used in dry cleaning and metal degreasing. Used to remove soot from industrial boilers. Used in manufacture of paint removers and printing inks.
1,2-Dibromo-3-chloropropane (DBCP)	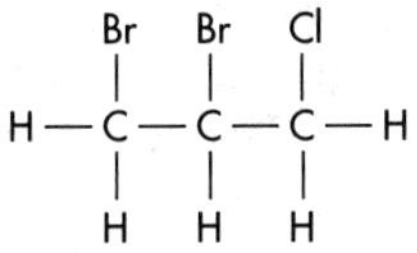 $H_2BrC-CHBr-CH_2Cl$	Soil fumigant to kill nematodes. Intermediate in organic synthesis.
o-Dichlorobenzene (1,2-dichlorobenzene)	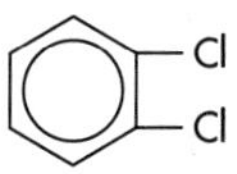 benzene ring with two adjacent Cl	Chemical intermediate. Solvent. Fumigant and insecticide. Used for industrial odor control. Found in sewage from odor control chemicals used in toilets.

FIGURE 7.4 Organic halides found in hazardous waste.

the structure and uses of some **organic halides** that are sometimes found as ground-water contaminants. Table 7.5 gives the physical properties of methane, ethane, ethene, and their chlorinated forms. These are widely used organic compounds that have often been found as contaminants in ground water. Tetrachloroethene (perchloroethene), trichloroethene, and 1,1,1-trichloroethane have a low flammability and a high vapor density, which make them very useful as solvents. They are frequently used for degreasing metal parts and as the fluid in dry cleaning. They are denser than water; if spilled on the ground in quantities great enough to overcome the residual saturation, the pure phase may migrate vertically downward through an aquifer. They are also quite soluble in water and can migrate as a dissolved phase with flowing ground water.

7.5.2 Alcohols

An **alcohol** has one or more hydroxyl groups, —OH, substituted for hydrogen atoms on an aliphatic hydrocarbon. Alcohols are named by finding the longest hydrocarbon chain that includes the carbon atom to which the hydroxyl group is attached. That becomes the base name, to which suffix -*ol* is added. The position of the carbon atom to which the hydroxyl group is attached is indicated by a numerical prefix. If there are two hydroxyl groups, then the suffix -*diol* is used, and so forth. Figure 7.5 lists some common alcohols that may find their way to a hazardous waste site. Alcohols are miscible with water and hence have a potential for significant mobility in ground water. However, many are also readily biodegraded.

7.5.3 Ethers

Ethers have an oxygen atom bonded between two carbon atoms. The common name is obtained from the names of the two hydrocarbons followed by the word ether. Methyl ethyl ether is:

$$CH_3-O-CH_2CH_3$$

In the IUPAC system, the *methoxy* functional group is CH_3O-, so the preceding compound is also known as 1-methoxyethane.

Two cyclical ethers have been found to be rather persistent contaminants in ground water. Tetrahydrofuran is used in large quantities as a chemical intermediate and as a solvent for resins such as PVC, adhesives, and various coatings. It is miscible with water, is resistant to biodegradation, and has the following structure:

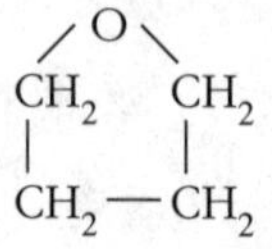

$$\begin{array}{c} \diagup O \diagdown \\ CH_2 \quad CH_2 \\ | \quad\quad | \\ CH_2 - CH_2 \end{array}$$

1,4-Dioxane is also miscible with water and is extremely resistant to biodegradation. At the Seymour Recycling Corporation Hazardous Waste Site in Indiana, more than 80 chemicals have been detected in the plume of contaminated ground water. The compound that has traveled the greatest distance is 1,4-dioxane, with tetrahydrofuran being the second-most mobile compound. 1,4-Dioxane is used as a solvent in a wide variety of lacquers, paints, varnishes, inks, dyes, emulsions, and so on. It is added to chlorinated solvents as a stabilizer to prevent degradation. Note that it is not a dioxin, which is a

TABLE 7.5 Common chlorinated alkanes and alkenes with their physical properties.

Alkanes	Synonym	Structure	Specific Gravity[a]	Melting Point[a]	Boiling Point[a]	Vapor Pressure[a]	Water Solubility[a]	Henry's Law Constant ($m^3 \times atm/mol$)[b]
Methane								
Methane		CH_4						
Methyl chloride	(Chloromethane)	CH_3Cl		−97.7°C	−24°C	5.0 atm at 20°C	4,000 cu cm/L	0.00584 at 17.5°C
Methylene chloride	(Dichloromethane)	CH_2Cl_2		−97°C	40−52°C	349 mm at 20°C	20,000 mg/L at 20°C	0.00131 at 17.5°C
Chloroform	(Trichloromethane)	$CHCl_3$		−64°C	62°C	160 mm at 20°C	8,000 mg/L at 20°C	0.00246 at 17.5°C
Carbontetrachloride	(Tetrachloromethane)	CCl_4	1.489 at 20°C	−23°C	76.7°C	90 mm at 20°C	800 mg/L at 20°C	0.0211 at 17.5°C
Ethanes								
Ethane	(Dimethyl)	CH_3CH_3		−172°C	−89°C	38.5 atm at 20°C	60.4 mg/L at 20°C	
Chloroethane	(Ethylchloride)	CH_2ClCH_3	0.92 at 20°C	−138.3°C	12.4°C	1,000 mm at 20°C	5,740 mg/L at 20°C	0.00846 at 17.5°C
1,1-Dichloroethane		$CHCl_2CH_3$	1.174 at 20°C	−97.4°C	57.3°C	180 mm at 20°C	5,500 mg/L at 20°C	0.00389 at 17.5°C
1,2-Dichloroethane	(Ethylenedichloride)	CH_2ClCH_2Cl	1.25 at 20°C	−35.4°C	83.5°C	61 mm at 20°C	8,690 mg/L at 20°C	
1,1,1-Trichloroethane	(Methylchloroform)	CCl_3CH_3	1.35 at 20°C	−32°C	71/81°C	100 mm at 20°C	4,400 mg/L at 20°C	0.0120 at 17.5°C
1,1,2-Trichloroethane	(Vinyltrichloride)	$CHCl_2CH_2Cl$	1.44 at 20°C	−35°C	113.7°C	19 mm at 20°C	4,500 mg/L at 20°C	
1,1,2,2-Tetrachloroethane		$CHCl_2CHCl_2$	1.60 at 20°C	−42.5°C	146.4°C	5 mm at 20°C	2,900 mg/L at 20°C	
1,2,2,2-Tetrachloroethane		CH_2ClCCl_3	1.60		138°C			
Pentachloroethane		CCl_3CHCl_2	1.67 at 25°C	−29°C	162°C	3.4 mm at 20°C	50 mg/L at 22°C	
Hexachloroethane		CCl_3CCl_3	2.09 at 20°C	187.4°C				
Ethenes								
Ethene	(Ethylene)	$CH_2{=}CH_2$		−169°C	−104°C	> 40 atm at 20°C	131 mg/L at 20°C	0.0193 at 17.5°C
Vinyl chloride	(Chloroethene)	$CH_2{=}CHCl$		−153°C	−13.9°C	2,660 mm at 25°C	1.1 mg/L at 25°C	0.0191 at 17.5°C
1,1-Dichloroethene		$CCl_2{=}CH_2$	1.218 at 20°C	−122.5°C	31.9°C	500 mm at 20°C		0.00265 at 17.5°C
cis-1,2-Dichloroethene		$CHCl{=}CHCl$	1.28	−81°C	60°C	200 mm at 25°C	800 mg/L at 20°C	
trans-1,2-Dichloroethene		$CHCl{=}CHCl$	1.26	−50°C	48°C	200 mm at 14°C	600 mg/L at 20°C	0.00660 at 17.5°C
Trichloroethene	(TCE)	$CHCl{=}CCl_2$	1.46 at 20°C	−87°C	86.7°C	60 mm at 20°C	1,100 mg/L at 25°C	0.00632 at 17.5°C
Tetrachloroethene	(Perchloroethene)	$CCl_2{=}CCl_2$	1.626 at 20°C	−22.7°C	121.4°C	14 mm at 20°C	150 mg/L at 25°C	0.0117 at 17.5°C

Sources: [a] K. Verschueren, *Handbook of Environmental Data on Organic Chemicals*, 2d ed. (New York: Van Nostrand Reinhold Company, 1983), 1310 pp.
[b] J. M. Gossett, "Measurement of Henry's Law Constants for C_1 and C_2 Chlorinated Hydrocarbons," *Environmental Science and Engineering* 21, no. 2 (1987):202−8.

Name	Structure	Uses and Sources
Methanol (wood alcohol)	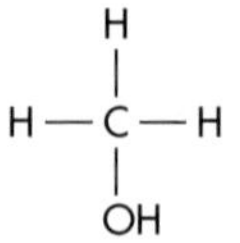 $\text{H}-\overset{\overset{\text{H}}{\mid}}{\underset{\underset{\text{OH}}{\mid}}{\text{C}}}-\text{H}$	Solvent. May be added to gasoline. Manufacture of formaldehyde and methyl halides.
Ethanol (grain alcohol) (ethylalcohol)	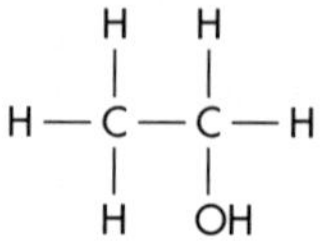 $\text{H}-\overset{\overset{\text{H}}{\mid}}{\underset{\underset{\text{H}}{\mid}}{\text{C}}}-\overset{\overset{\text{H}}{\mid}}{\underset{\underset{\text{OH}}{\mid}}{\text{C}}}-\text{H}$	Preparation of distilled spirits. Solvent. Manufacture of acetaldehyde, acetic acid, ethyl ether, etc. Preparation of lacquers, perfumes, cosmetics, over-the-counter medicines. Degradation of ethyl acetate in excess of water.
Ethylene glycol (1,2-ethanediol)	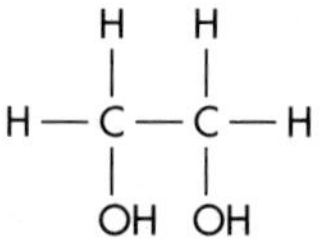 $\text{H}-\overset{\overset{\text{H}}{\mid}}{\underset{\underset{\text{OH}}{\mid}}{\text{C}}}-\overset{\overset{\text{H}}{\mid}}{\underset{\underset{\text{OH}}{\mid}}{\text{C}}}-\text{H}$	Antifreeze (engine coolent) compound. Manufacture of polyester fiber and film. Deicing compound for airplanes at gate. Solvent base.
Propanol	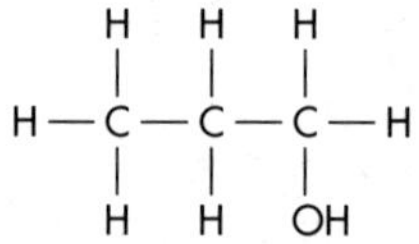 $\text{H}-\overset{\overset{\text{H}}{\mid}}{\underset{\underset{\text{H}}{\mid}}{\text{C}}}-\overset{\overset{\text{H}}{\mid}}{\underset{\underset{\text{H}}{\mid}}{\text{C}}}-\overset{\overset{\text{H}}{\mid}}{\underset{\underset{\text{OH}}{\mid}}{\text{C}}}-\text{H}$	Released from fermentation of whisky and during sewage treatment and decomposition of organic matter. Solvent in printing, used in nail polish, brake fluid, lacquers, cleaners, polishes.
1,2-Propanediol (propylene glycol)	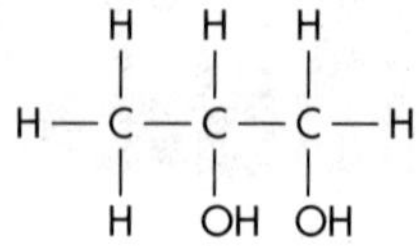 $\text{H}-\overset{\overset{\text{H}}{\mid}}{\underset{\underset{\text{H}}{\mid}}{\text{C}}}-\overset{\overset{\text{H}}{\mid}}{\underset{\underset{\text{OH}}{\mid}}{\text{C}}}-\overset{\overset{\text{H}}{\mid}}{\underset{\underset{\text{OH}}{\mid}}{\text{C}}}-\text{H}$	Solvent used in paints, inks, and coatings. Antifreeze formulations.
2-Methy-2-butanol	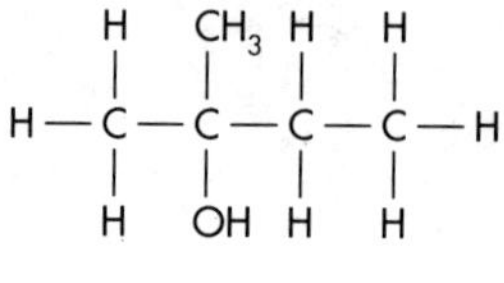 $\text{H}-\overset{\overset{\text{H}}{\mid}}{\underset{\underset{\text{H}}{\mid}}{\text{C}}}-\overset{\overset{\text{CH}_3}{\mid}}{\underset{\underset{\text{OH}}{\mid}}{\text{C}}}-\overset{\overset{\text{H}}{\mid}}{\underset{\underset{\text{H}}{\mid}}{\text{C}}}-\overset{\overset{\text{H}}{\mid}}{\underset{\underset{\text{H}}{\mid}}{\text{C}}}-\text{H}$	Solvent.
t-Butanol (2-methyl-2-propanol)	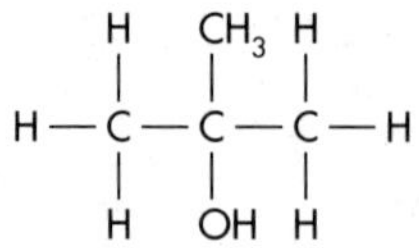 $\text{H}-\overset{\overset{\text{H}}{\mid}}{\underset{\underset{\text{H}}{\mid}}{\text{C}}}-\overset{\overset{\text{CH}_3}{\mid}}{\underset{\underset{\text{OH}}{\mid}}{\text{C}}}-\overset{\overset{\text{H}}{\mid}}{\underset{\underset{\text{H}}{\mid}}{\text{C}}}-\text{H}$	Manufacture of flotation agents, flavors, and perfumes. Solvent. Paint removers. Octane booster in gasoline. Lacquer. Solvent for pharmaceuticals.
4-Methyl-2-pentanol	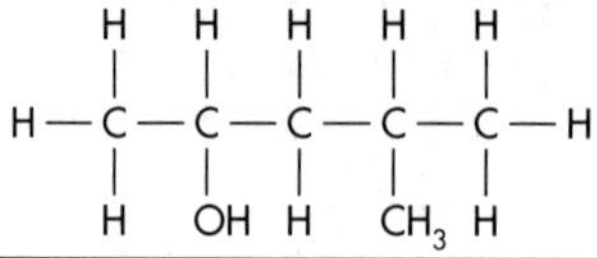 $\text{H}-\overset{\overset{\text{H}}{\mid}}{\underset{\underset{\text{H}}{\mid}}{\text{C}}}-\overset{\overset{\text{H}}{\mid}}{\underset{\underset{\text{OH}}{\mid}}{\text{C}}}-\overset{\overset{\text{H}}{\mid}}{\underset{\underset{\text{H}}{\mid}}{\text{C}}}-\overset{\overset{\text{H}}{\mid}}{\underset{\underset{\text{CH}_3}{\mid}}{\text{C}}}-\overset{\overset{\text{H}}{\mid}}{\underset{\underset{\text{H}}{\mid}}{\text{C}}}-\text{H}$	Solvent.

FIGURE 7.5 Alcohols found in hazardous waste.

Organic Compounds in Ground Water

class of compounds thought to be highly toxic. The structure of 1,4-dioxane is

$$\begin{array}{c} \text{CH}_2\text{—O—CH}_2 \\ | \quad\quad\quad | \\ \text{CH}_2\text{—O—CH}_2 \end{array}$$

7.5.4 Aldehydes and Ketones

Aldehydes and ketones contain the **carbonyl group,** which has this structure:

$$\diagdown \!\! C\!=\!O \diagup$$

An **aldehyde** has at least one hydrogen atom bonded to the carbon. The IUPAC name is obtained by finding the name of the hydrocarbon and adding the suffix -*al*. The two most simple aldehydes are methanal (formaldehyde) and ethanal (acetaldehyde):

Formaldehyde (methanal)   Acetaldehyde (ethanal)

Ketones have the carbonyl group bonded to two hydrocarbons. They are named by indicating the name of the hydrocarbon and ending with the suffix -*one*. The most simple ketone is propanone, which is commonly known as acetone. Acetone is a widely used solvent, as are other ketones such as 2-butanone (methyl ethyl ketone), 4-methyl-2-pentanone (methyl isobutyl ketone) and 2-pentanone (methyl propyl ketone). Structures of some of these ketones include the following:

2-Pentanone   2-Butanone   Propanone (acetone)

7.5.5 Carboxylic Acids

Carboxylic acids have the **carboxyl group,** —COOH. They are named by taking the stem name and adding the suffix -oic acid. The structure of the functional group is

$$-\!\!\overset{\overset{\displaystyle O}{\|}}{C}\!\!-\!\!OH$$

Acetic acid (ethanoic acid) is a common industrial chemical, which isn't thought to pose health threats. One carboxylic acid that is widely used is 2-propenic acid (acrylic acid). Large amounts are used during the manufacture of acrylic esters. Its structure is

$$\text{H}_2\text{C}=\text{CH}-\underset{\underset{\text{O}}{\|}}{\text{C}}-\text{OH}$$

Another carboxylic acid used in large quantities for the manufacture of synthetic fibers, plasticizers, resins, and foams is adipic, or 1,6-hexanedioic acid, which has two carboxyl groups

$$\text{HO}-\underset{\underset{\text{O}}{\|}}{\text{C}}-\text{CH}_2-\text{CH}_2-\text{CH}_2-\text{CH}_2-\underset{\underset{\text{O}}{\|}}{\text{C}}-\text{OH}$$

Carboxylic acids are weak acids and do not strongly dissociate in water.

7.5.6 Esters

Esters are the result of the combination of a carboxylic acid with an alcohol. The functional group is:

$$\text{R}-\underset{\underset{\text{O}}{\|}}{\text{C}}-\text{O}-\text{R}'$$

where R represents the remainder of the carboxylic acid and R' is the alcohol. The ester is named by the name of the alcohol group followed by the name of the carboxylic acid group with the suffix *-ate*. Ethanol and formic acid join to create ethyl formate. Some esters have odors that are associated with many everyday substances. For example, ethyl formate smells like rum.

$$\text{H}-\underset{\underset{\text{O}}{\|}}{\text{C}}-\text{O}-\text{CH}_2-\text{CH}_3 \quad \text{Ethyl formate}$$

Esters that are used in industry for flavorings, in perfumes, and as solvents, especially for paints, include ethyl formate, pentyl acetate (*n*-amyl acetate), butyl acetate, ethyl acetate, isobutyl acetate, and 3-methylbutyl acetate (isoamyl acetate).

One class of esters that is very common in industrial use is the phthalates. They are used as plasticizers to improve the flexibility of various plastics. Some are used as solvents and in insect repellents. Figure 7.6 gives the structural formulas for several common phthalates. Figure 7.6(a) shows the base structure of the phthalates, which is derived from phthalic acid.

7.5.7 Phenols

Phenols are based on a hydroxyl radical bonded to a benzene ring. Phenol can occur naturally in ground water, usually in low amounts associated with decomposing organic

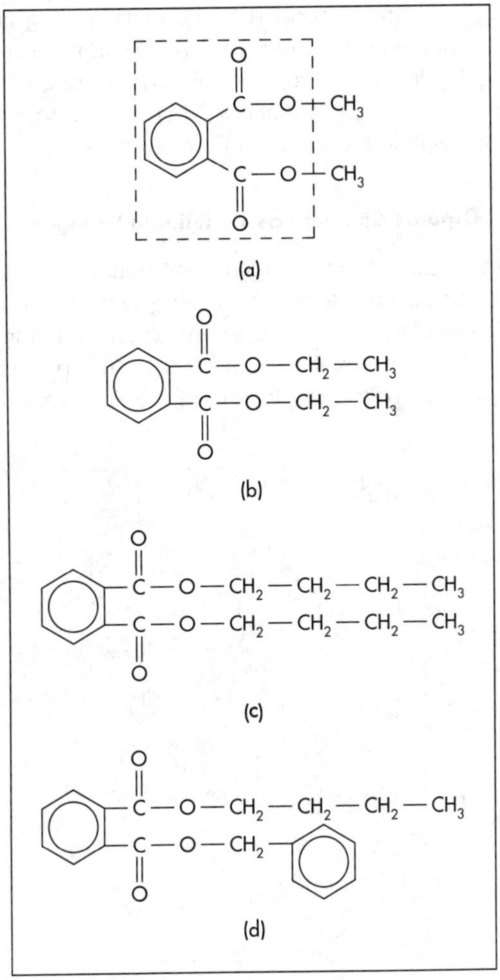

FIGURE 7.6 Structure of some phthalates. (a) Dimethyl phthalate; (b) diethyl phthalate; (c) di-n-butyl phthalate; (d) butyl benzyl phthalate.

matter. Phenol is also a common ground-water contaminant, due to its use in many industrial processes.

There are many phenol-based compounds used in industry that can occur in ground water. Cresols have a methyl group attached to the benzene ring of toluene. They are used in industry and are also released during coal-tar refining. Chlorophenols have one or more chloride ions on the benzene ring of a phenol. Chlorophenol is a synthetic intermediate for the manufacture of dye and more highly chlorinated phenols. Trichlorophenol is used to preserve wood and leather and as a biocide and antimildew agent. Pentachlorophenol is used as a wood preservative. It is a solid and is usually dissolved in a carrier solvent, such as diesel fuel, in making treated wood. Nitrophenols have a nitrate group bonded to the benzene ring of the phenol. Some phenols found in hazardous waste include phenol, *p*-chloro-*m*-cresol, 2-chlorophenol, 2,4-dichlorophenol, 2,4-dimethylphenol, 4,6-dinitro-*o*-cresol, 2,4-dinitrophenol, 2-nitrophenol, 4-nitrophenol, 2,4,6-trichlorophenol, and pentachlorophenol. The structure of some of these compounds is found in Figure 7.7.

7.5.8 Organic Compounds Containing Nitrogen

We have already seen that organic compounds can include nitrogen by attaching a nitrate ion to a carbon atom, for example nitrophenols. Nitrotoluenes are also common industrial chemicals and include 2- and 4-nitrotoluene, 2,4-dinitrotoluene, and 2,4,6-trinitrotoluene (TNT). TNT is an explosive and has been reported as a soil contaminant in areas of waste disposal from manufacture of munitions and explosives.

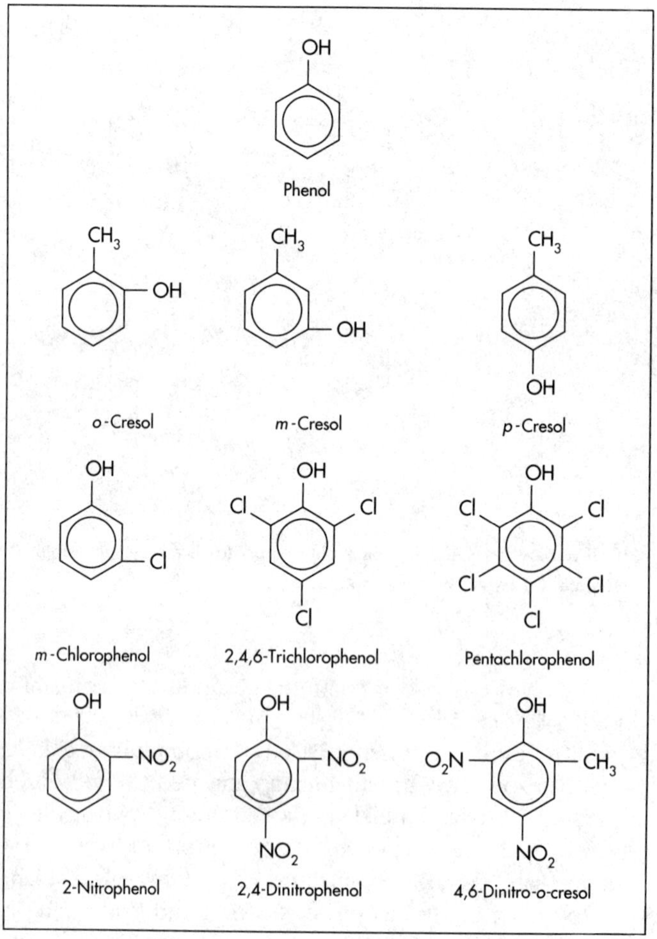

FIGURE 7.7 Structure of phenols.

Organic Compounds in Ground Water

Amines are organic compounds where the base molecule is ammonia and one or more of the hydrogen atoms is replaced by a hydrocarbon group. Some simple amines include the following:

$$CH_3-NH_2 \qquad CH_3-CH_2-NH_2 \qquad CH_3-NH-CH_3$$
$$\text{Methyl amine} \qquad \text{Ethyl amine} \qquad \text{Dimethyl amine}$$

Amides have the following group present:

$$R'-\underset{\underset{H}{|}}{N}-\underset{}{\overset{\overset{O}{\|}}{C}}-H$$

Acrylamide (2-propeneamide) is used in the synthesis of dyes and the manufacture of polymers, adhesives, and permanent-press fabrics. It has this structure:

$$H-\underset{\underset{H}{|}}{C}=\underset{\underset{H}{|}}{C}-\overset{\overset{O}{\|}}{C}-NH_2$$

Nitriles have a cyanide group, in which there is a triple bond between a carbon atom and a nitrogen atom. An example of a nitrile is acrylonitrile (2-propenenitrile). This chemical is produced in very large amounts and is used in the manufacture of acrylic fibers, polyacrylonitrile plastics, ABS resins, and other products. It has this structure:

$$H-C=C\begin{smallmatrix}\nearrow C\equiv N\\ \searrow H\end{smallmatrix}$$

Atrizine, a complex molecule containing a six-sided ring structure with alternating carbon and nitrogen atoms, is a very widely used herbicide. It is used as a preemergent weed control on corn in the midwestern United States and for weed control on sugar cane and pineapple fields in Hawaii. Its structure is given in Figure 7.8. Although atrazine is fairly quickly degraded in the environment by chemical and microbial processes it has been found in a number of wells in rural areas of Wisconsin. During biodegradation the parent compound is transformed to other compounds, called **metabolites.** Two pathways by which the degradation of atrizine begins are shown in Figure 7.8. Biodegradation is discussed in detail later in this chapter.

7.5.9 Organic Compounds Containing Sulfur and Phosphorus

Organic compounds may also contain sulfur and phosphorus molecules. Organosulfur compounds tend to have offensive odors. Examples of simple organosulfur compounds include mercaptans, the active ingredients in the spray of a skunk.

Methyl mercaptan — Ethyl mercaptan — Dimethyl mercaptan

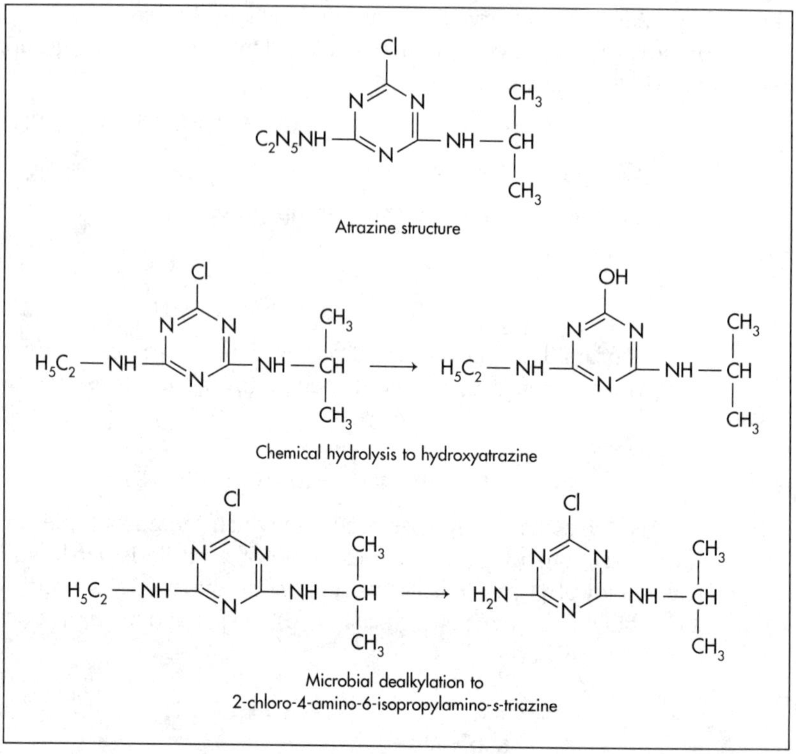

FIGURE 7.8 Structure of atrizine and metabolic pathways leading to degradation.

Many pesticides include sulfur, phosphorus, or both in their composition. Figure 7.9 gives the structure of some common organophosphorus and organosulfur pesticides. Most of these have been associated with ground-water contamination either in areas of their use or where they were manufactured.

7.6 Degradation of Organic Compounds

7.6.1 Introduction

It is well known that straight chain and aromatic hydrocarbons associated with petroleum products can undergo biological degradation (Barker, Patrick, and Major 1987; J. T. Wilson et al. 1986). Hydrogeologists have recently observed that halogenated organic solvents dissolved in ground water undergo transformations under natural conditions with the compounds undergoing progressive dehalogenation (Roberts, Schreiner, and Hopkins 1982; Parsons, Wood, and De Marco 1984; Cline and Viste 1985). Benzene, toluene and xylene have also been observed to undergo degradation in aquifers (Chiang et al. 1989; Barker and Patrick 1985; Barker, Patrick, and Major 1987).

Name	Structure	Use
Captan	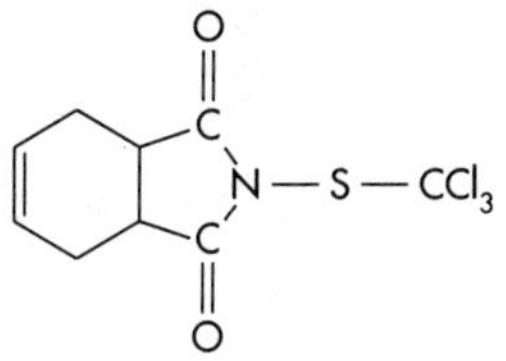 (cyclohexene-dicarboximide)N—S—CCl$_3$	Fungicide used on foliage
Disulfone	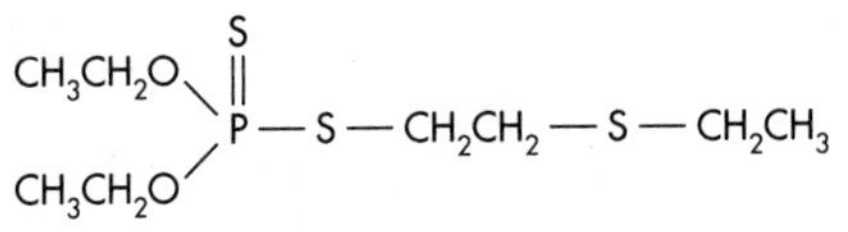 (CH$_3$CH$_2$O)$_2$P(=S)—S—CH$_2$CH$_2$—S—CH$_2$CH$_3$	Systemic insecticide used on plants
Aldicarb	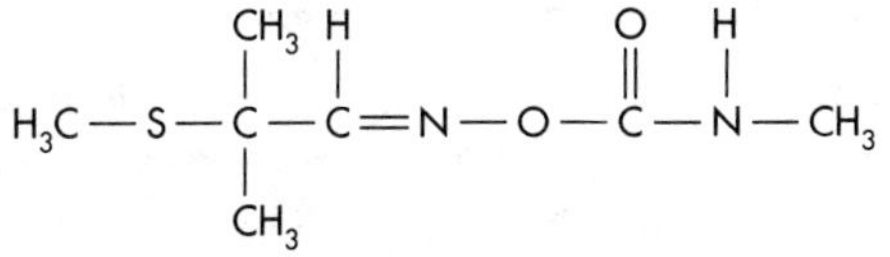 H$_3$C—S—C(CH$_3$)$_2$—C(H)=N—O—C(=O)—N(H)—CH$_3$	Systemic insecticide and nematicide
Endosulfan	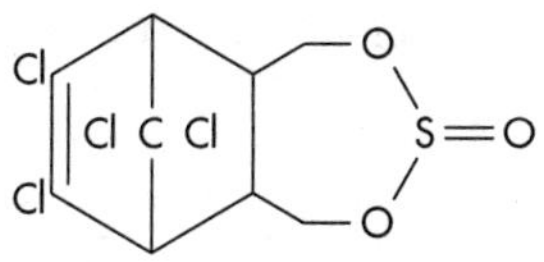 (chlorinated bicyclic with cyclic sulfite, S=O)	Insecticide
Malathion	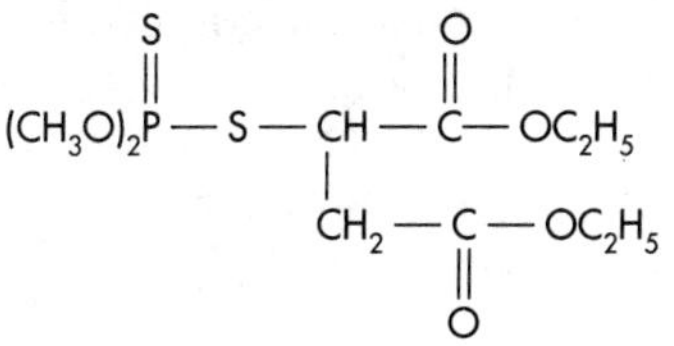 (CH$_3$O)$_2$P(=S)—S—CH(C(=O)OC$_2$H$_5$)—CH$_2$—C(=O)OC$_2$H$_5$	Insecticide
Parathion	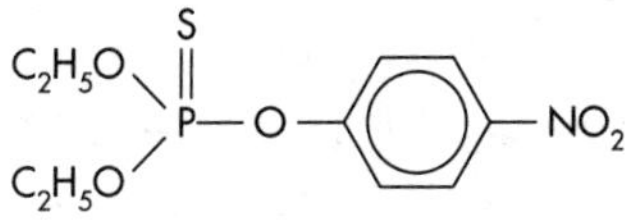 (C$_2$H$_5$O)$_2$P(=S)—O—C$_6$H$_4$—NO$_2$	Insecticide and acaricide
Asulam	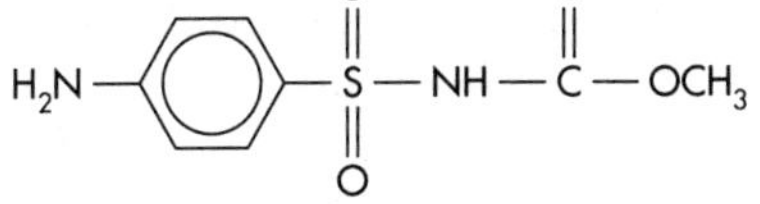 H$_2$N—C$_6$H$_4$—S(=O)$_2$—NH—C(=O)—OCH$_3$	Herbicide

FIGURE 7.9 Structure and uses of organosulfur and organophosphorus pesticides.

In this section we will examine the types of organic compounds that can undergo degradation, the means by which such degradation occurs, and the conditions under which it occurs. Degradation is defined as the process of an organic molecule becoming smaller by chemical or biological means. A molecule might have a halide ion replaced with a hydrogen ion, thus being transformed into a compound with a lower molecular weight. Carbon atoms can be broken off the molecule, leaving it with fewer atoms. The

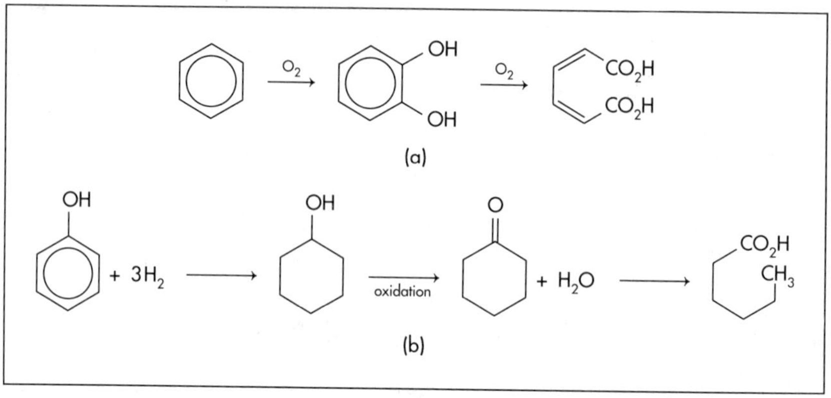

FIGURE 7.10 Degradation of a benzene ring. (a) Aerobic degradation of benzene in the presence of oxygen. (b) Anaerobic degradation of phenol in the presence of nitrate.

ultimate product of degradation of a hydrocarbon is methane or carbon dioxide and water.

7.6.2 Degradation of Hydrocarbons

Alkanes can be degraded under aerobic conditions by microbes (Manahan, 1984). Bacteria that can do this include *Micrococcus, Pseudomonas, Mycobacterium,* and *Nocardia*. The first step in the process is the conversion of a terminal CH_3 group to a carboxyl group. The microbes then attack the second carbon of the newly formed carboxylic acid and remove a two-carbon fragment, forming carbon dioxide. The oxidation of n-hexane follows these steps:

$$CH_3CH_2CH_2CH_2CH_2CH_3 + O_2 \longrightarrow CH_3CH_2CH_2CH_2CH_2CO_2H$$
$$CH_3CH_2CH_2CH_2CH_2CO_2H + 3\,O_2 \longrightarrow CH_3CH_2CH_2CO_2H + 2\,CO_2 + H_2O$$
$$CH_3CH_2CH_2CO_2H + 3\,O_2 \longrightarrow CH_3CO_2H + 2\,CO_2 + H_2O$$

The resulting acetic acid can easily be further degraded to carbon dioxide and water. Branched-chain hydrocarbons are more resistant to microbial degradation than normal alkanes.

Aromatic hydrocarbons can also be degraded under aerobic conditions. The first step in the cleavage of the ring is to replace two hydrogens on adjacent carbon atoms with hydroxyl groups. The benzene ring is then cleaved between these two rings to form a bicarboxylic acid. Figure 7.10(a) illustrates the ring-cleavage process, which requires molecular oxygen.

It is also possible for aromatic hydrocarbons to undergo anaerobic degradation (Evans 1977). The benzene ring can be degraded in the presence of nitrate by *Pseudomonas* sp. and *Moraxella* sp.* The benzene ring is first saturated to cyclohexane,

---

* The designation *Pseudomonas* sp. means that the bacterium has been identified to the level of the genus *Pseudomonas,* but not to the species level.

then oxidized to a ketone, and finally cleaved by hydrolysis to form a carboxylic acid. Figure 7.10(b) shows the degradation of phenol, first to cyclohexanol, then an oxidation to cyclohexanate and then ring cleavage to form hexanoic acid. During this process the nitrate is reduced to nitrogen.

Major et al. (1988) made a study of biodegradation of benzene, toluene, o-xylene, and m-xylene in the laboratory using nitrate as the electron acceptor under denitrifying conditions. They found that degradation of all four compounds occurred and suggested that bioremediation using added nitrate to ground water might be possible.

Polycyclic aromatic hydrocarbons (PAHs) have also been reported to be degraded by microbes. Bumpus (1989) demonstrated that under aerobic conditions in the laboratory, a fungus, *Phanerochaete chrysosporium*, was able to degrade all the major PAHs in anthracene oil, which is a complex mixture of PAH compounds obtained from the fractional distillation of coal oil. PAHs reported to be degraded included fluorene, phenanthrene, anthracene, carbazole, fluoranthene, pyrene, benzo[a]fluorene, 1-methylfluorene, and acenaphthalene.

The initial steps in ring cleavage of PAH compounds naphthalene, phenanthrene, anthracene, 1,2-benzanthracene, and benzo[a]pyrene are shown on Figure 7.11 (Guerin and Jones 1988). Heitkamp et al. (1988) have identified some of the products of ring oxidation and ring fusion of pyrene, a four-ring PAH, that were produced by the bacterium *Mycobacterium* sp. in an oxygenated lab environment that had additional organic nutrients. The identified metabolites of pyrene are shown in Figure 7.12. Heitkamp and Cerniglia (1989) have shown that this bacterium can also degrade a mixture of 2-methylnaphthalene, phenanthrene, pyrene, and benzo[a]anthracene in a laboratory culture made with natural sediment and ground water, demonstrating that the bacteria could compete with native soil bacteria for available nutrients.

These laboratory studies are valuable for identifying microbes that can degrade organic compounds. The microbes themselves are initially found at sites where PAH compounds have been released to the soil. The *Mycobacterium* sp. bacterium was isolated from hydrocarbon-contaminated sediment samples from an oil field near Port Aransas, Texas (Heitkamp and Cerniglia 1988). Mueller et al. (1990) isolated a bacterium, *Pseudomonas paucimoblis*, that is capable of using fluoranthene as the sole source of carbon for energy and growth. This strain was discovered at a site where creosote, which is rich in PAH, had been discarded. The ability of these naturally occurring strains of microbes to degrade PAH compounds bodes well for bioremediation efforts at PAH waste sites.

The previously cited studies were conducted under aerobic conditions. However, many sites with organic wastes have anaerobic conditions, because the soil bacteria consume oxygen in degrading the organic waste. Mihelcic and Luthy (1988) found that microbial populations could degrade acenaphthene and naphthalene under anaerobic conditions if sufficient nitrate was present to act as a source of oxygen. The PAH compounds and soil organic carbon were mineralized with the production of nitrogen. However, no mineralization occurred after the nitrate was depleted.

7.6.3 Degradation of Chlorinated Hydrocarbons

The following classes of organic compounds have been found to undergo either biotic or abiotic degradation: chlorinated methanes, chlorinated ethanes, chlorinated propanes,

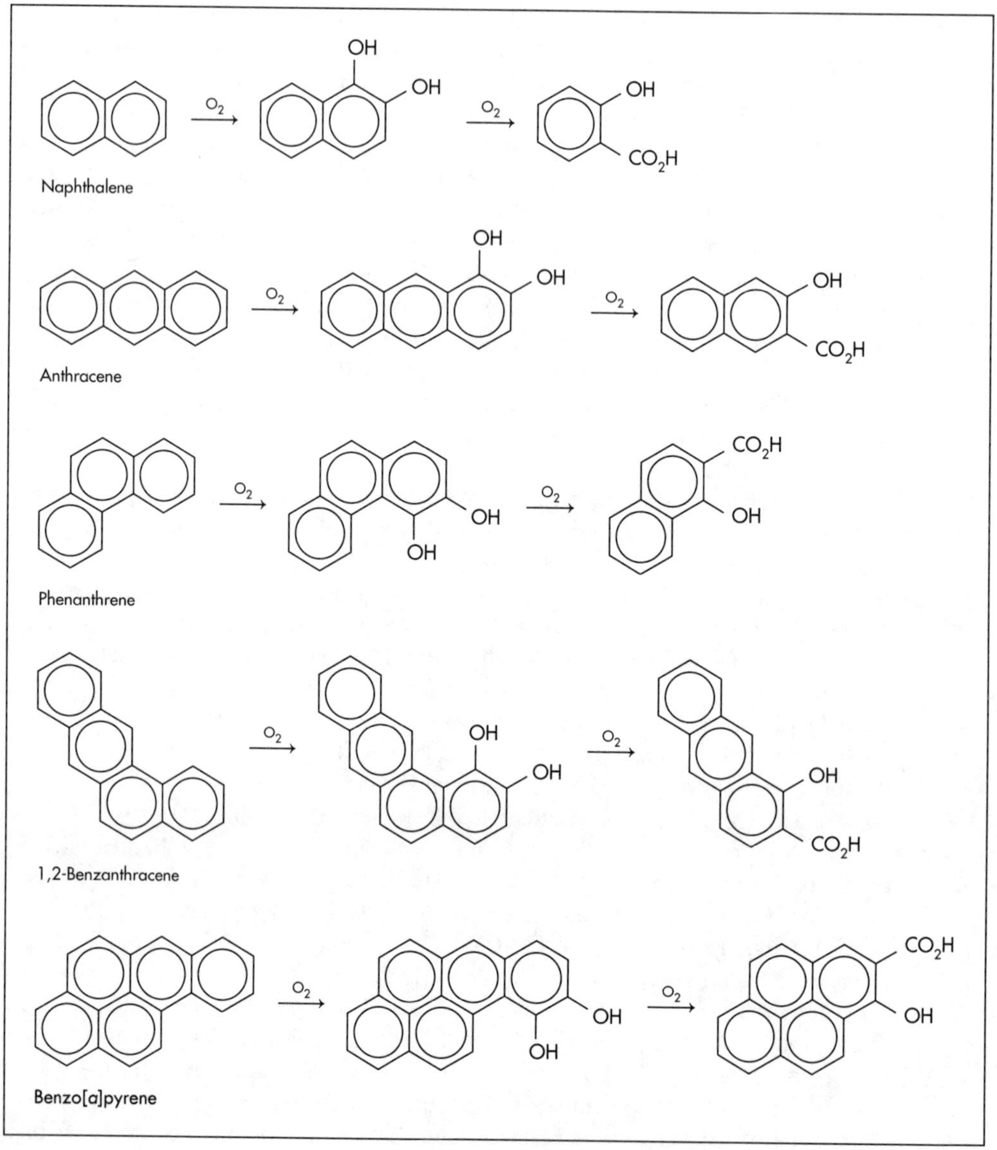

FIGURE 7.11 Cleavage of ring structure of some polycyclic aromatic hydrocarbons.

chlorinated butanes, chlorinated ethenes, brominated methanes, brominated ethenes, bromochloropropanes, chlorinated phenols, chlorinated benzenes, and halogenated acetates (e.g. Mikesell and Boyd 1990; Harker and Kim 1990; Topp and Hanson, 1990; Kohring, Rogers, and Wiegel 1989; Galli and McCarty 1989a, 1989b; Freedman and Gossett 1989; Tsien et al. 1989; Little et al. 1988; Fliermans et al. 1988; Nelson, Montgomery, and Pritchard 1988; Haigler, Nishino, and Spain 1988; Fathepure and Boyd 1988; Egli et al.

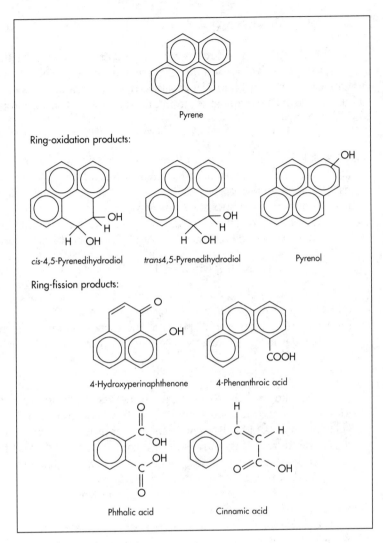

FIGURE 7.12 Structure of identified pyrene metabolites produced by *Mycobacterium* sp. *Source:* M. A. Heitkamp et al., *Applied and Environmental Microbiology* 54, no. 10, (1988): 2556–65.

1988; Vogel, Criddle, and McCarty 1987; Vogel and McCarty 1987a, 1987b, 1985; Barrio-Lage, Parsons, and Nassar 1987; Barrio-Lage, Parsons, Nassar, and Lorenzo 1986, 1987; Nelson et al. 1986, 1987; Vogel and Reinhard 1986; Grbic-Galic and Vogel 1987; Pignatello 1986; Strand and Shippert 1986; B. H. Wilson, Smith, and Rees 1986; Janssen et al. 1985; Schwarzenbach, Giger, Schaffner 1985; Wilson and Wilson 1985; Kleopfer et al. 1985; LaPat-Polasko, McCarty, and Zehnder 1984; Bouwer and McCarty 1983a, 1983b; Burlinson, Lee, and Rosenblatt 1982).

The chlorinated ethanes and ethenes have been well studied and are common ground-water contaminants. There are a number of reactions that can occur abiotically to break them into lower-molecular-weight compounds (Vogel, Criddle, and McCarty 1987).

Substitution is a reaction where water reacts with the halogenated compound to substitute an OH^- for an X^-, creating an alcohol. This can occur in water without either inorganic or biological catalysts, but the reaction rates are slow. The reaction is also called **hydrolysis** and can be illustrated with 1-bromopropane:

$$CH_3-CH_2-CH_2Br + H_2O \longrightarrow CH_3-CH_2-CH_2OH + HBr$$

Substitution reactions proceed most rapidly for monohalogenated compounds. Monohalogenated compounds have reaction half-lives of about 1 mo. As the number of halogen ions increases, the half-life for reactions due to substitution increases rapidly into the range of years to hundreds of years.

Other groups can be substituted as well, such as the reaction with an HS^- radical under reducing conditions to release an X^- ion and form a mercaptan:

$$CH_3-CH_2Br + HS^- \longrightarrow CH_3-CH_2SH + Br^-$$

Dehydrohalogenation is a reaction where an alkane loses a halide ion from one carbon atom and then a hydrogen ion from an adjacent carbon. The result is the formation of a double bond between the carbon atoms, thus creating an alkene. Dehydrohalogenation can transform 1,1,1-trichloroethane to 1,1-dichloroethene:

$$CCl_3-CH_3 \longrightarrow Cl_2=CH_2 + HCl$$

The rate of dehydrohalogenation increases with increasing numbers of halogen ions; hence compounds that undergo substitution most slowly undergo dehydrohalogenation most rapidly. Bromine ions are more rapidly removed than chlorine ions in these abiotic reactions. Burlinson, Lee, and Rosenblatt (1982) found that bromine is removed from dibromochloropropane six time faster than chlorine.

Oxidations and reductions are typically biologically mediated and require external electron donors or acceptors. In some cases polyhalogenated aliphatic compounds will act as electron acceptors and become reduced.

Oxidations include α-hydroxylation, which is the addition of an OH^- radical to an alkane in place of an H atom on a carbon that also contains a halogen ion. The result is the formation of a chlorinated alcohol. The α-hydroxylation of 1,1-dichloroethane forms 1,1-dichloroethanol:

$$CH_3CHCl_2 + H_2O \longrightarrow CH_3CCl_2OH + 2H^+ + 2e^-$$

The halogenated alcohol can then undergo the further loss of hydrogen from the hydroxyl group and a halide to form an aldehyde:

$$\begin{array}{c} H \quad Cl \\ | \quad\; | \\ H-C-C-OH \\ | \quad\; | \\ H \quad Cl \end{array} \longrightarrow \begin{array}{c} H \quad Cl \\ | \quad\; | \\ H-C-C=O + H^+ + Cl^- \\ | \\ H \end{array}$$

Oxidation of a carbon-carbon double bond can result in the formation of an epoxy; the process is called **epoxidation:**

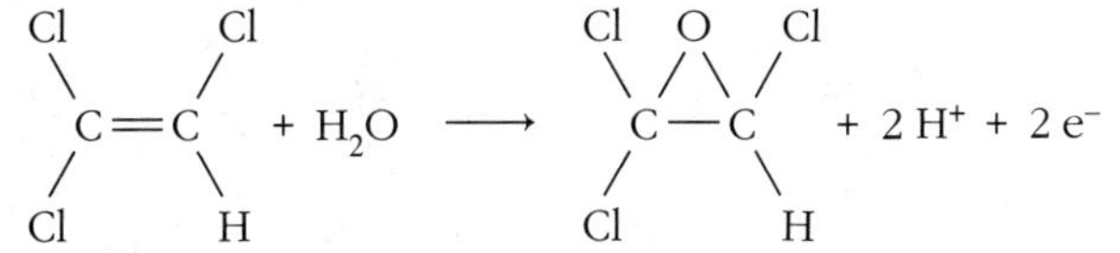

$$\begin{array}{c}Cl\\ \diagdown\\ C=C\\ \diagup\diagdown\\ ClH\end{array}\begin{array}{c}Cl\\ \diagup\end{array} + H_2O \longrightarrow \begin{array}{c}ClOCl\\ \diagdown\diagup\diagdown\diagup\\ C-C\\ \diagup\diagdown\\ ClH\end{array} + 2\,H^+ + 2\,e^-$$

The epoxy is short-lived and under neutral pH conditions can be oxidized to a carboxylic acid:

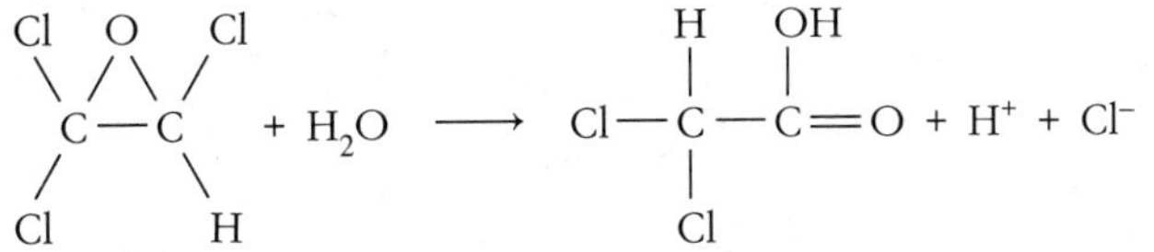

$$\begin{array}{c}ClOCl\\ \diagdown\diagup\diagdown\diagup\\ C-C\\ \diagup\diagdown\\ ClH\end{array} + H_2O \longrightarrow \begin{array}{c}HOH\\ ||\\ Cl-C-C=O\\ |\\ Cl\end{array} + H^+ + Cl^-$$

Reductions start with the removal of a halide ion by a reduced species, such as a reduced transition metal or a transition metal complex. The reduced species is thus oxidized. The alkyl radical thus formed can react with a H^+ ion, which substitutes for the departed X^-. This process is called **hydrogenolysis.**

$$CH_3-CCl_3 + H^+ + e^- \longrightarrow CH_3CHCl_2 + Cl^-$$

Reductions can also occur if there are halides on adjacent carbon atoms. In this case the loss of a halogen from each carbon atom creates an alkene by formation of a double bond between the carbon atoms. This reduction is called **dihaloelimination** and can transform a chlorinated alkane into a chlorinated alkene. The process is illustrated by the transformation of hexachloroethane to tetrachloroethene:

$$CCl_3-CCl_3 + 2\,e^- \longrightarrow CCl_2=CCl_2 + 2\,Cl^-$$

Figure 7.13 illustrates the biologically mediated reductions that can occur with chlorinated ethane and ethenes under hydrogenolysis and dihaloelimination. Also shown in the figure are the relative half-reaction reduction potentials calculated from the free energy of the reaction by using the Nernst equation.

Environmental conditions influencing the type and rate of the preceding reactions include pH, temperature, state of oxidation or reduction, microorganisms present, and types of other chemicals present. Reaction kinetics also play an important role in the determination of the abiotic and biotic fate of organic contaminants. Theoretically, the end products of the abiotic reactions are ethane and ethene, which should be amenable to further biodegradation. However, under field conditions such a favorable outcome might require many years to occur. Before it does, the contaminant might well flow from the point of origin to contaminate a large area of the aquifer system.

7.6.4 Degradation of Organic Pesticides

One of the first organic pesticides developed was DDT, or dichloro-diphenyl-trichloroethane. This proved to be a very effective agent against a wide variety of insects, and it was inexpensive. However it was found to be very resistant to degradation and hence persisted in the environment. Damage to nontargeted wildlife, especially birds, was

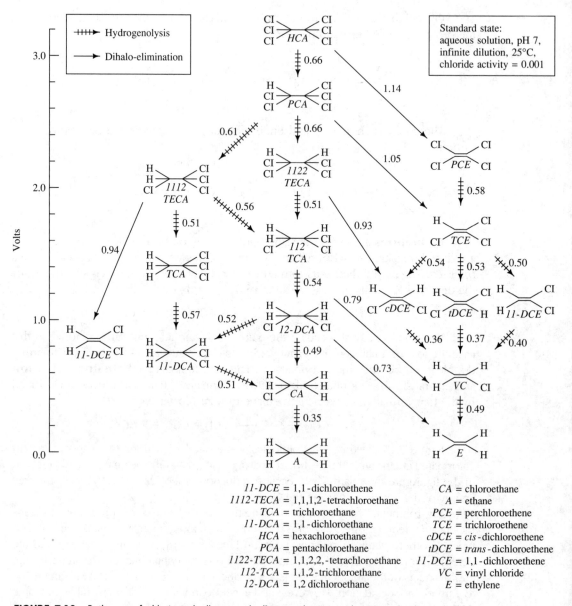

FIGURE 7.13 Pathways of chlorinated alkane and alkene reduction and estimated relative half-life reduction potentials in volts. *Source:* Reprinted with permission from T. M. Vogel, C. S. Criddle, and P. L. McCarty, *Environmental Science and Technology* 21, no. 8, (1987):722–36. Copyright 1987 American Chemical Society.

Organic Compounds in Ground Water

reported, and the use of this compound was eventually banned in the United States. Although DDT does degrade, the first metabolite formed, dichloro-diphenyl-dichloro-ethane (DDD), proved to be at least as toxic as DDT. DDD forms from DDT by the removal of a chlorine atom by hydrolysis. Figure 7.14 illustrates a metabolic pathway for the degradation of DDT.

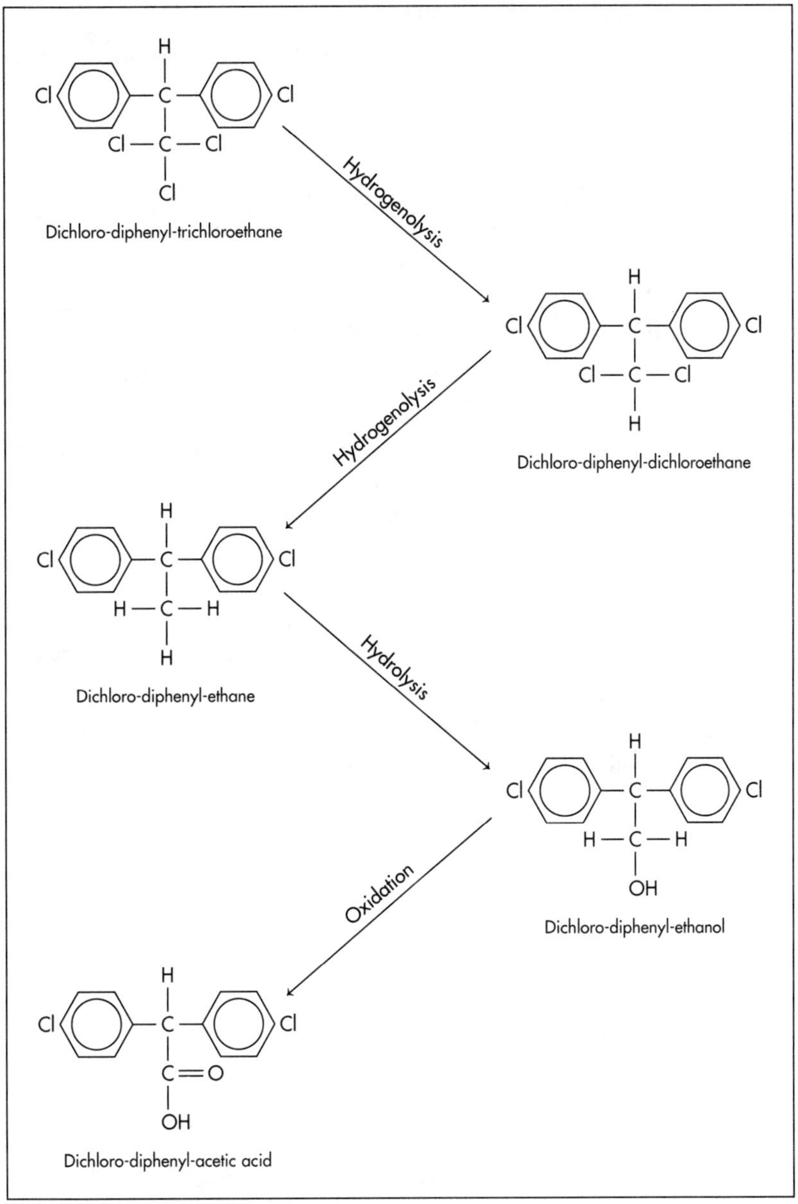

FIGURE 7.14 Metabolic pathway for the degradation of dichloro-diphenyl-trichloroethane (DDT).

Newer pesticides are formulated so that they will not persist in the environment. They have structures that will more readily undergo abiotic, or biologically mediated, degradation. Smith (1988) lists the following pathways by which herbicides are degraded. Naturally, the same reactions will apply to other pesticides.

Dealkylation: The biological removal of a methyl or other alkyl group from a nitrogen atom (see Figure 7.8)

Dealkoxylation: The biological removal of a methoxy (methyl ether) group from a nitrogen atom.

Decarboxylation: The biological or abiotic removal of one carbon and two oxygen atoms from a carboxyl group.

Dehalogenation: Biological replacement of a chlorine atom with a hydrogen atom. This is especially important in degradation of insecticides, many of which are chlorinated hydrocarbons.

Ether cleavage: The biological cleaving of an ether by breaking the bond between oxygen and carbon atoms.

Hydrolysis: The chemical or biological cleavage of molecules by the addition of water. The products usually include an alcohol or carboxylic acid.

Hydroxylation: The biological introduction of hydroxyl groups into either aliphatic or aromatic compounds.

Methylation: The biological addition of a methyl group to an alcohol or phenol to form a methyl ether.

Oxidation: Oxidation of an alcohol to an aldehyde, which can be further oxidized to a carboxylic acid, done by either chemical or biological processes. Epoxide formation is a major step in many oxidation processes.

Beta-oxidation: Biological removal of two carbon atoms from an alkane chain linked to an aromatic ring structure.

Reduction: Biological reduction of a nitrate group linked to a herbicide to form an amine group.

Ring cleavage: One of many processes by which microbes can break the structure or an aromatic string.

7.7 Field Examples of Biological Degradation of Organic Molecules

7.7.1 Introduction

A vast body of information of biological and abiotic degradation of organic molecules has accumulated, based on laboratory studies using microcosms, the microbial equivalent of an aquarium. Under such carefully controlled conditions, the disappearance of a compound can be determined and the appearance of metabolites noted. However, there have been far fewer field studies of microbial degradation. It is far more difficult to make field studies and interpret the results. At the most basic level, is a compound disappearing due to degradation, or is some other process—such as volatilization, sorption or dilu-

Organic Compounds in Ground Water

tion—involved? However, enough of a body of information has accumulated to document the natural degradation of a variety of organic compounds in soil and ground water. In Chapter 9 we will examine how enhanced biodegradation can be used as a means of remediating aquifers and soils contaminated with organic compounds.

7.7.2 Chlorinated Ethanes and Ethenes

Cline and Viste (1985) noted that at a field study of a solvent-recovery facility in Wisconsin, wells downgradient from the site contained di- and mono-chloro ethanes and ethenes, including vinyl chloride. However, the facility had never accepted these products, which were interpreted to be the metabolites of trichloroethane and trichloroethene.

Fetter (1989) reported the distribution of chlorinated ethanes and ethenes in ground water at the Seymour Recycling Center Superfund site, a former solvent-recycling facility in Indiana. Table 7.6 contains the analyses of ground water from three wells: 203a, located beneath the facility; 206a, located 50 ft down-gradient from the site boundary; and 207b, located 300 ft down-gradient from the site boundary. (Data from Fetter (1989) have been updated in Table 7.6.) The well beneath the site contained trichloroethene, 1,1,1-trichloroethane, 1,2-dichloroethane, 1,1-dichloroethane, 1,1-dichloroethene, *trans*-1,2-dichloroethene, chloroethane, and vinyl chloride. Chloroethane, 1,2-dichloroethane, 1,1-dichloroethane and *trans*-1,2-dichloroethene have similar K_{oc} values, ranging from 36 to 45. They would migrate at just about the same rate. Well 206a, located 50 ft from the site boundary, had higher chloroethane, lower 1,1-dichloroethane and *trans*-1,2-dichloroethene, variable vinyl chloride, and no detectable hits on the other compounds. Well 207b, located 250 ft down-gradient from well 206a, had elevated chloroethane and no significant hits on any of the other compounds. Over a 6-yr monitoring period the chloroethane in this down-gradient well increased from 41 μg/L to a maximum of 18,000 μg/L. This is convincing evidence that the chlorinated ethanes and ethenes are being degraded, with chloroethane being one end product. The mechanism of degradation is not known, but the ground water at the site is reducing, having large amounts of dissolved iron. The fate of the vinyl chloride is unclear. It may be being mineralized, it may be escaping through volatilization, or it may not be migrating due to a high K_{oc} value.

Jackson, Priddle, and Lesage (1990) investigated the transport and fate of CFC-113 (1,1,2-trichloro-1,2,2-trifluoroethane, or Freon®), a compound used as a solvent to clean circuit boards and semiconductors in the electronics industry. Field studies of a landfill where this compound had been disposed showed that it was present in the ground water along with two metabolites, indicating that biotransformation was occurring. Reductive dechlorination transforms it to 1,2-dichloro-1,2,2-trifluoroethane (CFC-123a), which can undergo dihalide elimination to form 1-chloro-1,2,2-trifluoroethene (CFC-1113). The latter compound is quite toxic, a property not shared with the more halogenated forms.

$$\underset{\text{CFC-113}}{\overset{\begin{array}{cc}Cl & Cl\\ | & |\end{array}}{F-C-C-F}\underset{\begin{array}{cc}| & |\\ Cl & F\end{array}}{}} \longrightarrow \underset{\text{CFC-123a}}{\overset{\begin{array}{cc}H & Cl\\ | & |\end{array}}{F-C-C-F}\underset{\begin{array}{cc}| & |\\ Cl & F\end{array}}{}} \longrightarrow \underset{\text{CFC-1113}}{\overset{F\quad\quad F}{\underset{Cl\quad\ \ F}{C=C}}}$$

TABLE 7.6 Seymour Recycling Corporation summary of well analyses for chlorinated ethanes and ethenes.

Compound	TCE	111TCA	11DCE	11DCA	12DCA	t-DCE	CA	VC
K_{oc}	152	155	217	45	36	39	42	8,400
Well 203a (Monitoring well under site)								
Aug. 84	nd	1,500	500	9,600	nd	16,000	2,300	1,100
Dec. 84	nd	nd	nd	800	200	200	4,000	2,700
Jun. 85	500	16,000	600	13,000	400	15,000	1,500	1,000
Jun. 90	54	690	190	4,500	<250	6,500	1,600	300
Well 206a (Monitoring well 50 ft down-gradient from site)								
Aug. 84	nd	nd	nd	330	nd	75	9,500	nd
Dec. 84	nd	nd	nd	nd	nd	60	3,000	<60
Jun. 85	nd	nd	nd	5,600	nd	1,700	17,000	3,500
Feb. 88	nd	nd	nd	51	nd	52	1,600	nd
Jan. 89	nd	nd	nd	12J	nd	nd	800	nd
Aug. 89	nd	nd	nd	nd	nd	12J	6,200	nd
Jun. 90	nd	nd	nd	28	nd	22	1,500	nd
Well 207b (Monitoring well 300 ft down-gradient from site)								
Aug. 84	nd	nd	nd	7	nd	nd	41	nd
Dec. 84	nd	nd	nd	nd	nd	nd	200	nd
Jun. 85	nd	nd	nd	nd	nd	nd	800	nd
Feb. 88	nd	nd	nd	nd	nd	nd	5,800	nd
Jan. 89	nd	nd	nd	nd	nd	nd	1,400	nd
Aug. 89	nd	nd	nd	nd	nd	2J	18,000	nd
Jun. 90	nd	nd	13J	nd	nd	nd	13,000	nd

Data from U.S. EPA.

Key: TCE = trichloroethene
111TCA = 1,1,1-trichloroethane
11DCE = 1,1-dichloroethene
11DCA = 1,1-dichloroethane
12DCA = 1,2-dichloroethane
t-DCE = trans-1,2-dichloroethene
CA = chloroethane
VC = vinyl chloride
All analyses in micrograms/liter.
nd = not detected
J = estimated value

7.7.3 Aromatic Compounds

There is special interest in the biodegradation of aromatic compounds, because they are the most water-soluble fraction of gasoline, which has leaked from underground storage tanks into soil and ground water all over the world. Benzene, toluene, *o*-, *m*-, and *p*-xylene, trimethylbenzene, and ethylbenzene constitute the major aromatic components of gasoline (Kreamer and Stezenbach 1990).

Barker and Patrick (1985) and Barker, Patrick, and Major (1987) injected ground water spiked with 2.36 mg/L benzene, 1.75 mg/L toluene, 1.08 mg/L *p*-xylene, 1.09 mg/L *m*-xylene, 1.29 mg/L *o*-xylene, and 1280 mg/L chloride into a sand aquifer. The aquifer was aerobic, with 7 to 8 mg/L of dissolved oxygen at the start of the experiment. The plume was sampled using a dense network of monitoring wells over a period of 434 days. All the aromatic compounds degraded, with the *m*- and *p*-xylenes being the most readily degraded, then the *o*-xylene, then toluene, and finally the benzene. Complete removal of benzene occurred within 1.2 yr.

Rifai et al. (1988) observed that at a site where there was a spill of aviation gasoline, those monitoring wells with high concentrations of dissolved oxygen in the ground water had low amounts of dissolved benzene, toluene, and xylene (BTX), and those with high BTX had low dissolved oxygen. This was interpreted to mean that aerobic decomposition of the BTX was occurring, with the dissolved oxygen in the ground water being consumed in the process. Chiang et al. (1989) also made a study of the relationship of dissolved oxygen in ground water to BTX in ground water. They found a statistically significant relationship between the amount of dissolved oxygen and the concentration of dissolved BTX. They observed that a minimum dissolved-oxygen concentration of 0.9 mg/L in ground water was necessary for complete aerobic degradation of BTX.

Degradation of phenol and naphthalene in ground water has been reported by Ehrlich et al. (1982). Concentrations of phenolic compounds in a contaminant plume decreased from 30 mg/L to less than 0.2 mg/L over a distance of 430 m, whereas naphthalene decreased from 20 mg/L to 2 mg/L over the same distance. Sodium, a conservative element, decreased only from 430 to 120 mg/L along the flow path, indicating that dilution was responsible for some—but by no means all—of the decline. Methane formation was occurring in the contaminated portion of the aquifer and methanogenic bacteria were isolated there, indicating anaerobic decomposition was a factor.

Aerobic degradation of phenol and naphthalene has also been observed in the field (Klecka et al. 1990). Ground water beneath a site used for the disposal of waste from a charcoal-making operation contained 220 μg/L phenol, 570 μg/L 2-methylphenol (cresol), 860 μg/L 2,4-dimethyl phenol, and 220 μg/L naphthalene. The native ground water was aerobic, with dissolved oxygen present in all areas except the most contaminated part of the contaminant plume. Concentrations of phenols and naphthalene decreased to background levels within 100 m down-gradient of the site. Laboratory studies in microcosms using sediment from the aquifer demonstrated that biodegradation of phenol, creosol, and naphthalene could occur under aerobic conditions. Computer modeling confirmed that the major factor in reducing the phenols over a 100-m flow path was biodegradation and not sorption.

7.8 Analysis of Organic Compounds in Ground Water

There are a number of methods by which organic compounds dissolved in ground water can be identified and quantified. In **gas chromatography** a mixture of volatile materials is transported by a carrier gas through a column packed with either an absorbing solid phase or an absorbing liquid phase that is coated on a solid material. The volatile component will partition between the carrier gas and the absorbing phase, and the length of time that it takes for the component to traverse the column will be characteristic. If there are several compounds in a mixture, the column will separate them and they will arrive at different times. A detector, such as a **photoionization detector** or a **flame-ionization detector,** located at the end of the column can determine the quantity and identity of each volatile component. In **mass spectrometry** the compound is ionized by an electrical discharge and the ions are then separated based on their charge-to-mass ratio. The output is a **mass spectrum,** which can be compared to the mass spectra of a large number of standard compounds stored in an electronic data base called a mass-spectra library. The most commonly used laboratory method is gas chromatography combined with mass spectrometry, frequently known as **GC/MS.** In GC/MS the mass spectrometry is preceded by gas chromatography to separate the compounds. Figure 7.15 shows a library spectrum of benzene matched against the mass spectrum of a sample identified as benzene.

The U.S. EPA has developed a series of standard methods for the analysis of organic chemicals dissolved in water. These are known as the 600 series methods (40 CFR Part 136, *Federal Register,* October 26, 1984). Table 7.7 lists the 600 series and the target compounds.

Method 601 for purgeable halocarbons and method 602 for purgeable aromatics are gas-chromatograph methods using a purge-and-trap procedure to isolate volatile organics. Nitrogen is bubbled through a column of water containing the dissolved organics, so that the organics are purged from the water and carried with the gas. The organics are trapped on Tenax, a solid sorbent. The Tenax is then heated, and the volatile organics are swept into a gas chromatograph for separation and then detected. Method 601 uses a halide-specific detector and method 602 specifies a photoionization detector. The results are reported as volatile organics or purgeable organics.

Methods 603 through 612 are gas-chromatographic methods for substances other than purgeable hydrocarbons and aromatics. Method 613 is a GC/MS method for dioxin.

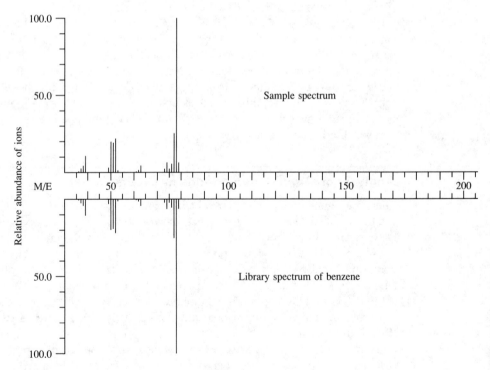

FIGURE 7.15 Comparison of the mass spectrum of an unknown with the mass spectrum of benzene from the library of mass spectra. *Source:* B. L. Roberts, *Ground Water Monitoring Review* 5, no. 4, (1985):41–43. Used with permission. Copyright © 1985 Water Well Journal Publishing Co.

TABLE 7.7 U.S. EPA 600 series analytical methods for organic compounds.

Method Number	Analytical Technique	Target Compounds
601	GC	Purgeable halocorbons
602	GC	Purgeable aromatics
603	GC	Acrolein and acrylonitrile
604	FIDGC	Phenols
605	HPLC	Benzidines
606	GC	Phthalate ester
607	GC	Nitrosamines
608	GC	Organochlorine pesticides and PCBs
609	GC	Nitroaromatics and isophorone
610	GC & HPLC	Polycyclic aromatic hydrocarbons
611	GC	Haloethers
612	GC	Chlorinated hydrocarbons
613	GC/MS	2,3,7,8-TCDD (dioxin)
624	GC/MS	Purgeable organics
625	GC/MS	Acid and base/neutral extractable organics

Key: GC = gas chromatography
FIDGC = flame ionization detector gas chromatography
HPLC = high-performance liquid chromatography
GC/MS = gas chromatography/mass spectrometry

Many target compounds can be determined by either a GC or a GC/MS method. Method 624 is a GC/MS method for purgeable organic compounds that includes compounds also determined by GC methods 601, 602, 603, and 612. Method 625 is a GC/MS method for semivolatile organics that includes compounds also detected by GC methods 604, 606, 607, 609, 610, 611, and 612.

Because of the larger number of target compounds, methods 624 and 625 are usually specified for organic analysis of contaminated water. These are supplemented by method 608 for pesticides and polychlorinated biphenols and method 613 for 2,3,7,8-TCDD (dioxin). If certain compounds are detected by method 625, they should then be confirmed with a GC method specific to that compound.

In method 624 an inert gas, such as nitrogen, is bubbled through a 5-mL water sample so that the volatile organics are transferred from the water phase to the vapor phase. The vapors are swept to a sorbent trap, where they are trapped on Tenax. After purging is complete, the Tenax is heated and flushed with an inert gas to desorb the volatiles and drive them into a gas-chromatographic column. The purgeables are separated in the column and then detected with a mass spectrometer.

Not all the organic compounds are purged from the water by the nitrogen gas. These are the semivolatile compounds. Method 625 is for the analysis of extractable semivolatiles. The pH of a 1-L sample is adjusted so that it is greater than 11. The sample is then extracted with methylene chloride. This is known as the **base/neutral fraction.** The pH of the sample is then adjusted to less than 2. The sample is again extracted with methylene chloride to form the acid fraction. Each fraction is chemically dried to remove the water and concentrated by distillation to 1 mL. The fractions are then analyzed separately by GC/MS.

The U.S. EPA has created a target compound list of organic chemicals for its contract laboratory program. The 126 compounds on this list are also known as **priority pollutant organic compounds.** Laboratories are required to identify and quantify any compounds on this list through methods 608, 624, and 625. Table 7.8 presents the target compound list. The first 34 compounds are volatiles determined by method 624. Compounds 35 to 99 are semivolatiles determined through method 625. Pesticides and PCBs constitute compounds 100 to 126 and are detected through method 608.

There are a number of additional EPA methods for organic compounds. One example is **total petroleum hydrocarbons** by method 418.1. In this method a liter of sample is extracted with fluorocarbon 113 at a pH less than 2. Interfering compounds are then removed with a silica gel absorbent. An infrared analysis of the sample is made by direct comparison to a calibration curve made from prepared samples with a known concentration.

There are also specific EPA methods for analysis of organic compounds on a soil matrix. This involves an extraction procedure followed by GC/MS analysis.

The analysis of water samples will frequently result in the analyst determining that a particular compound was not found. The answer that is reported, however, is not zero. It is instead reported as less than detection, with the **detection limit** given. The detection limit is the lowest concentration level that can be determined to be statistically different than a blank sample (American Chemical Society, 1983). A **method detection limit** is specified for each of the EPA 600 series analytical procedures. The **limit of quantification** is the level above which the quantitative results may be expressed with a specified degree of confidence. Values between the detection limit and the limit of quantification may be reported as **estimated values.**

Swallow, Shifrin, and Doherty (1988) point out some pitfalls in the analysis of organic compounds in ground water. The GC/MS instruments are designed for automatic operation, with the mass spectra matched electronically to a library spectrum. This match may not be accurate. Swallow, Shifrin, and Doherty report one case where the same compound was reported as eight different compounds in 13 samples, even though the spectra were virtually identical. The only differences were small and could be attributed to background interferences. If an experienced analyst had not examined the spectra, the error would not have been found. Reference standards are run on a limited set of potential compounds, and the accuracy of matching is high with compounds that have been used as a reference. However, for many compounds for which a reference has not been run, the automated procedure may not get a good match between the sample spectrum and a library spectrum. If no match is found, then the sample will be reported as an unknown. However, this could be a significant pollutant at the site under investigation and very important, even though a library match could not be made. In that case reference standards of unusual compounds that might have been used in the area would need to be prepared to obtain additional spectra for the library.

The instruments are sensitive, and large concentrations of a compound may not be tolerated. **Sample dilution** may be necessary for samples where one or more compounds are in high concentration. These are reported as diluted samples. However, dilution raises the detection limits for all compounds. A sample may require dilution because of a high concentration of a more-or-less benign compound, and the resulting

TABLE 7.8 Target compound list.

Number	Chemical Name	Number	Chemical Name	Number	Chemical Name
1	Chloromethane	43	bis(2-Chloroisopropyl) ether	85	di-n-Butyl phthalate
2	Bromomethane	44	4-Methylphenol	86	Fluoranthene
3	Vinyl chloride	45	n-Nitrosodi-n-Propylamine	87	Pyrene
4	Chloroethane	46	Hexachloroethane	88	Butyl benzyl phthalate
5	Methylene chloride	47	Nitrobenzene	89	3,3'-Dichlorobenzidine
6	Acetone	48	Isophorone	90	Benz[a]anthracene
7	Carbon disulfide	49	2-Nitrophenol	91	Chrysene
8	1,1-Dichloroethene	50	2,4-Dimethylphenol	92	bis(2-Ethylhexyl) phthalate
9	1,1-Dichloroethane	51	Benzoic acid	93	di-n-Octyl phthalate
10	1,2-Dichloroethene	52	bis(2-Chloroethoxy) methane	94	Benzo[b]fluoranthene
11	Chloroform	53	2,4-Dichlorophenol	95	Benzo[k]fluoranthene
12	1,2-Dichloroethane	54	1,2,4-Trichlorobenzene	96	Benzo[a]pyrene
13	2-Butanone	55	Naphthalene	97	Indeno(1,2,3-cd)pyrene
14	1,1,1-Trichloroethane	56	4-Chloroaniline	98	Dibenz[a,h]anthracene
15	Carbon tetrachloride	57	Hexachlorobutadiene	99	Benzo[g,h,i]perylene
16	Vinyl acetate	58	4-Chloro-m-cresol	100	BHC-alpha
17	Bromodichloromethane	59	2-Methylnaphthalene	101	BHC-beta
18	1,2-Dichloropropane	60	Hexachlorocyclopentadiene	102	BHC-gamma (lindane)
19	cis-1,3-Dichloropropene	61	2,4,6-Trichlorophenol	103	BHC-delta
20	Trichloroethene	62	2,4,5-Trichlorophenol	104	Heptachlor
21	Chlorodibromomethane	63	2-Chloronaphthalene	105	Aldrin
22	1,1,2-Trichloroethane	64	2-Nitroaniline	106	Heptachlor epoxide
23	Benzene	65	Dimethyl phthalate	107	Endosulfan 1 (alpha)
24	trans-1,3-Dichloropropene	66	Acenaphthylene	108	Dieldrin
25	Bromoform	67	2,6-Dinitrotoluene	109	4,4'-DDE
26	4-Methyl-2-pentanone	68	3-Nitroaniline	110	Endrin
27	2-Hexanone	69	Acenaphthene	111	Endosulfan 2 (beta)
28	Tetrachloroethene	70	2,4-Dinitrophenol	112	4,4'-DDD
29	Toluene	71	4-Nitrophenol	113	Endosulfan sulfate
30	1,1,2,2-Tetrachloroethane	72	Dibenzofuran	114	DDT-P,P' (4,4'-DDT)
31	Chlorobenzene	73	2,4-Dinitrotoluene	115	Methoxychlor
32	Ethylbenzene	74	Diethyl phthalate	116	Endrin ketone
33	Styrene	75	4-Chlorophenyl phenyl ether	117	Chlordane, alpha
34	Xylene, total	76	Fluorene	118	Chlordane, beta
35	Phenol	77	4-Nitroaniline	119	Toxaphene
36	bis(2-Chloroethyl) ether	78	4,6-Dinitro-2-methylphenol	120	PCB-1016 (Aroclor 1016)
37	2-Chlorophenol	79	N-Nitrosodiphenylamine	121	PCB-1221 (Aroclor 1221)
38	1,3-Dichlorobenzene	80	4-Bromophenyl phenyl ether	122	PCB-1232 (Aroclor 1232)
39	1,4-Dichlorobenzene	81	Hexachlorobenzene	123	PCB-1242 (Aroclor 1242)
40	Benzyl alcohol	82	Pentachlorophenol	124	PCB-1248 (Aroclor 1248)
41	1,2-Dichlorobenzene	83	Phenanthrene	125	PCB-1254 (Aroclor 1254)
42	2-Methylphenol	84	Anthracene	126	PCB-1260 (Aroclor 1260)

high detection limits may then mask the presence of an important compound. A 10:1 dilution will result in a 10-fold increase in the detection limit.

Quality assurance and quality control (QA/QC) are important parts of any analytical program for organic chemicals. A quality-assurance and quality-control program aims to determine the **accuracy,** or correctness, of the data as well as the **precision,** or repeatability, of the analyses.

Field and **method blanks** are used to detect if any organics are being inadvertently introduced during the sampling and analysis procedures. A field blank is a sample of very pure water that is run through the sampling equipment and then put into a bottle and returned to the lab for analysis. If all goes well, the field blank should not contain any organic compounds. Method blanks are also used to determine the purity of the solvents and reagents used in the analysis. Method blanks are samples of distilled water that are analyzed using the solvents and chemicals called for in the procedure.

If a field and/or method blank contains organic compounds, these compounds were introduced during the sampling or analytical procedures. Certain organic compounds that are used in laboratories are frequently detected in the method blanks. These compounds include acetone, methylene chloride, toluene, 2-butanone, di-*n*-butyl phthalate, di-*n*-octyl phthalate, and *bis*(2-ethylhexyl) phthalate. If one of these compounds is present in a field or method blank, then the U.S. EPA's Laboratory Data Validation Functional Guidelines specify that the detection limit for that compound be set at 10 times the greatest amount found in a blank. For all other organic compounds that might be found in a method or field blank, the detection limit is set at 5 times the greatest amount found in the blank.

Spiked samples have a known amount of an organic compound added to water and then run through the analytical process. If possible, uncontaminated ground water from the site is used to prepare the spike. This becomes a standard solution; since the initial concentration is known, a **percent recovery** of the analyte can be determined. A perfect analysis will have a 100% recovery. The QA/QC specifications for the lab contract will determine an acceptable percent recovery for valid data. If too much or too little of the spiked compound is recovered, the reported data are not valid.

Duplicate samples are used to validate the precision of the analysis. The lab may take a sample from the field, split it into two or more aliquots, and analyze them to make a **lab duplicate.** The hydrogeologist may also collect a duplicate or triplicate sample in the field and submit it as a field duplicate. Most often field duplicates are submitted as blind duplicates so that the analyst doesn't know that a split sample was submitted. Standard procedures generally call for 1 split sample to be collected for every 10 field samples.

7.9 Summary

There are a number of important physical properties of organic chemicals that influence their behavior. These include the melting point and boiling point, density, vapor pressure, vapor density, octanol-water partition coefficient, and Henry's law constant. Organic compounds are based on carbon, which may be bonded to hydrogen, halides, oxygen, nitrogen, phosphorus, and sulfur. The most simple organic compounds are hydrocarbons, which have only carbon and hydrogen. Carbon atoms have four bonding locations and may form single, double, and triple bonds with other carbon atoms. There is a formal system for naming organic compounds, although many compounds have common names that arose before the formal system was developed. Many organic compounds may be degraded by chemical and microbial means that reduce the molecular weight and complexity of the compounds. The end product of degradation of hydrocarbons is carbon dioxide or methane, depending upon the conditions under which degradation occurs. Organic compounds dissolved in water are most commonly analyzed by gas chromatography, mass spectrometry, or a combination of both.

References

American Chemical Society Committee on Environmental Improvement. 1983. *Analytical Chemistry,* vol. 55, pp. 2210–18.

Barker, J. F., and G. C. Patrick. 1985. Natural attenuation of aromatic hydrocarbons in a shallow sand aquifer. *Proceedings of the Conference on Petroleum Hydrocarbons and Organic Chemicals in Ground Water: Prevention, Detection and Restoration,* 160–177. Dublin, Ohio: National Water Well Association.

Barker, J. F., G. C. Patrick, and D. Major. 1987. Natural attenuation of aromatic hydrocarbons in a shallow sand aquifer. *Ground Water Monitoring Review* 7, no. 1; 64–71.

Barrio-Lage, G., F. Z. Parsons, and R. S. Nassar. 1987. Kinetics of the depletion of trichloroethene. *Environmental Science and Technology* 21, no. 4:366–70.

Barrio-Lage, G., F. Z. Parsons, R. S. Nassar, and P. A. Lorenzo. 1986. Sequential dehalogenation of chlorinated ethenes. *Environmental Science and Technology* 20, no. 1:96–99.

———. 1987. Biotransformation of trichloroethene in a variety of subsurface materials. *Environmental Toxicology and Chemistry* 6:571–78.

Bouwer, E. J., and P. L. McCarty. 1983a. Transformations of halogenated organic compounds under denitrification conditions. *Applied and Environmental Microbiology* 45, no. 4:1295–99.

———. 1983b. Transformations of 1- and 2-carbon aliphatic organic compounds under methanogenic conditions. *Applied and Environmental Microbiology* 45, no. 4:1286–94.

Bumpus, J. A., 1989. Biodegradation of polycyclic aromatic hydrocarbons by *Phanerochaete chrysosporium*. *Applied and Environmental Microbiology* 55, no. 1:154–58.

Burlinson, N. E., L. A. Lee, and D. H. Rosenblatt. 1982. Kinetics and products of hydrolysis of 1,2-dibromo-3-chloropropane. *Environmental Science and Technology* 16, no. 9:627–32.

Chiang, C. Y., J. P. Salanitro, E. Y. Chai, J. D. Colhart, and C. L. Klein. 1989. Aerobic biodegradation of benzene, toluene, and xylene in a sandy aquifer—data analysis and computer modeling. *Ground Water* 27, no. 6:823–34.

Cline, P. V., and D. R. Viste. 1985. Migration and degradation patterns of volatile organic compounds. *Waste Management & Research* 3:351–60.

Egli, C., T. Tschan, R. Scholtz, A. M. Cook, and T. Leisinger. 1988. Transformation of tetrachloromethane to dichloromethane and carbon dioxide by *Acetobacterium woodi*. *Applied and Environmental Microbiology* 54, no. 11:2819–24.

Ehrlich, G. G., D. F. Goerlitz, E. M. Godsy, and M. F. Hult, 1982. Degradation of phenolic contaminants in ground water by anaerobic bacteria: St. Louis Park, Minnesota. *Ground Water* 20, no. 6:703–10.

Evans, W. C. 1977. Biochemistry of the bacterial catabolism of aromatic compounds in anaerobic environments. *Nature* 270:17–22.

Fathepure, Babu Z., and S. A. Boyd. 1988. Dependence of tetrachloroethylene dechlorination on methanogenic substrate consumption by *Methanosarcina* sp. Strain DCM. *Applied and Environmental Microbiology* 54, no. 12:2976–80.

Fetter, C. W. 1989. Transport and fate of organic compounds in ground water, in *Recent Advances in Ground-Water Hydrology*. Edited by J. E. Moore, A. A. Zaporozec, S. C. Csallany, and T. C. Varney, 174–84. St. Paul, Minn.: American Institute of Hydrology.

Fliermans, C. B., T. J. Phelps, D. Ringelberg, A. T. Mikell, and D. C. White. 1988. Mineralization of trichloroethylene by heterotrophic enrichment cultures. *Applied and Environmental Microbiology* 54, no. 7:1709–14.

Freedman, D. L. and J. M. Gossett. 1989. Biological reductive dechlorination of tetrachloroethylene and trichloroethylene to ethylene under methanogenic conditions. *Applied and Environmental Microbiology* 55, no. 9:2144–51.

Galli, Rene, and P. L. McCarty. 1989a. Biotransformation of 1,1,1-trichloroethane, trichloromethane and tetrachloromethane by a *Clostridium* sp. *Applied and Environmental Microbiology* 55, no. 4:837–44.

———. 1989b. Kinetics of biotransformation of 1,1,1-tri-chloroethane by *Clostridium* sp. strain TCAAIIB.

Applied and Environmental Microbiology 55, no. 4: 844–51.

Grbic-Galic, D., and T. M. Vogel. 1987. Transformation of toluene and benzene by mixed methanogenic cultures. *Applied and Environmental Microbiology* 53, no. 2:254–60.

Guerin, W. F., and G. E. Jones. 1988. Two-stage mineralization of phenanthrene by estuarine enrichment cultures. *Applied and Environmental Microbiology* 54, no. 4:929–36.

Haigler, B. E., S. F. Nishino, and J. C. Spain. 1988. Degradation of 1,2-dichlorobenzene by a *Pseudomonas* sp. *Applied and Environmental Microbiology* 54, no. 2:294–301.

Hall, A. E. 1988. Transformations in soil. In *Environmental Chemistry of Herbicides*. Edited by R. Glover, vol. I, 171–200. Boca Raton, Fla.: CRC Press.

Harker, A. R., and Young Kim. 1990. Trichloroethylene degradation by two independent aromatic-degrading pathways in *Alcaligenes eutrophus* JMP134. *Applied and Environmental Microbiology* 56, no. 4:1179–81.

Heitkamp, M. A., and C. E. Cerniglia. 1988. Mineralization of polycyclic aromatic hydrocarbons by a bacterium isolated from sediment below an oil field. *Applied and Environmental Microbiology* 54, no. 6:1612–14.

———. 1989. Polycyclic aromatic hydrocarbon degradation by a *Mycobacterium* sp. in a microcosm containing sediment and water from a pristine ecosystem. *Applied and Environmental Microbiology* 55, no. 8:1968–73.

Heitkamp, M. A., J. P. Freeman, D. W. Miller, and Carl Cerniglia. 1988. Pyrene degradation by a *Mycobacterium* sp.: Identification of ring oxidation and ring fission products. *Applied and Environmental Microbiology* 54, no. 10:2556–65.

Hunt, J. M. 1979. *Petroleum Geochemistry and Geology*. New York: W. H. Freeman and Company, 617 pp.

Jackson, R. E., M. W. Priddle, and S. Lesage. 1990. Transport and fate of CFC-113 in ground water. *Proceedings of Petroleum Hydrocarbons and Organic Chemicals in Ground Water: Prevention, Detection and Restoration*, 129–42. Dublin, Ohio: National Water Well Association.

Janssen, D. B., A. Scheper, L. Dijkhuizen, and B. Witholt. 1985. Degradation of halogenated aliphatic compounds by *Xanthobacter autotrophicus* GJ10. *Applied and Environmental Microbiology* 49, no. 3:673–77.

Klecka, G. M., J. W. Davis, D. R. Gray, and S. S. Madsen. 1990. Natural bioremediation of organic contaminants in ground water: Cliffs-Dow superfund site. *Ground Water* 28, no. 4:534–43.

Kleopfer, R. D., D. M. Easley, B. B. Hass, T. G. Deihl, D. E. Jackson, and C. J. Wurrey. 1985. Anaerobic degradation of trichloroethylene in soil. *Environmental Science and Technology* 19, no. 3:277–80.

Kohring, G-W, J. E. Rogers, and J. Wiegel. 1989. Anaerobic biodegradation of 2,4-dichlorophenol in freshwater lake sediments at different temperatures. *Applied and Environmental Microbiology* 55, no. 2:348–52.

Kreamer, D. K., and K. J. Stezenbach. 1990. Development of a standard, pure-compound base gasoline mixture for use as a reference in field and laboratory experiments. *Ground Water Monitoring Review* 10, no. 2:135–45.

LaPat-Polasko, L. T., P. L. McCarty, and A. J. B. Zehnder. 1984. Secondary substrate utilization of methylene chloride by an isolated strain of *Pseudomonas* sp. *Applied and Environmental Microbiology* 47, no. 4:825–30.

Little, C. D., A. V. Palumbo, S. E. Herbes, M. E. Lidstrom, R. L. Tyndale, and P. J. Gilmer. Trichloroethylene biodegradation by a methane-oxidizing bacterium. *Applied and Environmental Microbiology* 54, no. 4:951–56.

Major, D. W., C. I. Mayfield, and J. F. Barker. 1988. Biotransformation of benzene by denitrification in aquifer sand. *Ground Water* 26, no. 1:8–14.

Manahan, S. E. 1984. *Environmental Chemistry*, 4th ed. Boston: PWS Publishers, 612 pp.

McCarty, P. L., M. Reinhard, and B. E. Rittman. 1981. Trace organics in groundwater. *Environmental Science and Technology* 15, no. 1:40–51.

Mihelcic, J. R., and R. G. Luthy. 1988. Microbial degradation of acenaphthene and naphthalene under denitrification conditions in soil-water systems. *Applied and Environmental Microbiology* 54, no. 5:1188–98.

Mikesell, M. D., and S. A. Boyd. 1990. Dechlorination of chloroform by *Methanosarcina* strains. *Applied and Environmental Microbiology* 56, no. 4:1198–1201.

Mueller, J. G., P. J. Chapman, B. O. Blattmann, and P. H. Pritchard. 1990. Isolation and characterization of a fluoranthene-utilizing strain of *Pseudomonas paucimobis*. *Applied and Environmental Microbiology* 56, no. 4:1079–86.

Nelson, M. J. K., S. O. Montgomery, W. R. Mahaffey, and P. H. Pritchard. 1987. Biodegradation of trichloroethylene and involvement of an aromatic biodegradative pathway. *Applied and Environmental Microbiology* 53. no. 5:949–54.

Nelson, M. J. K., S. O. Montgomery, E. J. O'Neill, and P. H. Pritchard. 1986. Aerobic metabolism of trichloroethylene by a bacterial isolate. *Applied and Environmental Microbiology* 52, no. 2:383–84.

Nelson, M. J. K., S. O. Montgomery, and P. H. Pritchard. 1988. Trichloroethylene metabolism by microorganisms that degrade aromatic compounds. *Applied and Environmental Microbiology* 54, no. 2:604–6.

Parsons, F., P. R. Wood, and J. DeMarco. 1984. Transformations of tetrachloroethene and trichloroethene in microcosms and groundwater. *Journal of American Water Works Association* 76, no. 2:56–59.

Pignatello, J. J. 1986. Ethylene dibromide mineralization in soils under aerobic conditions. *Applied and Environmental Microbiology* 51, no. 3:588–92.

Rifai, H. S., P. B. Bedient, J. T. Wilson, K. M. Miller, and J. M. Armstrong. 1988. Biodegradation modeling at aviation fuel spill site. *Journal of Environmental Engineering* 114, no. 5:1007–29.

Roberts, P. V., J. Schreiner, and G. D. Hopkins. 1982. Field study of organic water quality changes during groundwater recharge in the Palo Alto baylands. *Water Research* 16:1025–35.

Schwarzenbach, R. P., W. Giger, C. Schaffner, and O. Wanner. 1985. Groundwater contamination by volatile halogenated alkanes: Abiotic formation of volatile sulfur compounds under anaerobic conditions. *Environmental Science and Technology* 19, no. 4:322–27.

Smith, Allen E. 1988. Transformations in soil. In *Environmental Chemistry of Herbicides*, edited by R. Grover. 171–200. Boca Raton, Fla.: CRC Press.

Strand, S. E., and L. Shippert. 1986. Oxidation of chloroform in an aerobic soil exposed to natural gas. *Applied and Environmental Microbiology* 52, no. 1:203–5.

Swallow, K. C., N. S. Shifrin, and P. J. Doherty. 1988. Hazardous organic compound analysis. *Environmental Science and Technology* 22, no. 2:136–42.

Topp, Edward, and R. S. Hanson. 1990. Degradation of pentachlorophenol by a *Flavobacterium* species grown in continuous culture under various nutrient limitations. *Applied and Environmental Microbiology* 56, no. 2:541–44.

Tsien, H-C, G. A. Brusseau, R. S. Hanson, and L. P. Wackett. 1989. Biodegradation of trichloroethylene by *Methylosinus trichosporium* OB3b. *Applied and Environmental Microbiology* 55, no. 12:3155–61.

Vogel, T. M., C. S. Criddle, and P. L. McCarty. 1987. Transformations of halogenated aliphatic compounds. *Environmental Science and Technology* 21, no. 8: 722–36.

Vogel, T. M, and P. L. McCarty. 1985. Biotransformation of tetrachloroethylene to trichloroethylene, dichloroethylene, vinyl chloride, and carbon dioxide under methanogenic conditions. *Applied and Environmental Microbiology* 49, no. 5:1080–83.

———. 1987a. Abiotic and biotic transformations of 1,1,1-trichloroethane under methanogenic conditions. *Environmental Science and Technology* 21, no. 12:1208–13.

———. 1987b. Rate of abiotic formation of 1,1-dichloroethylene from 1,1,1-trichloroethane in groundwater. *Journal of Contaminant Hydrology* 1:299–308.

Vogel, T, M. and M. Reinhard. 1986. Reaction products and rates of disappearance of simple bromoalkanes, 1,2-dibromopropane and 1,2-dibromoethane in water. *Environmental Science and Technology* 20, no. 10:992–97.

Wilson, J. T., L. E. Leach, M. Henson, and J. N. Jones. 1986. *In situ* biorestoration as a ground water remediation technique. *Ground Water Monitoring Review* 6, no. 4:56–64.

Wilson, B. H., G. B. Smith, and J. F. Rees. 1986. Biotransformations of selected alkylbenzenes and halogenated aliphatic hydrocarbons in methanogenic aquifer material: A microcosm study. *Environmental Science and Technology* 20, no. 10:997–1002.

Wilson, J. T., and B. H. Wilson. 1985. Biotransformation of trichloroethylene in soil. *Applied and Environmental Microbiology* 49, no. 1:242–43.

Chapter Eight
Ground-Water and Soil Monitoring

8.1 Introduction

Methods of installing monitoring wells and collecting ground-water samples have been developed with the specific intention of obtaining a representative sample of water from an aquifer. These methods minimize the potential for the introduction of contaminants into the ground through the process of installing a monitoring well. Wells and sampling devices can be constructed of materials that have a minimum tendency to leach materials into and sorb compounds from the water sample. Ground-water samples can be collected in such a manner that dissolved gases are not lost or exchanged with the atmospheric gases. Soil samples can also be collected for classification and chemical analysis.

Methods of collecting samples of soil water are also available. Soil gas sampling can be done to give an indication of areas where volatile organic compounds are contained in the soil or ground water.

8.2 Monitoring Well Design

8.2.1 General Information

Monitoring wells are installed for a number of different purposes. During the installation of a monitoring well, a soil boring may be made or rock-core samples may be collected to determine the basic geology of the site. Prior to the design of a well, it is necessary to determine what its use will be. Some purposes of monitoring wells include the following:

- Measuring the elevation of the water table
- Measuring a potentiometric water level within an aquifer
- Collecting a water sample for chemical analysis
- Collecting a sample of a nonaqueous phase liquid that is less dense than water
- Collecting a sample of a nonaqueous phase liquid that is more dense than water
- Testing the permeability of an aquifer or aquiclude
- Providing access for geophysical instruments
- Collecting a sample of soil gas

Ground-Water and Soil Monitoring   339

The use for which the well is intended will dictate the design. For example, if a well is to be used for the collection of water samples, the casing must be large enough to accommodate the water-sampling device. However, the diameter should not be much larger than the minimum size, because prior to the sampling of a well, stagnant water must be removed from the casing; the larger the diameter of the casing, the greater the volume of water that must be pumped and properly disposed. The factors that should be included in the design of a monitoring well include

- Type of casing material
- Diameter of the casing
- If there will be a well screen or an open borehole
- Length of casing
- Depth of the well
- Setting and length of the well screen
- Diameter of well screen
- Type of material for well screen
- Slot opening of well screen
- If an artificial filter pack (gravel pack) is necessary
- Gradation of filter pack (gravel pack) material
- Method of installation of well and screen
- Material used to seal annular space between casing and borehole wall
- Protective casing or well vault

8.2.2 Monitoring Well Casing

All monitoring wells have a **casing,** whether they have a screen or terminate in an open borehole in bedrock. The casing is a piece of solid pipe that leads from the ground surface to the well screen or open borehole and is intended to keep both soil and water from entering the well other than through the screen or open borehole. Casing also prevents water from flowing from one aquifer horizon to another.

The diameter of the casing for a monitoring well is determined by the use for which the monitoring well is planned. If the only purpose of the monitoring well is to measure water levels, then a 1-in.-inside-diameter casing is all that is needed. An electric probe to measure water level or a pressure transducer will fit inside the 1-in. casing. Figure 8.1 shows an electric probe being lowered into a 2-in. casing.

If a well is to be used to collect a ground-water sample, the diameter of the well needs to be such that standard well-sampling equipment can fit inside. The common standard for well-sampling equipment is a nominal 2-in. diameter. This can accommodate a wide variety of pumps that can withdraw water at rates of 0.5 to 2 or 3 gal/min. Specially designed borehole geophysical equipment can also fit inside a 2-in. diameter casing. Some states mandate the casing diameter for monitoring wells. For example, the Wisconsin Department of Natural Resources requires a minimum inside diameter of 1.9 in. and a maximum inside diameter of 4.0 in., whereas the New Jersey Department of Environmental Protection requires a 4-in.-diameter well under all conditions.

For some applications, monitoring wells may be intended for several functions such as measuring water levels, collecting water samples, pumping to remove

FIGURE 8.1 Electric probe used to measure water levels in monitoring wells. Photo credit: Jim Labre.

contaminated water, and perhaps floating nonaqueous phase liquids and as a part of a vapor-extraction system. These wells generally have diameters larger than 2 in. to accommodate pumping equipment with a higher-flow capacity. The actual equipment to be used determines the casing diameter.

Casing diameter can also be influenced by the depth of the well. The deeper the well, the stronger the casing and screen must be to resist the lateral pressure at the final depth and the crushing force of the weight of the length of casing. Larger diameter casing can be made with thicker walls to have greater strength. It is easier to have a straight well with stronger casing. Straight wells are important in accommodating bailers and pumps.

The outside diameter of casing is standard; however, the inside diameter is a function of the wall thickness. Table 8.1 lists the wall thickness and inside diameter for various schedules of casing. Heavier-schedule casing is stronger because it has a thicker wall. The strength of a casing also depends upon the material from which it is constructed. A schedule 5 casing made of stainless steel is stronger than a schedule 40 casing made of polyvinyl chloride (PVC), yet leaves a greater inside diameter.

TABLE 8.1 Dimensions of inside and outside diameters of well casings.

Pipe Size	Outside Diameter	Schedule 5		Schedule 10		Schedule 40		Schedule 80	
		Wall Thickness	Inside Diameter	Wall Thickness	Inside Diameter	Wall Thickness	Inside Diameter	Wall Thickness	Inside Diameter
Nominal 2"	2.375"	0.065"	2.245"	0.109"	2.157"	0.154"	2.067"	0.218"	1.939"
Nominal 3"	3.500"	0.083"	3.334"	0.120"	3.260"	0.216"	3.068"	0.300"	2.900"
Nominal 4"	4.500"	0.083"	4.334"	0.120"	4.260"	0.237"	4.026"	0.337"	3.826"
Nominal 5"	5.563"	0.109"	5.345"	0.134"	5.295"	0.258"	5.047"	0.375"	4.813"
Nominal 6"	6.625"	0.109"	6.407"	0.134"	6.357"	0.280"	6.065"	0.432"	5.761"

There are a number of materials used to make well casings and screens. These materials vary in chemical inertness, strength, durability, ease of handling, and cost. One must always consider the intended use of the monitoring well before selecting a material. What is the chemistry of the ground water and associated contaminants? Will any compounds present in the ground water react with any of the possible casing materials? How deep will the well be; what are the strength requirements? Is the well intended for a short-term monitoring project or will it remain in service for many years?

Well casings are available in the following materials: fluoropolymers, such as PTFE, or polytetrafluoroethylene (Teflon® is the brand name of one manufacturer of PTFE), mild steel, stainless steel, galvanized steel, fiberglass, PVC, and polypropylene. Mild or galvanized steel is often used for water-supply well casings but is not as frequently found in monitoring wells because it may react with the ground water to leach metals from the casing (Barcelona, Gibb, and Miller 1983). Polypropylene is not widely available. Most monitoring wells are made of stainless steel or PVC, with PTFE being less common. PVC casing is the least expensive. Relative casing costs for other materials, compared with PVC, are mild steel = 1.1, polypropylene = 2.1, type 304 stainless steel = 6.9, type 316 stainless steel = 11.2, and PTFE = 20.7. Type 316 stainless steel is more resistant to corrosion than type 304 under reducing conditions (Aller et al. 1989).

Stainless steel has the greatest strength, followed by mild steel. Both are also resistant to heat, but they are heavier than the plastics and are, therefore, more difficult to install. The lower strength of the plastics is compensated for by using a heavier-schedule casing that necessary with steel. Most monitoring wells are shallow enough that schedule 40 or 80 PVC has sufficient strength. PTFE is more brittle and has less wear resistance than PVC or polypropylene and is hence less durable. PTFE also has a low tensile strength and high weight per unit length, which limits its use to shallow depths. Even there, PTFE casing tends to bow under its weight when installed in monitoring wells and may not be straight and plumb. Although its nonstick properties are good in frying pans, the neat cement grout used to seal the annular space between the casing and the borehole may not bond to the PTFE casing (Nielsen 1988).

In the selection of casing material for ground-water monitoring wells, we must consider the potential chemical reactions between the casing material and the ground water. Ideally, casing material should neither leach matter into water nor sorb chemicals from water.

Reynolds and Gillham (1985) studied the sorption from aqueous solution of five halogenated organic compounds by several polymer materials. The organic compounds used were 1,1,1-trichloroethane, 1,1,2,2-tetrachloroethane, hexachloroethane, perchloroethene, and bromoform. The materials tested were PVC, PTFE, nylon, polypropylene, polyethylene, and latex rubber. Nylon, polypropylene, polyethylene, and latex rubber rapidly absorbed all five compounds. PVC absorbed all the compounds but 1,1,1-trichloroethane, although the rate of absorption was low. PTFE absorbed all the compounds but bromoform; although the rate of adsorption of three of the four compounds was low, PTFE absorbed 50% of the perchloroethylene in 8 hr.

Parker, Hewitt, and Jenkins (1990) evaluated the suitability of PVC, PTFE, stainless steel type 304 (SS 304), and stainless steel type 316 (SS 316) as casing material for monitoring metals in ground water. They evaluated the interaction of four trace elements

Ground-Water and Soil Monitoring

that are of concern in ground-water studies: arsenic, cadmium, chromium, and lead. The metals were tested at concentrations of 50 and 100 µg/L dissolved in ground water. Figure 8.2 shows the results of this study. If the concentration relative to control remains at 1.0, there is no interaction; if it drops to less than 1.0, then the element is sorbing onto the casing material; and if it rises above 1.0, the element is being leached from the casing. The PTFE was the most inert with respect to the metals, and the PVC was much better than either SS 304 or SS 316.

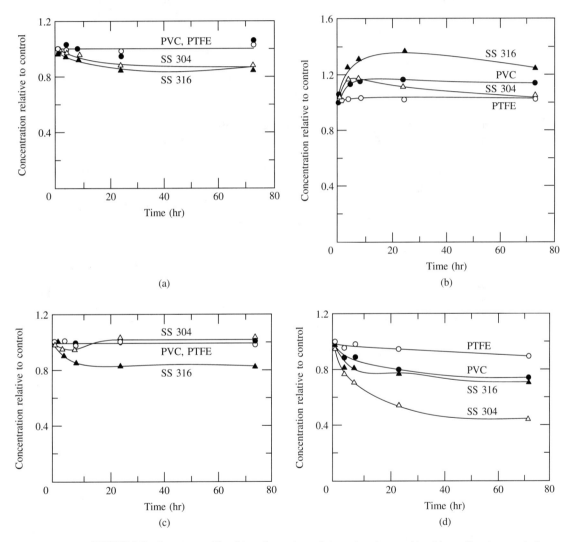

FIGURE 8.2 Sorption and leaching of arsenic, cadmium, chromium, and lead by well casings made from PVC, PTFE, type 304 stainless steel, and type 316 stainless steel. *Source:* L. V. Parker, A. D. Hewitt, and T. F. Jenkins, *Ground Water Monitoring Review* 10, no. 2 (1990):146–56. Used with permission. Copyright © 1990 Water Well Journal Publishing Co.

The interaction of several organic compounds with the same well-casing materials was also studied by Parker, Hewitt, and Jenkins (1990). Ten organic compounds were tested, including chlorinated ethenes, chlorobenzenes, nitrobenzenes, and nitrotoluenes. None of the compounds was sorbed onto either type of stainless steel. Many of the compounds were sorbed by the plastic casings, with the PTFE sorbing at a greater rate than the PVC. The amount and rate of sorption varied by compound. Figure 8.3 shows the sorption of trichloroethene by the four casing types. Clearly, stainless steel is the material of choice for monitoring organics, and PTFE is to be avoided. For a compromise material for monitoring both organics and inorganics, PVC appears to be the best. It also has the appeal of having the lowest cost. PVC manufactured specifically for well casing should be used, and it should carry the designation *NSF wc,* which indicates that the casing conforms to National Sanitation Foundation Standard 14 for potable water supply (National Sanitation Foundation 1988).

However, PVC should be avoided if certain organic compounds are present in the ground as nonaqueous phase liquids. It is reportedly soluble in low-molecular-weight ketones, aldehydes, amines and chlorinated alkanes, and alkenes (Barcelona, Gibb, and Miller 1983). Likewise, PVC casing should also never be joined with solvent-glued joints. These solvents include compounds such as methylethylketone and tetrahydrofuran and they may leach into ground water samples. Threaded joints that are machined directly onto the PVC are the preferred method of joining casing sections and casing to screen. Joints should be flush on the inside of the casing to prevent equipment being lowered into the casing from hanging up in a projecting joint.

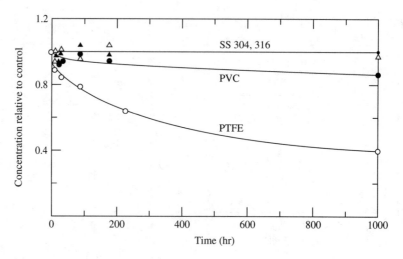

FIGURE 8.3 Sorption of trichloroethene from ground water by PVC, PTFE, type 304, and type 316 stainless steel well casings. *Source:* L. V. Parker, A. D. Hewitt, and T. F. Jenkins, *Ground Water Monitoring Review* 10, no. 2 (1990):146–56. Used with permission. Copyright © 1990 Water Well Journal Publishing Co.

Ground-Water and Soil Monitoring   345

8.2.3 Monitoring Well Screens

If the monitoring well terminates in an unconsolidated formation, a **screen** is necessary to allow the water to enter while keeping the sediment out. In most monitoring well applications, the well screen is the same diameter as the casing to which it is attached by a threaded coupling. Likewise, the well screen is normally made of the same material as the casing. The considerations that go into deciding the material to use for the casing also apply to the screen.

The screen will have openings to permit the water to enter. Manufactured well screen should always be used rather than hand-cut slots or drilled holes in plastic pipe. The two common screens for monitoring wells are slotted pipe, which is available in PVC and PTFE, and continuous wire wrap, which is available in stainless steel. Figure 8.4 illustrates these two screen types.

The width of the slot or wire-wrap opening is precisely controlled during the manufacture of the screen; the screen is available in a variety of opening sizes, generally ranging from 0.008 to 0.250 in. A screen with an opening of 0.010 is referred to as a 10-

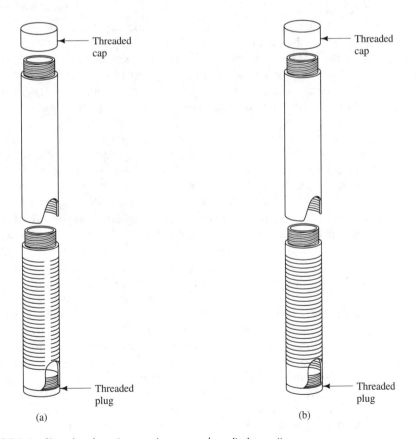

FIGURE 8.4 Slotted and continuous wire-wrapped monitoring well screens.

slot screen. Many manufacturers carry only a limited number of slot sizes in stock—for example, 10- and 20-slot. Since the casing and screen are typically ordered in advance of the well construction, the hydrogeologist usually has settled on a standard design prior to going on the job.

8.2.4 Naturally Developed and Filter-Packed Wells

The casing and screen may be placed in the borehole and the native sediment allowed to cave around the screen. This is called a **naturally developed well** and is often used in sandy sediment with very limited amounts of silt and clay present. At least 90% of the sediment should be retained on a 10-slot screen before a naturally developed well is considered (Aller et al. 1989). When water is withdrawn from such a well, it may initially be cloudy due to suspended silt and clay, but the water should eventually clear as the fines near the screen are removed by a process called well development. In a naturally developed well the slot size is selected to allow some of the fine sediment to enter the well during development; this leaves only the coarser sediment outside the screen.

In designing a water well, it is very important that the well be hydraulically effective—i.e., there should be a minimal loss of energy as the water flows into the well. The selection of the slot opening for naturally developed water wells is very important and is based on a grain-size distribution curve of the sediment opposite the well screen. Monitoring wells are designed to retain much more of the natural formation than water wells because they are much more difficult to develop (Driscoll 1986). Monitoring wells are not usually designed with the precision necessary for a water-supply well. The well should be hydraulically efficient as well as being as clear of silt and clay as possible. If preliminary investigations indicate that the aquifer to be monitored has reasonably coarse sand or gravel and few fines, a standard slot size may be preselected for all the monitoring wells. Ten-slot screen is frequently used under these conditions.

If the formation is cohesive—that is, has a high clay content—or if it is sandy with a high silt content, it will be necessary to use an **artificial filter pack.** Filter-pack material is medium to coarse sand that is predominately silica with no carbonates. It is mined and graded to have a specific grain-size distribution. Manufactured filter-pack material comes washed and bagged and is far preferable to native sand as artificial filter pack. The filter-pack material is placed in the borehole opposite the well screen. Its purpose is to stabilize the natural formation and keep it out of the screen. This will reduce the amount of silt and clay that enters the well when it is developed.

The grain size of the filter-pack material is based on the nature of the formation opposite the screen. If the formation is fine sand, then the grain-size distribution is determined. The filter pack material should have an average grain size that is twice the average grain size of the formation and have a uniformity coefficient (ratio of 40% retained size to 90% retained size) between 2 and 3 (Driscoll 1986). The screen-slot opening is then selected to retain 90% of the filter pack. The minimum practical slot size for monitoring well screens is 0.008 in. Figure 8.5 shows a grain-size distribution curve for a filter-pack material designed for an eight-slot screen. If the monitoring well is in silt or clay, all one can do is install an 8 slot screen and appropriate filter pack.

Ground-Water and Soil Monitoring

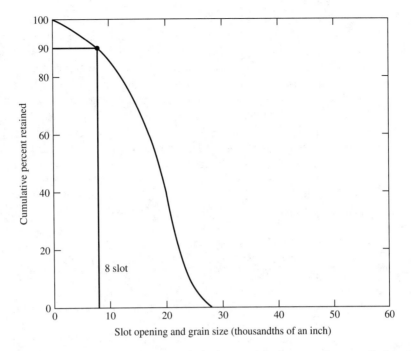

FIGURE 8.5 Grain-size distribution curve used to select an eight-slot screen for a monitoring well.

The filter-pack material should be 2 to 3 in. thick. This means that a 2-in.-diameter well screen should be installed in a borehole 6 to 8 in. in diameter. The filter-pack material is normally extended 2 or 3 ft above the top of the well screen to allow for settlement of the material during development.

8.2.5 Annular Seal

The **annular space** in the borehole above the filter pack must be sealed to prevent the movement of surface water downward to the filter pack. It may also be sealed to prevent vertical movement of ground water from one zone to another or to isolate a discrete sampling zone. The seal should be made of a material that has a low permeability, bonds well to the natural formation and the casing, and expands after it has been emplaced to ensure a tight seal. It should set up within a day or so and be durable and permanent.

Materials typically used for an annular seal are bentonite pellets, granular bentonite slurry, neat cement grout, bentonite-sand slurry, and neat cement grout with a powdered bentonite additive.

Neat cement grout is a mixture of 94 lb of type I Portland cement with 5 to 6 gal of water. Granular bentonite slurry is a mixture of 30 lb of untreated bentonite powder mixed with 125 lb of untreated bentonite granules with 100 gal of water. Bentonite-cement grout is a mixture of 5 lb of untreated powdered bentonite with 94 lb of type

I Portland cement and 8.5 gal of water. Bentonite-sand slurry is a mixture of 55 lb of untreated powdered bentonite with 100 gal of water and 10 to 25% sand by volume to make a slurry that weighs 11 lb/gal. All water used to make these slurries should be from a source that is fresh and known to be uncontaminated and free from floating oil.

Bentonite is a clay containing at least 85% sodium montmorillonite; it will swell to several times its original volume when thoroughly hydrated. This hydration takes place below the water table. However, bentonite has a high cation-exchange capacity and can affect the chemistry of water that comes into contact with it. Portland cement is used to make cement grout. When Portland cement cures, it is highly alkaline and can affect the pH of ground water that comes into contact with it. Neat cement grout will shrink by at least 17% when it cures. The addition of bentonite to make a bentonite-cement grout significantly reduces the shrinkage problem. If neat cement grout or bentonite cement grout is used, the casing material should be either stainless steel or schedule 80 PVC due to the heat generated as the cement cures.

The materials available for an annular seal are not ideal. Although they can be used to make an impermeable seal, there is a chance they might affect ground-water quality in their immediate vicinity. This problem is mitigated if 2 ft of fine sand is placed in the annular space above the filter-pack material or native sand opposite the screen. This keeps the annular seal material from coming into contact with the water entering the well screen.

Many hydrogeologists place a 2- or 3-ft layer of bentonite pellets above the fine sand if the pellets will be below the water table. The pellets will swell and keep the grout material from entering the filter-pack material. If the top of the 2-ft fine-sand seal is above the water table, then 2 ft of granular bentonite may be placed prior to the addition of the annular seal.

8.2.6 Protective Casing

In order to provide physical protection for the investment in a costly monitoring well, as well as to protect from vandalism by individuals accidentally or intentionally putting foreign fluids and objects into a monitoring well, a locking protective steel casing or well vault is needed.

A protective casing extends several feet above the ground surface. It extends above the top of the monitoring well and has an inside diameter sufficiently large so that the hydrogeologist can reach inside and unscrew a cap from the monitoring well. It is set into a surface cement seal. For monitoring wells installed in freezing climates, a drain hole at the bottom of the surface casing is desirable to prevent accumulation of moisture that could freeze in the annular space between the protective casing and the monitoring well. (The author has seen a stainless-steel monitoring well casing pinched shut by water that accumulated in a protective casing without a drain hole and then froze!)

In some applications, it is not practical to have a monitoring well that extends above ground—for example, in the driveway at a gas station. There are small well vaults available that can be used for protection for monitoring wells. However, they should be in places that are not going to flood; otherwise floodwaters could enter the aquifer via the monitoring well. If a well vault is used in a gas station or similar location, it should be clearly marked and should be distinctive from the fillers for underground storage

Ground-Water and Soil Monitoring

tanks so that an inattentive person doesn't try to fill it with gasoline! A locking well cap without a vent hole should also be used.

8.2.7 Screen Length and Setting

The hydrogeologist must decide on the length of the screen and the depth to which it will be set, based on the objectives of the monitoring program. Objectives could include monitoring the position of the water table, measuring the potentiometric head at some depth in the aquifer, collecting representative water samples from various depths in the aquifer, and detecting both light and dense nonaqueous phase liquids. Moreover, monitoring might be intended to detect the migration of ground water containing contaminants into an aquifer or evaluating the effectiveness of removing contaminants from an aquifer. All might require different approaches.

To monitor the position of the water table or to detect the presence of LNAPLs, the screen must be set so that it intersects the water table. The screen must be long enough to intersect the water table over the range of annual fluctuation. In addition, the screen must be long enough so that when the water table is at its greatest depth below the land surface, there is enough of the screen remaining below the water table to contain sufficient water for a water sample. A water table monitoring well will also be able to detect the presence of light nonaqueous phase liquids. In most applications the minimum length of the screen for a water table–monitoring well is 10 ft with 5 ft above and 5 ft below the water table. If the water table has more than 5 ft of annual fluctuation, a longer well screen is needed. However, some states specify a maximum screen length of 10 ft. Figure 8.6 shows examples of incorrect (a and b) and correct (c) placement of a multipurpose monitoring well intended to measure the position of the water table, detect floating nonaqueous phase liquids, and collect water samples from the upper part of the aquifer.

If the purpose of a monitoring well is to measure the potentiometric pressure at some depth in the aquifer, then the well is called a **piezometer.** A piezometer should have a relatively short screen length, 2 to 5 ft, so that the pressure that is recorded is representative of only a small vertical section of the aquifer. A piezometer can also be used to collect ground-water samples that are representative of a small vertical section of the aquifer.

Monitoring wells utilized to collect ground-water samples should be designed with respect to a specific ground-water monitoring goal. The concentration of ground-water contaminants can vary vertically. If a monitoring well has a long well screen, it has a greater probability of intersecting a plume of contamination. However, a water sample taken from such a well may draw water from both contaminated and uncontaminated parts of the aquifer, resulting in a reported concentration that is less than that of the ground water in the plume. This is illustrated in Figure 8.7.

The collection of such unrepresentative water samples may have serious implications for the implementation of ground-water regulations. In monitoring ground water in order to find the actual concentration of contaminants in a plume, it may be necessary to use several piezometers screened at different depths at the same location. This is expensive, not only due to the initial cost of the wells but also due to the costs of multiple chemical analyses for each round of sampling. However, such a configuration

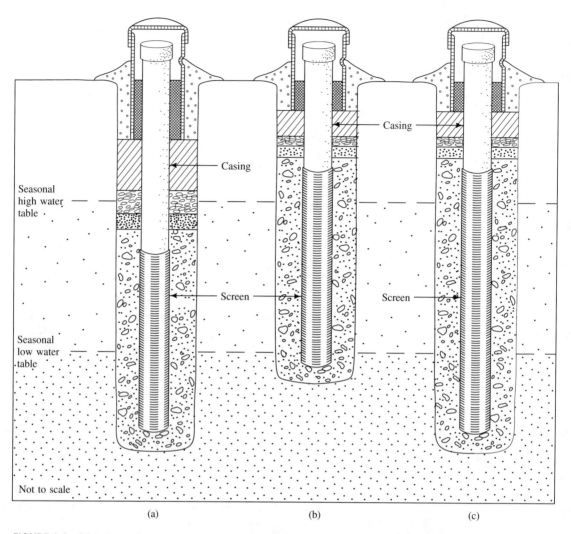

FIGURE 8.6 (a) Incorrect placement of water table–monitoring well screen. Seasonally high water table is above the top of the screen and floating, nonaqueous phase liquids would be above the screen and not detected. (b) Incorrect placement of water table–monitoring well screen. Seasonally low water table is so far down in well that there is not enough water in well to collect a sample for chemical analysis. (The water table elevation could still be determined.) (c) Correct length and placement of water table–monitoring well screen.

will yield the greatest amount of information about the hydraulic head as well as the water quality.

If a monitoring well is intended to serve as warning that a plume of contamination is escaping from a potential source, then it should be screened in the most permeable parts of the aquifer. Ground water and contaminants that it may be carrying not only

Ground-Water and Soil Monitoring

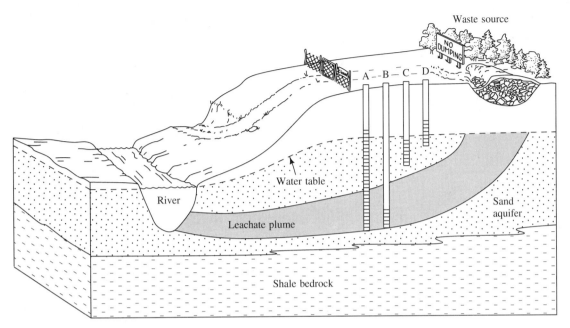

FIGURE 8.7 Effect of monitoring well–screen length on water-quality sampling. Monitoring well A is fully screened through the thickness of the aquifer. It intersects the plume of leachate but the reported concentration will be less than the actual concentration as water is withdrawn from both contaminated and uncontaminated parts of the aquifer. Piezometer B is also screened to intersect the plume of leachate. The reported concentration will be representative of the leachate. Piezometer C and water table monitoring well D don't intersect the plume, indicating that it is deep in the aquifer.

preferentially travel through the most permeable material but travel faster there as well. Hence, the leading edge of a plume of contamination will follow the most permeable pathway.

If the plume of contaminated water is following a zone or direction of high hydraulic conductivity, it may flow in a direction that is not parallel to grad h. This may mean that the location of the plume is not exactly down-gradient from the source.

On the other hand, if an aquifer is contaminated and a monitoring well has been installed to monitor the progress of a remediation effort, the well should not be screened in the most permeable part of the aquifer. In pump and treat systems, the water will preferentially travel through and flush out the more permeable zones. A well screened in a permeable zone may indicate that the aquifer is rapidly being cleaned, but in fact less permeable zones located nearby may still have high concentrations of contaminants that have yet to be removed.

8.2.8 Summary of Monitoring Well Design

Figure 8.8 illustrates details of the final design of a water table observation well and a piezometer illustrating all the design elements discussed in this section.

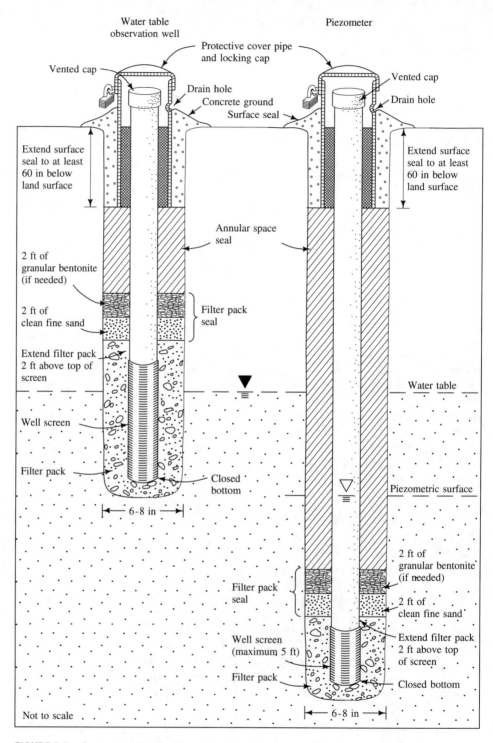

FIGURE 8.8 Construction details of a water table observation well and piezometer. *Source:* Wisconsin Department of Natural Resources.

8.3 Installation of Monitoring Wells

8.3.1 Decontamination Procedures

Because the purpose of drilling a monitoring well is to collect a sample of water and analyze it for very small concentrations of chemicals, it is highly desirable not to introduce any chemicals into the aquifer as a part of the well-drilling and installation procedure. The process of cleaning the equipment and supplies that will be used is called **decontamination.**

When materials are manufactured, they may become coated with substances such as grease and oil. Therefore, unless the manufacturer specifically guarantees that the article has been decontaminated and has shipped it in a well-sealed wrapper, it should be decontaminated. Equipment that has been used at a contaminated site should be assumed to be contaminated and should be decontaminated before it is used at another site. Even at the same site, if a drill rig or a bailer is used at different wells, decontamination is required to prevent cross contamination (contamination from one area being introduced into a clean area) that could occur.

There is wide variability in required and recommended decontamination procedures between the USEPA and the various states (Mickham, Bellandi, and Tifft 1989). The hydrogeologist must consult with the appropriate regulatory authority to determine if a specific decontamination procedure is required. In the absence of a specific requirement, the following generic procedure should adequately clean equipment and supplies. In some cases not all the steps are required. For cleaning large equipment such as a drilling rig, a specific area must be set aside and a decontamination pad must be constructed to capture all the fluids used in the process. If the rig is contaminated, wash water from it may also become contaminated. Be careful that any solvents used aren't accidentally released to the environment. Small tools such as bailers and Shelby tubes can be cleaned in buckets set on a polyethylene sheet. Sampling pumps can be cleaned by running various wash solutions through them, as well as washing the exterior.

The following steps are used to clean drilling and soil sampling equipment including drill rig, augers, drill rod, tools, sampling tubes, etc.:

1. Use a wire brush or similar equipment to remove all dried sediment and thick accumulations of grease.
2. Wash the equipment with a soft brush and water with phosphate-free detergent.
3. In extreme conditions organic residues can be removed by washing the equipment with an organic solvent such as methanol or propanol. Don't use solvents such as trichloroethene that might be expected to be found at a hazardous waste site.
4. Clean and rinse the equipment with potable water.
5. Rinse the equipment with deionized water.

Steam cleaning with a pressure sprayer can be used in step 4 for equipment that can withstand the heat and force of the spray.

When equipment has been decontaminated, it should not be placed on the ground. It can be wrapped in clean paper or aluminum foil or set on polyethylene sheets.

Sampling equipment should also be decontaminated between uses. If the equipment has not come into contact with nonaqueous phase liquids, rinsing with potable water and washing thoroughly with phosphate-free detergent, including scrubbing the inside of tubes with a bottle brush, followed by a potable-water wash and then a deionized-water rinse, should suffice. If the equipment has come into contact with nonaqueous phase organic liquids, then an initial solvent wash may be necessary.

The cost of decontamination of sampling equipment and the uncertainty introduced by solvent washing has led many hydrogeologists to specify dedicated sampling equipment in each well. Disposable bailers are also available that are less expensive than the cost of labor involved in cleaning reusable bailers.

8.3.2 Methods of Drilling

There are a number of methods of drilling that are appropriate for installation of monitoring wells. When working in shallow unconsolidated formations, hollow-stem augers are commonly used. If a well is to be drilled deeper than about 100 ft or into bedrock, a rotary drilling method may be appropriate. Cable-tool drilling is an excellent way of installing monitoring wells in both unconsolidated and consolidated formation, but it is slow and may be expensive. Hollow-stem augers with a bit that contains carbide teeth can also be used in weak, indurated rock.

Hollow-Stem Augers

Hackett (1987, 1988) presents an authoritative discussion of all aspects of drilling with hollow stem augers. A hollow-stem auger looks a little like a large, untapered screw (Figure 8.9). The auger flights are constructed around a hollow pipe. A drilling rig rotates the augers and a bit on the end of the auger loosens the sediment, which is then brought up to the surface by the rotating auger flights. The cuttings accumulate at the surface and must be shoveled away from the augers. Figure 8.10 pictures an auger drilling rig. The auger is advanced into the ground as it is rotated. A plug on the end of a rod inserted through the hollow stem may be screwed into the bit to seal the end of the opening and prevent sediment from coming up inside the hollow stem. Alternatively, a nonretrievable plug can be placed in the end of the bit. This plug is knocked out of the end of the augers when the final depth is reached and it is no longer needed. However, knock-out plugs preclude the collection of soil and water samples during drilling.

One advantage of drilling with hollow-stem augers is that drilling fluids and mud are normally not required. Circulation of drilling fluids has the potential to spread contaminants throughout the borehole. Drilling mud is a viscous liquid needed in mud-rotary drilling that can line the borehole and partially seal it. However, when drilling in formations with cohesive layers, the auger bit may smear clay from the cohesive layers so that it mixes with sand and gravel layers at the perimeter of the borehole. Auger drilling typically can advance about 50 to 100 ft per day if samples are being taken.

Augers usually come in lengths of 5 ft. One flight is advanced into the ground and then the drill stem is disconnected and another flight is attached to the augers in the ground. At this time samples of the formation ahead of the auger bit may be taken. The plug on the end of the bit must be removed before sampling can occur. The maximum depth at which hollow-stem auger drilling can normally be used is 150 ft (Hackett 1987); as a practical matter, it rarely exceeds 100 ft.

Ground-Water and Soil Monitoring

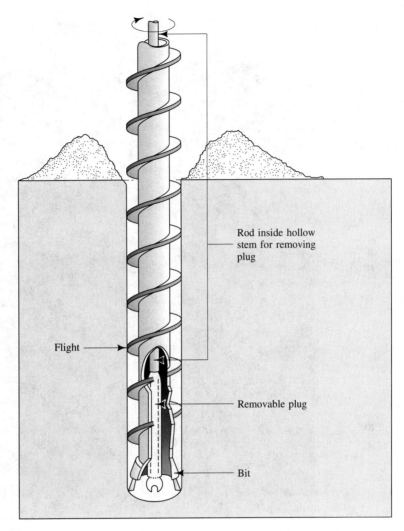

FIGURE 8.9 Hollow-stem auger drill rod and bit. *Source:* M. L. Scalf et al. *Manual of Ground Water Sampling Procedures,* 1981. National Water Well Association. Used with permission.

Some loosely consolidated sands, called heaving sands, can enter into the borehole when the plug is being removed. This problem can be avoided by keeping the hollow stem filled with potable water when the plug is removed. The potable water should not contain any contaminants that might be found in the ground water. If no formation sampling is planned, then a knock-out plug can be used to prevent the introduction of heaving sands into the hollow stem during drilling. The knock-out plug can be removed after the final depth is reached by pushing down with the well casing if it is stainless steel; otherwise it will be necessary to use a rod prior to the time that the well is installed.

FIGURE 8.10 Hollow-stem auger drilling rig. *Photo credit*: C. W. Fetter.

Hollow-stem auger drilling can also be used to sample water quality at various depths during drilling. At a selected horizon the plug at the end of the hollow stem is removed. A well point on the end of a rod is lowered to the bottom of the augers and then is driven ahead of the bit by hammering or hydraulic pressure. The well point is developed by pumping until clear water is obtained. A sample of ground water at that depth is then obtained. It is best if potable water is not used in the borehole during this procedure, because it could interfere with the ground water quality. If there is heaving sand inside the augers, the well point can be driven through it.

Keeley and Boateng (1987) suggested a modification of the hollow-stem auger drilling technique, in which a temporary casing that is larger in diameter than the auger bit is employed. The auger is advanced several feet and then the temporary casing is driven to the depth of the auger bit by repeatedly dropping a heavy weight on the top of the casing. The advantage of this modification is that it prevents mixing of soil horizons as the augers rotate.

Mud-Rotary Drilling

Mud-rotary drilling can be used in both unconsolidated and consolidated formations. It is fairly rapid, up to 100 ft/day, and can be used to depths far in excess of any that might be required for ground-water–contamination studies. A heavy drilling fluid, made by mixing various additives to water, is circulated in the borehole by pumping it down the inside of hollow drill rods. The mud rises back to the surface in the annular space between the borehole wall and the drill pipe. The rising mud carries with it the drill cuttings, which settle out in a mud tank at the surface. Figure 8.11 shows the circulation pattern for mud-rotary drilling.

One advantage of mud-rotary drilling is that the borehole will remain open after the string of drill pipe and the bit are removed. This means that a complete suite of geophysical logs can be run on the hole, which is kept open by the weight of the mud inside it. However, the fluid in the drilling mud can penetrate the native formations and alter the ground-water geochemistry. The coating of drilling mud on the borehole walls may be difficult to remove. This can impede the hydraulic connection between the well and the formation. Bentonite-based drilling muds may remove metals from the ground water and affect the chemistry. Because of the potential problems with drilling mud, rotary drilling may not be as suitable for ground-water contamination studies as hollow-stem augers. However, under many hydrogeologic conditions it is the only drilling method that is practical. Good well development to remove the residual drilling mud in the screen zone is very important.

Air-Rotary Drilling

If the monitoring well is to be installed in bedrock, then air-rotary drilling may be considered. First a surface casing needs to be installed through any unconsolidated material. Typically this is done by using mud-rotary drilling. The surface casing is large enough that the air-rotary bit can fit inside of it.

The fluid used in air rotary is compressed air, which is blown down the inside of the drill pipe. The air then blows the cuttings back up the annular space, where they accumulate around the borehole. When the water table is encountered, the air may blow ground water out of the borehole as well. If this occurs, it is possible to determine when the water table is encountered and the relative yield of the well. However, the air may also force the water back into the formation. Air-rotary drilling using a down-hole percussion bit can drill up to 60 ft/hr. Samples are collected as chips, which are brought to the surface with the return flow of air and water.

Air-rotary drilling is fast and can go to depths of a thousand feet of more. Because air is the drilling fluid, contamination problems are minimized. However, the drillers may want to add a foaming agent to the air as it goes down the hole. The foam helps to float the chips to the surface, but it consists of organic chemicals such as isopropyl alcohol, ethyl alcohol, and alcohol ether sulfate, and its use should be avoided. The air

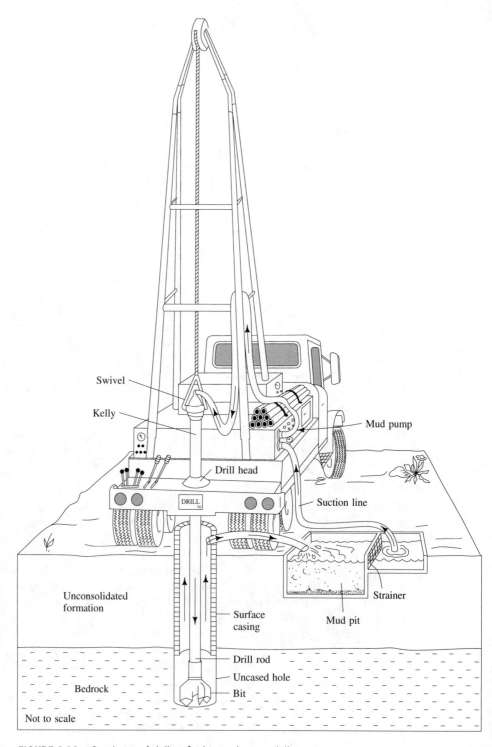

FIGURE 8.11 Circulation of drilling fluid in mud-rotary drilling.

compressors have air filters, which need to be in good working condition; otherwise, lubricating oil and other contaminants may be introduced into the borehole with the compressed air. Percussion hammers used for air-rotary drilling may also need lubricants.

Air-rotary drilling may introduce volatile organic compounds into the atmosphere as well as blowing contaminated dust out of the borehole.

Reverse-Rotary Drilling

In reverse-rotary drilling the circulating drilling fluid drains down the annular space and then is pulled up the center of the drill stem by a suction pump located on the drill rig. Because the drilling fluid rises with a much greater velocity in reverse-rotary than in mud-rotary drilling, a much less viscous drilling fluid is used. In many cases clear water mixed with the drill cuttings is all that is necessary. This gives the reverse-rotary method an advantage over mud-rotary drilling, since it is much easier to develop the well because there is no mud wall on the borehole to break down. However, reverse-rotary drilling is more expensive than the mud-rotary method, and the minimum borehole diameter is 12 in.

Cable-Tool Drilling

Cable-tool drilling is one of the oldest drilling methods and has been used widely for the installation of water wells. Although the drilling equipment is less expensive than for some other methods, the drilling is slow and overall costs may be expensive due to high labor costs.

In cable-tool drilling a heavy bit is located at the end of a tool string hanging from a cable. The drill rig repeatedly lifts and drops the hammer, which breaks up consolidated rock or loosens unconsolidated sediment. A steel casing is driven into the formation behind the bit. When the bottom of the casing fills with broken rock and sediment, the tool string and bit are removed, and a bottom-loading bailer is used to remove the accumulated cuttings. Below the water table, the ground water and cuttings make a slurry. Above the water table, water must be added to make a slurry so that the bailer can be used. Drive casing is needed only until bedrock is reached. In most bedrock formations the hole will stay open without drive casing. Figure 8.12 shows the tools used for cable-tool drilling.

Advantages of cable-tool drilling include the fact that no drilling fluids are used and that nothing is circulated through the well. Both factors serve to limit contamination problems. It is easy to collect representative samples of the formation during bailing of the casing. Well points can be driven ahead of the casing in unconsolidated formations for the collection of water quality samples. Cable-tool drilling can be used to depths in excess of 1000 ft.

8.3.3 Drilling in Contaminated Soil

When drilling at a contaminated site, the cuttings that are brought to the surface may be contaminated. Drilling personnel should wear appropriate protective clothing and, if necessary, use breathing apparatus. A large, heavy sheet of plastic should be placed in the work area, and the drill bit should be advanced through a hole in the center of the plastic sheet.

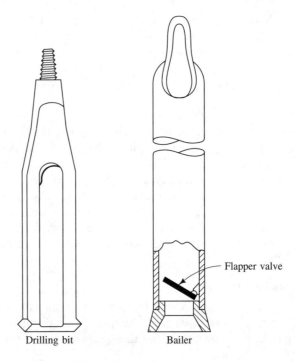

FIGURE 8.12 Tools used for cable-tool drilling.

The cuttings augered to the surface can be collected from the plastic sheet and put into containers for proper disposal.

8.4 Sample Collection

Samples of unconsolidated materials and rock are needed to delineate the geology of a site. They are collected by drilling **borings.** Borings may be made just for sample collection, or they may be made as a part of the process of installing a monitoring well. Borings may be made by any of the methods of drilling discussed in Section 8.3.

During the drilling process, earth materials are brought to the surface. During the augering process, soil and sediment ride up the augers; in the mud-rotary process, earth materials come up mixed with the mud; in air-rotary drilling there is a slurry of rock and water brought to the surface by air pressure; and cuttings are brought to the surface with a bailer in cable-tool drilling. In all cases the samples are disturbed, some (such as the samples on the augers) more than others. In addition, it may be difficult to tell the exact depth represented by the sample. Fine layering of sedimentary materials cannot be distinguished in such samples. An imprecise model of the geology can be constructed from these samples, but for more details, undisturbed samples should be collected.

A **core sample** is collected in a special sampling tube that is driven into unconsolidated formations or by the drilling of a **rock-core sample** with a special rock-core bit studded with industrial diamonds.

There are two main types of sampling tubes for unconsolidated samples. Both types of tubes can be used with hollow-stem augering and mud-rotary drilling if the drill rod and bit have a provision for a sample tube to be extended through them into the formation ahead of the bit. Figure 8.13 shows the sequence for the extension of a sampler through the end of the bit of a hollow-stem auger. The sampler is driven or pushed into undisturbed formation ahead of the bit. In cable-tool drilling the drill bit is removed from the borehole and the sampler is lowered on a rod or cable.

A **Shelby tube** is a thin-walled tube that can be screwed to the end of a rod, lowered to the bottom of the drilled hole, and pressed into cohesive sediments by using hydraulic pressure reacting against the weight of the drill rig. These samples are said to be *undisturbed*, although they are in fact minimally disturbed. The precise method of collecting a Shelby tube sample is described in method ASTM D1587 (American Society for Testing and Materials 1983). The sample can be extruded from the Shelby tube in the lab and trimmed into a permeameter for a permeability test. Details of the microstratigraphy can be examined as well. Shelby tube sampling does not work with noncohesive sediments.

A **split-spoon sampler** can be used for the collection of samples of both cohesive and noncohesive sediments. The split-spoon sampler consists of a split tube with thicker walls than a Shelby tube. The two halves are placed together and joined by screwing a circular drive shoe on the bottom and a head assembly on the top. The assembled split-spoon sampler is screwed to a rod and lowered to the bottom of the drill hole. A pipe-like weight of 140 lb is placed on the top of the rod. The weight is repeatedly raised and dropped a distance of 30 in. in order to drive the split-spoon sampler into the formation. The number of blows necessary to drive the sampler every 6 in. is recorded as the sampler is driven 18 in. into the formation. The more dense the formation, the greater the number of blows needed to drive it 6 in. The process is described by ASTM 1586 (American Society for Testing and Materials 1984).

After the split-spoon sampler is driven 18 in., or after refusal, it is brought to the surface and opened. Frequently, less than 18 in. of sediment have been collected. Fine, noncohesive sediment such as sand may fall out of the sampler as it is being retrieved. Sediment greater in diameter than one-third of the diameter of the sampler may not enter it at all. A pebble may lodge in the barrel and not allow any other sediment to enter. Sediments may compact in the sampler so that a full 18-in. sample may actually occupy less than 18 in. of the core barrel. The hydrogeologist examining the split-spoon sample must use his or her judgment in making a log based on the split-spoon samples. Figure 8.14 is a photograph of a split-spoon sample.

A standard for core samples is to collect one 18-in. sample every 5 ft. This frequency of sampling is suitable for relatively homogeneous formations. However, if the microstratigraphy of the formations is important—for example, if there are permeable sand seams in a clay formation—then continuous-core samples should be collected. Continuous cores are made by advancing the drill bit to the full depth that was sampled by a split spoon and then immediately taking another core sample of the fresh formation

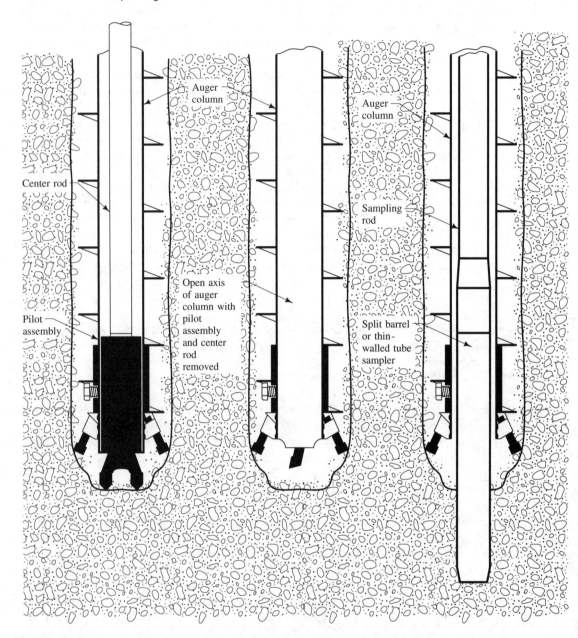

FIGURE 8.13 Sequential steps for the collection of a core sample through a hollow-stem auger. *Source:* Glen Hackett, *Ground Water Monitoring Review* 7, no. 4 (1987): 51–62. Used with permission. Copyright © 1987 Water Well Journal Publishing Co.

FIGURE 8.14 Hydrogeologist describing a split-spoon sample. *Photo credit:* C. W. Fetter.

ahead of the previous core. Continuous-core samples can also be collected in cohesive soils by using a special core barrel that collects a 5-ft-long sample inside a hollow-stem auger as the auger is being advanced. If a sequence consisting of a 3-in split spoon followed by a 2.5-in split spoon followed by a 2-in split spoon is used, 4.5 to 5 ft of continuous sample core can be collected before the augers need to be advanced.

In consolidated formation a rock-core sample is collected by use of a core barrel with a diamond-studded bit. The rotating bit grinds up rock in an annular pattern, leaving an undisturbed center of rock that enters the core barrel. There is a core lifter just behind the bit to keep the core from falling out of the core barrel when the drill rods and bit are retrieved from the borehole.

8.5 Installation of Monitoring Wells

Following the collection of samples during the installation of borings, a monitoring well can be installed in the borehole. Boreholes drilled with mud should stay open with the drill rod removed. Hollow-stem augers are generally left in the ground and the well is installed through them, as is drive casing in cable-tool drilling. There must be a sufficient

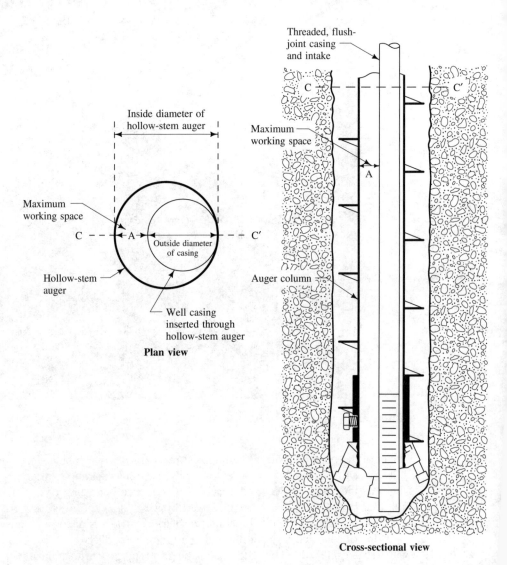

FIGURE 8.15 Casing offset inside hollow-stem auger to give greatest working opening. *Source:* Glen Hackett, *Ground Water Monitoring Review* 8, no. 1 (1988):60–68. Used with permission. Copyright © 1988 Water Well Journal Publishing Co.

Ground-Water and Soil Monitoring

working opening inside the casing or augers. For a 2-in. nominal monitoring well, this means a minimum $4\frac{1}{4}$-in. opening is needed. The casing can be offset within the auger to give the largest working opening (Figure 8.15). With a $4\frac{1}{4}$-in. inside diameter auger and a 2-in. nominal casing, this creates a 1.875-in. working opening.

The first step in the installation of a monitoring well is to screw the well screen to the casing and then lower the assembly through the inside of the augers or temporary casing. Prior to installation the casing and screen should be thoroughly decontaminated. The casing and screen may be wrapped in white butcher paper after it has been decontaminated and then kept wrapped until just before it is lowered into the augers. Figure 8.16 shows a 20-ft casing and screen being lowered by hand. Longer casings need to be lowered on a cable using the drilling rig.

FIGURE 8.16 Lowering well screen and casing into hollow-stem augers. Note the white wrapping paper around the decontaminated casing. *Photo credit:* C. W. Fetter.

Once the casing and screen have been lowered into the well, the filter-pack material needs to be placed. The volume of filter-pack material necessary to fill the annular space between the screen and casing and the borehole wall from the bottom of the borehole to a point 2 ft above the top of the screen should be computed. At least this much material must be on hand before starting the filter-pack installation. A weighted measuring tape is lowered into the working opening between the casing and the hollow-stem auger and the total depth of the borehole is measured and recorded.

If the formation is cohesive and can stand open for a short while, the augers are withdrawn 1 or 2 ft from the bottom. Filter-pack material is then poured into the working opening, and the annular space is filled to the levels of the auger bit. Care should be taken that the filter-pack sand doesn't fill the space between the casing and the augers, because it can lock the casing and hollow-stem augers together. The weighted tape is used to determine the position of the top of the filter pack. The augers are then withdrawn another 1 to 2 ft, and the process is repeated until the entire filter pack is placed. Figure 8.17 illustrates what is known as the free-fall method of filter-pack emplacement.

When filter-pack material drops through a water column, it may separate according to size. It may also bridge the space between the casing and the auger and create a void below. To avoid these problems, a **tremmie pipe** should be used wherever possible (Figure 8.18). A tremmie pipe is a pipe that extends from the surface and through which the filter-pack sand may be poured. After the augers are withdrawn a few feet, the annular space is filled with sand being poured down the tremmie pipe. The tremmie pipe is raised as the level of sand rises. The tremmie pipe can be used to tamp down the sand, and the weighted tape is used to measure the position of the top of the filter pack.

If the formation is noncohesive, it will collapse as the augers are withdrawn. Under these conditions, the withdrawal of the augers and the addition of the filter-pack material must occur simultaneously (Hackett 1988). A cable must be attached to the top of the auger string so that the working opening is accessible at all times. The hollow stem of the augers is filled with clean water, and a positive hydraulic head is thus maintained throughout the operation. The augers are very slowly withdrawn, and at the same time filter-pack material is added so that the top of the filter pack is within an inch or so of the bottom of the augers. This requires precise coordination of the rate of addition of filter-pack sand and the rate of withdrawal of the augers.

The final depth of the top of the filter pack is confirmed by measurement with the weighted tape. It should be 2 ft above the top of the screen. The augers are then withdrawn another 2 ft and 2 ft of pure bentonite clay are placed by free fall through the working space. If the area is below the water table, bentonite pellets are used; if the area is above the water table, granular bentonite is used. Pellets should be dropped a few at a time so that they aren't caught in the working space as they start to swell by hydration. The weighted tape is used to confirm the final thickness of the bentonite layer after enough time has elapsed to allow the bentonite to hydrate.

Placement of the annular seal should take place by use of a tremmie pipe. The tremmie pipe should be lowered to the top of the bentonite seal. The augers are withdrawn 2 ft or so and the annular space is filled from the bottom with grout, which is either pumped down the tremmie pipe or is fed by gravity (Figure 8.19). The weighted tape is used to confirm the position of the top of the grout. The augers are then repeatedly

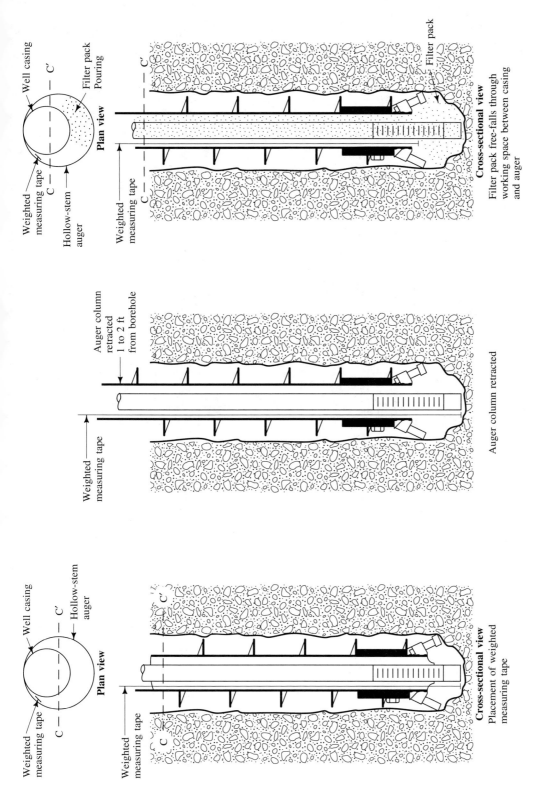

FIGURE 8.17 Free-fall method of filter-pack emplacement with a hollow-stem auger. The method also works with drive casing. Source: Glen Hackett, *Ground Water Monitoring Review* 8, no. 1 (1988):60–68. Used with permission. Copyright © 1988 Water Well Journal Publishing Co.

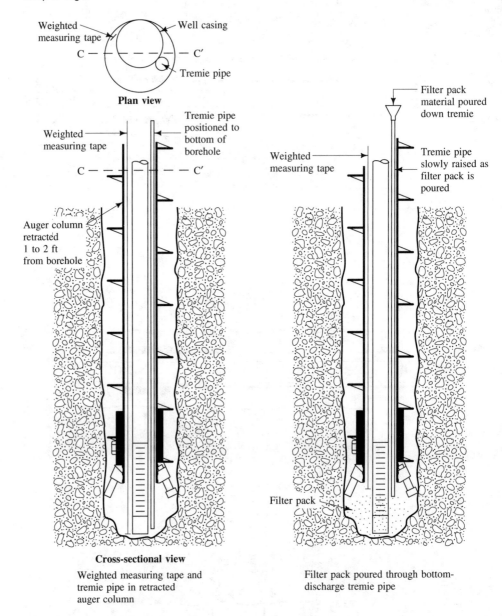

FIGURE 8.18 Use of a tremmie pipe for emplacement of filter-pack material. *Source:* Glen Hackett, *Ground Water Monitoring Review* 8, no. 1 (1988):60–68. Used with permission. Copyright © 1988 Water Well Journal Publishing Co.

Ground-Water and Soil Monitoring

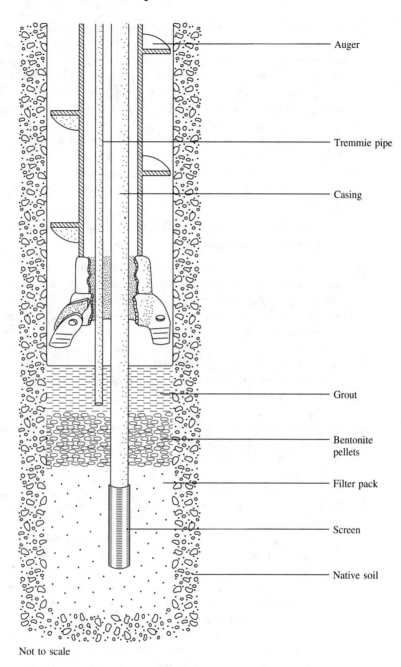

Not to scale

FIGURE 8.19 Use of a tremmie pipe for emplacement of grout above the bentonite pellets.

withdrawn and grout is emplaced until it is brought close to the surface. The tremmie pipe can be left at the bottom until the grout is brought to the surface, or it can be raised as the augers are withdrawn. For slumping sediments, the grout must be emplaced simultaneously and at the same rate as the augers are withdrawn.

The final step is the installation of a locking protective cap. Bentonite-cement can be brought all the way to the surface. If bentonite grout or bentonite-sand grout is used, the final few feet filling the annular space must be neat cement or bentonite-cement grout. The locking protective casing can be pushed into the cement grout when it is still soft. If the grout settles overnight, its level should be brought to the surface with additional material. A stronger surface seal can be obtained if the top 2 ft of the annular space are filled with concrete, as opposed to bentonite-cement grout.

8.6 Monitoring Well Development

Once a monitoring well is constructed, it is necessary that it undergo development. This is a process of removing fine sand, silt, and clay from the aquifer around the well screen. If drilling mud was used, then vigorous development may be needed to break down the mud pack on the borehole wall. Development is needed to create a well that ideally will not pump silt and clay when it is sampled. It also may create a zone around the well screen that is more permeable than the native soil and that stabilizes the native soil so that the fine sediments do not enter the filter pack.

Aller et al. (1989) made the following observations about monitoring well development:

1. Using compressed air for well development may alter native water chemistry, crack the casing or blow the bottom cap off the screen.
2. Adding water to the well for flushing the well or surging can alter the ground-water chemistry, at a minimum by dilution.
3. Breaking down a mud wall left in the borehole from mud-rotary drilling is very difficult.
4. Developing a well when the screen is in a clean, homogeneous, high-permeability aquifer is relatively easy.
5. Developing a well when the screen is in a fine-grained, stratified, low-permeability formation is difficult.
6. Developing a large-diameter well is easier than developing a small-diameter well.
7. Shallow monitoring wells are easier to develop than deeper monitoring wells.
8. Monitoring wells that can be bailed dry tend to be turbid because of the steep hydraulic gradients that are developed.
9. In the final analysis, many monitoring wells cannot be developed to the point where a nonturbid, ground-water sample can be collected. This is especially true if the formation does not yield very much water, so that extensive development is not possible.

If the borehole is drilled into a stable, consolidated formation, especially if mud is used during the drilling, it may be advantageous to flush the borehole with potable, fresh water to wash out as much of the mud as possible prior to installation of the well and filter pack. This will greatly cut the time needed for well development. In some cases it will not be permissible to add water to the borehole, since this might alter the ground-water chemistry.

There are three procedures used for monitoring well development: bailing, surge-block surging, and pumping/overpumping/backwashing. These may be used alone or in combination.

A **bailer** for a monitoring well is a section of pipe that is open on the top end and has a foot valve on the bottom end. It is attached to a line so that it can be lowered into the well. Water fills it from the bottom; then when it is raised by the line, the foot valve closes and the water inside is trapped. Figure 8.20 is a diagram of a bailer.

When developing a well by bailing, the bailer should be allowed to free-fall to the water surface. When it strikes the water, a pressure wave results, which pushes water from the well screen out into the formation. After the bailer is filled with water, it is withdrawn and water from the formation enters the screen. This back-and-forth motion of water through the filter pack loosens fine sediment so that it can be drawn into the well and removed by the bailer. The bailer should be allowed to sink to the screen area so that the water that fills it contains the fine sediment that is brought into the well from the filter pack. The bailer can also be raised and lowered when it is submerged to force water to move back and forth through the screen area. Bailing can take some hours to develop a monitoring well effectively; this can translate into significant labor cost. Bailing for development can be undertaken by hand or by using a cable attached to a power-operated drum on a drill rig or truck. Some well-development outfits have an arm that can go up and down like a walking beam to create a surging action. Care should be taken so that the surging action is not vigorous enough to collapse the well screen.

A **surge block** is a device that fits inside the well with a flexible gasket that is close in size to the inside diameter of the well. Figure 8.21 shows the design of a surge block for small-diameter monitoring wells. It is attached to a rod that is raised and lowered with a stroke of about 3 ft. Most of the water is moved up or down by the action of the surge block, although some fraction of the water bypasses the surge block. The surging is initiated with the surge block at the top of the well screen, and the block is gradually lowered until the entire screened area has been surged. Every so often the surge block is removed and the well bailed to remove the sediment that has been brought into the well. If too much sediment accumulates above the surge block, it can bind between the surge block and the casing wall and lock up the surge block. To avoid this, the surge block must be removed and the well bailed frequently. Surging begins with a gentle action; as development progresses it becomes more vigorous, with a more rapid plunging action. Again, too vigorous a surging action might cause the screen to collapse.

Pumping the well can aid in development. A number of different types of pumps are suitable. However, some pumps might be damaged by the sediment that must necessarily be removed during development. The ideal pump for development is capable of a wide range of flow velocities and doesn't have a valve that prevents backflow. The pump intake should be in the screen zone so that it will immediately pick up sediment that is brought into the well. The pump is started at a low velocity and is shut off every so often. When it is shut down, the water in the pump column between the water surface and the pump will flow back into the well and out into the filter pack. When the pump is started again, this water will be drawn into the well and will loosen fine sediment in

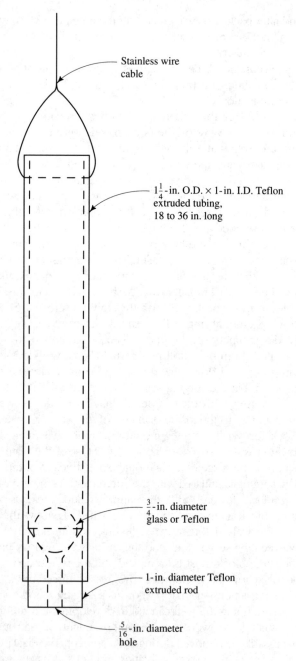

FIGURE 8.20 Diagram of a bottom-loading bailer.

Ground-Water and Soil Monitoring

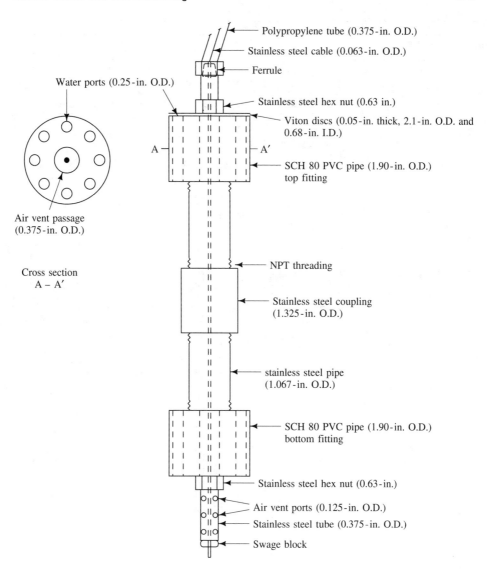

FIGURE 8.21 Design of a specialized surge block for monitoring wells. *Source:* Ronald Schalla and R. W. Landick, *Ground Water Monitoring Review* 6, no. 2 (1986):77–80. Used with permission. Copyright © 1986 Water Well Journal Publishing Co.

the filter pack. With time the rate at which the well is pumped is increased so that the water velocity through the filter pack into the screen is increased. Eventually, the well will be overpumped—that is, pumped at a rate that is greater than the flow into the well through the well screen. There will be a rapid decrease in the water level in the well during overpumping, and it can't be sustained for very long. Eventually the water from the well should clear. Monitoring wells should be periodically redeveloped.

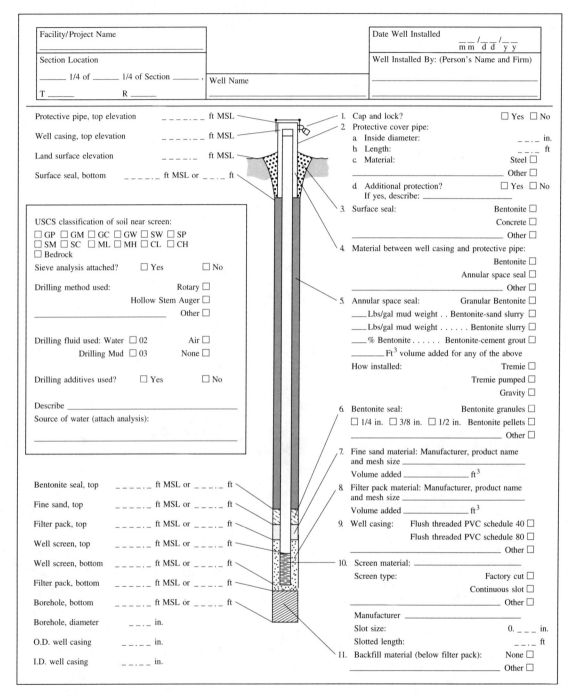

FIGURE 8.22 Form for recording information about construction details of a monitoring well. *Source:* Modified from Wisconsin Department of Natural Resources.

8.7 Record Keeping During Monitoring Well Construction

Many states have detailed record-keeping requirements for monitoring well construction. Even when detailed records aren't required by statute, they should be kept as a matter of sound professional practice. Figure 8.22 shows a one-page form that can be used to record important information on monitoring well construction. Many firms have their own forms, and many states have a required form. The form should be filled out in the field as the information is collected.

Records should also be kept of well development, including the date that it occurred, the method used, the water level at the start of development, the water level at the end of development, the time spent developing the well, and the volume of water removed. The thickness of sediment on the bottom of the well can be determined by measuring the depth to the bottom of the well with a weighted tape. This measured distance is subtracted from the measured length of well casing and screen that was installed. The difference between the two dimensions is the thickness of sediment inside the well. This thickness should be measured and recorded both before and after development. If possible, sediment should be removed with a bottom-loading bailer.

8.8 Monitoring Well and Borehole Abandonment

Sometimes difficulties are encountered during the construction of a monitoring well that prevent its completion. Monitoring wells may be installed for a specific time period, after which they must be removed. Test borings may be made with no intention of using the borehole for construction of a monitoring well. In all such cases, proper abandonment of the well should be undertaken. Many states have specific well-abandonment codes. In the absence of specific requirements, monitoring wells and boreholes should be abandoned in such a manner that surface water cannot drain into the aquifer. Otherwise a direct connection for contaminated water from the surface to the aquifer can result.

If a casing and screen have been installed, they should be removed if possible. This can be accomplished by pulling if the annular seal has not been filled with a cement-type grout. If a plastic casing breaks while being pulled, it can be removed by drilling it out with hollow-stem augers. Following the removal of the casing, a tremmie pipe should be used to fill the resulting borehole from the bottom with an appropriate material, such as neat cement or bentonite grout. The grout is placed while the augers are being pulled out of the hole. Material removed from a monitoring well may be contaminated and should be properly disposed of.

If the casing and screen have been grouted into place, it may not be possible to remove them. If this is the case, the casing should be cut off below grade; then the screen and casing must be filled from the bottom using a tremmie pipe and an appropriate material such as neat cement or bentonite grout.

Boreholes in sediments can be filled with grout or native soils mixed with bentonite. Boreholes into bedrock should be grouted with a cement-type grout.

8.9 Multiple-level Devices for Ground-Water Monitoring

Usually monitoring wells are installed in nested configurations, with a water table monitoring well and one or more piezometers screened below the water table. The best situation is to install each monitoring well or piezometer in an individual borehole. Under these circumstances it is possible to obtain an excellent seal to prevent vertical movement of water along the casing. These individual wells can be placed within 5 ft of each other.

Under some circumstances it may be possible to begin with a large-diameter borehole, place two or more monitoring wells or piezometers in the borehole, and provide a seal between them. However, the construction of such a well nest might be quite expensive. For example, it could be done in an unconsolidated formation by (1) driving a 10- or 12-in.-diameter casing with a cable tool rig to the depth of the bottom piezometer, (2) installing the deepest piezometer, (3) pulling back the casing and installing the filter pack and seal for the deepest piezometer, (4) pulling back the casing and grouting up to the level of the next piezometer, (5) allowing the grout to harden, (6) installing the next piezometer, (7) pulling back the casing and installing the filter pack and seal for the second piezometer, (8) pulling back the casing and grouting up to the level of the third piezometer, etc. Moreover, it is difficult to install a seal between piezometers in the same borehole that can positively prevent vertical movement of ground water along the casing. Such a design may make sense when multiple piezometers are to be put into a bedrock borehole hundreds of feet deep where the drilling costs are very high. For shallow wells, especially in sediment, the design usually is not cost-effective.

It is possible to install inexpensive multilevel sampling devices in a sandy aquifer. One such device is shown in Figure 8.23. It consists of a rigid PVC tube, inside of which are multiple tubes of flexible tubing. Each tube leads from the surface and ends at a different depth. Each has a port into which a ground-water sample can be drawn. Sampling ports can be spaced vertically at distances as close as 1 ft or less, so that very detailed vertical sampling can be accomplished. Water is withdrawn from the tubing by applying a suction, so the water table must be less than 25 ft below the surface.

Multilevel samplers also can be constructed by fastening a bundle of flexible tubes, each of a different length, to the outside of a rigid PVC pipe that acts as a spine.

The multilevel sampler is installed by using a hollow-stem auger. The device is constructed at the surface and lowered to the desired depth through the augers. It is not possible to develop the sampling ports, so this device can be installed only in clean, sandy sediment. No filter pack or grout is used in this construction. The augers are withdrawn and the native sand is allowed to slump around it. One disadvantage of this device is that it is usually not possible to measure water levels with it.

A second type of multilevel sampler can be used in bedrock boreholes. In this design (Figure 8.24) packers are located above and below each sampling port; when they are inflated, they seal off that part of the borehole. This is a permanently installed device, but it can be removed and reused if desired.

Barker et al. (1987) evaluated the bias in samples that can be introduced by the use of multilevel piezometers constructed out of flexible tubing. Leaching of plastics from the plastic tubing is one problem. Another problem is the sorbing of organics by the tubing. Both these problems can be minimized by using Teflon® tubing and thor-

Ground-Water and Soil Monitoring

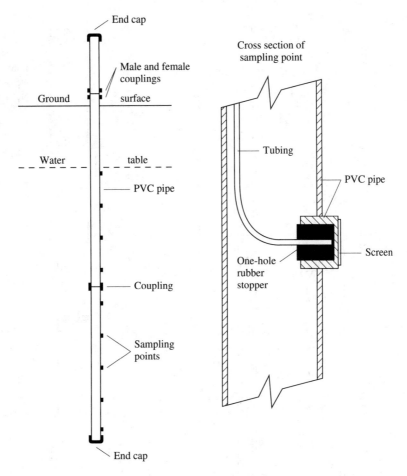

FIGURE 8.23 Multilevel ground-water sampling device for use in sandy soil. *Source:* J. F. Pickens et al. *Ground Water Monitoring Review* 1, no. 1 (1981):48–51. Used with permission. Copyright © 1981 Water Well Journal Publishing Co.

oughly purging the tube prior to sampling. The tendency of various plastics to leach or sorb organics has been studied. However, flexible tubing can transmit organics from ground water to the water in the tubing. This is especially true for polyethylene tubing. Samples drawn from below an organic plume may indicate contamination, when in fact the organics are being transmitted across the plastic tubing from the contaminated ground water. This can apparently occur even with Teflon® tubing. For this reason multilevel piezometers designed so that the flexible tubing is exposed to the ground water may not be appropriate for monitoring plumes of organic contaminants. If the flexible tubing is contained within a casing, such as in Figure 8.23, this should not be an issue because the tubing doesn't come into contact with the ground water.

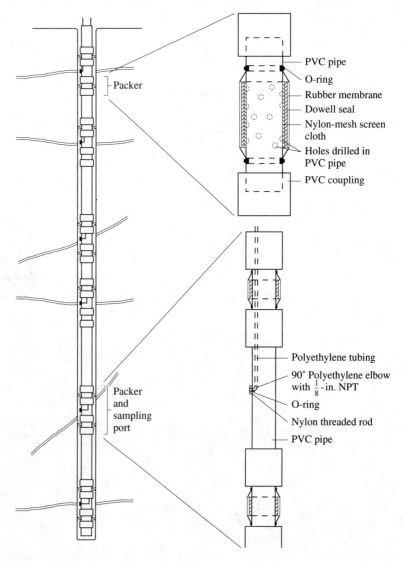

FIGURE 8.24 Multilevel ground-water sampling device for use in fractured rock aquifers. *Source:* J. A. Cherry and P. E. Johnson, *Ground Water Monitoring Review* 2, no. 3 (1982):41–44. Used with permission. Copyright © 1982 Water Well Journal Publishing Co.

8.10 Well Sampling

8.10.1 Introduction

After a monitoring well has been designed, installed, and developed, the next step is to collect a water sample. The water sample should be representative of the water in the formation; that is, the sampling techniques should collect water from the aquifer and not from water that has been standing in the well casing or screen. In addition, the sampling device should provide a sample that has not been altered by the sampling process and should not cause cross contamination.

8.10.2 Well Purging

Water that has been standing in the well has been in contact with atmospheric gases and the well casing and screen. This contact can affect the water chemistry. Oxygen can diffuse into the water and dissolved gases can volatilize or oxidize. Trace elements may be leached from the well casing. Organics may be sorbed by the well casing. In order to be sure that the water being drawn in to the sampling device comes from the aquifer, the well must be purged of standing water prior to sampling. The goal of purging is to remove all the water that has been standing in the well. The volume of water that must be removed to accomplish that goal depends upon the method of purging and formation permeability.

The first step in well purging is to measure the depth of water in the well, the total well depth, and the inside diameter of the well casing. These measurements are used to compute the volume of water standing in the casing. If a well is purged by a method that withdraws water from the top of the water column, then theoretically only one well volume needs to be withdrawn. For example, purging with a bailer that is lowered slowly into the well to a depth no greater than the length of the bailer will remove water only from the top of the column. If a pump is used to purge the well, the pump intake should be as close as possible to the top of the water column. Although one well volume would theoretically remove all the standing water, good practice suggests that at least three well volumes should be removed to be sure that the standing water in the casing and screen is totally removed. This also removes water from the filter-pack area. If a well is bailed dry, or nearly so, it is not necessary to attempt to remove multiple well volumes. As soon as the well has recovered enough to contain sufficient sample volume, the sample should be collected.

If the pump intake is lowered to the level of the screen in the well during purging, then most of the water will come from the screen area, and an area of stagnant water will develop in the water column above the pump intake. Under such conditions up to five well volumes need to be pumped to remove all of the stagnant water in the well (Gibb, Schuller, and Griffin 1981). Keeley and Boateng (1987) advocate a staged technique when purging with a pump. The pump intake is lowered to just below the water surface at the beginning of the purging process and then is gradually lowered through the water column until it is at the screen zone, when purging is complete. Purging three well volumes with this technique should be adequate.

Electrical conductivity and pH can be monitored during the well-development procedure. If they vary widely during the well-purging process, this may mean that water from different sources is being withdrawn. If these values don't stabilize, this doesn't necessarily mean that the stagnant water hasn't been withdrawn from the well. There may be instrument drift, or the water quality in the aquifer may be changing as water from different parts of the aquifer is being withdrawn. If possible, the well should be purged until it is not turbid.

The water being purged from the monitoring well may be contaminated. If so, it must be properly disposed of in a treatment facility. For this reason, purging techniques that limit the amount of water withdrawn are desirable.

8.10.3 Well-Sampling Devices

There are a large number of sampling devices available for monitoring wells. They operate under different physical principles and designs and have different applications. Most are available in a variety of materials. The following is a partial list of available devices (Nielsen and Yeates 1985; Pohlmann and Hess 1988):

1. *Open bailer:* This device is a rigid tube with an open top and either a closed bottom or a check valve on the bottom. It is attached to a line and is lowered and raised by hand. It withdraws the sample from the top of the water column.

2. *Point-source bailer:* This device has a check valve on both the top and bottom and can be lowered on a line to a given depth below the surface, where the valves can be closed by a cable. It can collect grab samples from any depth in the water column.

3. *Syringe sampler:* A medical syringe or similar device is attached to a length of tubing and is lowered to a selected depth in the water column. A suction is applied to the tubing and the syringe, which was lowered in the "empty" position. The syringe fills as the water comes into the needle because of the vacuum being developed by the suction on the tubing.

4. *Gear-drive pump:* This device is similar to a traditional submersible electrical pump. There is a miniature electrical motor attached to the pump, which rotates a set of gears to drive the sample up the discharge line via positive displacement. A continuous flow of water under positive pressure is developed.

5. *Bladder pump:* This sampler has a rigid tube containing an internal flexible bladder. There are check valves on either end of the rigid tube. When the bladder is deflated, water enters the lower end of the tube through the check valve. When the tube is full, the bladder is inflated with an inert gas pumped down from the surface. The bottom check valve closes, the top check valve opens, and the water sample flows up the discharge line. When the bladder deflates, the water in the discharge line can't drain back into the rigid tube because of the check valve. The water is under positive pressure at all times and doesn't come into contact with the gas.

6. *Helical-rotor pump:* This pump has a submersible electrical motor. It rotates a helical rotor-stator, which drives water up the discharge line under positive pressure.

7. *Gas-drive piston pump:* A piston that pumps the water is driven up and down by gas pressure from the surface. The gas does not contact the sample.

8. *Submersible centrifugal pump:* A submersible electrical motor drives an impeller in the pump, which creates a pressure and forces the water up a discharge line.

9. *Peristaltic pump:* Unlike the others, this is a pump located at the land surface. It is a self-priming vacuum pump that can draw a water sample up tubing under suction. Loss of volatile compounds and dissolved gases may occur due to the vacuum developed.

Ground-Water and Soil Monitoring

10. *Gas-lift pump:* A constant stream of gas or air is used to force the water up a discharge tube. The water comes into contact with the gas or air and oxidation and loss of volatiles can occur.

11. *Gas-driven pump:* An inert gas is used to alternately pressurize and depressurize a sample chamber. The sample chamber has check valves to create a one-way flow of water up a discharge line. This is similar to a bladder pump except that there is no bladder, so the gas comes into contact with the water sample.

Figure 8.25 is a matrix of sampling devices and the suitable applications for collection of ground water samples.

Pearsall and Eckhardt (1987), Yeskis et al. (1988), and Tai, Turner, and Garcia (1991) evaluated the efficacy of a number of different sampling devices in collecting representative samples of volatile organic compounds in ground water. The first two studies were performed by withdrawing water from ground-water–monitoring wells with different devices and comparing the results. The authors had no controls—that is, they did not know the actual concentration of the contaminants in the aquifer.

Yeskis et al. (1988) tested a bailer, a bladder pump, a piston pump, a submersible electric impeller pump, and a submersible electric helical stator-rotor pump. They found very little variation in the results except for the bailer. The bailer samples were in general less than half of the values obtained by the other devices. Pearsall and Eckhardt (1987) evaluated two 2-in. submersible helical stator-rotor pumps made of different materials, a 4-in. submersible electrical water-well pump, a cast-iron centrifugal suction pump, a peristaltic pump, and a bailer. The submersible pumps all gave similar results; the peristaltic and centrifugal suction pumps gave somewhat lower results. Silicon and PVC tubing appeared to give lower results than Teflon® tubing when used with the same pump. The bailer gave variable results, depending upon the concentration.

Tai, Turner, and Garcia (1991) built a 100-ft-tall, 5-in.-diameter, stainless-steel, vertical standpipe with sampling ports at various depths. They were able to use a sampling device to collect a sample from the standpipe, at the same time drawing a control sample from that depth through a sample port. They tested a Teflon® point-source bailer, a manual-driven piston pump, a motor-driven piston pump, a submersible helical-rotor pump, a peristaltic pump, and a bladder pump. Teflon® tubing was used to convey samples to the surface from the pumps. Because they had control samples, they were able to calculate a percent recovery for each sampling device. Table 8.2 shows the percent recovery for each sampling device at different depths. Examination of this table shows that submersible pump, the peristaltic pump, and the bladder pump all had excellent recoveries, ranging from 98.5 to 100.5%. The other three devices were not as accurate, with the bailer having the lowest recovery.

The low recovery using the bailer was probably due to agitation of the sample as it was being transferred from the bailer to a 40-mL sample bottle for volatile organic analysis. Although it is a simple piece of equipment, an experienced operator is needed for correct use of a bailer.

On the basis of available research, the piston pump, the bladder pump, the electrical submersible helical-rotor pump, and the peristaltic pump appear to be the best devices for collection of water samples for volatile organic samples. Although not tested, they

Portable Sampling Devices					Ground Water Parameters													
					Inorganic							Organic				Radioactive		Biol.
	Device	Approximate Maximum Sample Depth	Minimum Well Diameter	Sample Delivery Rate or Volume	EC	pH	Redox	Major Ions	Trace metals	Nitrate, Fluoride	Dissolved Gases	Non-volatile	Volatile	TOC	TOX	Radium	Gross Alpha & Beta	Coliform Bacteria
Grab	Open bailer	no limit	$\frac{1}{2}$ in.	Variable	•			•	•	•		•				•		•
Grab	Point-source bailer	no limit	$\frac{1}{2}$ in.	Variable	•	•	•	•	•	•		•	•	•	•	•		•
Grab	Syringe sampler	no limit	$1\frac{1}{2}$ in.	0.01–0.2 gal	•	•	•	•	•	•		•				•	•	•
Positive displacement (submersible)	Gear-drive	200 ft	2 in.	0–0.5 g/min									•					
Positive displacement (submersible)	Bladder pump	400 ft	$1\frac{1}{2}$ in.	0–2 g/min	•	•	•	•	•	•	•	•	•	•	•	•	•	•
Positive displacement (submersible)	Helical rotor	160 ft	2 in.	0–1.2 g/min	•	•	•	•	•	•	•	•	•	•	•	•	•	
Positive displacement (submersible)	Piston pump (gas-drive)	500 ft	$1\frac{1}{2}$ in.	0–0.5 g/min	•			•	•	•		•				•	•	
Positive displacement (submersible)	Centrifugal	variable	3 in.	variable	•			•		•						•	•	
Suction lift	Peristaltic	26 ft	$\frac{1}{2}$ in.	0.01–0.3 g/min	•			•		•		•				•		•
Gas contact	Gas-lift	variable	1 in.	Variable														
Gas contact	Gas-drive	150 ft	1 in.	0.2 g/min	•			•		•		•				•		

FIGURE 8.25 Matrix showing applications of a number of ground-water sampling devices. *Source:* K. F. Pohlmann and J. W. Hess, *Ground Water Monitoring Review* 8, no. 4 (1988): 82–84. Used with permission. Copyright © 1988 Water Well Journal Publishing Co.

TABLE 8.2 Percent recovery of sampling devices compared with control samples.

	\multicolumn{2}{c}{17.5 ft}	\multicolumn{2}{c}{54 ft}	\multicolumn{2}{c}{92 ft}
Sampler	Low[a] Conc.	High[b] Conc.	Low[a] Conc.	High[b] Conc.	Low[a] Conc.	High[b] Conc.
Submersible pump	97.8	99.7	99.5	98.5	100.1	100.1
Peristaltic pump	97.7	98.7	98.7	98.5	97.7	100.0
Bladder pump		101.5		100.5		99.9
Teflon bailer	93.9	96.3	92.4	93.9	92.7	91.4
Manual-driven piston pump	102.2	—	102.5	101.1	102.3	103.5
Motor-driven piston pump	97.4	96.8	99.8	98.4	100.9	100.7

[a] 10 to 30 μg/L of methylene chloride, 1,1-dichlorethene, *trans*-1,2-dichloroethene, 1,2-dichloroethane, and 1,1,1-trichloroethane.
[b] 100 to 200 μg/L of methylene chloride, 1,1-dichlorethene, *trans*-1,2-dichloroethene, 1,2-dichloroethane, and 1,1,1-trichloroethane.
Source: D. Y. Tai, K. S. Turner, and L. A. Garcia. The use of a standpipe to evaluate ground water samplers. *Ground Water Monitoring Review* 11, no. 1:125–132. Used with permission. Copyright © 1991 Water Well Journal Publishing Company.

would most likely also be excellent for semivolatile organics and inorganics. Other devices might also prove to be acceptable if subjected to rigorous testing. The bailer is also acceptable for collection of samples if used properly to give a minimum of agitation to the sample. The slightly lower recovery for the bailer is still in excess of 90% of the control. This is well within the precision of many analytical methods.

Wherever possible, dedicated sampling equipment should be used in each well. This prevents the possibility of cross contamination between wells. It also eliminates the cost of decontaminating the equipment between wells and the possibility of interference from solvents used for decontamination. Inexpensive disposable bailers are available and may be an acceptable sampling device for many studies where the cost of dedicated sampling equipment in each well is not justified.

8.11 Soil-Gas Monitoring

8.11.1 Introduction

Volatile organic compounds can partition into a vapor phase in the vadose zone. The volatile organic compound may be pure product adhering to a mineral surface or forming a nonaqueous phase layer on top of the capillary zone. It may also be dissolved in soil water or ground water.

Soil-gas monitoring can be valuable for several reasons. Pure product present in the vadose zone is a reservoir for ongoing contamination of ground water. Infiltrating precipitation can dissolve the product and carry it down to the water table. Restoration of a site may involve soil remediation if organic compounds are detected in the vadose zone.

Soil-gas monitoring can also be used to determine the extent of a layer of a volatile, nonaqueous phase liquid, such as gasoline, floating on the water table. Vapors from the gasoline can partition into the vadose zone, so that if they are detected by soil-gas monitoring in an area where product spills or leaks are not likely, this is an indication that the product may be migrating into the area.

Soil-gas monitoring has also been used as a screening method to evaluate the extent of a plume of ground water contaminated with volatile organic compounds. The soil gas above the plume may contain volatile organics if the plume is at the water table.

Under certain conditions soil-gas monitoring can be done much more quickly and inexpensively than installing ground-water monitoring wells.

8.11.2 Methods of Soil-Gas Monitoring

There are at least three ways of monitoring the volatile organic compounds in the vadose zone: (1) direct sampling of soil, (2) installation of a soil probe to withdraw soil gas for analysis, and (3) installation of passive soil-gas collectors.

Test borings can be installed and soil samples can be collected directly with a split-spoon sampler. The split spoon is opened, and samples of soil can be collected with a small-diameter, thin-walled tube, such as a cork borer. These samples are placed in a 40-mL VOA vial. Volatile organics can be measured in the headspace in the vial. The rate at which volatiles are lost during the sampling process is unknown, but the measured concentration will be less than the actual value.

In sand and other types of loose soil, a steel soil-probe tube can be hydraulically pushed into the vadose zone for dynamic soil-gas monitoring. A vacuum is applied to the tube and soil gas is drawn up it. The soil gas can be trapped in a container for later analysis, or it can be analyzed directly in the field by gas chromatography (Robbins et al. 1990a, 1990b). Soil gas from a probe can also be sorbed onto an activated carbon trap, which is then taken back to a laboratory for analysis (Wallingford, DiGiano, and Miller 1988).

Passive, soil-gas monitoring uses an *in situ* absorbent, which is placed in the vadose zone for a period of days to weeks. A shallow soil boring is made in the vadose zone and an activated-carbon, organic-carbon, vapor monitor is suspended in the boring. The top of the boring is sealed and the monitor left in place for days to weeks. The boring is then unsealed; the vapor monitor is removed and sealed in a container for shipping to the lab, where it is analyzed (Kerfoot and Meyer 1986).

Soil-gas monitoring to find ground-water contamination plumes works best in areas where the vadose zone is comprised of dry, coarse-grained soils. The depth to water cannot be too deep, but at least 15 ft is preferable. If the water table is too shallow, the concentration gradients are very steep, and a slight difference in depth of measurement may give a great difference in measured values (Marrin 1988). It is not possible to find an exact correlation between the soil-gas concentration and contaminant concentration in the underlying ground water. At best an order-of-magnitude correlation is possible (Thompson and Marrin 1987). Soil-gas measurements are affected not only by the soil gas but also by the sampling technique and the soil-air permeability. If there are several organic compounds in the ground water, they will partition into the vadose zone according to their individual Henry's law constants. The organic compound with the greatest concentration in the soil gas may be the compound with the greatest Henry's law constant and not the one with the greatest concentration in the ground water. Although care must be taken when interpreting the results of a soil-gas survey, these surveys are valuable screening techniques and have significant qualitative value.

8.12 Soil-Water Sampling

8.12.1 Introduction

Contamination moving from the surface toward the water table passes through the vadose zone. Monitoring of soil-water quality in the vadose zone beneath hazardous-waste land-treatment systems is required under Subtitle C of the Resource Conservation and Recovery Act. States may also require vadose zone monitoring beneath other types of hazardous-waste facilities.

In order to determine the chemical composition and quality of soil moisture in the vadose zone, a sample must be collected. Because the soil water in the vadose zone is under tension, it cannot flow into a well under gravity the way that ground water flows into a well. Soil water must be collected with a suction lysimeter (Wilson 1990).

8.12.2 Suction Lysimeters

A **suction lysimeter** is a porous cup located on the end of a hollow tube. The tube can be PVC or even stainless steel. The porous cup can be ceramic, nylon, PFTE, or fritted stainless steel. Tubing connects the suction lysimeter with the surface.

A suction is applied to the hollow tube and held for a period of time. If the suction is greater (more negative) than the soil-moisture tension in the soil, a potential gradient will develop from the soil to the porous cup. Soil water will flow into the porous cup, from which it can then be directed through the tubing to the surface for collection. The flow of soil moisture can be slow, and it may be necessary to hold the vacuum overnight to supply a sufficient volume.

A vacuum lysimeter simply has a porous tip on the end of a hollow tube with a stopper that extends to the surface. The vacuum is applied to the lysimeter by means of a hand-vacuum pump attached to a small tube that extends down to the porous tip, and a sample is drawn to the surface when vacuum is applied (Figure 8.26). The practical depth of this type of sampler is about 6 ft due to the awkwardness and cost of installing long tubes (Wilson 1990).

A pressure-vacuum lysimeter has a hollow tube that is about 2 in. in diameter and 1 ft long. Two tubes run from the lysimeter to the surface (Figure 8.27). One of the tubes, the discharge line, extends to the bottom of the lysimeter and the other tube, the pressure-vacuum line, ends near the top. A vacuum is applied to the pressure-vacuum line with a vacuum pump while the discharge line is shut off with a pinch clamp. The pressure-vacuum line is then sealed with a pinch clamp and the lysimeter is allowed to sit overnight so that the sample can be drawn into the cup. The pinch clamps are then removed. A hand-pressure pump is then attached to the pressure-vacuum line, and when pressure is applied, the water is forced up the discharge line to the surface. (A single pressure-vacuum hand pump can be used for this operation.) The maximum practical operational depth for the pressure vacuum pump is about 50 ft. At depths greater than

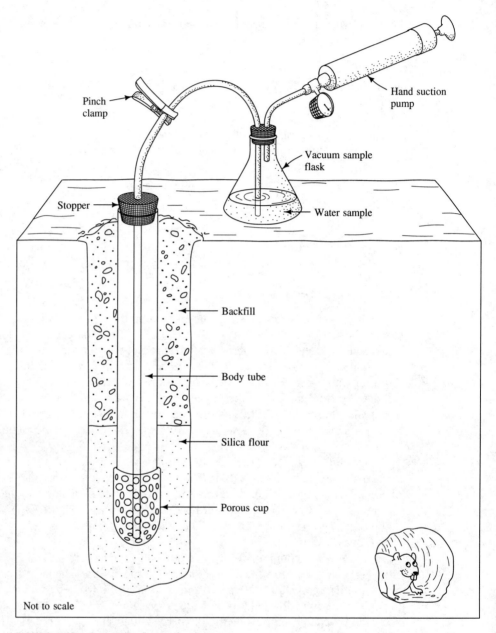

FIGURE 8.26 Operation of a vacuum lysimeter.

Ground-Water and Soil Monitoring

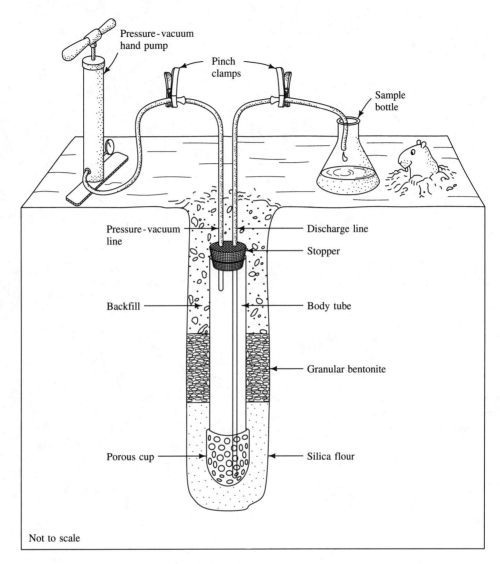

FIGURE 8.27 Operation of a pressure-vacuum lysimeter.

this, the pressure needed to force the sample to the surface tends to drive it back out of the porous cup. This can be avoided by using a lysimeter with a check valve and an internal reservoir (Figure 8.28). When the vacuum is applied the water is drawn into the reservoir. When the pressure is then applied, the check valve prevents backflow into the porous cup and the sample must go to the surface.

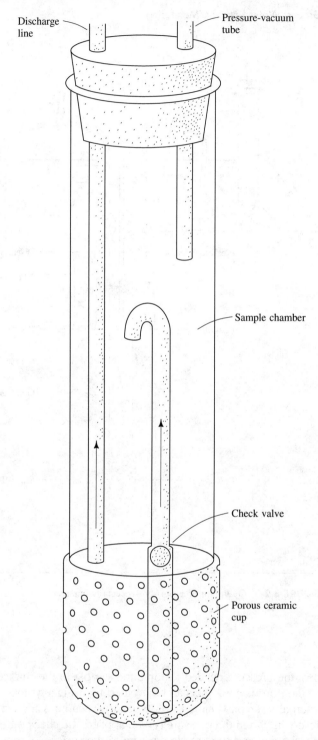

FIGURE 8.28 Design of a pressure-vacuum lysimeter with a sample chamber and a check valve.

8.12.3 Installation of Suction Lysimeters

In order for suction lysimeters to work properly, they must be carefully installed (Wilson 1991). Prior to use, new porous ceramic cups should be leached with a 10% hydrochloric acid solution for 24 hr to remove dust from manufacturing. They should then be thoroughly rinsed by passing distilled water through them. The lysimeter is then pressure tested before being installed by submerging it in water and applying a pressure of 30 lb/in.^{2} to test for leaks. No leaks should appear at any fittings, and air should bubble evenly through the porous cup.

The suction lysimeter is installed in an augered hole that has a greater diameter than the hollow tube. If the suction lysimeter is placed at a shallow depth, less than about 10 ft, then the access hole can be made by a hand auger. If it is greater than that depth, then a hollow-stem auger-drilling rig is needed to auger a hole.

The porous tip must be well bedded in slurry made of 200-mesh silica flour or native soil material that has been dried and screened to remove the fraction larger than coarse sand. There needs to be good hydraulic contact with the porous cup, the bedding material, and the native soil. Prior to installation the porous tip should be well hydrated by soaking in distilled water.

If a hollow-stem auger is being used, the augers should first be pulled back about 2 ft to expose the native soil. A slurry made from silica flour and distilled water (1 lb of 200-mesh silica flour and 150 mL of distilled water) is put into the hole with a tremmie pipe to fill the hole with about 6 in. of slurry. The suction lysimeter is then lowered into the hole and centered. The tremmie pipe is then used carefully to place the slurry around the lysimeter up to an elevation of about 1 ft above the top of the lysimeter. The unit should be held in place until the water drains from the slurry. The unit is then tested to see if it will hold a vacuum of 0.6 bars. If it does, the augers are pulled back another 3 ft; then about 1 ft of sieved native soil and 2 ft of bentonite granules are added. The tremmie pipe is used to add distilled water to the bentonite to hydrate it. Finally, the augers are pulled and the borehole is backfilled with native soils that are tamped down to ensure compaction. Since a lot of distilled water was added to the soil during installation, the lysimeter needs to be purged until consistent water quality is obtained.

8.13 Summary

In order to collect data for ground-water contamination studies, it is usually necessary to install monitoring wells. Such wells can permit one to measure water levels, collect water samples for analysis, and detect the presence of nonaqueous phase liquids.

Monitoring well design considers the type and diameter of casing and screen, the material from which the casing and screen is constructed, and whether the well will have an artificial-filter pack or be naturally developed. The annular space between the casing and the borehole must be properly sealed, and the well needs a protective covering. Monitoring wells may be installed by hollow-stem auger drilling, mud-rotary drilling, air-rotary drilling, reverse-rotary drilling, and cable-tool drilling. The equipment used for monitoring well installation and the material used for the well must be properly decontaminated. Once installed, the well needs to be developed to remove fine material from the area outside the well screen. Water table wells are installed to monitor the water table and piezometers are used to monitor formations below the water table. Frequently, a water table well and several piezometers are installed at the same location to form a nest of wells to sample the formations at different depths. In sandy aquifers multilevel monitoring devices may be used to collect samples from very closely spaced vertical intervals.

There are a number of sampling devices that have been developed to withdraw water samples from monitoring wells. These include bailers, bladder pumps, stator-rotor pumps, piston pumps, and peristaltic pumps. The ability of these devices to collect an unbiased sample varies with the design, technique used, and material from which they are constructed. Prior to sampling a well, it is necessary to purge it to remove stagnant water from the casing.

Soil-gas vapor can be sampled through active probes that withdraw soil-gas samples for analysis or passive devices that sorb soil gas onto activated charcoal devices. Monitoring soil gas can indicate areas where the soil has been contaminated and requires remediation. It can also be used to delineate areas where shallow ground water is contaminated by volatile organic compounds and where there is a layer of a volatile, nonaqueous phase liquid floating on the top of the capillary zone.

Soil water must be collected through the use of a suction lysimeter. These are installed in boreholes above the water table.

References

Aller, Linda, T. W. Bennett, Glen Hackett, R. J. Petty, J. H. Lehr, Helen Sedoris, D. M. Nielsen, and J. E. Denne. 1989. *Handbook of Suggested Practices for the Design and Installation of Ground-Water Monitoring Wells*. Dublin, Ohio: National Water Well Association, 398 pp.

American Society for Testing and Materials. 1983. Standard practice for thin-wall tube sampling of soils: D1587. 1986 Annual Book of American Society for Testing and Materials Standards. Philadelphia, Penn. pp. 305–7.

American Society for Testing and Materials. 1984. Standard method for penetration test and split barrel sampling of soils: D1586. 1986 Annual Book of American Society for Testing and Materials Standards. Philadelphia, Penn. pp. 298–303.

Barcelona, M. J., J. P. Gibb, and R. A. Miller. 1983. *A Guide to the Selection of Materials of Monitoring Well Construction and Ground-Water Sampling*. Illinois State Water Survey, Contract Report 327, 78 pp.

Barker, J. F., G. C. Patrick, L. Lemon, and G. M. Travis. 1987. Some biases in sampling multilevel piezometers for volatile organics. *Ground Water Monitoring Review* 7, no. 2:48–54.

Driscoll, F. G. 1986. *Ground Water and Wells*, St. Paul, Minn.: Johnson Division, 1089 pp.

Gibb, J. P., R. M. Schuller, and R. A. Griffin. 1981. *Procedures for the Collection of Representative Water Quality Data from Monitoring Wells*. Illinois State Geological Survey and Illinois State Water Survey Cooperative Report 7, 61 pp.

Gilbert, R. O., and R. R. Kinnison. 1981. Statistical methods of estimating the mean and variance from radionuclide data sets containing negative, unreported or less than values. *Health Physics* 40:377–90.

Hackett, Glen. 1987. Drilling and constructing monitoring wells with hollow-stem augers—Part I: Drilling considerations. *Ground Water Monitoring Review* 7, no. 4:51–62.

———. 1988. Drilling and constructing monitoring wells with hollow-stem augers—Part II: Monitoring well installation. *Ground Water Monitoring Review* 8, no. 1:60–68.

Keeley, J. F., and Kwasi Boateng. 1987. Monitoring well installation, purging and sampling techniques—Part I: Conceptualizations. *Ground Water* 23, no. 3:300–13.

Kerfoot, H. B., and C. L. Meyer. 1986. The use of industrial hygiene samplers for soil-gas surveying. *Ground Water Monitoring Review* 6, no. 4:88–93.

Marrin, D. L. 1988. Soil-gas sampling and misinterpretation. *Ground Water Monitoring Review* 8, no. 2:51–54.

Mickam, J. T., Robert Bellandi, and E. C. Tifft, Jr. 1989. Equipment decontamination procedures for ground water and vadose zone monitoring programs: Status and prospects. *Ground Water Monitoring Review* 9, no. 2:100–21.

National Sanitation Foundation. 1988. National Sanitation Foundation Standard 14. Ann Arbor, Mich., 65 pp.

Nielsen, D. M. 1988. Much ado about nothing: The monitoring well construction materials controversy. *Ground Water Monitoring Review* 8, no. 1:4–5.

Nielsen, D. M., and G. L. Yeates. 1985. A comparison of sampling mechanisms available for small-diameter ground-water monitoring wells. *Ground Water Monitoring Review* 5, no. 2:83–99.

Parker, L. V., A. D. Hewitt, and T. F. Jenkins. 1990. Influence of casing materials on trace-level chemicals in ground water. *Ground Water Monitoring Review* 10, no. 2:146–156.

Pearsall, Kenneth, and D. A. V. Eckhart. 1987. Effects of selected sampling equipment and procedures on the concentrations of trichloroethylene and related compounds in ground water samples. *Ground Water Monitoring Review* 7, no. 2:64–73.

Pohlmann, K. F., and J. W. Hess. 1988. Generalized ground-water sampling-device matrix. *Ground Water Monitoring Review* 8, no. 4:82–84.

Reynolds, G. W. and R. W. Gillham. 1985. Absorption of halogenated organic compounds by polymer materials commonly used in ground-water monitors. In *Proceedings of the Second Canadian/American Conference on Hydrogeology*, 125–32. Banff, Alberta, Canada, June. 1985. National Water Well Association.

Robbins, G. A., B. G. Deyo, M. R. Temple, J. D. Stuart, and M. J. Lacy. 1990. Soil-gas surveying for subsurface gasoline contamination using total organic vapor detection instruments—Part 1: Theory and laboratory experimentation. *Ground Water Monitoring Review* 10, no. 3:122–31.

———. 1990. Soil-gas surveying for subsurface gasoline contamination using total organic vapor detection instruments—Part 2: Field experimentation. *Ground Water Monitoring Review* 10, no. 4:110–17.

Schalla, Ronald, and R. W. Landick. 1986. A new valved and air-vented surge plunger for developing small-diameter monitor wells. *Ground Water Monitoring Review* 6, no. 2:77–80.

Tai, D. Y., K. S. Turner, and L. A. Garcia. 1991. The use of a standpipe to evaluate ground water samplers. *Ground Water Monitoring Review* 11, no. 1:125–32.

Thompson, G., and D. Marrin. 1987. Soil gas contaminant investigations: A dynamic approach. *Ground Water Monitoring Review* 7, no. 3:88–93.

Wallingford, E. D., F. A. DiGiano, and C. T. Miller. 1988. Evaluation of a carbon absorption method for sampling gasoline vapors in the subsurface. *Ground Water Monitoring Review* 8, no. 4:85–92.

Wilson, L. G. 1990. Methods for sampling fluids in the vadose zone. In *Ground Water and Vadose Zone Monitoring*, ASTM STP 1053. Edited by D. M. Nielsen and A. I. Johnson, 7–24. Philadelphia: American Society for Testing and Materials.

———. 1991. Characterization and Modeling of the Vadose Zone. Lecture Notes, University of Arizona.

Yeskis, D., K. Chiu, S. Meyers, J. Weiss, and T. Bloom. 1988. A field study of various sampling devices and their effect on volatile organic compounds. *Proceedings, Second National Outdoor Action Conference on Aquifer Restoration, Ground Water Monitoring and Geophysical Methods*, 471–79. Dublin, Ohio: National Water Well Association.

Chapter Nine
Site Remediation

9.1 Introduction

During the past decade ground-water scientists and engineers have developed a number of techniques for both containing and remediating soil and ground-water contamination. In the United States this effort has been driven by increased environmental awareness as well as newly enacted legislation such as the Resource Conservation and Recovery Act (RCRA) and the Comprehensive Environmental Response, Compensation, and Liability Act (CERCLA). This is still an area of intense research and there is likely to be much additional progress in frontier fields such as genetic engineering of microbes to bioremediate specific organic compounds.

In general, the remediation of a site must address two issues. If there is an ongoing source of contamination, source control will be necessary to prevent the continuing release of contaminants to the subsurface or from the subsurface to ground water. Examples of sources include leaking landfills, spills of soluble substances that have been sorbed in the vadose zone, and mobile nonaqueous phase liquids (NAPLs) floating on the top of the capillary zone. The second part of remediation is the treatment of contaminated ground water and/or soil to remove or greatly reduce the concentrations of the contaminants.

The goals of a soil and ground-water remediation effort can range over a continuum from isolating the waste and halting any further spread of contamination to completely treating the soil and ground water to remove all traces of contamination. The latter goal is generally neither technically nor economically feasible. However, the means are available to remediate contaminated soil and ground water at many sites to a level where the contamination has been reduced by as much as 90 to 99%.

9.2 Source-Control Measures

9.2.1 Solid Waste

Solid waste may have been buried in an unsecure landfill, placed in an open excavation, or simply spread on the land surface. If contaminants continue to be leached from these sources by infiltrating precipitation or even ground water if the solid wastes have been buried below the water table, then a source-control measure is necessary before ground-water remediation is attempted. If the source continues to leach contaminants, then ground-water remediation may be futile.

Source-control measures include physical removal of the waste and transportation to a secure landfill or incinerator, construction of impermeable covers or low-permeability caps to eliminate or minimize infiltration of precipitation, and construction of physical barriers around the waste source.

9.2.2 Removal and Disposal

If the source is removed, then wastes can no longer migrate from it. Solid waste that has been spread on the land surface can easily be removed by conventional earth-moving equipment. Waste that has been buried in a landfill can also be exhumed and transported to secure landfill. A hazardous-waste landfill was operated near Wilsonville, Illinois, from 1976 to 1981. Hazardous wastes, including liquids, were buried in drums that were placed in 26 trenches, each approximately 10 to 20 ft deep, 50 to 100 ft wide, and 175 to 400 ft long. In 1981 it was found that hazardous wastes had migrated up to 50 ft from the trenches over a 3-yr period, a rate 100 to 1000 times greater than predicted prior to construction of the landfill. Following a court order, the site owner exhumed and removed all the drums and transported them to a more secure landfill. The process took 4 yr and many millions of dollars (Herzog, et al. 1989).

Excavation and removal of hazardous materials must be done in a manner that protects the health and safety of workers and the public. The materials may be hazardous if one comes into contact with them, give off toxic or dangerous vapors, or be harmful if ingested. The risk of moving material as opposed to leaving it in place must always be evaluated prior to a removal action.

Furthermore, the final disposition must be environmentally sound. In at least one case, spent solvents were moved from an abandoned hazardous-waste site to a solvent-recycling facility. The latter facility eventually went bankrupt and became a hazardous waste site itself. Under CERCLA, the courts have found that the generator of hazardous wastes is responsible for their cleanup and disposal costs if the disposal-site operator becomes bankrupt. In this case, the generator had to pay twice for the disposal of wastes.

Soil contaminated with organic compounds may be remediated by excavation and incineration. This is a technique that has been proposed for highly refractory organic compounds such as PCBs and some pesticides. The organic matter of the soil is incinerated while a supplemental fuel is burned in the incinerator. One concern about this method has been the fate of inorganic elements in the soil (Wall and Richards 1990). It is also very expensive. Compounds with a high Btu (heat) value, such as hydrocarbons, can also be incinerated. This is less costly, since supplemental fuel is not needed. However, permits are needed for incineration, and these may be difficult to obtain from state and local authorities. Incineration is usually a very expensive option.

9.2.3 Containment

If the waste can't economically or technically be excavated, then it may be possible to contain it. If the waste is below the water table, flowing ground water can pass through it and create leachate as illustrated in Figure 9.1(a). Such a landfill can be surrounded by a **ground-water cutoff wall.** The purpose of the ground-water cutoff wall is to

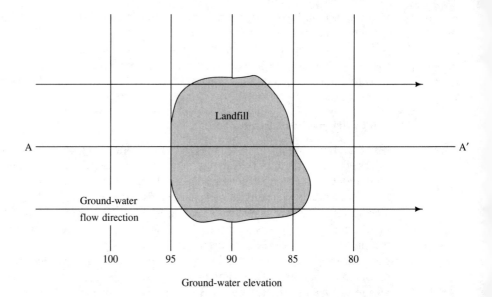

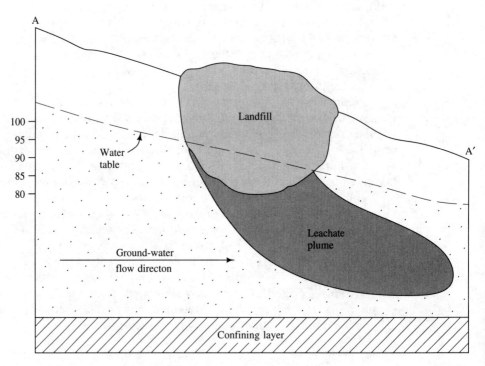

FIGURE 9.1(a) Top view and cross section of a landfill that was constructed with an excavation that extends below the water table. Ground water can flow through the waste and create leachate.

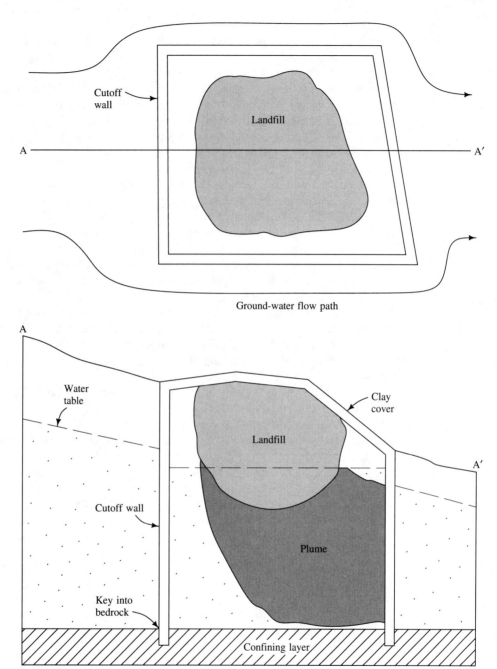

FIGURE 9.1(b) Top and side view of a cutoff wall that completely surrounds the landfill in (a), opposite page. Cutoff wall is keyed into underlying low permeability formation.

divert ground-water flow from passing through the waste so that it can't form leachate (Lynch et al. 1984; Need and Costello 1984). The cutoff wall needs to be deep enough to key into an impermeable layer so that ground water can't pass under the wall.

There are several ways that cutoff walls can be used. The wall can extend all around the waste (Figure 9.1(b)). If this is done, the ground water will flow around the wall and be diverted from the waste. The water table will rise on the up-gradient wall and fall on the down-gradient wall. If there is no recharge or flow through the cutoff walls, the water table within the cutoff walls will be flat. However, there is generally leakage through the cover or cutoff walls, so some extraction wells will be needed within the cutoff walls to prevent build up of water within walls.

If the cutoff walls are extended far enough to surround both the waste body and the plume of contamination, then remediation may proceed without worry that it will spread further. One of the problems that has arisen with the remediation of ground water is that the plume may spread rapidly, whereas legal action to assess blame for the plume proceeds through the courts with glacial speed. If a cutoff wall is installed as an emergency action, then the plume can wait until the courts have spoken.

Cutoff walls have also been used to stop the spread of a contaminant plume. At the Rocky Mountain Arsenal near Denver, Colorado, a cutoff wall was constructed across a bedrock valley containing higher-permeability, unconsolidated deposits that acted as a pathway for a contaminant plume. The contaminated water is pumped from the up-gradient side of the cutoff wall, treated to remove the contamination, and injected into the aquifer on the down-gradient side of the cutoff wall (Shukle 1982; Konikow and Thompson 1984). This system is illustrated in Figure 9.2.

A number of different types of materials have been suggested for use in cutoff walls. Most are constructed of soil-bentonite slurries, but concrete and concrete/polymer mixtures are also used. A trench is excavated with an excavator or backhoe. The trench is held open by a slurry of bentonite and water. The slurry acts in the same manner as the drilling mud that holds open a borehole. The bentonite slurry penetrates into the more permeable formations and forms a low-permeability filter cake. As the working end of the trench is built around the site, the opposite end of the trench is back-filled with a soil-bentonite slurry. The soil-bentonite slurry has a very low permeability and minimizes the movement of most ground water through it. The trench is typically 2 to 3 ft wide and can be up to 60 ft deep if excavated with a backhoe or up to 120 ft deep if dug with a clamshell shovel (Need and Costello 1984). Technology to install cutoff walls as deep as 400 ft is available (Valkenburg 1991). Preconstruction design studies are needed to determine if the waste that is to be contained is compatible with the bentonite-soil slurry.

In most cases it will also be necessary to construct a cover over the waste material to prevent the infiltration of precipitation. If the waste material is above the water table, a cover without a slurry wall might be all that is needed. For a waste material buried below the water table, a low-permeability cover is needed in association with a slurry wall. In the absence of a cover, the infiltrating water will fill the area within the slurry wall like a bathtub.

Covers may be constructed of native soils, synthetic membranes, or a combination of both. Covers are typically sloped in order to promote runoff of precipitation. It may even be necessary to bring in fill material to create the necessary elevation at the center of the cover to form the needed slope.

Site Remediation

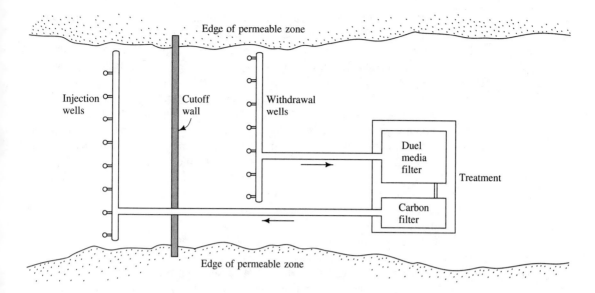

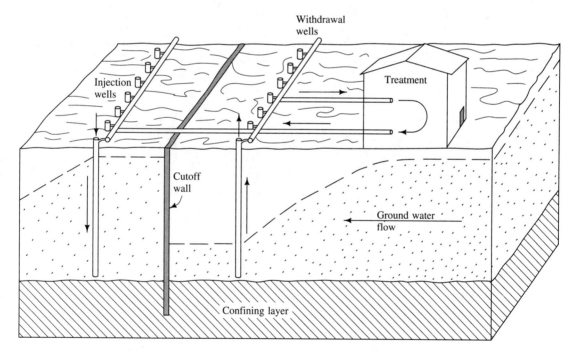

FIGURE 9.2 Cutoff wall used with extraction and injection wells at the Rocky Mountain Arsenal, Denver, Colorado, to isolate and treat a plume of contaminated ground water.

A multilayer cover constructed of natural materials might have the following elements, starting from the surface: (1) 2 ft of topsoil as a rooting medium for vegetation with shallow roots, such as grass or crown vetch, which is needed to prevent erosion; (2) a 2-ft-thick compacted layer of cobbles mixed with clay to stop burrowing animals from breaching the cap; (3) a 1-ft-thick low-permeability zone to separate the cobble layer from the next-lower zone; (4) a 1-ft-thick layer of very permeable sand, which acts as a capillary break to the movement of soil moisture and provides for lateral drainage of any infiltrating precipitation; and (5) a bottom layer that consists of 2 to 3 ft of recompacted clay with a maximum permeability less than 10^{-7} cm/sec to act as a final barrier to ground-water recharge. Geotextile fabric would be needed between several of the layers to keep the materials from mixing during construction and to distribute the stresses evenly. Figure 9.3 shows the details of this design. Erosion of the soil and penetration of the cover by tree roots are the main concerns for the long-term integrity of this design.

Synthetic geomembranes can also be used in cover designs as low-permeability layers. One advantage of using native soil materials is that their long-term performance is assured, whereas the long-term behavior of plastic membranes has not been tested. However, for many applications, plastic membrane covers are suitable. They can easily be installed so that monitoring and extraction wells extend through the cap.

Infiltration of precipitation into the ground can also be reduced by paving the surface with a material such as asphalt. However, an asphalt seal is more permeable than a multilayer cover and would require extensive maintenance to seal cracks that form.

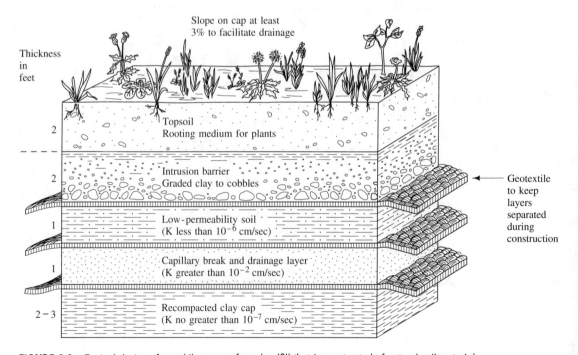

FIGURE 9.3 Typical design of a multilayer cap for a landfill that is constructed of natural soil materials.

Site Remediation

Diversion ditches and drains might be used to prevent surface runoff from entering an area where it can infiltrate into the soil and come into contact with buried waste material.

9.2.4 Hydrodynamic Isolation

Hydraulic controls can also be used to isolate a zone where the ground water has been contaminated. An extraction well positioned at the leading edge of a contaminant plume can be used to stabilize the position of the plume (Figure 9.4). The plume-stabilization well will pump contaminated water, which may require treatment before disposal. It will prevent the encroachment of the plume of contamination to uncontaminated parts of the aquifer. With the contamination thus isolated, work on source control and other remediation measures can progress at the most expedient pace.

Wilson (1984) described the use of a pair of injection and withdrawal wells to create a hydraulic isolation zone around a plume of gasoline contamination. The withdrawal well draws contaminated water to it; the water is then reinjected into the ground up-gradient of the position of the plume (Figure 9.5(a)). The withdrawn water may be treated prior to injection, or nutrients can be added to promote bioremediation. In addition, treatment systems may need periodic maintenance necessitating a shutdown.

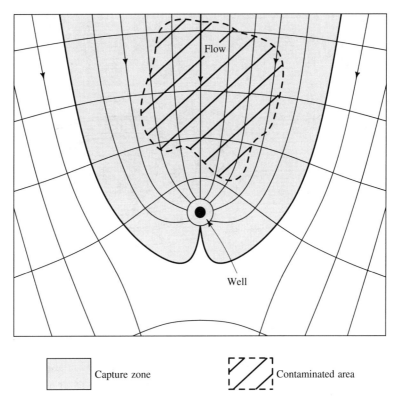

FIGURE 9.4 Plume stabilization well used to isolate plume of contaminated ground water.

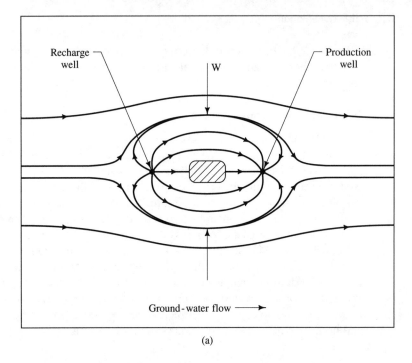

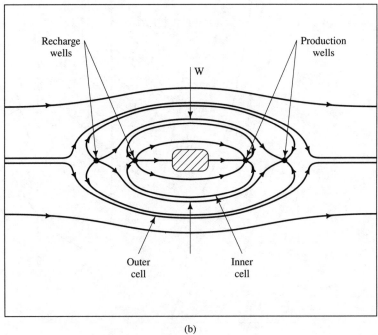

FIGURE 9.5 Plan view of (a) single-cell and (b) double-cell hydraulic containment of contaminated groundwater. *Source:* J. L. Wilson, *Proceedings of the Fourth National Symposium and Exposition on Aquifer Restoration and Ground Water Monitoring,* 1984, 65–70. National Water Well Association. Used with permission.

Site Remediation

With a reinjection system the untreated water can still be injected, and the plume-stabilization well can be kept pumping. Otherwise, if the plume-stabilization well is shut down, the plume might be able to spread beyond its hydraulic boundary. Wilson (1984) also described a double-cell hydraulic containment system with two pairs of injection and production wells, as shown in Figure 9.5(b). This system provides further isolation of the plume and also creates a smaller inner cell so that the production well pumping contamination will pump a smaller volume of water, which will lower treatment costs. With two injection and pumping wells, there is a possibility of shutting down one well periodically for maintenance and still having the system operational.

9.3 Pump-and-Treat Systems

9.3.1 Overview

Ground water that contains dissolved inorganic and organic chemicals can be removed from the ground so that the water can be treated at the surface to remove the contaminants. The advantage of this method of treatment is that conventional methods of wastewater treatment can be employed. The treated water can be discharged to a receiving water body, passed to a publicly owned wastewater treatment plant for further treatment and dilution, or reinjected into the ground. A pump-and-treat system for the plume of contaminants works best in combination with source-control measures to stop the release of material into the plume.

The disposal of the treated wastewater may require state and local permits. Discharge to a publicly owned wastewater-treatment plant may require local permits for industrial discharge and there might be pretreatment standards. Water discharged to a surface-water body requires state permits, including an NPDES permit. Many states require a permit to inject water into the ground. There may be concentration limits specified as a part of the permit. At least one state, Wisconsin, prohibits the injection of water or liquid wastes into the ground.

If there are NAPLs present, the situation is much more complex than if all contaminants are in a dissolved form. As long as an NAPL is present, it will partition between the NAPL phase and the dissolved phase. Thus, as contaminated water is withdrawn from the aquifer for treatment, the clean water that is drawn into the aquifer eventually becomes contaminated with material partitioning from the remaining NAPL. If some of the NAPL is mobile, it may be captured by pumping. LNAPLs that are floating on the water table are relatively easy to locate and remove. However, DNAPLs that sink to the bottom of the aquifer are very difficult even to locate, much less recover (Mackay and Cherry 1989, Freeze and Cherry 1989). Because considerable amounts of residual NAPL will remain even if the mobile NAPL is removed, a great many years may be required for pump-and-treat systems to remove all the residual NAPL by partitioning into the dissolved phase, which can be recovered. In the case of contamination by DNAPLs, especially in fractured rock aquifers, it may be impossible to remediate a contaminated aquifer (Mackay and Cherry 1989, Freeze and Cherry 1989).

If the dissolved phase sorbs onto the mineral matter of the soil, that phase may desorb as the contaminated water is flushed from the pores. The greater the distribution coefficient, the more slowly the sorbed phase will be released and the longer it will take to remediate the aquifer. The kinetics of desorption dictate that many pore volumes of uncontaminated water might be needed to remove completely the sorbed phase of both organic and inorganic contaminants. Whiffin and Bahr (1984) observed that the observed rate which organic compounds desorbed from an aquifer were slower than the rate predicted by transport equations that assumed equilibrium conditions between the sorbed phase and the dissolved phase (Figure 9.6). The advection-dispersion equation results were initially fairly accurate; however, when the concentration dropped to about half the initial concentration, the actual rate of removal was slower than the predicted rate.

Contaminants that have been in the ground for a long period of time have been able to diffuse into the less permeable zones of porous media aquifers and into the bedrock matrix of fractured rock aquifers. Pump-and-treat systems are inefficient in removing these contaminants, since the majority of the water being removed by pumping will come from the most permeable zones of the aquifer.

For example, a porous media aquifer might consist of coarse sand layers with interbedded fine sands and silts. Over time, the contaminated water will diffuse from the highly permeable coarse sands into the fine sands and silts, which have very limited permeability. Some of the contamination will remain in the dissolved phase and some will partition onto the surface of the fine sediment and associated soil organic carbon. The fine sediment will have a larger surface area per unit volume of the aquifer than the coarse sediment and thus will sorb more contamination.

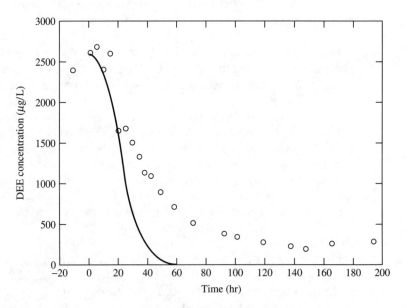

FIGURE 9.6 Measured desorption values for diethyl ether for water pumped from a purge well (open circles) versus calculated desorption curve based on advection-dispersion equation (solid line). *Source:* R. B. Whiffin and J. M. Bahr, *Proceedings of the Fourth National Symposium and Exposition on Aquifer Restoration and Ground Water Monitoring*, 1984, pp. 75–81. National Water Well Association. Used with permission.

Water removed by pumping will come primarily from the coarse sand layers. The contaminated water in the pores of these layers will be fairly quickly removed; perhaps only a few pore volumes of water will need to be flushed through to accomplish this.

Contaminants sorbed onto the coarse sand grains will then be removed by partitioning into the clear water that then occupies the pores in the coarse sand layers. Meanwhile, the contaminated water occupying the pores of the fine sand and silt layers will be flushed from these layers very slowly. It will take the movement of many pore volumes of water through the coarse sand layers to flush just a few pore volumes of water through the less permeable layers. It will take much longer to remove the contaminants that are desorbing from the fine sediments after the contaminated pore water is removed.

There will be an initial rapid decline in the concentration of contaminants in the water being removed by a pump-and-treat system. This decline represents the removal of the contaminated water contained in the pores of the aquifer. Initially, when the larger pores are being flushed, the drop in concentration will be rapid. As the smaller pores are being flushed, the rate at which the concentration declines will decrease. The concentration will eventually approach some constant value, which represents a steady-state condition where the rate at which the contaminants are being removed by the pump-and-treat system is equal to the rate at which they are being released into the ground water by desorption and/or partitioning from residual NAPLs.

At this point in the remediation process contaminant mass is still being removed from the aquifer, but the ground water is not becoming any cleaner. The concentration at which this occurs may be higher than the cleanup goals for the remediation. It may take many years of pumping at this stage to reduce the amount of sorbed contaminants and/or the residual NAPLs sufficiently for the ground-water quality to improve to a point where it meets cleanup goals.

Figure 9.6 illustrates this phenomenon. The initial concentration of diethyl ether, illustrated by open circles, was about 2600 μg/L. After 100 hr of pumping, the concentration dropped by 86% to about 360 μg/L. After an additional 100 hr of pumping, the concentration had declined only an additional 5% to about 230 μg/L (91% total decline) and was indicating a constant trend.

A study of pump-and-treat cleanup at 16 sites in the United States where the ground water was contaminated with organics revealed that if the initial concentration of contamination was high (>1000 μg/L), pumping could achieve reductions of contaminant concentrations of 90 to 99% before leveling occurred. At sites where the initial concentration was less than 1000 μg/L, leveling occurred before a 90% reduction was accomplished (Doty and Travis 1991). Moreover, if remediation is halted before the sorbed phase is completely removed, the dissolved concentration will eventually rise above the level detected at the end of the remedial period as additional material desorbs to come to equilibrium with the dissolved phase.

9.3.2 Capture Zones

In order to capture a plume of contaminated water, it is necessary to have one or more pumping wells located down-gradient of the source area. Each well will have what is known as a **capture zone,** which is the area contributing flow to that particular well.

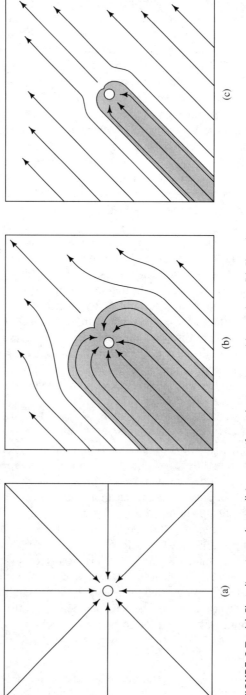

FIGURE 9.7 (a) Flow lines toward a well in an aquifer with no water table gradient. (b) Flow lines toward a well and the capture zone with uniform flow to the upper left of the figure. (c) Flow lines toward a well and the capture zone with uniform flow to the upper left at a rate 10 times the rate of (b).

Site Remediation

If the water table is flat, there is no regional flow. The capture zone of a well is radially symmetrical, centered on the well and extending as far as the edge of the cone of depression. See Figure 9.7(a). If there is a slope to the water table, the ground water flows and the capture zone is asymmetrical, with the greatest extent in the up-gradient direction (Figure 9.7(b)). As the ground-water velocity increases, the width of the capture zone decreases for a given pumping rate. See Figure 9.7(c).

The shape of the capture zone is a function of the average linear ground-water velocity, the quantity of the water being pumped from the aquifer, and the distribution of hydraulic conductivity. The up-gradient extent of the capture zone depends upon the length of time over which the pumping occurs. Shafer (1987) developed a numerical model to determine the time-related capture zone around extraction wells. Figure 9.8(a) shows the initial hydraulic head for a homogeneous, isotropic, unconfined aquifer. The hydraulic-head distribution after 20 yr of pumping from a single well is shown in Figure 9.8(b). During this time period the cone of depression has not spread all the way to the up-gradient end of the diagram. The extent of the capture zone is shown superimposed on the model grid in Figure 9.8(c).

Even if the hydrogeology is not homogeneous, the numerical model (Shafer 1987) can still be used to determine the capture zone. Figure 9.9(a) shows the hydraulic conductivity distribution of a nonuniform aquifer. Figure 9.9(b) shows the hydraulic-head distribution in this aquifer after 20 yr of pumping at the same rate as the previous example. The shape of the 20-yr capture zone is shown in Figure 9.9(c). Its irregular shape is the result of the irregular distribution of hydraulic conductivity. This numerical model can be used to determine the capture zone of an extraction well used for site remediation or it can be used to determine the area contributing flow to a production well. In the latter mode it is useful for developing management plans to protect the water quality in the recharge area of a well field.

9.3.3 Computation of Capture Zones

A method for determining the optimum number of extraction wells, their locations, and the rate at which each should be pumped has been developed by Javandel and Tsang (1986). Their solution has yielded simple type curves for single-well and multiple-well applications. Equations are also available to compute the maximum extent of the capture zone. The aquifer is assumed to be confined and have a uniform thickness, B. However, for unconfined aquifers if the drawdown is small relative to the total saturated thickness of the aquifer, the resulting error from using the equations for confined flow should not be large. Prior to pumping there is a uniform regional specific discharge (Darcian velocity), U. All extraction wells are fully penetrating. The well discharge rate is Q. All parameters need to be in consistent units, i.e., B in meters, U in meters per second, and Q in cubic meters per second.

Let us locate the position of a single pumping well at the origin. The equation of the line that divides the area that will be captured by the well from the rest of the flow field is

$$y = \pm \frac{Q}{2BU} - \frac{Q}{2\pi BU} \tan^{-1} \frac{y}{x} \qquad (9.1)$$

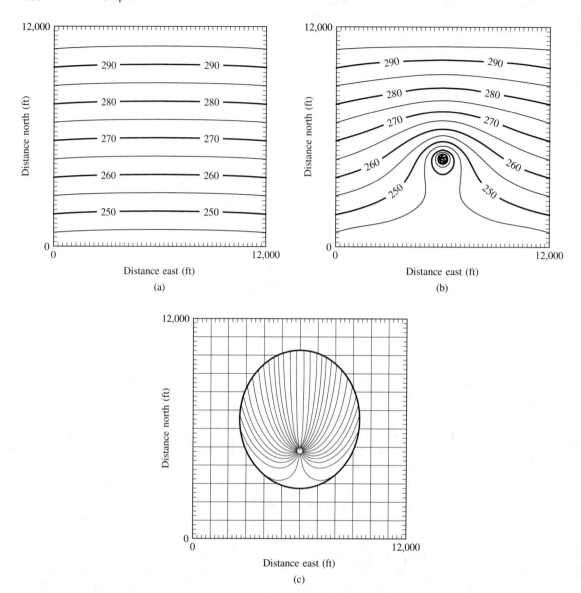

FIGURE 9.8 (a) Hydraulic head distribution with no wells pumping. (b) Hydraulic head distribution in a homogeneous, isotropic aquifer with one well pumping. (c) Shape of the 20-yr capture zone based on hydraulic head distribution of Figure 9.8(b). *Source:* J. M. Shafer, *Ground Water* 25, no. 3 (1987): 283–89. Used with permission. Copyright © 1987 Water Well Journal Publishing Co.

The only parameter in Equation 9.1 is the ratio Q/BU. Figure 9.10 shows a type curve for Equation 9.1 for different values of Q/BU with the units in meters and seconds. All the water within the given type curve will flow to the well. Figure 9.11 shows some of the flow paths for the capture zone for $Q/BU = 2000$ m. All the flow paths end at

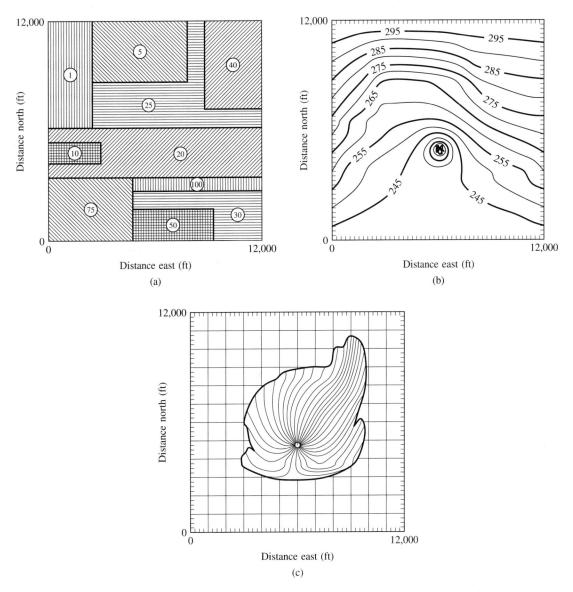

FIGURE 9.9 (a) Distribution of hydraulic conductivity in nonhomogeneous aquifer with the initial head distribution of Figure 9.8(a). (b) Hydraulic head distribution in the nonhomogeneous aquifer with one well pumping. (c) Shape of the 20-yr capture zone based on hydraulic head distribution of Figure 9.9(b). *Source:* J. M. Shafer, *Ground Water* 25, no. 3 (1987): 283–89. Used with permission. Copyright © 1987 Water Well Journal Publishing Co.

the well, and a stagnation point develops downgradient of the extraction well. Figure 9.12 shows a water table profile along the y axis. It can be seen that the stagnation point forms a ground-water divide between flow toward the well and flow in the regional direction.

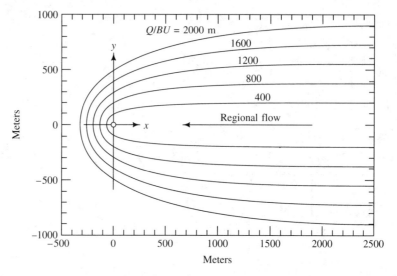

FIGURE 9.10 Type curve for analytical solution to capture zone analysis for a single extraction well. *Source:* I. Javandel and C-F. Tsang, *Ground Water* 24, no. 5 (1986): 615–25. Used with permission. Copyright © 1986 Water Well Journal Publishing Co.

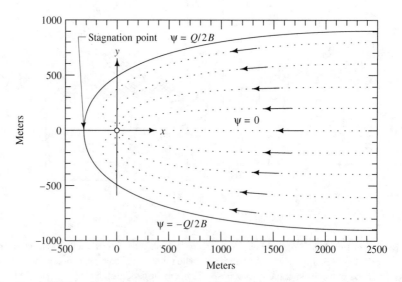

FIGURE 9.11 Flow lines for capture zone with a single extraction well for $Q/BU = 2000$ m. *Source:* I. Javandel and C-F. Tsang, *Ground Water* 24, no. 5, (1986): 615–25. Used with permission. Copyright © 1986 Water Well Journal Publishing Co.

Site Remediation

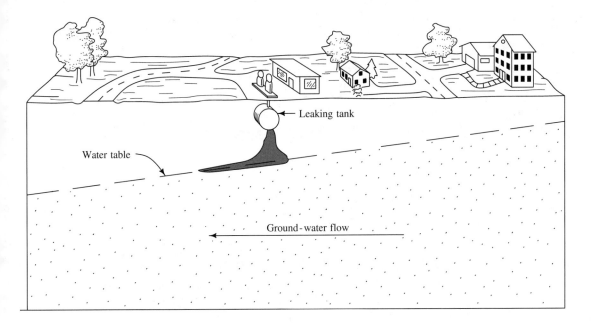

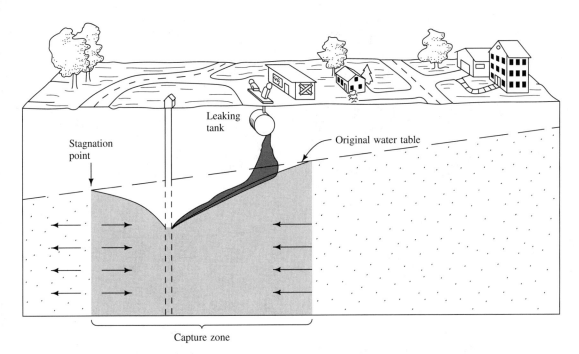

FIGURE 9.12 Cross section along the x axis showing the cone of depression for a single extraction well superimposed on the regional water table.

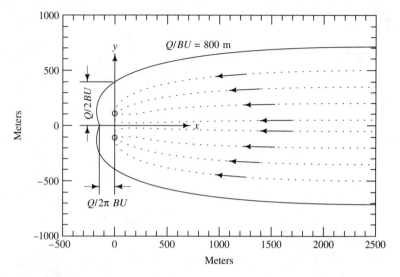

FIGURE 9.13 Capture zones of two wells properly located to prevent the leakage of any water in the space between them when $Q/BU = 800$ m. *Source*: I. Javandel and C-F. Tsang, *Ground Water* 24, no. 5 (1986): 615–25. Used with permission. Copyright © 1986 Water Well Journal Publishing Co.

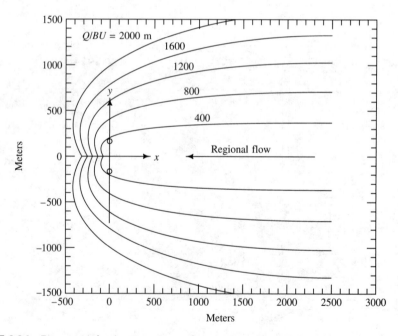

FIGURE 9.14 Type curves for the capture zone for two extraction wells located on the y axis with the distance between them equal to $Q/\pi BU$. *Source*: I. Javandel and C-F. Tsang, *Ground Water* 24, no. 5 (1986): 615–25. Used with permission. Copyright © 1986 Water Well Journal Publishing Co.

Site Remediation

There is a maximum quantity of water that can be pumped from a single extraction well. If the plume is wider than the capture zone developed by the maximum pumping rate, then multiple extraction wells are needed. One concern with multiple extraction wells is that their capture zones must overlap, or ground-water flow can pass between them. If the distance between extraction wells is less than or equal $Q/\pi BU$, then the capture zones will overlap. The optimum well spacing is thus $Q/\pi BU$. Figure 9.13 shows the flow lines for a two-extraction-well setup with a value of Q/BU of 800 m.

If $2d$, the distance between two extraction wells, is less than or equal $Q/\pi BU$, then the capture zone is defined as

$$y + \frac{Q}{2\pi BU}\left(\tan^{-1}\frac{y-d}{x} + \tan^{-1}\frac{y+d}{x}\right) = \pm\frac{Q}{BU} \qquad (9.2)$$

Figure 9.14 shows a type curve for the two-extraction-well problem for a variety of values of Q/BU, with the extraction wells separated by the optimum distance $(Q/\pi BU)$.

If there are three wells, then we locate one with an origin on the x axis and also at distances $+d$ and $-d$ from the origin on the y axis. The optimum well spacing is $1.26Q/\pi BU$. The equation for the capture zone boundary is

$$y + \frac{Q}{2\pi BU}\left(\tan^{-1}\frac{y}{x} + \tan^{-1}\frac{y-d}{x} + \tan^{-1}\frac{y+d}{x}\right) = \pm\frac{3Q}{2BU} \qquad (9.3)$$

Figure 9.15 shows the type curves for this situation.

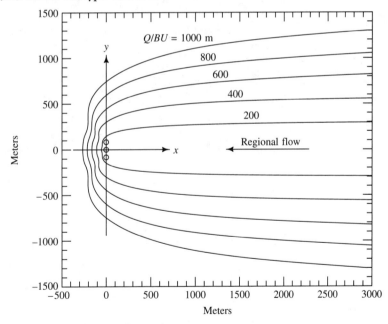

FIGURE 9.15 Type curves for the capture zone for three extraction wells located on the y axis with the distance between them equal to $1.26Q/\pi BU$. Source: I. Javandel and C-F. Tsang, *Ground Water* 24, no. 5 (1986): 615–25. Used with permission. Copyright © 1986 Water Well Journal Publishing Co.

If there are four wells, the optimal well spacing is approximately $1.2Q/\pi BU$; the type curve for this extraction-well spacing is given in Figure 9.16.

There are four steps to be followed in implementing the use of these type curves. It should be stressed that these type curves assume uniform conditions, which generally don't occur in nature. Moreover, if the aquifer is unconfined, which it most likely would be, and if the drawdown is a significant fraction of the total saturated thickness, then significant errors might be introduced by the assumption that the aquifer is confined.

Step 1. Prepare a site map at the same scale as the type curves. Indicate the regional flow direction and the plume shape on the map. The plume consists of all areas where ground-water contamination is in excess of some allowable value. This depends upon the type of contamination and the regulatory climate.

Step 2. Superimpose the site map on the one-well type curve. The direction of regional flow must be parallel to the x axis. Place the leading edge of the plume just beyond the location of the withdrawal well. Select the type curve that completely captures the plume. This is the required value of Q/BU.

Step 3. Compute the regional value of the specific discharge, where $U =$ hydraulic conductivity times hydraulic gradient. Find the pumping rate by multiplying the value of Q/BU found in step 2 by BU. If the resulting pumping rate can be achieved by the use of one well, then the problem is solved. If one well will not produce the required volume, then go on to step 4.

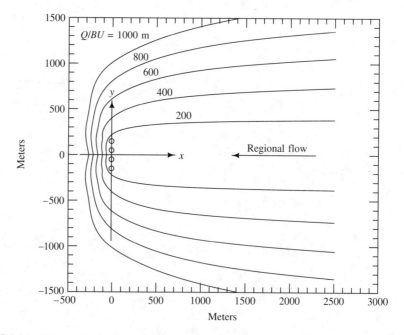

FIGURE 9.16 Type curves for the capture zone for four extraction wells located on the y axis with the distance between them equal to $1.2Q/\pi BU$. Source: I. Javandel and C-F. Tsang, Ground Water 24, no. 5 (1986): 615–25. Used with permission. Copyright © 1986 Water Well Journal Publishing Co.

Site Remediation

Step 4. Place the site map on the type curve diagram for two withdrawal wells, orienting it as before. Select the type curve that completely encloses the plume. Find the required pumping rate for each well, Q, by multiplying the value of Q/BU from the type curve by BU. If this value could reasonably be pumped from each withdrawal well, then the problem is solved. If not, then a three-well or a four-well configuration must be evaluated.

EXAMPLE PROBLEM

The plume shown in Figure 9.17 is found in an aquifer that has a hydraulic conductivity of 3×10^{-5} cm/sec, a storativity of 0.003, and a hydraulic gradient of 0.003. The aquifer is 20 m thick. When overlain on Figure 9.10, the plume is contained within the $Q/BU = 800$-m type curve.

The regional seepage velocity is found from

$$U = 3 \times 10^{-5} \text{ m/sec} \times 0.003 = 9 \times 10^{-8} \text{ m/sec}$$

The pumping rate is

$$Q = \left(\frac{Q}{BU}\right) BU = (800 \text{ m})(20 \text{ m})(9 \times 10^{-8} \text{m/sec})$$

$$= 1.44 \times 10^{-3} \text{ m}^3/\text{sec} \ (23 \text{ gal/min})$$

The nonequilibrium equation for large values of time can be approximated by (Fetter 1988, eq. 6-7)

$$h_0 - h = \left(\frac{2.3 Q}{4\pi T}\right) \log\left(\frac{2.25 Tt}{Sr_2}\right)$$

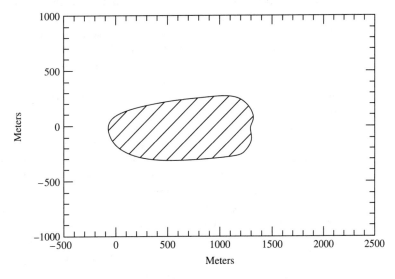

FIGURE 9.17 Contaminant plume for capture zone computation problem.

The transmissivity is hydraulic conductivity times thickness:

$$T = 3 \times 10^{-5} \text{ m/sec} \times 20 \text{ m} = 6 \times 10^{-4} \text{ m}^2/\text{sec}$$

The parameter r is the well radius. For a 10-in.-diameter well, $r = 0.127$ m. If the well is pumped for 1 yr (3.154×10^7 sec) without any recharge to the system, then the drawdown, $h_0 - h$, is

$$h_0 - h = \frac{(2.3)(1.44 \times 10^{-3} \text{ m}^3/\text{sec})}{(4)(3.14)(6 \times 10^{-4} \text{ m}^2/\text{sec})} \log \frac{(2.25)(6 \times 10^{-4} \text{ m}^2/\text{sec})(3.154 \times 10^7 \text{ sec})}{(0.003)(0.127)(0.127)}$$

$$h_0 - h = \frac{3.31 \times 10^{-3} \text{ m}^3/\text{sec}}{7.53 \times 10^{-3} \text{ m}^2/\text{sec}} \log \frac{4.258 \times 10^4 \text{ m}^2/\text{sec} \times \text{sec}}{4.84 \times 10^{-5} \text{ m}^2}$$

$$h_0 - h = (0.439 \text{ m}) \log 8.798 \times 10^8$$

$$h_0 - h = (0.439 \text{ m})(8.94)$$

$$h_0 - h = 3.93 \text{ m}$$

The aquifer, which is 20 m thick, can support a drawdown of 3.93 m at the pumping rate of 1.44×10^{-3} m^3/sec. If there were recharge to the confined aquifer, then the actual drawdown would be even less. Therefore, the plume can be captured with one withdrawal well.

9.3.4 Optimizing Withdrawal-Injection Systems

The rate at which ground-water restoration can be accomplished by pump-and-treat systems depends in part on how many pore volumes of water can be withdrawn from the contaminated zone. If the entire plume falls within the capture zone of one or more withdrawal wells, then we know that the plume won't spread and eventually maximum feasible restoration will occur. However, by increasing the rate (number of pore volumes of water per year) at which contaminated water is pumped, the restoration time can be decreased. For optimal conditions we also want to minimize the volume of contaminated water that is pumped, because that also must be treated.

Satkin and Bedient (1988) used a contaminant-transport model to investigate the use of various pumping and injection patterns to remediate a plume of contamination. They examined the effectiveness of seven different well patterns for various combinations of hydraulic gradient, maximum drawdown, and aquifer dispersivity. The patterns are shown on Figure 9.18.

If a single pumping well is used, it must be placed so that the capture zone encompasses the plume. The closer the well can be to the center of mass of the contaminant, the faster the contamination can be removed. If the plume can be captured by a single withdrawal well, then multiple pumping wells aligned along the axis of the plume will increase the rate of cleanup over a single well by pumping a greater volume of water. The use of pumping wells without injection wells may create a problem if there isn't a receiving body of water in which to dispose of treated ground water.

If injection wells are used in combination with withdrawal wells, cleanup time can be reduced, because steeper hydraulic gradients can be created. These steeper gradients will produce more water flowing to the withdrawal well(s) than occurs if extraction wells alone are used. We have already seen how a system of a down-gradient withdrawal well

Site Remediation

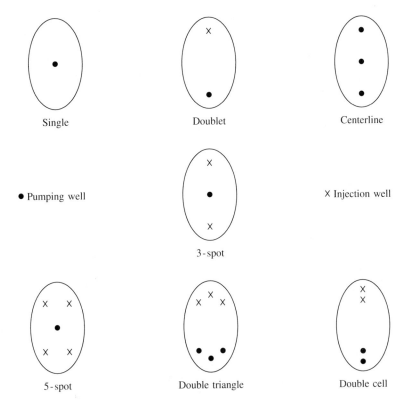

FIGURE 9.18 Possible patterns for extraction and extraction-injection well systems. *Source*: R. L. Satkin and P. B. Bedient, *Ground Water* 26, no. 4 (1988): 488–98. Used with permission. Copyright © 1988 Water Well Journal Publishing Co.

in conjunction with an up-gradient injection well can be used to create a circulation cell to isolate the plume. This is called a doublet on Figure 9.18. Other injection-withdrawal combinations tested included the double cell and double triangle, which are variations of the doublet. Two patterns tested were based on one extraction well and multiple injection wells: the three-spot and the five-spot. When injection wells are used in conjunction with extraction wells, the treated water is disposed via the reinjection. However, ground-water injection wells are prone to clogging and may need periodic maintenance (Fetter 1988). Additionally, states may have water-quality standards for any water that is reinjected; most states require a permit for injection wells.

Satkin and Bedient (1988) found that the best well pattern for cleanup was highly site-specific. They also found that even with the same well pattern, variation in the placement of the wells yielded different cleanup times. When the hydraulic gradient is low, the doublet, double cell, and three-spot patterns were very effective. Under conditions of high hydraulic gradient, the centerline was most effective. In this pattern the down-gradient injection well, which must be located beyond the leading edge of the plume, creates a hydraulic barrier to further migration of the plume. The five-spot pattern was not found to be very effective under any conditions.

9.3.5 Permanent Plume Stabilization

In some cases it may not be feasible or even technically possible to fully remediate a badly contaminated aquifer by pump-and-treat technology. This is especially true if the source cannot be located and removed—for example, if there are nonbiodegradable residual DNAPLs present in a fractured rock aquifer. For such aquifers the only feasible technology might be permanent plume stabilization. One or more plume-stabilization wells would be installed and just enough water pumped to capture the plume. The volume of water pumped could be less than that needed for a full site remediation, since there is no need to maximize the number of pore volumes drawn through the aquifer. The pumped water could either be treated and discharged into a surface-water body or reinjected up-gradient to recirculate through the aquifer. This arrangement would be necessary in perpetuity.

9.4 Treatment of Extracted Ground Water

9.4.1 Overview

Water withdrawn during a pump-and-treat restoration project is treated according to the type of contamination. Different types of treatment are needed for water contaminated with heavy metals vis-a-vis that contaminated by dissolved organic compounds (Nyer 1985). Most of the treatment techniques that are used were developed for wastewater and have been adapted to contaminated ground water.

The design of a pump-and-treat system should be cost effective. To this end the designer must consider the trade-off between capital costs and operating costs. Extraction systems can be designed by the hydrogeologist to withdraw the maximum volume of water in the shortest period of time. This yields the fastest, but not necessarily the most cost-effective, cleanup of the aquifer. The size of the treatment plant is dictated by the maximum rate that water is pumped for treatment. The capital costs for the treatment plant include the treatment vessels, pumps, piping, and tanks. The greater the flow rate at which contaminated water is pumped through the treatment system, the larger these items must be and the greater the initial capital costs. Operating costs include the electricity to run the plant, the cost of chemicals, the labor to operate the plant, and the cost of repairs. Some of the operating costs will be the same no matter how long the project lasts—for example, the cost of chemicals used to treat the water. If the same total volume of water is treated, the same amount of chemicals will be needed, no matter how long or short the treatment period. Other costs, such as labor, depend primarily upon the length of time of the operation. If the total volume of water is treated over a very short period of time, there will be high initial capital costs for the large-capacity plant and low operating costs because of the short time period. If the same volume of water is pumped over a longer period of time, the capital costs will be lower, since a smaller treatment plant is needed, but the operating costs will be higher. The smallest treatment plant possible is that needed to treat the quantity of water generated by the minimum pumping rate, which is just high enough to capture the plume. There will be some optimum treatment rate that minimizes the combined capital and operating costs.

9.4.2 Treatment of Inorganic Contaminants

The majority of inorganic contamination needing treatment consists of metals, which can be removed by precipitation. Many metal hydroxides precipitate at a specific alkaline pH value. For these metals, adjustment of pH by adding lime can cause precipitation of the metal hydroxide, which is removed via a clarifier followed by filtration. Ferrous iron can be removed by aeration to create ferric iron, which will precipitate at a slightly alkaline pH. Hexavalent chromium must first be reduced to the trivalent state by lowering the pH to 3 and then adding a reducing agent such as sulfur dioxide. The trivalent chromium can be precipitated as a hydroxide by raising the pH above the neutral value. Arsenic can be coprecipitated with iron by adding dissolved iron at a pH of 5 to 6 and then raising the pH with lime to between 8 and 9.

Any inorganic compound can be removed by ion exchange. This is used commonly for nitrate, which can't be removed by precipitation. Inorganic contaminants can also be removed by reverse osmosis and electrodialysis.

9.4.3 Treatment of Dissolved Organic Contaminants

Many of the organic contaminants found in ground water are volatile. They can be stripped from the water by exposing the water to a flow of air. This is accomplished in an air-stripping tower (Figure 9.19). The tower is a tall cylinder filled with an inert packing material, typically made of polypropylene. The packing material is designed to have a very high porosity and a large total surface area. The contaminated water is sprayed into the top of the tower onto the packing material. A blower attached to the bottom of the tower forces air up the tower at the same time that the water, which was broken up into droplets by the spray nozzles, trickles down the packing material. The volatile organic chemicals vaporize from the water into the air and are expelled out of the top of the tower. Care must be taken that emissions from the air-stripping tower do not create an air-pollution problem. The volatile organic compounds are eventually degraded by exposure to sunlight in the atmosphere. If air-quality modeling indicates that the air-stripper emissions will cause a health risk, the vapors can be captured by sorption onto activated carbon.

A number of organic compounds have low volatility and, hence, a low removal rate in an air-stripping tower. Many of these can be sorbed onto activated carbon. The contaminated water is pumped through a reaction vessel filled with activated carbon. These compounds partition onto the carbon and are hence removed from the water. The ability of the carbon to remove organics is eventually exhausted, and it must be replaced. Spent activated carbon can be regenerated by heating to drive off the sorbed organics.

Other methods of treating dissolved organic matter include the many biological methods that have been developed for wastewater treatment. A few organics, such as 1,4-dioxane, are resistant to air stripping, carbon absorption, or biological treatment and prove to be very difficult to remove from contaminated ground water.

Ground water pumped for treatment of organics may also contain natural dissolved iron. The iron will precipitate from solution as the water becomes aerated during treatment, for example in an air stripper. Mechanical filters, such as sand beds, are needed to filter the precipitated iron from the wastewater stream.

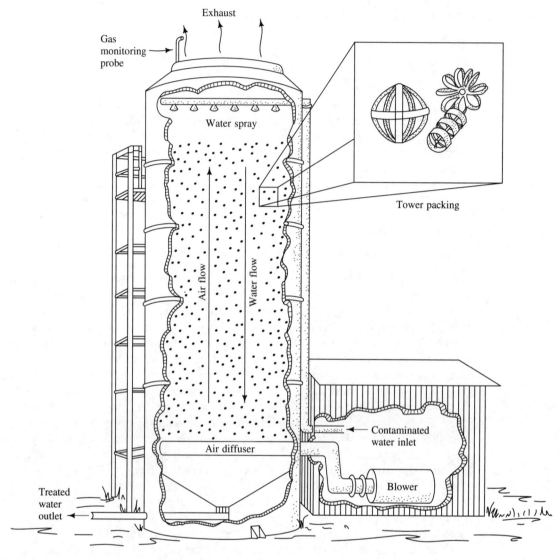

FIGURE 9.19 Design of an air-stripping tower.

9.5 Recovery of Nonaqueous Phase Liquids

If a mobile layer of a light nonaqueous phase liquid (LNAPL) has accumulated on the water table or the top of the capillary zone, it will flow in the direction in which the water table is sloping. The floating product can be recovered by depressing the water table with extraction wells or trenches. The product then flows to the well or trench, where it can be captured.

Figure 9.20 shows a simple recovery well for LNAPLs. The recovery well has a continuously slotted screen, which extends from above the top of the floating product

Site Remediation

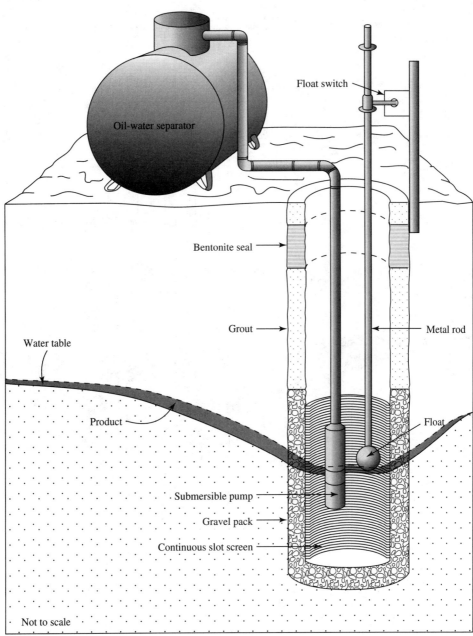

FIGURE 9.20 Single-pump system for recovery of light nonaqueous phase liquid. *Source*: S. B. Blake and R. W. Lewis, *Proceedings of the Second National Symposium on Aquifer Restoration and Ground Water Monitoring*, 1982, pp 69–76. National Water Well Association. Used with permission.

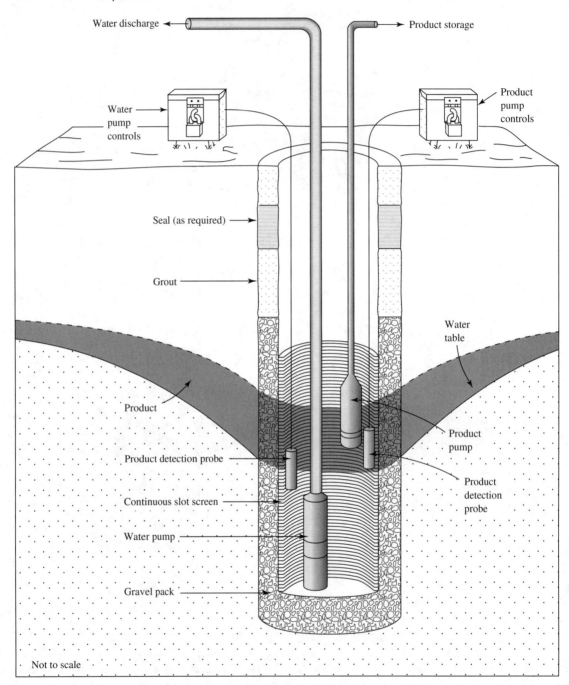

FIGURE 9.21 Double-pump, single-well system for recovery of light nonaqueous phase liquid. *Source:* S. B. Blake and R. W. Lewis, *Proceedings of the Second National Symposium on Aquifer Restoration and Ground Water Monitoring,* 1982, pp. 69–76. National Water Well Association. Used with permission.

to below the planned drawdown of the water table. A single pump is positioned so that it can pump both water and the LNAPL. The pump is activated by a float switch set so that the pumping level is maintained close to the pump intake so that both water and LNAPL will be withdrawn. If a floating skimmer pump is used, the switch isn't needed. Such a system is relatively inexpensive and easy to operate. However, the pump may emulsify the water and oil, so that an oil-water separator is needed to recover the product (Blake and Lewis 1982). In addition, soluble organics may be introduced into the water during the mixing process. If this occurs, then the water may also need treatment. However, the water may already contain soluble organics from the floating-product layer, so water treatment is already required.

Use of a two-pump system avoids the problem of the oil-water emulsion forming. A water pump is used to depress the water table. This is set some distance below the pumping-water level. A product-recovery pump set at the pumping level of the water table recovers the product in a condition allowing it to be sent directly to storage for later disposal. Two-pump systems can be installed in a single well (Figure 9.21). The casing and screen must have a large-enough diameter to hold both pumps and some float switches. A continuous-slot screen that extends from a point above the LNAPL layer to well below the water-pumping level is used. The water pump is set near the bottom of the well. A product-detection probe is located just above the water pump. If the product level drops to that depth, it is detected, and a signal is sent to shut down the water pump so that product is not drawn into the water pump. This isolates the water discharge so that it doesn't become contaminated with product. The product pump is located at the planned pumping level and has a switch activated by a product-detection probe to turn it on and off. The advantage of this system is that the water and LNAPL are not mixed. The recovered LNAPL can often be used, so some cost recovery is possible.

If wells already exist that are not suitable to hold two pumps—for example, their diameters are too small or the screen doesn't intercept the water table—then two pumps in two wells can be used. The deeper well, with a screen set below the water table, can be used as the extraction well to depress the water table. A second well for the product pump is constructed so that the screen extends from above the top of the product layer to below the pumping level (Figure 9.22). Product-detection probes are used to turn the pumps on and off.

Care needs to be taken when the product-recovery wells are first installed to be sure the pumps are set at the proper elevation and pumping rate. It will take several days of adjustment to determine the stable pumping level and the proper setting of the product-recovery pump. These systems can be set up to operate automatically and need only periodic checking to determine that the pumps and controls are still operating properly.

Skimming trenches can also be used to recover floating product. The trench is excavated to a depth below the water table and extends beyond the limits of the product plume. If possible, the trench should be down-gradient from the plume so that a minimal amount of water needs to be withdrawn from the trench to capture the plume, as shown in Figure 9.23(a). Drawdown in the trench needs to be great enough to reverse the ground-water gradient on the down-gradient side of the trench so that the floating product cannot flow out. A floating skimmer pump is used to lower the water table and remove

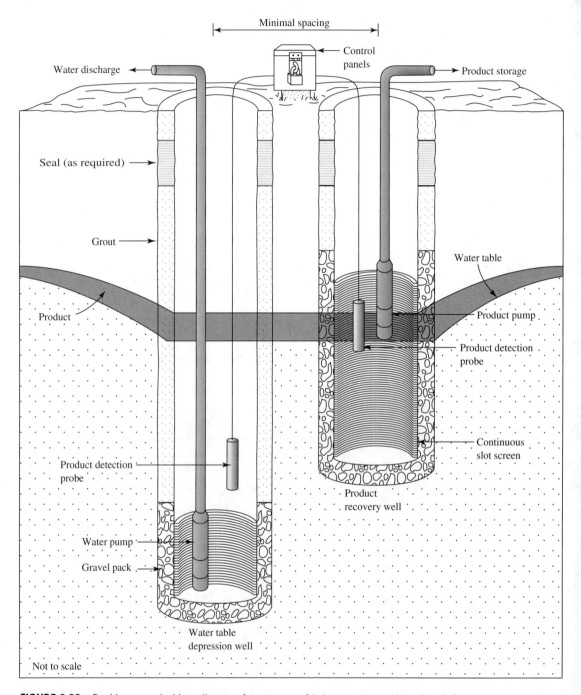

FIGURE 9.22 Double-pump, double-well system for recovery of light nonaqueous phase liquid. *Source:* S. B. Blake and R. W. Lewis, *Proceedings of the Second National Symposium on Aquifer Restoration and Ground Water Monitoring*, 1982, pp. 69–76. National Water Well Association. Used with permission.

Site Remediation

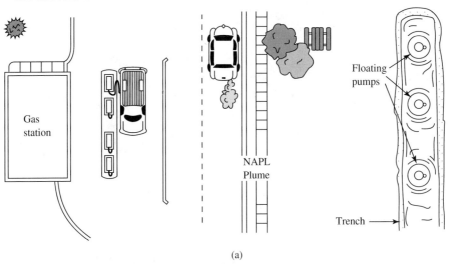

(a)

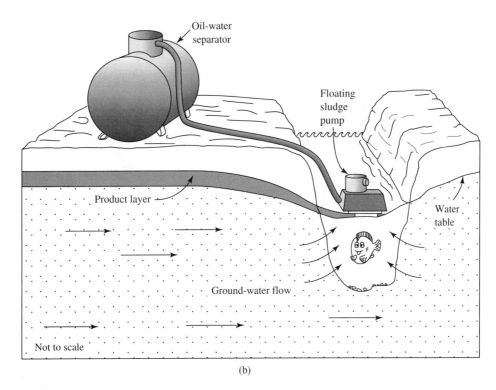

(b)

FIGURE 9.23 (a) Location of an interceptor trench used to capture a floating plume of a light nonaqueous phase liquid. (b) Cross section of trench and floating pump to capture the floating product and depress the water table.

the product. The mixture of product and water is sent to an oil-water separator (Figure 9.23(b)).

Floating product can also be captured by buried drains. A trench is excavated below the lowest expected position of the water table at a location down-gradient of the floating plume. Six inches of coarse stone is placed at the bottom of the trench and then a perforated plastic pipe is laid on the stone. The plastic material of the pipe must be compatible with the product to be recovered. The pipe drains into a sump, from which product and ground water are pumped for treatment. The fluid level in the sump is kept low enough that the water table falls to the elevation of the pipe, causing the product layer to drain into the pipe. This type of system can be installed in urban areas where an open trench would present a safety hazard. It can also be used with gasoline, which can present an explosion hazard. If gasoline is being recovered, explosion-proof pumps and motors must be used.

9.6 Removal of Leaking Underground Storage Tanks

Underground storage tanks have been used for many types of products, particularly petroleum distillates. The underground storage-tank system consists of the tank, fittings, and piping to add product to the tank as well as to remove it. Leaks can develop either in the tank or in the associated fittings and pipes. Steel tanks can corrode to the point where holes develop. Fittings may not have been properly tightened when installed. Ruptures may develop due to settling, and the tank may simply overflow if it is overfilled. Leaky tanks are generally identified by means of a "tightness" test performed by a qualified contractor.

The remedy for an underground tank that is known or suspected of leaking is to remove it. The removal process is performed by a contractor, but the process should be monitored by an environmental professional. The following steps are taken in tank removal:

1. Notify the local fire marshal and obtain all necessary permits.
2. If the tank holds an unknown liquid, analyze the liquid to determine the U.S. EPA hazardous-waste classification of the contents. This should be the procedure at sites where the tanks have not been used for some time. If the tank is in use up to the point of abandonment, then the nature of the product is probably known.
3. Pump the product from the tank and properly dispose of it.
4. Remove any sludges from the tank and properly dispose of them.
5. Purge vapors from the tank using an inert gas such as carbon dioxide or nitrogen.
6. Steam-clean the interior of the tank to remove any toxic residue. Pump the water used in steam cleaning from the tank and properly dispose of it.
7. Remove the tank from the ground by excavating the overlying soil and lifting the tank with a backhoe or excavator.
8. Cut up the tank for scrap or otherwise dispose of it.
9. Examine soil underlying the tank for contamination. Visually inspect badly-contaminated soil for stains and/or a distinctive odor. Then use an organic-vapor analyzer to find other areas of less severe soil contamination.

Site Remediation

10. Remove all contaminated soil. In some cases the soil may be disposed of at a landfill, where it can be used for daily cover material. Petroleum-contaminated soil can also be biologically treated by landfarming (Lynch and Genes 1989; Czarnecki 1989) or used in the manufacture of asphalt (Kosetecki Calabrese, and Fleischer 1989) and bituminous concrete (Eklund 1989).
11. Backfill the excavation with clean soil.
12. Complete all necessary reports and file them with the proper authorities.

Figure 9.24 shows a leaking underground storage tank being removed. This tank is still leaking, because the product wasn't removed before the tank was lifted from the excavation. This is an example of how not to "yank a tank."

If the contaminated soil extends to the water table, it is possible that ground water has become contaminated. If the tank held an LNAPL, then a floating-product layer might have formed. If the tank held a DNAPL, then the DNAPL may have sunk into underlying aquifers. In either case, a ground-water–contamination investigation is needed. Different types of investigations are used for sites with LNAPL contamination than for those with DNAPL contamination, because the NAPLs behave differently in the subsurface.

FIGURE 9.24 Removal of a leaking underground storage tank. Note the product pouring from holes in the tank! *Photo credit*: Kenneth Hawk.

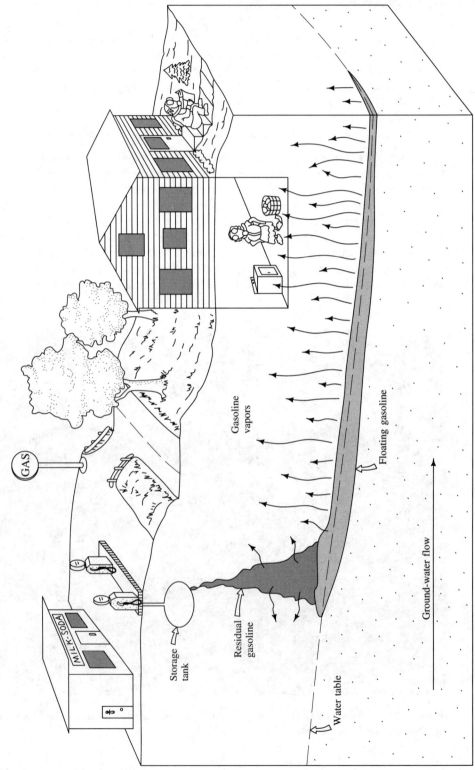

FIGURE 9.25 Release of organic vapors in vadose zone from residual saturation and floating product.

9.7 Soil-Vapor Extraction

Organic vapors are found in the unsaturated zone in association with spills and leaks of volatile organic compounds. When a volatile organic compound is discharged into the unsaturated zone, it will partition between the liquid and vapor state. Even if the soil absorbs all the spilled liquid before it reaches the water table, the vapors may migrate through the vadose zone. If there is a migrating plume of a mobile LNAPL that is volatile, then the LNAPL will continue to partition into the vapor phase, and the vadose zone above the plume will contain vapors (Figure 9.25). In addition, as the water table rises and falls with the floating-product layer, the product will be sorbed by the soil in a zone representing the annual cycle of water table rise and fall. The residual saturation in this zone will also contribute soil vapors.

Hydrocarbon vapors can migrate through the soil and accumulate in basements, where they can pose a threat of fire or explosion. Vapor-control measures may be needed to prevent explosions. Such measures can be accomplished by installing wells in the vadose zone and pumping air and vapors from them (Figure 9.26). This will keep the vapors from migrating into the basement. Another tactic is to place a fan so that it blows air into a basement. This pressurizes the basement and keeps the organic vapors out.

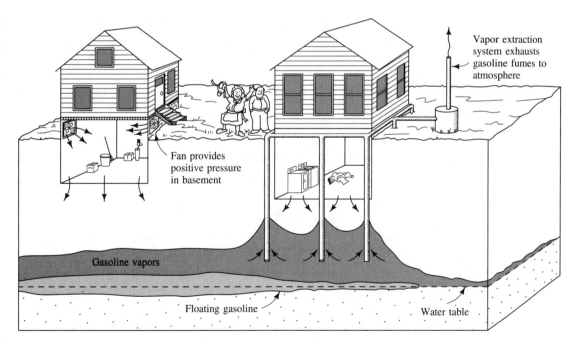

FIGURE 9.26 Control of organic vapors in the vadose zone. *Source:* Modified from M. J. O'Conner, J. G. Agar and R. D. King, *Proceedings of Conference on Petroleum Hydrocarbons and Organic Chemicals in Ground Water: Prevention, Detection and Restoration,* 1984, pp. 519–33. National Water Well Association. Used with permission.

The positive-pressure technique can't be used in climates where the outside air in the winter is below freezing; otherwise the cold air will freeze the pipes in the basement.

Site remediation can also be accomplished with soil-vapor extraction. If residual organic compounds remain in the vadose zone, infiltrating precipitation will continue to dissolve them and carry them in solution to the water table. Rather than sealing the surface to prevent infiltration, soil-vapor extraction can be used to remove the residual saturation of volatile organic compounds. Soil-vapor extraction can also be used to remove floating layers of very volatile hydrocarbons. Rather than the hydrocarbon being removed in liquid form, it is removed as a vapor through the vapor-extraction wells (Malot 1989; Trowbridge and Malot 1990).

Vapor-extraction systems can be constructed using wells in the vadose zone that are designed in much the same way as ground-water wells. Wells would be used in areas where the depth to the water table is 10 ft or more deep. The wells contain a slotted-plastic well screen. The wells will not be developed the way that water wells are, so the well screen is set in coarse gravel backfill for greatest air flow. The upper 5 ft or so of the well is solid plastic casing set in cement grout. It is important to seal the annular space so that the well doesn't just pull atmospheric air down the outside of the casing. A vapor-extraction well is designed to withdraw soil vapor from the vadose zone in a

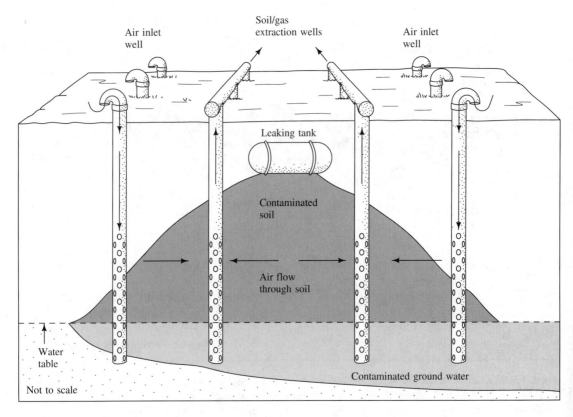

FIGURE 9.27 Soil-vapor extraction system consisting of vapor-extraction wells and air-vent wells.

circular area around the well. Air-vent wells are used in conjunction with the vapor-extraction wells. In order to extract the vapors, fresh air must be circulated through the pores containing the residual NAPL material. The air-vent wells provide a pathway of least resistance to ensure that air will circulate through the entire volume of the soil that is being remediated (Figure 9.27). Air-vent wells are constructed in the same way as vapor-extraction wells, but instead of being connected to a suction header, the upper end is open and capped with an inverted-U trap to keep out rain. If air-vent wells are not used, the soil-vapor–extraction wells must be pumped at varying rates to make sure permanent stagnation zones between wells do not develop.

If the water table is shallow, vapor-extraction wells are not effective. In this case, vapor-extraction trenches can be used. The trenches are excavated through the vadose zone to just above the seasonally high water table. A layer of gravel is put down and then a perforated plastic pipe is laid in the trench and covered with gravel. The rest of the trench is backfilled with low-permeability material to prevent short-circuiting of the vadose zone by atmospheric air circulating through the backfill material. If a low-permeability cap is to be constructed over the vapor-extraction system, then some type of venting system using air-inlet trenches is needed to allow air to sweep through the vadose zone in order to flush out the vapors and allow the residual organic compounds sorbed onto the soil to partition into the air. Even if a clay cap will not be placed over the vapor-extraction system, the air-inlet trenches will help to circulate air through the entire depth of the contaminated soil. Figure 9.28 shows a vapor-extraction system with air-inlet trenches. Air trenches alternate with soil-vapor–extraction trenches. Such a system can be designed so that a given trench could be used as either an air-inlet trench or a vapor-extraction trench. This maximizes the operational utility of the system.

Whether vapor-extraction wells or trenches are used, a vacuum system is needed to apply a suction to the wells or pipes. A suction header is attached to each extraction pipe or well and extended to a vacuum pump. The exhausted air with the vapors is passed through activated-carbon filters to remove the vapor if an air-pollution problem would result from the discharge of the vapors to the atmosphere.

9.8 *In Situ* Bioremediation

In Chapter 7 we saw that in some locations naturally occurring soil microbes are able to biodegrade hydrocarbons and other organic compounds. In some instances natural processes might be sufficient to remediate a contaminated aquifer or soil zone. However, in many additional cases it is possible to enhance the ability of the native soil microbes to degrade organic contaminants by the addition of nutrients and/or oxygen to the system (Thomas and Ward 1989; Lee et al. 1988; Mueller, Chapman, and Pritchard 1989; and Flathman, Jerger, and Bottomley 1989).

The subsurface is populated with microbes, which have a fairly consistent population density with depth. This subsurface includes not only the soil-moisture zone, but also the rest of the vadose zone and underlying aquifers. The majority of subsurface microbes are growing on a substrate—i.e., a soil particle. Low levels of microbes in ground water do not mean low levels of microbes in the ground. The majority of subsurface microbes are bacteria, but fungi and protozoa are also found.

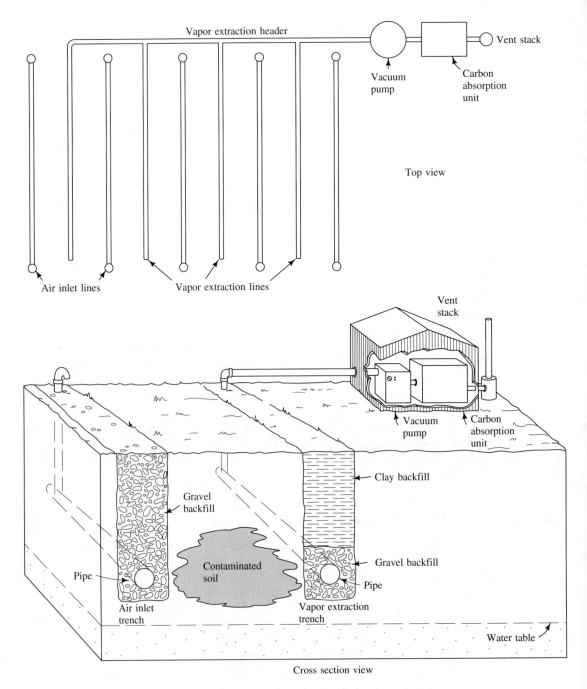

FIGURE 9.28 Soil-vapor extraction system consisting of trenches installed under a clay cap.

Many subsurface microbes can metabolize a wide variety of naturally occurring organic compounds such as sugars, starches, and amino acids. Microbes from soil that has been contaminated with various synthetic organic compounds have been found to be capable of degrading these compounds. In some cases microbes from uncontaminated soil have been unable to degrade the same compounds. However, other compounds can readily be degraded by microbes from uncontaminated soil. In some cases it is necessary for microflora to become acclimated to contamination before they can begin to metabolize it. The length of time that this acclimatization may take is not known. The amount of nutrients (nitrogen, phosphorus, potassium) and oxygen present, pH, and temperature influence the rate at which microbes can metabolize organic contaminants.

The earliest use of enhanced bioremediation in the United States was in remediating ground water contaminated with high-octane gasoline (Jamison, Raymond, and Hudson 1975). The process that was used was patented by Raymond (1974). Nutrients in the form of inorganic nitrogen and phosphorus were added via injection wells and air was supplied to the aquifer by bubbling (sparging) it into the water standing in the injection wells. After the addition of the nutrients, the number of bacteria in ground water increased by a factor of 1000. Ten months after the addition of the nutrients, gasoline could no longer be detected in the ground water.

Prior to the start of a biodegradation process, the nutrient requirements of the native soil bacteria are determined with a laboratory investigation. Microcosms of soil from the site are assembled and the contaminant is added. As much as possible, the microcosm should resemble the site conditions. If the ground water is to be remediated, the soil should be saturated with ground water from the site. Soil remediation should be evaluated in unsaturated microcosms. Various microcosms are given different amounts of nutrients, and the rates at which the contaminant disappears in the various microcosms are determined.

Before a bioremediation process is begun, any floating NAPLs should be removed if feasible. Figure 9.29 shows a schematic of *in situ* enhanced bioremediation using injection and withdrawal wells, which add the nutrients directly to the ground water. If there is considerable residual contamination in the soil above the water table, then nutrients must be added to the soil to stimulate soil bacteria. This can be done by injecting the nutrients directly into the soil below the rooting zone. (The nutrients need to get to the soil bacteria and not the plants growing on the surface.) An infiltration gallery located above the area of contaminated soil can be periodically flooded with water containing the necessary nutrients. The water should be charged with oxygen. Between flooding periods the infiltration gallery should be allowed to dry out so that oxygen can diffuse to the soil. The nutrients will wash into the ground water and promote bacterial growth there as well as remediate any dissolved hydrocarbons. Water pumped from extraction wells may be discharged or recirculated via injection wells and infiltration galleries.

The factors necessary for *in situ* bioremediation of organic compounds start with a geologic medium that is permeable enough to allow the introduction of oxygen and nutrients. The *in situ* biodegradation project will fail if the nutrients cannot reach the zone of contamination. Microbes that can metabolize the specific contaminants must be present. Atlas (1975) indicated that microorganisms that can degrade hydrocarbons are ubiquitous. However, much more specialized microbes may be needed for compounds

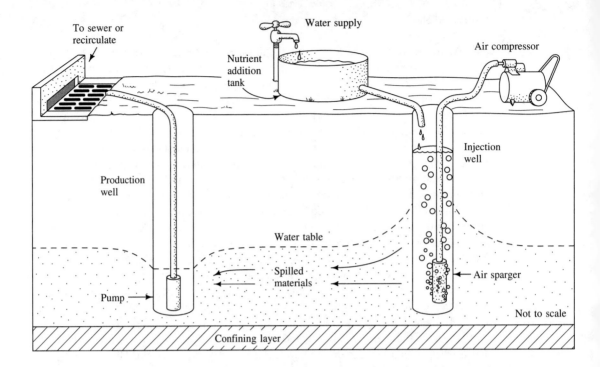
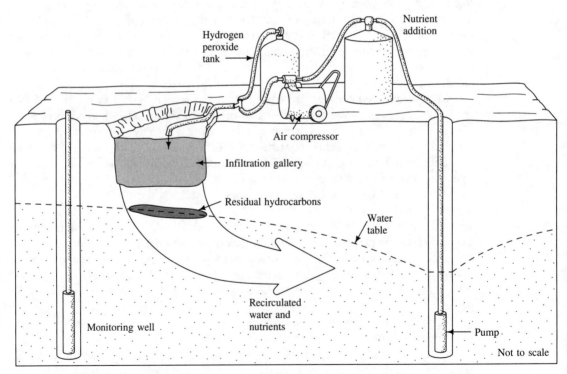

FIGURE 9.29 Enhanced *in situ* aerobic bioremediation. *Source*: Reprinted with permission from M. D. Lee et al. *Critical Reviews in Environmental Controls* 18, no. 1 (1988): 29–89. Copyright CRC Press, Inc., Boca Raton, Fla.

such as polynuclear aromatic hydrocarbons and heterocyclic compounds (ring compounds with nitrogen or sulfur in the ring) (Mueller, Chapman, and Pritchard 1989). It may be necessary to inoculate compound-specific microbes at a site if they are not naturally present.

Oxygen can be added to ground water by sparging either air or pure oxygen. Hydrogen peroxide can be used as a much more efficient way of adding oxygen to water, but it can be toxic to some microbes and hence does not have universal application (Raymond et al. 1986; Yaniga 1982). In addition, hydrogen peroxide is more expensive. Air-vent wells and vapor-extraction wells can be used to supply oxygen to the vadose zone if nutrients have been added by injection or infiltrating water.

Case Study: Enhanced Biodegradation of Chlorinated Ethenes

Chlorinated ethenes are among the more common ground-water contaminants due to their widespread use as solvents. In Section 7.6.3 methods of degradation of this class of compounds were discussed. A field experiment was conducted at the Moffett Naval Air Station, Mountain View, California, to see if biostimulation could be used to enhance their *in situ* degradation (Roberts, Hopkins, Mackay, and Semprini 1990; Semprini, Roberts, Hopkins, and McCarty 1990; Semprini, Hopkins, Roberts, Grbic-Galic, and McCarty 1991).

An uncontaminated, confined sand aquifer that was 1.5 m thick was instrumented with a line of injection and extraction wells located 6 m apart. The direction of ground-water flow from the injection to the extraction wells was parallel to the regional hydraulic gradient. Intermediate sampling wells were placed at distances of 1, 2.2, and 4 m from the injection wells. Depending upon the injection and withdrawal rates, travel times from the injection wells to the withdrawal wells were from 20 to 42 hr. When oxygenated water was injected into the aquifer prior to the biostimulation experiments, the oxygen was transported to the extraction wells with little loss.

The compounds that were selected for study were vinyl chloride (VC), *trans*-1,2-dichloroethene (t-DCE), *cis*-1,2-dichloroethene (c-DCE), and trichloroethene (TCE). When these were injected prior to biostimulation, they were retarded in the rank order of TCE $>$ t-DCE $>$ c-DCE $>$ VC. It was found that with a long period of injection prior to biostimulation, the sorption capacity of the aquifer could be saturated with respect to TCE, t-DCE, and c-DCE. After 1000 hr of injection, the concentration of these compounds in the monitoring well located 1 m from the injection well was found to be 90 to 95% of the injected concentration.

After the aquifer reached steady-state concentrations of the organic halides, it was biostimulated by injecting alternating pulses of dissolved oxygen and methane, along with continuous injection of the organic halides. The methane acted as the primary substrate (electron donor) for the growth of indigenous methane-utilizing bacteria, while the oxygen was the electron acceptor. The organic halides were degraded by cometabolism, a process by which the methantrophic bacteria that are utilizing the methane produce enzymes that are able to degrade the chlorinated ethenes. During the biostimulation experiments, decreases in concentration of both methane and the organic halides were observed. Within 2 m of travel through the aquifer, VC was reduced by 90 to 95%, t-DCE by 80 to 90%, c-DCE by 45 to 55%, and TCE by 20 to 30%. Residence times in the aquifer were only 1 to 2 days for this amount of biodegradation. It took about three weeks for the biostimulation experiment to reach these steady-state rates of reduction. An intermediate degradation product, *trans*-dichloroethene epoxide, was detected. When the injection of methane was halted, the concentration of the epoxide quickly decreased and the concentration of the halogenated ethenes slowly increased. However, when the rate of methane addition was increased beyond a certain concentration, it was shown to reduce the rate of transformation of VC and t-DCE. Thus, while methane was necessary for stimulating *in situ* aerobic biotransformation, there appears to be an optimal concentration beyond which it inhibits the process.

9.9 Combination Methods

If there is a considerable amount of residual product below the water table, nutrients can be added by circulating water. Then the water table can be depressed by dewatering. This creates a situation that allows the circulation of air through the zone of contamination,

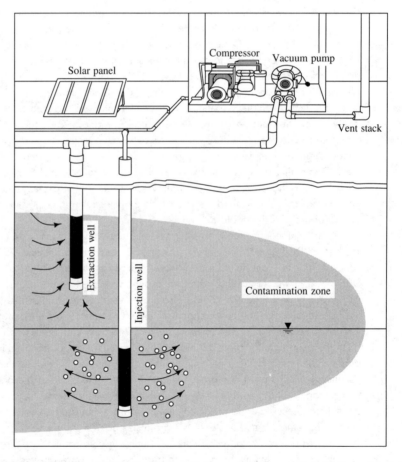

FIGURE 9.30 Subsurface volatilization and ventilation system. *Source:* C. P. Ardito and J. E. Billings, *Proceedings of Petroleum Hydrocarbons and Organic Chemicals in Ground Water: Prevention, Detection and Restoration,* 1990, pp 281–96. National Water Well Association. Used with permission.

Site Remediation

which can more effectively get oxygen to the microbes than circulating water. This is especially true if the formation is fine grained. The water removed by dewatering might need treatment before it is discharged. If the contaminant is volatile, a vapor-extraction system can also be used in conjunction with the lowered water table. This increases the rate of remediation by adding physical removal of the organic compound to the biological remediation.

Ardito and Billings (1990) described a hybrid method called the subsurface volatilization and ventilation system. The system consists of a number of well nests. Each nest consists of an air-injection well, which is screened below the water table, and a vapor-extraction well, which is screened in the vadose zone (Figure 9.30).

This approach combines physical removal of hydrocarbons with bioremediation. The vapor-extraction wells remove the volatile fraction of residual NAPL. The air sparged into the ground water removes some of the dissolved organics by air stripping. The oxygen added to the ground water promotes biological activity to degrade organics below the water table. Nutrients can also be introduced by batch addition to further bioremediation. Ardito and Billings (1990) reported the use of this system to treat a site where 40,000 gal of leaded gasoline leaked from an underground storage tank. A pump-and-treat system at the site was operated for 3 yr but had operational problems due to clogging of the injection well. A subsurface volatilization and ventilation system was installed and in 6 mo caused a significant reduction in the extent and concentration of benzene, toluene, ethylbenzene, and xylene. Figure 9.31 shows the extent of the plume, in milligrams per liter of benzene, before and after the 6-mo period. The reduction in benzene, toluene, xylene, and ethylbenzene in three monitoring wells is shown in Figure 9.32.

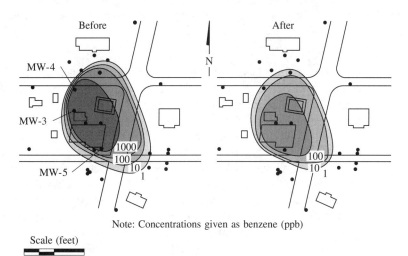

FIGURE 9.31 Extent of plume of benzene at a gasoline spill before and after treatment by subsurface volatilization and ventilation system. *Source:* C. P. Ardito and J. E. Billings, *Proceedings of Petroleum Hydrocarbons and Organic Chemicals in Ground Water: Prevention, Detection and Restoration*, 1990, pp 281–96. National Water Well Association. Used with permission.

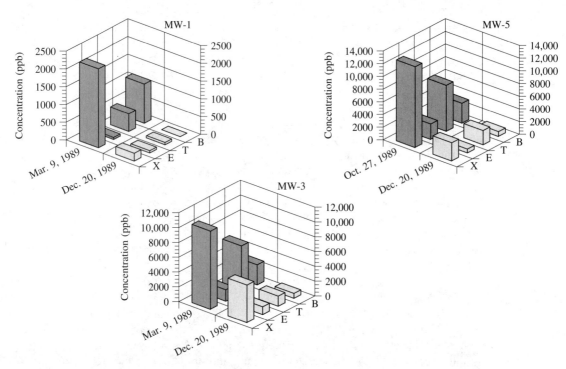

FIGURE 9.32 Reduction in benzene, ethylbenzene, xylene, and toluene measured in monitoring wells after 6 mo of operation of subsurface volatilization and ventilation system. *Source*: C. P. Ardito and J. E. Billings, *Proceedings of Petroleum Hydrocarbons and Organic Chemicals in Ground Water: Prevention, Detection and Restoration*, 1990, pp 281–96. National Water Well Association. Used with permission.

This system is particularly attractive for remediation of a plume that contains compounds such as chlorinated solvents, which are not as biodegradable as hydrocarbons. The air sparging provides an *in situ* air-stripping system, with the stripped material being captured by the vapor-extraction wells.

Vapor-extraction wells can be combined with product-recovery wells to provide remediation of both a floating NAPL and a residual product in the vadose zone. This is especially useful for volatile NAPLs such as gasoline. The product-recovery well has a well screen that extends up through much of the vadose zone. A single skimming pump is used; it pumps enough water to create a cone of depression. The product in the area of influence of the cone of depression flows into the product-recovery well, where it is also pumped by the skimming pump. If gasoline is being recovered, an explosion-proof pump is needed. Vacuum is applied to the well when the skimming pump is operating. The vacuum extracts the vapors from the vadose zone. Many times such systems are installed under pavement areas at service stations. The pavement prevents short circuiting of the air flow, so that it sweeps from outside the paved area and across the area where the soil is contaminated. The water and product are pumped to a separator, where the product is recovered and sent to storage. The water is then directed either to an air-stripping tower or, if only a few gallons per minute are being pumped, to a diffuser, which strips the volatile compounds from the water. A diffuser is a tank, which water enters at one end. There are a series of baffles in the tank that are each progressively lower, so the water cascades over each baffle as it flows toward the outlet. The volatiles are stripped as the water becomes turbulent going over the baffles. The volatiles are collected in a hood over the tank and exhausted with a blower to a stack. The exhausts from the vacuum-extraction pump and from the diffuser or air stripper are combined for dispersion through a stack. Figure 9.33 illustrates such a system.

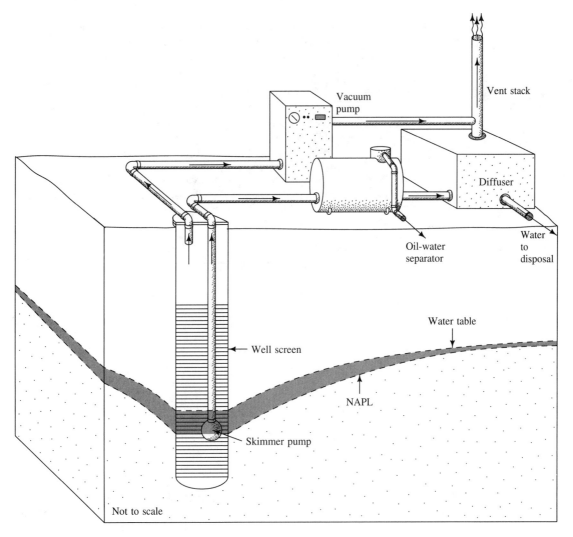

FIGURE 9.33 Combined vacuum extraction and product recovery well for remediation of a volatile nonaqueous phase liquid such as gasoline.

Case Study: Remediation of a Drinking Water Aquifer Contaminated with Volatile Organic Compounds

Cookson and Leszcynski (1990) describe the remediation of an aquifer supplying a well field for the city of South Bend, Indiana. In April 1980, water from the Olive-Sample Well Field developed an odor. Chemical testing showed that three of the six wells were contaminated with 100 to 225 μg/L of volatile organic compounds, including 1,1,1-trichloroethane and trichloroethene. The production wells were screened from about 120 to 190 ft below grade in a highly permeable sand and gravel aquifer. Total output from the well field was 3 to 6 million gallons per day. The well field was shut down in June 1980 due to the contamination.

A remedial investigation showed that the well field was at the leading edge of a plume of contamination. Figure 9.34 shows the isoconcentration lines for total volatile organic compounds in November 1980. An analysis of ground water from the most contaminated

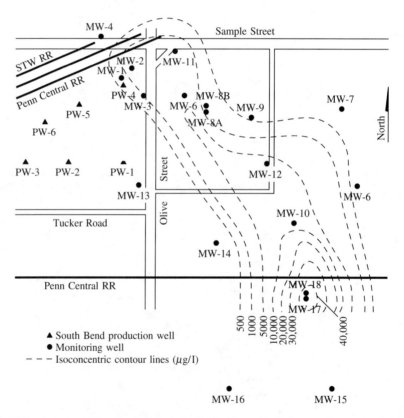

FIGURE 9.34 Isoconcentration lines in November 1980 for total volatile organic compounds in ground water at the Olive-Sample Well Field, South Bend, Indiana. Source: J. T. Cookson, Jr., and J. E. Leszcynski, *Proceedings of the Fourth National Outdoor Action Conference on Aquifer Restoration, Ground Water Monitoring and Geophysical Methods*, 1990, pp. 669–81. National Water Well Association. Used with permission.

Site Remediation

TABLE 9.1 Volatile organic compounds in a monitoring well located at the source area of a plume of volatile organic compounds.

Carbon tetrachloride	2,655 µg/L
1,1-Dichlorethane	1,480 µg/L
1,2-Dichloroethene	390 µg/L
trans-1,2-Dichloroethene	24,120 µg/L
Tetrachloroethene	2,605 µg/L
1,1,1-Trichloroethane	14,350 µg/L
Trichloroethene	1,555 µg/L
Vinyl chloride	4,360 µg/L
Total volatile organic compounds	51,785 µg/L

Source: Reprinted with permission from J. T. Cookson, Jr., and J. E. Leszcynski, *Proceedings of the Fourth National Outdoor Action Conference on Aquifer Restoration, Ground Water Monitoring, and Geophysical Methods.* (Dublin, Ohio: National Water Well Association 1990), 669–81.

monitoring well, No. 18, is shown in Table 9.1. The source of the contamination was found to be soil contaminated with organic compounds at the site of a chemical mixing and storage site.

Interceptor wells were installed in the aquifer between the contaminated soil source area and the well field. These wells were pumped at rates of about 1000 gal/min with the purged water being sent to a wastewater treatment plant. The contaminated soil was removed almost to the depth of the water table. The purge wells were operated for 5 yr after soil removal. The production wells were put into operation again in 1987. The water pumped from the well that was the most contaminated had total volatile organic compounds ranging from below detection to less than 10 µg/L/ This remediation was successful because of several factors: (1) The source could be contained by removing the contaminated soil; (2) a means existed to dispose of 1000 gal/min of contaminated ground water for a period of several years; (3) the contamination was contained in a highly permeable aquifer that apparently had mostly large pores. Contaminated ground water in an aquifer with pores of a similar size can be purged much more readily than from an aquifer with both large and small pores. In the latter case, the purge water flows through the large pores and the contaminated water is more or less trapped in the smaller pores.

Case Study: Ground-Water Remediation Using a Pump-and-Treat Technique Combined with Soil Washing

Thomsen et al. (1989) described a pump-and-treat system that combines soil washing with artificial recharge. The waste-disposal area for a manufacturing facility was located adjacent to a river. Over a period of time ground water beneath and adjoining the waste-disposal area became contaminated with solvents, primarily trichloroethene (TCE), perchloroethene (PCE), and 1,1,1-trichloroethane (TCA). A plume of contaminated ground water flowed into the river. Vertical contamination was limited primarily to a sand and gravel aquifer overlying an aquiclude, with limited contamination in the area below the aquiclude. A cross section of the site is shown in Figure 9.35.

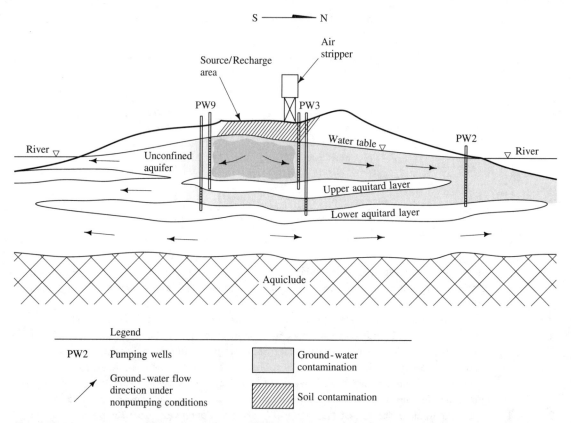

FIGURE 9.35 Cross section of stratigraphy showing the area of ground-water contamination at a solvent disposal facility site. *Source:* K. O. Thomsen et al., *Ground Water Monitoring Review* 9, no. 1(1989): 92–99. Copyright © 1989 Water Well Journal Publishing Co.

In all, 14 extraction wells were installed at nine locations. Five locations had two purge wells, each screened at a different depth, whereas at four locations there was only one purge well. The maximum extraction rate for all wells pumping at the same time was 600 gal/min. Figure 9.36 shows the location of the nine pumping locations. The cross section on Figure 9.35 shows three pumping locations, two of which have two wells and one with one well.

The extracted water was run through an air-stripping tower, which during the first season of operation had a 75% removal rate for trichloroethene with a flow rate of 210 gal/min and an influent TCE concentration of 4000 μg/L. The effluent from the air stripper was then spray-irrigated into a recharge basin that was constructed over the area of contaminated soil. The spray irrigation further reduced the TCE concentration by about 20%, so that the overall removal was about 95%. The recharge basin was surrounded with a berm to prevent runoff and was covered with gravel to promote infiltration. The maximum infiltration rate was 210 gal/min, which meant that not all the extraction wells could operate at once.

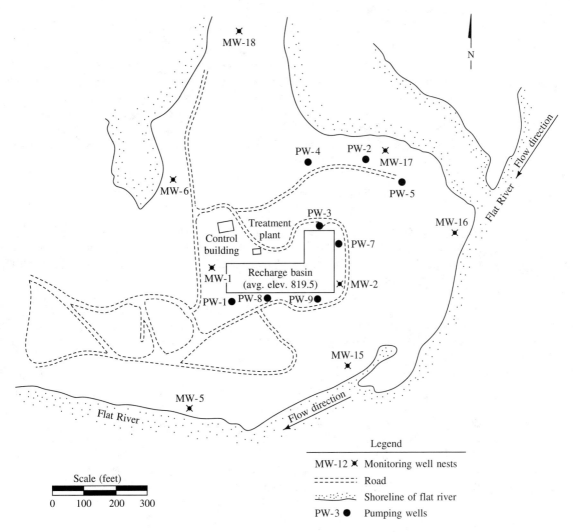

FIGURE 9.36 Plan view of location of recovery wells and recharge basin. *Source*: K. O. Thomsen et al. *Ground Water Monitoring Review* 9, no. 1 (1989): 92–99. Copyright © 1989 Water Well Journal Publishing Co.

The clean-up goal that was negotiated between the responsible party and the regulatory agency was 15 µg/L of TCE as an indicator parameter. This figure was agreed upon partly because of the TCE concentrations in the nearby river. The water being recharged into the basin was flushing the residual solvents out of the contaminated soil above the water table. The system operated from April through October, since spray irrigation could not be accomplished during the winter months at this location. During the initial season of operation, the water being pumped from the aquifer averaged 4000 µg/L of TCE and the water being recharged was about 200 µg/L of TCE. As remediation progresses, the TCE concentration in the purged water will decline, as will that in the recharging water. No estimate of the length of time necessary to effect remediation was given.

It is possible to remove 99% or more of TCE by air stripping. However, such a high removal rate would have resulted in a higher capital cost for a more efficient air-stripping column than the one that was used. Since the treated water was being recharged on site and recirculated through the system, it was not necessary to achieve the highest possible level of treatment. The system will probably need to be operated for a longer period of time than would have been necessary if the treated effluent being recharged were initially reduced to less than 15 μg/L of TCE. However, the longer operating costs are balanced by lower capital costs for the less efficient treatment system.

9.10   Summary

There is great interest in developing effective and efficient methods of remediating contaminated soils and ground water in order to meet the mandates of federal legislation such as RCRA and CERCLA. In order to have a successful remediation, it is necessary to first isolate or remove the source of the contamination. Sources can include such things as wastes spread on the land or improperly buried in the earth, leaking landfills, leaking underground storage tanks, or soils that have become contaminated with spills and leaks. If it is not possible to remove sources, they can be isolated by physical barriers such as slurry walls and impermeable covers or by hydrodynamic barriers created by pumping and injection wells.

Contaminated ground water can be pumped from the ground and treated. Any mobile NAPLs present should be removed before a pump-and-treat program is initiated. Residual NAPL and contaminants sorbed onto mineral surfaces and soil organic carbon will slowly partition into the clean ground water that replaces the contaminated ground water removed by pumping. This will greatly prolong the period of time that it takes to remediate the aquifer. If all the residual contamination is not removed from the aquifer, the concentration of contaminants will increase after the completion of a pump-and-treat operation. It may be impossible to remediate sites contaminated with DNAPLs by pump-and-treat methods. Such aquifers might require permanent plume-stabilization wells to prevent the spread of the plume.

Water that is extracted from the aquifer can be treated to remove both organic and inorganic contaminants. Methods of treatment of dissolved organics include air stripping and carbon absorption. Floating NAPLs can be removed by skimming pumps located in wells or trenches.

Underground tanks that are leaking should be removed and replaced. Soil and ground-water remediation is frequently necessary after a leaking tank has been removed.

Soil contaminated by residual organic compounds can be remediated by soil-vapor–extraction systems. Air containing organic vapors is drawn from the soil pores via vapor-extraction wells in the vadose zone and replaced with fresh air. Some methods exist that combine pumping water for treatment with soil-vapor extraction to treat contaminated soil and ground water simultaneously.

Soil and ground water contaminated by organic compounds may be amenable to bioremediation by both aerobic and anaerobic microbes. Nutrients and perhaps oxygen must be added to the soil or aquifer to encourage microbial activity. *In situ* bioremediation has the advantage of treating the contaminants dissolved in soil and ground water at the same time as residual contaminants.

Chapter Notation

B	Uniform thickness of a confined aquifer
U	Regional specific discharge (Darcy velocity)
Q	Pumping rate of plume-capture well
d	Distance between extraction wells
x	Distance along the x axis
y	Distance along the y axis
$h_0 - h$	Drawdown
T	Aquifer transmissivity
S	Aquifer storativity
r	Radius of a pumping well
t	Time since pumping began

References

Ardito, C. P., and J. F. Billings. 1990. "Alternative remediation strategies: The subsurface volatilization and ventilation system." *Proceedings of Petroleum Hydrocarbons and Organic Chemicals in Ground Water: Prevention, Detection and Restoration*, 281–96. Dublin, Ohio: National Water Well Association.

Atlas, R. M. 1975. Effects of temperature and crude oil composition on petroleum biodegradation. *Applied and Environmental Microbiology* 30:396–403.

Blake, S. B., and R. W. Lewis. 1982. "Underground oil recovery." *Proceedings of the Second National Symposium on Aquifer Restoration and Ground Water Monitoring*, 69–76. Dublin, Ohio: National Water Well Association.

Cookson, J. T., Jr., and J. E. Leszcynski. 1990. "Restoration of a contaminated drinking water aquifer." *Proceedings of the Fourth National Outdoor Action Conference on Aquifer Restoration, Ground Water Monitoring and Geophysical Methods*, 669–81. Dublin, Ohio: National Water Well Association.

Czarnecki, R. C. 1989. "Hot mix asphalt technology and the cleaning of contaminated soils." *Petroleum Contaminated Soils*, vol. 2. Edited by E. J. Calabrese and P. T. Kostecki, 267–78. Chelsea, Mich.: Lewis Publishers.

Doty, C. B., and C. C. Davis. 1991. *The effectiveness of groundwater pumping as a restoration technology*. Oak Ridge National Laboratory Report ORNL/TM-11866.

Eklund, K. 1989. "Incorporation of contaminated soils into bituminous concrete." In *Petroleum Contaminated Soils*, vol. 1. Edited by E. J. Calabrese and P. T. Kostecki, 191–200. Chelsea, Mich.: Lewis Publishers.

Fetter, C. W. 1988. *Applied Hydrology*, 2d. ed. New York: Macmillen Publishing Company, 588 pp.

Flathman, P. E., D. E. Jerger, and L. S. Bottomley. 1989. Remediation of contaminated ground water using biological techniques. *Ground Water Monitoring Review* 9, no. 1:105–19.

Freeze, R. A., and J. A. Cherry. 1989. What has gone wrong. *Ground Water* 27, no. 4:458–64.

Herzog, B. L., R. A. Griffin, C. J. Stohr, L. R. Follmer, W. J. Morse, and W. J. Su. 1989. Investigation of failure mechanisms and migration of organic chemicals at Wilsonville, Illinois. *Ground Water Monitoring Review* 9, no. 2:82–88.

Jamison, V. W., R. L. Raymond, and J. O. Hudson. 1975. Biodegradation of high octane gasoline in ground water. *Developments in Industrial Microbiology* 16:305 ff.

Javandel, I., and C-F Tsang. 1986. Capture zone type curves: a tool for aquifer cleanup. *Ground Water* 24, no. 5:616–25.

Konikow, L. F., and D. W. Thompson. 1984. "Groundwater contamination and aquifer reclamation at the Rocky Mountain Arsenal, Colorado." In *Groundwater Contamination*, 93–103. Washington, D. C.: National Academy Press.

Kostecki, P. T., E. J. Calabrese, and E. J. Fleischer. 1989. "Asphalt batching of petroleum contaminated soils as a viable remedial option." In *Petroleum Contaminated Soils*, vol. 1. Edited by E. J. Calabrese and P. T. Kostecki, 175–90. Chelsea, Mich.: Lewis Publishers.

Lee, M. D., J. M. Thomas, R. C. Borden, P. B. Bedient, C. H. Ward, and J. T. Wilson. 1988. Biorestoration of aquifers contaminated with organic compounds. *Critical Reviews in Environmental Control* 18, no. 1:29–89.

Lynch, E. R., S. W. Anagnost, G. A. Swenson, and R. K. Goldman. 1984. "Design and evaluation of in-place containment structures utilizing ground-water cutoff walls." *Proceedings of the Fourth National Symposium and Exposition on Aquifer Restoration and Ground Water Monitoring*, 1–7. Dublin, Ohio: National Water Well Association.

Lynch, J., and B. R. Genes. 1989. "Land treatment of petroleum contaminated soils." In *Petroleum Contaminated Soils*, vol. 1. Edited by E. J. Calabrese and P. T. Kostecki, 163–74. Chelsea, Mich.: Lewis Publishers.

Mackay, D. M., and J. A. Cherry. 1989. Groundwater contamination: pump and treat remediation. *Environmental Science and Technology* 23, no. 6:630–36.

Malot, J. J. 1989. "Cleanup of a gasoline contaminated site using vacuum extraction technology." In *Petroleum Contaminated Soils*, vol. 2. Edited by E. J. Calabrese and P. T. Kostecki, 283–301. Chelsea, Mich.: Lewis Publishers.

Mueller, J. G., P. J. Chapman, and P. H. Pritchard. 1989. Creosote contaminated sites: their potential for bioremediation. *Environmental Science and Technology* 23, no. 10:1197–1201.

Need, E. A, and M. J. Costello. 1984. "Hydrogeologic aspects of slurry wall isolation systems in areas of high downward gradients." *Proceedings of the Fourth National Symposium and Exposition on Aquifer Restoration and Ground Water Monitoring*, 18–26. Dublin, Ohio: National Water Well Association.

Nyer, E. K. 1985. *Groundwater Treatment Technology*. New York: Van Nostrand Reinhold, 188pp.

Raymond, R. L. 1974. Reclamation of Hydrocarbon Contaminated Ground Water. U.S. Patent 3,846,290, November 5, 1974.

Raymond, R. L., R. A. Brown, R. D. Norris, and E. T. O'Neill. 1986. Stimulation of Biooxidation Processes in Subterranean Formations. U.S. Patent 4,588, 506, May 13, 1986.

Roberts, P. V., G. D. Hopkins, D. M. Mackay, and L. Semprini. 1990. A field evaluation of in-situ biodegradation of chlorinated ethenes: part 1, methodology and field site characterization. *Ground Water* 28, no. 4:591–604.

Satkin, R. L., and P. B. Bedient. 1988. Effectiveness of various aquifer restoration schemes under variable hydrogeologic conditions. *Ground Water* 26, no. 4:488–98.

Schukle, R. J. 1982. "Rocky Mountain Arsenal ground-water reclamation program." *Proceedings of the Second National Symposium on Aquifer Restoration and Ground Water Monitoring*, 336–68. Columbus, Ohio, National Water Well Association.

Semprini, L., P. V. Roberts, G. D. Hopkins, and P. L. McCarty. 1990. A field evaluation of in-situ biodegradation of chlorinated ethenes: part 2, results of biostimulation and biotransformation experiments. *Ground Water* 28, no. 5:715–27.

Semprini, L., G. D. Hopkins, P. V. Roberts, D. Grbic-Galic, and P. L. McCarty. 1991. A field evaluation of in-situ biodegradation of chlorinated ethenes: part 3, studies of competitive competition. *Ground Water* 29, no. 2:239–250.

Shafer, J. M. 1987. Reverse pathline calculation of time-related capture zones in nonuniform flow. *Ground Water* 25, no. 3:283–89.

Thomas, J. M., and C. H. Ward. 1989. *In situ* biorestoration of organic contaminants in the subsurface. *Environmental Science and Technology* 23, no. 7:760–66.

Thomsen, K. O., M. A. Chaudhry, K. Dovantzis, and R. R. Riesing. 1989. Ground water remediation using an extraction, treatment and recharge system. *Ground Water Monitoring Review* 9, no. 1:92–99.

Trowbridge, B. E., and J. J. Malot. 1990. "Soil remediation and free product removal using *in situ* vacuum extraction with catalytic oxidation." *Proceedings of the Fourth National Outdoor Action Conference on Aquifer Restoration, Ground Water Monitoring and Geophysical Methods*, 559–70. Dublin, Ohio: National Water Well Association.

Valkenburg, Nicholas, 1991. Vice President, Geraghty and Miller, Inc. Personal communication.

Wall, H. O., and M. K. Richards. 1990. "The incineration of arsenic-contaminated soils related to the Comprehensive Environmental Response, Compensation and Liability Act (CERCLA)". In *Remedial Action, Treatment and Disposal of Hazardous Waste*, 50–55. U.S. EPA EPA/600/9-90 037.

Whiffin, R. B., and J. M. Bahr. 1984. "Assessment of purge well effectiveness for aquifer decontamination." *Proceedings of the Fourth National Symposium and Exposition on Aquifer Restoration and Ground Water Monitoring*, 75–81. Dublin, Ohio: National Water Well Association.

Wilson, J. L. 1984. "Double-cell hydraulic containment of pollutant plumes." *Proceedings of the Fourth National Symposium and Exposition on Aquifer Restoration and Ground Water Monitoring*, 65–70. Dublin, Ohio: Natonal Water Well Association.

Yaniga, P. M. 1982. "Alternatives in decontamination for hydrocarbons-contaminated aquifers." *Proceedings of the Second National Symposium on Aquifer Restoration and Ground Water Monitoring*, 47–57. Dublin, Ohio: National Water Well Association.

Appendix A
Error Function Values

Values of the error function, erf(x), and the complementary error function, erfc(x), for positive values of x

x	erf(x)	erfc(x)	x	erf(x)	erfc(x)
0	0	1.0	1.1	0.880205	0.119795
0.05	0.056372	0.943628	1.2	0.910314	0.089686
0.1	0.112463	0.887537	1.3	0.934008	0.065992
0.15	0.167996	0.832004	1.4	0.952285	0.047715
0.2	0.222703	0.777297	1.5	0.966105	0.033895
0.25	0.276326	0.723674	1.6	0.976348	0.023652
0.3	0.328627	0.671373	1.7	0.983790	0.016210
0.35	0.379382	0.620618	1.8	0.989091	0.010909
0.4	0.428392	0.571608	1.9	0.992790	0.007210
0.45	0.475482	0.524518	2.0	0.995322	0.004678
0.5	0.520500	0.479500	2.1	0.997021	0.002979
0.55	0.563323	0.436677	2.2	0.998137	0.001863
0.6	0.603856	0.396144	2.3	0.998857	0.001143
0.65	0.642029	0.357971	2.4	0.999311	0.000689
0.7	0.677801	0.322199	2.5	0.999593	0.000407
0.75	0.711156	0.288844	2.6	0.999764	0.000236
0.8	0.742101	0.257899	2.7	0.999866	0.000134
0.85	0.770668	0.229332	2.8	0.999925	0.000075
0.9	0.796908	0.203092	2.9	0.999959	0.000041
0.95	0.820891	0.179109	3.0	0.999978	0.000022
1.0	0.842701	0.157299			

Appendix B
Bessel Functions

Modified Bessel functions of the second kind and zero order, $K_0(x)$

x	$K_0(x)$	x	$K_0(x)$	x	$K_0(x)$
0.010	4.721	0.040	3.336	0.070	2.780
0.011	4.626	0.041	3.312	0.071	2.766
0.012	4.539	0.042	3.288	0.072	2.752
0.013	4.459	0.043	3.264	0.073	2.738
0.014	4.385	0.044	3.241	0.074	2.725
0.015	4.316	0.045	3.219	0.075	2.711
0.016	4.251	0.046	3.197	0.076	2.698
0.017	4.191	0.047	3.176	0.077	2.685
0.018	4.134	0.048	3.155	0.078	2.673
0.019	4.080	0.049	3.134	0.079	2.660
0.020	4.028	0.050	3.114	0.080	2.647
0.021	3.980	0.051	3.094	0.081	2.635
0.022	3.933	0.052	3.075	0.082	2.623
0.023	3.889	0.053	3.056	0.083	2.611
0.024	3.846	0.054	3.038	0.084	2.599
0.025	3.806	0.055	3.019	0.085	2.587
0.026	3.766	0.056	3.001	0.086	2.576
0.027	3.729	0.057	2.984	0.087	2.564
0.028	3.692	0.058	2.967	0.088	2.553
0.029	3.657	0.059	2.950	0.089	2.542
0.030	3.623	0.060	2.933	0.090	2.531
0.031	3.591	0.061	2.916	0.091	2.520
0.032	3.559	0.062	2.900	0.092	2.509
0.033	3.528	0.063	2.884	0.093	2.499
0.034	3.499	0.064	2.869	0.094	2.488
0.035	3.470	0.065	2.853	0.095	2.478
0.036	3.442	0.066	2.838	0.096	2.467
0.037	3.414	0.067	2.823	0.097	2.457
0.038	3.388	0.068	2.809	0.098	2.447
0.039	3.362	0.069	2.794	0.099	2.437

Source: Adapted from M. S. Hantush, "Analysis of Data From Pumping Tests in Leaky Aquifers," *Transactions, American Geophysical Union* 37 (1956):702–14.

Modified Bessel functions of the second kind and zero order, $K_0(x)$

x	$K_0(x)$	x	$K_0(x)$	x	$K_0(x)$
0.10	2.427	0.60	0.777	1.0	0.421
0.11	2.333	0.61	0.765	1.1	0.366
0.12	2.248	0.62	0.752	1.2	0.318
0.13	2.169	0.63	0.740	1.3	0.278
0.14	2.097	0.64	0.728	1.4	0.244
0.15	2.030	0.65	0.716	1.5	0.214
0.16	1.967	0.66	0.704	1.6	0.188
0.17	1.909	0.67	0.693	1.7	0.165
0.18	1.854	0.68	0.682	1.8	0.146
0.19	1.802	0.69	0.671	1.9	0.129
0.20	1.753	0.70	0.660	2.0	0.114
0.21	1.706	0.71	0.650	2.1	0.101
0.22	1.662	0.72	0.640	2.2	0.0893
0.23	1.620	0.73	0.630	2.3	0.0791
0.24	1.580	0.74	0.620	2.4	0.0702
0.25	1.541	0.75	0.611	2.5	0.0623
0.26	1.505	0.76	0.601	2.6	0.0554
0.27	1.470	0.77	0.592	2.7	0.0493
0.28	1.436	0.78	0.583	2.8	0.0438
0.29	1.404	0.79	0.574	2.9	0.0390
0.30	1.372	0.80	0.565	3.0	0.0347
0.31	1.342	0.81	0.557	3.1	0.0310
0.32	1.314	0.82	0.548	3.2	0.0276
0.33	1.286	0.83	0.540	3.3	0.0246
0.34	1.259	0.84	0.532	3.4	0.0220
0.35	1.233	0.85	0.524	3.5	0.0196
0.36	1.207	0.86	0.516	3.6	0.0175
0.37	1.183	0.87	0.509	3.7	0.0156
0.38	1.160	0.88	0.501	3.8	0.0140
0.39	1.137	0.89	0.494	3.9	0.0125
0.40	1.114	0.90	0.487	4.0	0.0112
0.41	1.093	0.91	0.480	4.1	0.0100
0.42	1.072	0.92	0.473	4.2	0.0089
0.43	1.052	0.93	0.466	4.3	0.0080
0.44	1.032	0.94	0.459	4.4	0.0071
0.45	1.013	0.95	0.452	4.5	0.0064
0.46	0.994	0.96	0.446	4.6	0.0057
0.47	0.976	0.97	0.440	4.7	0.0051
0.48	0.958	0.98	0.433	4.8	0.0046
0.49	0.941	0.99	0.427	4.9	0.0041
0.50	0.924			5.0	0.0037
0.51	0.908				
0.52	0.892				
0.53	0.877				
0.54	0.861				
0.55	0.847				
0.56	0.832				
0.57	0.818				
0.58	0.804				
0.59	0.791				

Appendix C
W(t, B) Values

Values of the function W(t,B) for various values of t

t \ B	0.002	0.004	0.006	0.008	0.01	0.02	0.04	0.06	0.08
0	12.7	11.3	10.5	9.89	9.44	8.06	6.67	5.87	5.29
0.000002	12.1	11.2	10.5	9.89	9.44	8.06	6.67	5.87	5.29
0.000004	11.6	11.1	10.4	9.88	9.44	8.06	6.67	5.87	5.29
0.000006	11.3	10.9	10.4	9.87	9.44	8.06	6.67	.87	5.29
0.000008	11.0	10.7	10.3	9.84	9.43	8.06	6.67	5.87	5.29
0.00001	10.8	10.6	10.2	9.80	9.42	8.06	6.67	5.87	5.29
0.00002	10.2	10.1	9.84	9.58	9.30	8.06	6.67	5.87	5.29
0.00004	9.52	9.45	9.34	9.19	9.01	8.03	6.67	5.87	5.29
0.00006	9.13	9.08	9.00	8.89	8.77	7.98	6.67	5.87	5.29
0.00008	8.84	8.81	8.75	8.67	8.57	7.91	6.67	5.87	5.29
0.0001	8.62	8.59	8.55	8.48	8.40	7.84	6.67	5.87	5.29
0.0002	7.94	7.92	7.90	7.86	7.82	7.50	6.62	5.86	5.29
0.0004	7.24	7.24	7.22	7.21	7.19	7.01	6.45	5.83	5.29
0.0006	6.84	6.84	6.83	6.82	6.80	6.68	6.27	5.77	5.27
0.0008	6.55	6.55	6.54	6.53	6.52	6.43	6.11	5.69	5.25
0.001	6.33	6.33	6.32	6.32	6.31	6.23	5.97	5.61	5.21
0.002	5.64	5.64	5.63	5.63	5.63	5.59	5.45	5.24	4.98
0.004	4.95	4.95	4.95	4.94	4.94	4.92	4.85	4.74	4.59
0.006	4.54	4.54	4.54	4.54	4.54	4.53	4.48	4.41	4.30
0.008	4.26	4.26	4.26	4.26	4.26	4.25	4.21	4.15	4.08
0.01	4.04	4.04	4.04	4.04	4.04	4.03	4.00	3.95	3.89
0.02	3.35	3.35	3.35	3.35	3.35	3.35	3.34	3.31	3.28
0.04	2.68	2.68	2.68	2.68	2.68	2.68	2.67	2.66	2.65
0.06	2.30	2.30	2.30	2.30	2.30	2.29	2.29	2.28	2.27
0.08	2.03	2.03	2.03	2.03	2.03	2.03	2.02	2.02	2.01
0.1	1.82	1.82	1.82	1.82	1.82	1.82	1.82	1.82	1.81
0.2	1.22	1.22	1.22	1.22	1.22	1.22	1.22	1.22	1.22
0.4	0.702	0.702	0.702	0.702	0.702	0.702	0.702	0.702	0.701
0.6	0.454	0.454	0.454	0.454	0.454	0.454	0.454	0.454	0.454
0.8	0.311	0.311	0.311	0.311	0.311	0.311	0.311	0.310	0.310
1	0.219	0.219	0.219	0.219	0.219	0.219	0.219	0.219	0.219
2	0.049	0.049	0.049	0.049	0.049	0.049	0.049	0.049	0.049
4	0.0038	0.0038	0.0038	0.0038	0.0038	0.0038	0.0038	0.0038	0.0038
6	0.0004	0.0004	0.0004	0.0004	0.0004	0.0004	0.0004	0.0004	0.0004
8	0	0	0	0	0	0	0	0	0

Source: After M. S. Hantush, "Analysis of data from Pumping Tests in Leaky Aquifers," *Transactions, American Geophysical Union*, 37 (1956):702–14.

W(t, B) Values

t \ B	0.1	0.2	0.4	0.6	0.8	1	2	4	6	8
0	4.85	3.51	2.23	1.55	1.13	0.842	0.228	0.0223	0.0025	0.0003
0.000002	4.85	3.51	2.23	1.55	1.13	0.842	0.228	0.0223	0.0025	0.0003
0.000004	4.85	3.51	2.23	1.55	1.13	0.842	0.228	0.0223	0.0025	0.0003
0.000006	4.85	3.51	2.23	1.55	1.13	0.842	0.228	0.0223	0.0025	0.0003
0.000008	4.85	3.51	2.23	1.55	1.13	0.842	0.228	0.0223	0.0025	0.0003
0.00001	4.85	3.51	2.23	1.55	1.13	0.842	0.228	0.0223	0.0025	0.0003
0.00002	4.85	3.51	2.23	1.55	1.13	0.842	0.228	0.0223	0.0025	0.0003
0.00004	4.85	3.51	2.23	1.55	1.13	0.842	0.228	0.0223	0.0025	0.0003
0.00006	4.85	3.51	2.23	1.55	1.13	0.842	0.228	0.0223	0.0025	0.0003
0.00008	4.85	3.51	2.23	1.55	1.13	0.842	0.228	0.0223	0.0025	0.0003
0.0001	4.85	3.51	2.23	1.55	1.13	0.842	0.228	0.0223	0.0025	0.0003
0.0002	4.85	3.51	2.23	1.55	1.13	0.842	0.228	0.0223	0.0025	0.0003
0.0004	4.85	3.51	2.23	1.55	1.13	0.842	0.228	0.0223	0.0025	0.0003
0.0006	4.85	3.51	2.23	1.55	1.13	0.842	0.228	0.0223	0.0025	0.0003
0.0008	4.84	3.51	2.23	1.55	1.13	0.842	0.228	0.0223	0.0025	0.0003
0.001	4.83	3.51	2.23	1.55	1.13	0.842	0.228	0.0223	0.0025	0.0003
0.002	4.71	3.50	2.23	1.55	1.13	0.842	0.228	0.0223	0.0025	0.0003
0.004	4.42	3.48	2.23	1.55	1.13	0.842	0.228	0.0223	0.0025	0.0003
0.006	4.18	3.43	2.23	1.55	1.13	0.842	0.228	0.0223	0.0025	0.0003
0.008	3.98	3.36	2.23	1.55	1.13	0.842	0.228	0.0223	0.0025	0.0003
0.01	3.81	3.29	2.23	1.55	1.13	0.842	0.228	0.0223	0.0025	0.0003
0.02	3.24	2.95	2.18	1.55	1.13	0.842	0.228	0.0223	0.0025	0.0003
0.04	2.63	2.48	2.02	1.52	1.13	0.842	0.228	0.0223	0.0025	0.0003
0.06	2.26	2.17	1.85	1.46	1.11	0.839	0.228	0.0223	0.0025	0.0003
0.08	2.00	1.94	1.69	1.39	1.08	0.832	0.228	0.0223	0.0025	0.0003
0.1	1.80	1.75	1.56	1.31	1.05	0.819	0.228	0.0223	0.0025	0.0003
0.2	1.22	1.19	1.11	0.996	0.857	0.715	0.227	0.0223	0.0025	0.0003
0.4	0.700	0.693	0.665	0.621	0.565	0.502	0.210	0.0223	0.0025	0.0003
0.6	0.453	0.450	0.436	0.415	0.387	0.354	0.177	0.0222	0.0025	0.0003
0.8	0.310	0.308	0.301	0.289	0.273	0.254	0.144	0.0218	0.0025	0.0003
1	0.219	0.218	0.213	0.206	0.197	0.185	0.114	0.0207	0.0025	0.0003
2	0.049	0.049	0.048	0.047	0.046	0.044	0.034	0.011	0.0021	0.0003
4	0.0038	0.0038	0.0038	0.0037	0.0037	0.0036	0.0031	0.0016	0.0006	0.0002
6	0.0004	0.0004	0.0004	0.0004	0.0004	0.0004	0.0003	0.0002	0.0001	0
8	0	0	0	0	0	0	0	0	0	0

Appendix D
Exponential Integral

Values of the exponential integral, $Ei(x)$

$$Ei(x) = \int_{-\infty}^{x} \frac{e^v}{v} dv$$

x	Ei(x)	x	Ei(x)	x	Ei(x)	x	Ei(x)
0.0	$-\infty$	3.0	9.93383	6.0	85.9898	9.0	1,037.88
0.1	−1.62281	3.1	10.6263	6.1	93.0020	9.1	1,132.04
0.2	−0.82176	3.2	11.3673	6.2	100.626	9.2	1,234.96
0.3	−0.30267	3.3	12.1610	6.3	108.916	9.3	1,347.48
0.4	0.10477	3.4	13.0121	6.4	117.935	9.4	1,470.51
0.5	0.45422	3.5	13.9254	6.5	127.747	9.5	1,605.03
0.6	0.76988	3.6	14.9063	6.6	138.426	9.6	1,752.14
0.7	1.06491	3.7	15.9606	6.7	150.050	9.7	1,913.05
0.8	1.34740	3.8	17.0948	6.8	162.707	9.8	2,089.05
0.9	1.62281	3.9	18.3157	6.9	176.491	9.9	2,281.58
1.0	1.89512	4.0	19.6309	7.0	191.505	10.0	2,492.23
1.1	2.16738	4.1	21.0485	7.1	207.863	10.5	3,883.74
1.2	2.44209	4.2	22.5774	7.2	225.688	11.0	6,071.41
1.3	2.72140	4.3	24.2274	7.3	245.116	11.5	9,518.20
1.4	3.00721	4.4	26.0090	7.4	266.296	12.0	14,959.5
1.5	3.30128	4.5	27.9337	7.5	289.388	12.5	23,565.1
1.6	3.60532	4.6	30.0141	7.6	314.572	13.0	37,197.7
1.7	3.92096	4.7	32.2639	7.7	342.040	13.5	58,827.0
1.8	4.24987	4.8	34.6979	7.8	372.006	14.0	93,193.0
1.9	4.59371	4.9	37.3325	7.9	404.701	14.5	147,866
2.0	4.95423	5.0	40.1853	8.0	440.380	15.0	234,955
2.1	5.33324	5.1	43.2757	8.1	479.322		
2.2	5.73261	5.2	46.6249	8.2	521.831		
2.3	6.15438	5.3	50.2557	8.3	568.242		
2.4	6.60067	5.4	54.1935	8.4	618.919		
2.5	7.07377	5.5	58.4655	8.5	674.264		
2.6	7.57611	5.6	63.1018	8.6	734.714		
2.7	8.11035	5.7	68.1350	8.7	800.749		
2.8	8.67930	5.8	73.6008	8.8	872.895		
2.9	9.28602	5.9	79.5382	8.9	951.728		

Appendix E
Unit Abbreviations

Length (L)		Radiation	
m	meter	Ci	curie
cm	centimeter	pCi	picocurie
mm	millimeter	Bq	becquerel
μm	micrometer (micron)	Gy	gray
in.	inch	rad	radiation dose
ft	foot	rem	roentgen equivalent man
mi	mile	Sv	sievert
Mass (M)		**Volume (M^3)**	
kg	kilogram	L	liter
g	gram	mL	milliliter
mg	milligram	gal	gallon
μg	microgram	m^3	cubic meters
Time (T)		**Pressure ($ML^{-1}T^{-2}$)**	
sec	second	atm	atmosphere
min	minute	N/m^2	newtons per square meter
hr	hour	Pa	pascal
mo	month	**Temperature and Heat**	
yr	year	°C	degrees Celsius
Weight (ML/T^2)		K	kelvin
lb	pound	Btu	British thermal unit
Acceleration (LT^{-2})		**Area (M^2)**	
mi/hr/sec	miles per hour per second	ha	hectares
m/sec^2	meters per second per second	m^2	square meters
Discharge (L^3T^{-1})		**Molecular Weight**	
gal/min	gallons per minute	mol	mole
Concentration (ML^{-3})		M	molar
mg/L	milligrams per liter	**Velocity (LT^{-1})**	
μg/L	micrograms per liter	cm/sec	centimeters per second
		ft/hr	feet per hour
		ft/day	feet per day
		mi/hr	miles per hour

Index

Above-ground
    storage tanks, 23
Absorption, 117
Accuracy (of chemical analysis), 334
Acenaphthalene, 319
Acetaldehyde, 311
Acetic acid, 311
Acetone, 311, 334
Acrylamide, 315
Acrylic acid, 311–12.
Acrylonitrile, 315
Activated carbon, 417
Activity coefficient, 246
Adepic acid, 312
Adsorption, 117
Advection (advective transport), 47–48
Advection-dispersion equation, derivation of, 52–54
Aerobic biodegradation, 145, 147, 289–90, 429–33
Air entry value. *See* Bubbling pressure
Air sparging, 431
Air stripping, 417–18, 440
Alcohols, 308
Aldehydes, 311
Aliphatic hydrocarbons, 297–300
Alkanes, 297–99
Alkenes, 300
Alpha particle, 281
Americium, 150
Amides, 315
Amines, 315

Anaerobic biodegradation, 146, 289–90
Animal burials, 23
Anion exclusion, 190–93
Anisotropic, 35
Anisotropy ratio, 83
Annular seal, 347–48
Anthracene, 319
Apparent longitudinal dispersivity, 91
Aromatic hydrocarbons, 300–301
Arsenic
    chemistry of, 274–76
    ground water contaminated with, 31–32
Asbestos fibers, 150
Atmospheric pollutants, 27
Atrazine, 26, 315
Auger drilling, 354–57
Autocorrelation function, 80
Autocovariance, 81
Average linear velocity, 47, 52, 119
Average solute front velocity, 119

Bacteria, 150
Bailer, 371, 380–83
Barium, 277
Base/neutral fraction, 331
Becquerel, 282
Bentonite, 348, 366
1,2-Benzanthracene, 319
Benzene, 18, 221, 223, 316, 319, 328, 329, 330, 435

Benzene ring, 297, 300–301, 318
Benzo[*a*]anthracene, 319
Benzo[*a*]fluorene, 319
Benzo[*a*]pyrene, 319
Beryllium, 277
Beta particle, 281
Bilinear adsorption model, 130
BIO1D model
    description, 147
    used to compare equilibrium sorption models, 123–26
    used to illustrate retardation and decay, 149
Biodegradation, 144–49
Biofilm, 145
Bioremediation, 399, 429–34
bis(2-Ethylhexyl)phthalate, 334
Bladder pump, 380–83
Boiling point
    definition, 295
    relationship to molecular weight, 304
Borden (Ontario) landfill, 96–102, 150–57
Borings
    abandonment, 375
    as source of contamination, 28
    defined, 360
Boundary conditions, 56–57
Bromine, 271
Bromoform, 342
1-Bromopropane, 322
Brooks-Corey pore-size distribution index, 172
Brooks-Corey soil parameters, 172, 226–28

Index

Bubbling pressure, 169–70, 205
Buckingham flux law, 182–83
Buckley-Leverett equation for two-phase flow, 237–38
2-Butanone, 334

Cadmium, 280
Cancer risk, 14–16
Capillary fringe, 217–18
Capillary potential, 168
Capillary pressure, 169, 204
Capillary zone, 218
Capture zones, 403–16
 computation of, 405–14
 defined, 403–5
 optimization, 414–15
Carbazole, 319
Carbon tetrachloride, 151, 155, 156
Carbonyl group, 311
Carboxyl group, 311
Carboxylic acids, 311–12
Casing, 339–44
Casing diameter, 339–41
Casing material, 342–44
Cation exchange, 117, 166–67
Center of mass, 82
Centrifugal pump, 380–83
CERCLA, 392, 393
Cesspools, 16
CFC-113, 327
CFC-123a, 327
CFC-1113, 327
Chelating agents, 267, 269
Chemical activity, 246
Chemical equilibrium, 141
Chemical kinetics, 141
Chemical reactions, classification of, 116–17
Chemisorption, 117
Chlorine, 271
Chlorite, 164
Chlorobenzene, 147
Chloroethane, 327
Chlorophenol, 313
Cholera, 15
Chromatographic effect, 151
Chromium, 277–78
Coal-tar, 32, 301
Cobalt, 269, 278–79

Colloidal transport, 149–50
Colloids, 149, 164–65
Cometabolism, 433
Complementary error function, 46
Complexation, 267–70
Comprehensive Environmental Response, Compensation and Liability Act, 392, 393
Concentration, units of, 244
Conservative solute, 53
Containment (of waste), 393–99
Continuous injection, 61–63
Convection, 47–48
Copper, 279
Core sample, 361
Correlation coefficient, 133
Correlation length, 81
Cover, landfill, 396, 398
Creosote, 32, 301
Cresols, 313, 329
Cross-contamination, 353
Crude oil, 301
Curie, 281
Cut-off wall, 393–96
Cycloalkanes, 299
Cyclobutane, 299
Cyclohexane, 223

Darcy flux, 35
Darcy's law
 defined, 35
 for two-phase flow, 211–12
DCPD, 22
DDT, 323, 325
Dealkoxylation, 326
Dealkylation, 326
Debye-Huckel equation, 246
Decarboxylation, 326
Decontamination, 353–54
Dehalogenation, 326
Dehydrohalogenation, 322
Deicing salt, 26
Denitrification, 272
Dense nonaqueous phase liquids (DNAPL). *See also* NAPL
 definition, 202
 migration in saturated zone
  horizontal, 235–38

Dense nonaqueous phase liquids (DNAPL), *continued*
 vertical, 233–35
 migration in vadose zone, 231–33
Derivatives, 32–35
Detection limit, 11, 332
Deterministic model, 77
Dibromochloropropane (DBCP), 322
1,2-Dichlorobenzene, 151, 156
1,4-Dichlorobenzene, 147, 148
1,1-Dichloroethane, 327
1,2-Dichloroethane, 327
1,1-Dichloroethene (1,2-Dichloroethylene), 322, 327
1,2-Dichloroethene (DCE, 1,2-Dichloroethylene), 300, 433–34
Diesel fuel, 202
Diethyl ether, 402, 403
Dihaloelimination, 323
2,4-Dimethylphenol, 329
di-n-Butyl phthalate, 313, 334
di-n-Octyl phthalate, 334
Differential equations, 32–41
Diffusion, 43–47
Diffusion-controlled rate law, 130
DIMP, 22
1,4-Dioxane, 308, 311
Dispersivity, 50
 field tests, 68–70
 lab tests, 66–67
Displacement pressure. *See* Bubbling pressure
Disposal, residential, 21
Distribution coefficient, 118
Distribution coefficient for soil organic carbon, K_{oc}, 132
 estimated from molecular structure, 138–40
 estimated from octanol-water partition coefficient, 133–34
 estimated from solubility data, 134–37
Divergence, 37
Dose equivalent, 282

Dot product, 37
Drainage curve. See Drying curve
Drilling, at contaminated sites, 359
Drilling methods, 354–59
   air rotary, 357–59
   cable tool, 359
   hollow stem augers, 354–57
   mud rotary, 357
   reverse rotary, 359
Drilling mud, 354
Drinking water
   risk, 14–15
   standards, 11–14
Drying curve, 175, 205
Dumps, open, 21
Dynamic dispersivity, 50

Effective diffusion coefficient, 44
Effective porosity, 47
Effective saturation, 172
Eh
   definition, 250
   Eh-pH diagrams
      defined, 254
         stability field for dissolved iron calculated, 257–67
         stability field for water calculated, 254–57
   relationship with pH, 253–54
Electron acceptor, 249
Electrostatic double layer, 165–67, 190
Ensemble mean concentration, 81
EPA analytical methods for organics, 330–32
Epoxidation, 323
Equilibrium constant, 141, 245
Equilibrium sorption isotherm, 117
Equilibrium surface reactions, 117–28
Equivalent weight, 244
Ergodic hypothesis, 80
Esters, 312
Estimated value (of chemical analysis), 332
Ether cleavage, 326
Ethers, 308–11

Ethylbenzene, 221, 223, 328, 435
Ethylenediaminetetraacetate, 269–70
Ethyl formate, 312

Fertilizer application, 26
Fick's first law, 43–44
Fick's second law, 44
Field blank, 334
Filter pack, 346–47
Fingering, 196
First-type boundary. See Fixed-concentration boundary
Fixed-concentration boundary, 56, 57–58
Fixed-gradient boundary, 56, 58–60
Fixed-step function, 58
Flame-ionization detector, 329
Flow equation, derivation of 37–40
Fluid potential, 212–17
Fluoranthene, 319
Fluorene, 319
Fluoride, 270–71
Fluorite, 271
Force vector, 214–15
Formaldehyde, 311
Fractal geometry approach to solute transport, 85–92
   fractal cutoff limit, 88, 89
   fractal dimension, 87, 89
   fractal mathematics, 85–88
   fractal streamtube, 89
   fractal tortuosity, 88
   fractal velocity, 88
Fracture flow, 99, 103–7
Fractured media
   monitoring for DNAPLs in, 238–39
   solute transport in, 103–7
Free energy, 251–52
Freundlich sorption isotherm, 119–22
Fulvic acids, 269
Funicular water, 217
Funneling, 196

Gamma ray, 281
Gas chromatography, 329

Gasoline, 202, 221, 223
Gaussian distribution, 44
GC/MS, 330, 332
Gear drive pump, 380
Geochemical zonation, 288–92
Geomembrane, 398
Geometric mean, 82
Geotextile, 398
Gibbs free energy, 251–52
Graveyards, 23
Grey (as a unit of radiation), 282
Ground water
   contaminants
      sources of, 15–31
      treatment of, 416–18
      types of, 2–10
   cost of analysis, 11
   cut-off wall, 393–96
   monitoring, 349–51, 376–78
   treatment, 416–18
   usage in U.S., 1–2
Grout, 347–48
Gypsum, 272

Head related to fluid potential, 216–17
Henry's law
   definition, 223
   effect on soil-gas distribution, 384
   equation, 296
Henry's law constant, 223
Hexachloroethane, 151, 153, 156, 157, 323, 342
Hexane, 223, 224
Hobson's formula, 233
Homologous series, 295
Humic acid, 269
Humic substances, 269
Humin, 269
Humus, 165
Hurst coefficient, 91
Hydraulic conductivity, 35, 36, 37
   salinity effects on, 167–68
   unsaturated. See Unsaturated hydraulic conductivity
Hydrocarbons
   aliphatic hydrocarbons, 297–300

Hydrocarbons, *continued*
 aromatic hydrocarbons, 300–301
 biologic degradation of, 318–19, 429–33
 petroleum distillates, 301–5
Hydrodynamic dispersion, 51–52
Hydrodynamic dispersion coefficient, 51–52
Hydrodynamic isolation, 399–401
Hydrogenolysis, 323
Hydrolysis, 322, 326
Hydrophobic effect, 132
Hydroxylation, 326
Hysteresis, 175–76, 180, 205, 209, 211

Illite, 164
Imbibition curve. *See* Wetting curve
Imbibition pressure. *See* Bubbling pressure
Impoundments, 21–22
Incineration, 24
Initial conditions, 56–57
Injection wells, 18
Ink bottle effect, 175
Inner product, 37
Integral scale, 81
Interception trench, 421, 423
Interfacial tension, 203–04
Ionizing radiation, 281
Iron
 chemistry of, 257
 stability field of, 257–67
Irreducible nonwetting-fluid saturation, 205, 209
Irreducible wetting-fluid saturation, 205, 209
Irreversible first-order kinetic sorption model, 129
Irrigation, 25
Isomers, 298
Isotherms, 117–26, 187
Isotropic, 35

Kaolinite, 164
Ketones, 311

Kinetic sorption model, 117, 129–31
Kjeldahl nitrogen, 290

Lab duplicate, 334
Lag, 80
Lagoons, 21–22
Land application, 19
Landfill cover, 396, 398
Landfills, 19–20, 392–96
Langmuir sorption isotherm, 122–23
Langmuir two-surface sorption isotherm, 123
Leachate, 19, 288
Lead, 280–81
Leaking underground storage tanks, 424–26
Ligand, 267
Light nonaqueous phase liquids (LNAPL), 202
 defined, 202
 effect of rise and fall of water table, 231
 estimating recoverable amounts, 226–31
 measurement of thickness, 225–31
 migration, 217–24
 recovery of, 418–24
 treatment of, 427–37
Limit of quantification, 332
Linear sorption isotherm, 117–19, 187
Lognormal distribution, 75
Longitudinal dispersion, 50, 83, 84, 85
Longitudinal hydrodynamic dispersion, 51, 53
Lysimeter, 385–89

Macrodispersion, 73
Macromolecules, 150
Macropores, 196
Mass action, law of, 245–49
Mass conservation, law of, 37
Mass spectrometry, 329
Mass spectrum, 329
Mass transport
 equilibrium models, 186–88

Mass transport, *continued*
 in fractured rocks, 103–7
 in saturated zone, 43–114
 in unsaturated zone, 185–96
Material stockpiles, 23
Mathematical notation, 40–41
 del notation, 41
 Einstein's summation notation, 40
 index notation, 40
 tensor notation, 40
 vector notation, 40
Matric potential, 168, 169
 measurement of, 177–79
Maximum contaminant level (MCL), 11, 14
Maximum contaminant level goal (MCLG), 11
Mean, 44, 78
Mechanical dispersion, 49–50
Melting point, 295
Mercaptans, 315, 322
Mercury, 280
Metabolite, 315
Metal chemistry, 276–81
Metal complexes, 267–70
Methane, 289–90
Method blank, 334
Method detection limit, 332
Methylation, 326
Methylene chloride, 18, 334
Methyl ethyl ketone, 311, 344
1-Methylfluorene, 319
Methyl isobutyl ketone, 311
2-Methylnaphthalene, 319
2-Methylphenol, 329
Methyl propyl ketone, 311
Michaelis-Menten function. *See* Monad function
Microbial degradation of organic compounds, 316–29
 of aromatic compounds, 328–29
 of chlorinated ethanes and ethenes, 327–28
 of chlorinated hydrocarbons, 319–23
 of hydrocarbons, 318–19
 of organic pesticides, 323, 325

Microorganisms, 145, 146, 147
Mine drainage, 27
Mine wastes, 22–23
Mobile oil, 225–26, 231
Moffit Naval Air Station, California, 433–34
Molal solution, definition, 244
Molarity, 136
Molar solution, definition, 244
Mole, 136
  defined, 244
Molecular connectivity index, 138
Molecular diffusion, 43–47
Molecular topology, 138–39
Mole fraction, 136
Molybdenum, 279
Monad function, 145, 146
Monitoring for DNAPLs and LNAPLs, 238–39
Monitoring wells
  abandonment, 375
  annular seal, 347–48
  as a source of contamination, 28
  casing, 339–44
  design, 338–52
  development, 370–73
  factors to consider in design, 339
  installation, 364–70
  installation of borehole, 353–60
  measuring DNAPL, 234, 238–39
  measuring LNAPL, 225–26, 238
  multiple-level, 376–78
  protective casing, 348–49
  purging, 379–80
  purposes of, 338
  sampling, 378–83
  screen design, 345–46
  screen length and setting, 349–51
Montmorillonite, 164
Multiple solute effect, 140

Naphthalene, 300, 319, 329
Nernst equation, 250
Nickel, 279
Nitrile, 315
Nitrogen
  denitification, 272
  forms of, 272
  groundwater contamination, 272–74
  isotope fractions, 274
  total reduced, 290
Nitrophenol, 313–14
Nitrotoluenes, 314
Nonaqueous phase liquids, 202
  monitoring for, 349–51
  pump and treat systems for, 401
  recovery of, 418–24
  See also Light nonaqueous phase liquids; Dense nonaqueous phase liquids
Nonequilibrium sorption models, 129–31

Octane, 223, 224
Octanol-water partition coefficient, 132–33
  definition, 296
  used to estimate K_{oc}, 133–34
Oil table, 219–21
Oil-water separator, 421, 424
Olefins, 300
Organic compounds
  degradation of, 316–26
    aromatic compounds, 328–29
    chlorinated ethanes and ethenes, 327–28
    chlorinated hydrocarbons, 319–23
    hydrocarbons, 318–19
    organic pesticides, 323, 325
Organic halides, 305–8, 309, 319–28
Oxidation, Beta, 326
Oxidation of organic compounds, 322–23, 326
Oxidation potential, 250–51
Oxidation-reduction reactions, 249–52

Paraffins, 297
Partial derivative, 35
Partitioning, 117, 132–33
PCBs. *See* Polychlorinated biphenyls
Peclet number, 54, 55, 58, 60, 61
Peds, 164
Pendular ring, 204
Pendular water, 217
Pentachlorophenol, 202, 313
Peristaltic pump, 380–84
Pesticide application, 25–26
Pesticides, 316, 317
Petroleum contaminated soil, 425
Petroleum distillates, 301–5
pH
  definition, 253
  relationship with Eh, 253–54
Phenanthrene, 319
Phenols, 312–14, 329
Phosphorus, 276
Photoionization detector, 329
Phthalates, 312–13
Picocurie, 281–82
Piezometer, 349
Pipelines, 25
Piston pump, 380–84
Pits, 21–22
Plug flow, 48
Plume stabilization, 399, 416
Plutonium, 150
Polarity correction factor, 138–39
Polychlorinated biphenyls (PCBs), 301, 303, 393
Polycyclic aromatic hydrocarbons, 300, 302–3, 319–20
Ponds, 21–22
Population mean, 79
Pore volume, 66
Precision (of chemical analysis), 334
Pressure-plate assembly, 177
Pressure potential, 169
Pressure-vacuum lysimeter, 385–88
Priority pollutants, 332
Probability-density function, 80

Index

Product-recovery wells, 418–22
Protective casing, 348–49
Pump and treat, 401–16
Pyrene, 319
Pyrite, 272

Quality assurance/quality control, 334

Rad, 282
Radioactive decay, 144
Radioactive isotopes, 281–88
Radioactive waste, 24–25
Radium, 286–87
Radon, 287–88
RCRA, 19, 21, 392
Recovery wells, 418–22
Redox reactions, 249–52
Reduction of organic compounds, 323, 326
Relative permeability, 206–11
Rem, 282
Representative elementary volume, 38, 52
Residential disposal, 21
Residual DNAPL, 235
Residual nonwetting fluid saturation. *See* Irreducible nonwetting-fluid saturation
Residual oil, 225–26, 231
Residual wetting fluid saturation. *See* Irreducible wetting-fluid saturation
Resource Conservation and Recovery Act (RCRA), 19, 21, 392
Retardation, 115
Retardation factor
  Freundlich, 122
  Langmuir, 123
  linear, 119
Return flow, 25
Reversible linear kinetic sorption model, 129
Reversible nonlinear kinetic sorption model, 130
Richards equation, 183–84

Ring cleavage, 326
Road salt. *See* Deicing salt
Rocky Mountain Arsenal, 22, 396–97

Salt water intrusion, 28
Sample dilution (for chemical analysis), 332
Saturation ratio, 172, 203, 205
Scalar, 35
Scale effect of dispersion, 71–77
Screens, 345–46
Second type boundary. *See* Fixed-gradient boundary
Selenium, 276
Semivariogram, 91
Separation, 80
Septic tanks, 16, 18
Sesquioxides, 164–65
Seymour Recycling Corporation, 10, 327, 328
Shelby tube, 361
Sievert, 282
Silver, 279
Size-exclusion effect, 150
Skimming trench, 421, 423
Slug injection, 63
Slurry walls, 396
Sodium adsorption ration, 167
Sodium ethylenediaminetetra-acetate, 269–70
Sodium nitriloacetate, 269–70
Sodium tripolyphosphate, 269
Soil, 164
Soil aggregates, 164
Soil colloids, 164–65
Soil-gas monitoring, 383–85
Soil peds, 164
Soil-vapor extraction, 427–29
Soil-water-characteristic curves, 169–75
Soil-water potential, 168–69
  measurement of, 177–79
  sampling, 385–89
Soil-water-retention curves, 169–77
  construction of, 176–77
Solid waste, 392–93
Solubility, water, 295

Solubility product, 246
Solute transport, 43–114
Sorption of hydrophobic (organic) compounds, 132–40
Sorption processes, 117
Source control, 392–401
South Bend, Indiana, 438–39
Specific discharge, 35
Specific gravity, 295
Spiked sample, 334
Split spoon sample, 361–63
Standard deviation, 78
Stochastic descriptions of heterogeneity, 78–81
Stochastic models of solute transport, 77–85
Strontium, 277
Submersible pump, 380–84
Substitution, 322
Suction lysimeter, 385–86, 389
Sulfur, 272
Surface impoundments, 21
Surge block, 371
Syringe sampler, 380

Target compound list, 332, 333
Tenad, 142–44
Tensiometer, 177–79
Tension plate assembly, 176–77
Tensor, 35–36
1,1,2,2-Tetrachloroethane, 342
Tetrachloroethene (perchloroethene, tetrachloroethylene, perchloroethylene, PCE), 151, 153, 154, 155, 156, 157, 308, 322, 342, 439–42
Tetrahydrofuran, 308, 344
Third type boundary. *See* variable flux boundary
Thorium, 285–86
Threshold value. *See* Bubbling pressure
Toluene, 221, 222, 223, 224, 300, 316, 319, 328, 329, 334, 435
Tortuosity, 44
Tortuosity factor, 44, 54

Total petroleum hydrocarbons, 332
Tracer tests, 69–70
Transverse dispersion, 50, 65–66, 83
Transverse hydrodynamic dispersion, 51, 53
Tremmie pipe, 366
1,1,1-Trichloroethane (TCA), 308, 322, 327, 342, 438–39, 439–42
Trichloroethene (Trichloroethylene, TCE), 15, 18, 147, 202, 206, 207, 209, 210, 308, 327, 344, 433, 438–39, 439–42
Trichlorophenol, 313
Trihalomethane, 15
Trimethylbenzene, 328
Trinitrotoluene, 314
Tritium, 288
Typhoid, 15

Underground storage tanks, 24
Undisturbed sample, 361
Unit vector, 36
Unsaturated hydraulic conductivity, 180–82
  estimated from soil parameters, 182

Unsaturated hydraulic conductivity, *continued*
  relationship to intrinsic permeability, 181
Unsaturated zone. *See* Vadose zone
Uranium, 282–85
Urban runoff, 27

Vadose zone, 163–98, 204
  distribution of water, 217
  mass transport in, 185–90
  migration of DNAPLs, 231–33
  preferential flowpaths, 196–98
  soil-gas monitoring, 383–85
  soil-water sampling, 385–89
  treatment of organic compounds contained in, 427–29, 434–37
Vanadium, 277
Van Genuchten soil parameters, 172–75, 182, 228, 231
Vapor density, 296
Vapor extraction, 427–29, 435–37
Vapor phase transport, 184–85
Vapor pressure, 296
Variable flux boundary, 56, 60–61
Variance, 51, 52, 80

Vector, 36
Vinyl chloride (VC), 327, 433, 434
Viruses, 150
Volumetric water content, 169

Water softeners, 26–27
Water solubility, 295
Wells
  casing, 339–44
  design, 338–52
  development, 370–73
  filter packed, 346–47
  installation, 364–70
  naturally developed, 346
  purging, 379–80
  sampling, 378–83
  sampling devices, 380–83
  screen, 345–46
Wettability, 203–4
Wetting curve, 175, 205
Wilsonville, Illinois, hazardous waste site, 393
Wood preservatives, 32

X-ray, 281
Xylene, 221, 223, 224, 300, 316, 319, 328, 329, 435

Zinc, 280

ISBN 0-02-337135-8